Learning to Understand Remote Sensing Images

Learning to Understand Remote Sensing Images

Special Issue Editor

Qi Wang

MDPI • Basel • Beijing • Wuhan • Barcelona • Belgrade

Special Issue Editor
Qi Wang
Northwestern Polytechnical University
China

Editorial Office
MDPI
St. Alban-Anlage 66
4052 Basel, Switzerland

This is a reprint of articles from the Special Issue published online in the open access journal *Remote Sensing* (ISSN 2072-4292) from 2017 to 2019 (available at: https://www.mdpi.com/journal/remotesensing/special_issues/rsimages)

For citation purposes, cite each article independently as indicated on the article page online and as indicated below:

LastName, A.A.; LastName, B.B.; LastName, C.C. Article Title. *Journal Name* **Year**, *Article Number*, Page Range.

Volume 2
ISBN 978-3-03897-698-1 (Pbk)
ISBN 978-3-03897-699-8 (PDF)

Volume 1-2
ISBN 978-3-03897-700-1 (Pbk)
ISBN 978-3-03897-701-8 (PDF)

Contents

About the Special Issue Editor

Qi Wang, Professor, received his B.E. degree in automation and the Ph.D. degree in pattern recognition and intelligent systems from the University of Science and Technology of China, Hefei, China, in 2005 and 2010, respectively. He is currently a Professor with the School of Computer Science and the Center for OPTical IMagery Analysis and Learning (OPTIMAL), Northwestern Polytechnical University, Xi'an, China. His research interests include computer vision and pattern recognition.

Preface to "Learning to Understand Remote Sensing Images"

Accurate and efficient understanding of remote sensing data is an increasingly important issue which can make significant contributions to global environmental analysis and economic development. In this book, we introduce the challenges and advanced techniques in the field of remote sensing image understanding. This area has attracted a lot of research interest, and significant progress has been made during the past years, particularly in the optical, hyperspectral, and microwave remote sensing communities.

Our topic mainly focuses on learning to understand remote sensing images. We discuss some critical problems in major practical applications including image classification, object detection, image segmentation, image correction, hyperspectral unmixing, change detection, etc. We report the state-of-the-art of machine learning techniques and statistical computing methods to analyze remote sensing data, such as deep learning, graphical models, sparse coding, and kernel machines.

Throughout this book, it is assumed that the readers have a basic background in machine learning and remote sensing. We believe the reported advanced techniques can provide considerable value for researchers in teaching and scientific research.

This book is published with the tireless efforts of countless contributors. We thank each author for sharing their research findings with us. We thank the editors and the publishers for their time and support. We hope that through our efforts, more people can contribute to the development of remote sensing.

Qi Wang
Special Issue Editor

remote sensing

MDPI

Article

Road Segmentation of Remotely-Sensed Images Using Deep Convolutional Neural Networks with Landscape Metrics and Conditional Random Fields

Teerapong Panboonyuen [1], Kulsawasd Jitkajornwanich [2], Siam Lawawirojwong [3], Panu Srestasathiern [3] and Peerapon Vateekul [1,*]

1 Chulalongkorn University Big Data Analytics and IoT Center (CUBIC), Department of Computer Engineering, Faculty of Engineering, Chulalongkorn University, Phayathai Rd., Pathumwan, Bangkok 10330, Thailand; teerapong.pan@student.chula.ac.th

2 Data Science and Computational Intelligence (DSCI) Laboratory, Department of Computer Science, Faculty of Science, King Mongkut's Institute of Technology Ladkrabang, Chalongkrung Rd., Ladkrabang, Bangkok 10520, Thailand; kulsawasd.ji@kmitl.ac.th

3 Geo-Informatics and Space Technology Development Agency (Public Organization), 120, The Government Complex, Chaeng Wattana Rd., Lak Si, Bangkok 10210, Thailand; siam@gistda.or.th (S.L.); panu@gistda.or.th (P.S.)

* Correspondence: peerapon.v@chula.ac.th; Tel.: +6-62-218-6989

Academic Editors: Qi Wang, Nicolas H. Younan, Carlos López-Martínez and Prasad S. Thenkabail
Received: 1 June 2017; Accepted: 26 June 2017; Published: 1 July 2017

Abstract: Object segmentation of remotely-sensed aerial (or very-high resolution, VHS) images and satellite (or high-resolution, HR) images, has been applied to many application domains, especially in road extraction in which the segmented objects are served as a mandatory layer in geospatial databases. Several attempts at applying the deep convolutional neural network (DCNN) to extract roads from remote sensing images have been made; however, the accuracy is still limited. In this paper, we present an enhanced DCNN framework specifically tailored for road extraction of remote sensing images by applying landscape metrics (LMs) and conditional random fields (CRFs). To improve the DCNN, a modern activation function called the exponential linear unit (ELU), is employed in our network, resulting in a higher number of, and yet more accurate, extracted roads. To further reduce falsely classified road objects, a solution based on an adoption of LMs is proposed. Finally, to sharpen the extracted roads, a CRF method is added to our framework. The experiments were conducted on Massachusetts road aerial imagery as well as the Thailand Earth Observation System (THEOS) satellite imagery data sets. The results showed that our proposed framework outperformed Segnet, a state-of-the-art object segmentation technique, on any kinds of remote sensing imagery, in most of the cases in terms of *precision*, *recall*, and *F*1.

Keywords: deep convolutional neural networks; road segmentation; conditional random fields; satellite images; aerial images; THEOS

1. Introduction

Extraction of terrestrial objects such as buildings and roads, from remotely-sensed images has been employed in many applications in various areas, e.g., urban planning, map updates, route optimization, and navigation. For road extraction, most primary research is based on unsupervised learning, such as graph cut and global optimization techniques [1]. These unsupervised methods, however; have one common limitation, color-sensitivity, since they rely on only the color features.

That is, the segmentation algorithms will not perform well if the road colors presented in the suburban remotely-sensed images contain more than one color (e.g., yellowish brown roads in the countryside regions and cement-grayed roads in the suburban regions). This, in fact, has become a motivation of this work, that is, to overcome the color sensitivity issues.

Deep learning, a large convolutional neural network with performance that can be scaled depending on the size of training data and model complexity as well as processing power, has shown significant improvements in object segmentation from images as seen in many recent works [2–13]. Unlike unsupervised learning, more than one feature—other than color—can be extracted: line, shape, and texture, among others. The traditional deep learning methods such as the deep convolutional neural network (DCNN) [3,14], deep deconvolutional neural network (DeCNN) [5], recurrent neural network, namely reSeg [15], and fully convolutional networks [4]; however all suffer from accuracy performance issues.

A deep convolutional encoder-decoder (DCED) architecture, one of the most efficient newly developed neural networks, has been proposed for object segmentation. The DCED network is designed to be a core segmentation engine for pixel-wise semantic segmentation, and has shown good performance in the experiments tested using PASCAL VOC 2012 data—a well-known benchmark data set for image segmentation research [6,8,16]. In this architecture, the rectified linear unit (ReLU) is employed as an activation function.

In the road extraction task, there are many issues that can cause limited detection performance. First, based on [6,8], although the most recent DCED approach for object segmentation (or SegNet) showed promising detection performance on overall classes, the result for road objects is still limited as it fails to detect many road objects. This could be caused by the rectified linear unit (ReLU) which is sensitive to the gradient vanishing problem. Second, even when we apply Gaussian smoothing at the last step to connect detected roads together, this still yields excessive detected road objects (false road objects).

In this paper, we present an improved deep convolutional encoder-decoder network (DCED) for segmenting road objects from aerial and satellite images. Several aspects of the proposed method are enhanced, including incorporation of exponential linear units (ELUs), as opposed to ReLUs that typically outperform ELU in most object classification cases; adoption of landscape metrics (LMs) to further improve the overall quality of results by removing falsely detected road objects; and lastly, combination with the traditional fully-connected conditional random field (CRF) algorithms used in semantic segmentation problems. Although the ELU-SegNet-LM network may suffer a performance issue due to the loss of spatial accuracy, it can be alleviated by the conditional random fields algorithm, which takes into account the low-level information captured by the local interactions of pixels and edges [17–19]. The experiments were conducted using well-known aerial imagery, a Massachusetts roads data set (Mass. Roads), which is publicly available, and satellite imagery (from the Thailand Earth Observation System (THEOS) satellite) which is provided by GISTDA. The results showed that our method outperforms all of the baselines including SegNet in terms of *precision*, *recall*, and $F1$ scores. The paper is organized as follows. Related work is discussed in Section 2. Section 3 describes our proposed methodology. Experimental data sets and evaluations are described in Section 4. Experimental results and discussions are presented in Section 5. Finally, we conclude our work and discuss future work in Section 6.

2. Related Work

Deep learning is one of the fast-growing fields in machine learning which has been successfully applied to remotely-sensed data analysis, notably land cover mapping on urban areas [20]. It has increasingly become a promising tool for accelerating image recognition process with high accuracy results [4], [6], [21]; new architectures are proposed constantly on a weekly basis. This related work is divided into three subsections: we first discuss deep learning concepts for semantic segmentation, followed by a set of road object segmentation techniques using deep learning, and finally activation functions and post processing technique of deep learning are discussed.

Note that this paper only focuses on approaches built around deep learning techniques. Therefore, prior attempts at semantic segmentation [22,23] are not included and compared here since they are not based on a deep learning approach.

2.1. Deep Learning for Semantic Segmentation

Semantic segmentation algorithms are often formulated to solve structured pixel-wise labeling problems based on the deep convolutional neural network (DCNN), and are state-of-the-art supervised learning algorithms for modeling and extracting latent feature hierarchies. Noh et al. [5] proposed a novel semantic segmentation technique utilizing a deconvolutional neural network (DeCNN) and the top layer from DCNN adopted from VGG16 [24]. The DeCNN structure is composed of upsampling layers and deconvolution layers, describing pixel-wise class labels and predicting segmentation masks, respectively. Their proposed deep learning methods yield high performance in the PASCAL VOC 2012 data set [16], with a 72.5% accuracy in the best case scenario (this was the highest accuracy—at the time of writing this paper—compared to other methods that were trained without requiring additional or external data). Long et al. [4] proposed an adapted contemporary classification network incorporating Alex, VGG and Google networks into a full DCNN. In this method, some of the pooling layers were skipped: layer 3 (FCN-8s), layer 4 (FCN-16s), and layer 5 (FCN-32s). The skip architecture reduces the potential over-fitting problem and has shown improvements in performance ranging from 20 to 62.2% in the experiments tested using PASCAL VOC 2012 data. Ronneberger et al. [12] proposed U-Net, a DCNN for biomedical image segmentation. The architecture consists of a contracting path and a symmetric expanding path that capture context and consequently, enable precise localization. The proposed network claimed to be capable of learning despite the limited number of training images, and performed better than the prior best method (a sliding-window DCNN) on the ISBI challenge for segmentation of neuronal structures in electron microscopic stacks. In this work, VGG16 is selected as our baseline architecture since it is the most popular architecture used in various networks for object recognition. Furthermore, we will investigate the effect of the skipped layer technique, especially FCN-8s, since it is the top-ranking architecture as shown in Long et al. [4].

There is a new research area called "instance-aware semantic segmentation" which is slightly different from "semantic segmentation." Instead of labeling all pixels, it focuses on the target objects and labels only pixels of those objects. FCIS [25] is a technique developed based on fully convolutional networks (FCN). Mask R-CNN [26] is also created on top of FCN but incorporates with a proposed joint formulation. Even though their results are promising, they are not directly related to our scope on "semantic segmentation." In the future, we can extend these works and compare them to our proposed technique.

2.2. Deep Learning for Road Segmentation

There are many approaches to road network extraction in very-high-resolution (VHR) aerial and satellite imagery literature. Wand et al. [14] proposed a DCNN and finite state machine (FSM)-based framework to extract road networks from aerial and satellite images. DCNN recognizes patterns from a sophisticated and arbitrary environment while FSM translates the recognized patterns to states such that their tracking behaviors can be captured. The results showed that their approach is more accurate compared to the traditional methods. The extension of the method for automatic road point initialization was left for future work. DCNN for multiple object extraction from aerial imagery was proposed in [3] by Saito et al. Both features (extractors and classifiers) of DCNN were automated in that a new technique to train a single DCNN for extracting multiple kinds of objects simultaneously was developed. Two objects were extracted: buildings and roads, thus a label image consists of three channels: buildings, roads, and background. Finally, the results showed that the proposed technique not only improved the prediction performance but also outperformed the cutting-edge method tested on a publicly available aerial imagery data set. Muruganandham et al. [2] designed an automated framework to extract semantic maps of roads and highways, so the urban growth of cities from remote sensing images could be tracked. They used the VGG16 model—a simplistic architecture with homogeneous 3 × 3 convolution kernels and 2 × 2 max pooling throughout the pipeline—as a baseline for a fixed feature extractor. The experimental results showed that their proposed technique for the prediction performance was improved with $F1$ scores of 0.76 on the Mass. Roads data set.

2.3. Recent Techniques in Deep Learning

Activation function is an important factor for the accuracy of DCNN. While the most popular activation function for neural networks is the rectified linear unit (ReLU), Clevert et al. [21] have just proposed the exponential linear unit (ELU), which can speed up the learning process in DCNN and therefore lead to higher classification accuracies as well as overcoming the previously unsolvable problem, i.e., the vanishing gradient problem. Compared to other methods with different activation functions, ELU has greatly improved many of the learning characteristics. In the experiments, ELUs enable fast learning as well as more effective generalization performance than the ReLUs and the leaky rectified linear units (LReLUs) in networks with five layers or more. In ImageNet, ELU networks substantially increased the learning time compared to ReLU networks with the identical architecture; less than 10% classification error was presented for a single crop, model network.

Recently, there have been some efforts to enhance the performance of DCNN by combining it with other classifier as a post-processing step. Conditional random fields (CRFs) has been reported successful in increasing the accuracy of DCNN, especially in the image segmentation domain. CRFs have been employed to smooth maps [7,17–19]. Typically these models contain energy terms that couple neighboring nodes, favoring same-label assignments to spatially proximal pixels. Qualitatively, the primary function of these short-range CRFs has been used to clean up the spurious predictions of weak classifiers built on top of local hand-engineered features.

3. Proposed Methodology

In this section, we propose an enhanced, improved DCED network (or SegNet) to efficiently segment road objects from aerial and satellite images. Three aspects of the proposed method are enhanced: (**1**) modification of DCED architecture; (**2**) incorporation of landscape metrics (LMs); and (**3**) adoption of conditional random fields (CRFs). An overview of our proposed method is shown in Figure 1.

Figure 1. A process in our proposed framework.

3.1. Data Preprocessing

Data preparation is required when working with neural network and deep learning models. In addition, data augmentation is often required in more complex object recognition tasks. Thus, we increased the size of our data sets to improve the method efficiency by rotating them incrementally with eight different angles. All images on Massachusetts road data sets are standardized and cropped into 1500 × 1500 pixels with a resolution of 1 m^2/pixel. The data sets consist of 1108 training images, 49 test images, and 14 validation images. The original training images were further extended to 8864 training images.

On the THEOS data sets, we also increased the size of data sets in a similar fashion. Each image has 1500 × 1500 pixels with a resolution of 2 m^2/pixel.

3.2. Object Segmentation (ELU-SegNet)

SegNet, one of the deep convolutional encoder-decoder architectures, consists of two main networks encoder and decoder, and some outer layers. The two outer layers of the decoder network are responsible for feature extraction task, the results of which are transmitted to the next layer adjacent to the last layer of the decoder network. This layer is responsible for pixel-wise classification (determining which pixel belongs to which class). There is no fully connected layer in between feature extraction layers. In the upsampling layer of decoder, pool indices from encoder are distributed to the decoder where the kernel will be trained in each epoch (training round) at the convolution layer. In the last layer (classification), softmax is used as a classifier for pixel-wise classification. The encoder network consists of convolution layer and pooling layer. A technique, called batch normalization (proposed by Ioffe and Szegedy [27]), is used to speed up the learning process of the DCNN by reducing internal covariate shift. In the encoder network, the number of layers is reduced to 13 (VGG16) by removing the last three layers (fully connected layers) [6,8,28,29] for the following two reasons: to maintain the high-resolution feature maps in the encoder network, and to minimize the countless number of parameters from 134 million features to 14.7 million features compared to the traditional deep learning networks such as DCNN [4] and DeCNN [5], where the fully connected layer remains intact. In the activation function of feature extraction, ReLU, max-pooling, and 7 × 7 kernels are used in both encoder and decoder networks. For training images, three-channel images (RGB) are used. The exponential linear unit (ELU) was introduced in [21], which can speed up learning in deep neural networks, offer higher classification accuracies, and give better generalization performance than ReLUs and LReLUs on networks. In SegNet architecture, to perform optimization for training networks,the stochastic gradient descent (SGD) [30] with a fixed learning rate of 0.1 and momentum of 0.9 is used. In each training round (epoch), a mini-batch (a set of 12 images) is chosen such that each image is used once. The model with the best performance on the validation data set in each epoch will be selected. Our architecture (see Figure 2) is enhanced from SegNet, consisting of two main networks responsible for feature extraction. In each network, there are 13 layers, with the last layer being the classification based on softmax supporting pixel-wise classification.

In our work, an activation function called ELU is used as opposed to ReLU based on its performances. For the network training optimization, stochastic gradient descent (SGD) is used and configured with a fixed learning rate of 0.001 and momentum of 0.9 to delay the convergence time and so, can avoid local optimization trap.

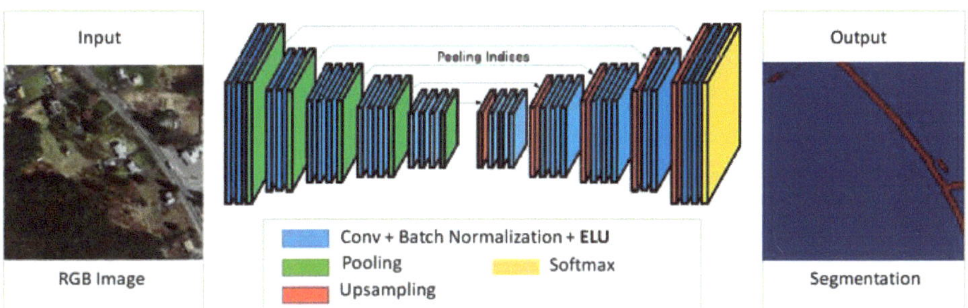

Figure 2. A proposed network architecture for object segmentation (exponential linear unit (ELU)-SegNet).

3.3. Gaussian Smoothing

Gaussian smoothing [31] is a 2-D convolution operator that is used to 'blur' images and remove unnecessary details and noises by utilizing the Gaussian function. The Gaussian function is used to determine the transformation needed for each pixel, resulting in a more complete extended road objects. We applied the Gaussian function first in the post-processing step in order to expand and prepare objects that are close to each other to be combined into components in the next step (as we shall see in Section 3.4).

The 1-D and 2-D Gaussian functions are described in Equations (1) and (2), respectively.

$$G(x) = \frac{1}{2\pi\sigma^2} e^{-\frac{x^2}{2\sigma^2}} \quad (1)$$

$$G(x) = \frac{1}{2\pi\sigma^2} e^{\frac{-x^2-y^2}{2\sigma^2}} \quad (2)$$

where x represents the distance from the origin in the X-axis, y represents the distance from the origin in the Y-axis, and σ represent the standard deviation of the Gaussian distribution.

3.4. Connected Component Labeling (CCL)

In connected components labeling (CCL) [31], all pixels are scanned and adjacent pixels with similar connectivity values are combined. Eight neighbors of each pixel were considered when analyzing connected components.

The expanded and overlapped objects from the Gaussian smoothing were actually grouped together in this step. The labeled objects will be further calculated using geometric attributes (e.g., area and perimeter) based on landscape metrics (LMs) as described in the next section.

3.5. False Road Object Removal (LMs)

After smoothing and labeling the objects, we compute the shape complexity of the objects through the shape index (as seen in Equation (3)), one of the landscape metrics for measuring arrangement and composition property of spatial objects. The resulting objects along with their shape scores are shown in Figure 3. As seen in Figure 3, the geometrical characteristics of roads were captured and differentiated from other spatial objects in the given image. Other geometry metrics can also be used such as rectangular degree, aspect ratio, etc. More information on other landscape metrics can be found in [32,33].

$$shape\ index = \frac{e(i)}{4x\sqrt{A(i)}} \quad (3)$$

where $e(i)$ and $A(i)$ denote the perimeter and area for object i, respectively.

3.6. Road Object Sharpening (CRFs)

Conditional random fields (CRFs) have traditionally been implemented to sharpen noisy segmentation maps [18]. These models are generally composed of energy terms comprising nodes in the neighborhood, causing false assignments of pixels that are in close proximity. To resolve these spatial limitations of short-range CRFs, the fully connected CRFs are integrated into our system [19]. Equation (4) expresses the energy function of the dense CRFs.

In the last step, we extended the ELU-SegNet-LMs model to ELU-SegNet-LMs-CRFs to enhance the network performance by adding explicit dependencies among the neural network outputs. Particularly, we added smoothness terms between neighboring pixels to our model, which can eliminate the need to learn smoothness from remotely-sensed images. Using the resulting models as part of the post-processing significantly increases the overall performance of the network over unstructured deep neural networks.

$$E(x) = \sum_{i} \theta_i(x_i) + \sum_{ij} \theta_{ij}(x_i, x_j) \tag{4}$$

where x denotes the label assignment for pixels. A unary potential used is $\theta_i(x_i)) = -logP(x_i)$, while $P(x_i)$ denotes the label assignment probability at pixel i as computed by a DCNN.

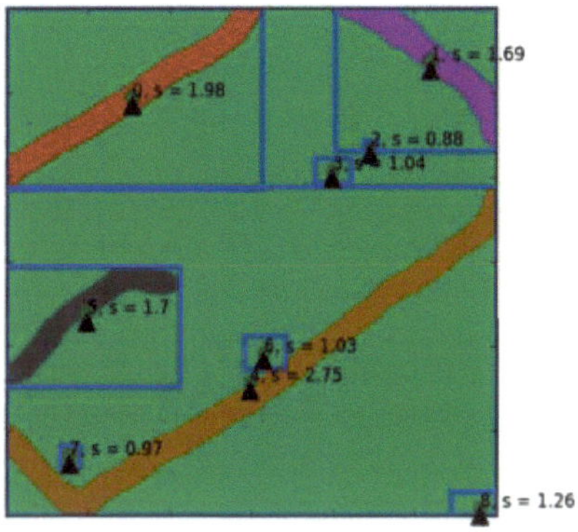

Figure 3. Illustration of shape index scores on each extracted road object. Any objects with shape index score lower than 1.25 are considered as noises and subsequently removed.

The inference can be efficiently established in the pair-wise potentials when using the fully connected graph. We treated the unary potential as local classifiers which are defined by the output of the ELU-SegNet-LMs model, which is a probability map for each class in each of the pixels. The pairwise potentials depict the interaction of pixels in the neighborhood and are influenced by the color similarity. In the DeepLab CRF model [19], the dense CRFs (instead of neighboring information) are used as a means to identify relationships between pixels. Furthermore, they define the following pairwise potentials as shown in Equation (5).

$$\theta_{ij}(x_i, x_j) = \mu(x_i, x_j)[w_1 \exp(-\frac{\| p_i - p_j \|^2}{2\sigma_\alpha^2} - \frac{\| I_i - I_j \|^2}{2\sigma_\beta^2}) + w_2 \exp(-\frac{\| p_i - p_j \|^2}{2\sigma_\gamma^2})] \tag{5}$$

where $\mu(x_i, x_j) = 1\ if\ x_i \neq x_j$ and zero otherwise, which, as in the Potts model, means that only nodes with distinct labels are penalized. The remaining expression uses two Gaussian kernels in different feature spaces; the first, 'bilateral' kernel depends on both pixel positions (denoted as p) and red-green-blue (RGB) color (denoted as I), and the second kernel only depends on pixel positions. The hyperparameters σ_α, σ_β *and* σ_γ control the scale of Gaussian kernels. The first kernel forces pixels to similar color and position to have similar labels, while the second kernel only considers spatial proximity when enforcing smoothness.

In summary, the first term of pairwise potentials depends on both pixel positions and color intensities whereas the second term depends solely on the pixel positions [18,19]. Although the dense CRFs can have billions of edges (which is technically infeasible to solve), it was recently found that the inference/maximum posterior can be approximated by the mean-field algorithm.

4. Experimental Data Sets and Evaluation

In our experiments, two types of data sets are used: aerial images and satellite images. Table 1 shows one aerial data set (Massachusetts) and five satellite data sets (Nakhonpathom,

Chonburi, Songkhla, Surin, and Ubonratchathani). All experiments are evaluated based on *precision*, *recall*, and *F1*.

Table 1. Numbers of training, validation, and testing sets.

	Training Set	Validation Set	Testing Set
Massachusetts	1108	14	49
Nakhonpathom	200	14	49
Chonburi	100	14	49
Songkhla	100	14	49
Surin	70	14	49
Ubonratchathani	70	14	49

4.1. Massachusetts Road Data Set (Aerial Imagery)

This data set (made publicly available by [7]) consists of 1171 aerial images of the state of Massachusetts. Each image is 1500×1500 pixels in size, covering an area of 2.25 square kilometers. We randomly split the data into a training set of 1108 images, a validation set of 14 images and a testing set of 49 images. The samples of this data set are shown in Figure 4. The data set covers a wide variety of urban, suburban, and rural regions with a total area of over 2600 square kilometers. With our test set alone, it covers more than 110 square kilometers which is by far the largest and most challenging aerial image labeling data set.

(**a**) (**b**)

Figure 4. Two sample aerial images from the Massachusetts road corpus, where a row refers to each image (**a**) Aerial image and (**b**) Binary map, which is a ground truth image denoting the location of roads.

4.2. THEOS Data Sets (Satellite Imagery)

In this type of data, the satellite images were separated into five data sets—one for each province. The datasets were obtained from the Thailand Earth Observation System (THEOS), also known as Thaichote, an Earth observation satellite of Thailand developed by EADS Astrium SAS, France. This data set consists of 855 satellite images covering five provinces: 263 images of Nakhonpathom, 163 images of Chonburi, 163 images of Songkhla, 133 images of Surin, and 133 images of Ubonratchathani. Some samples of these images are shown in Figure 5.

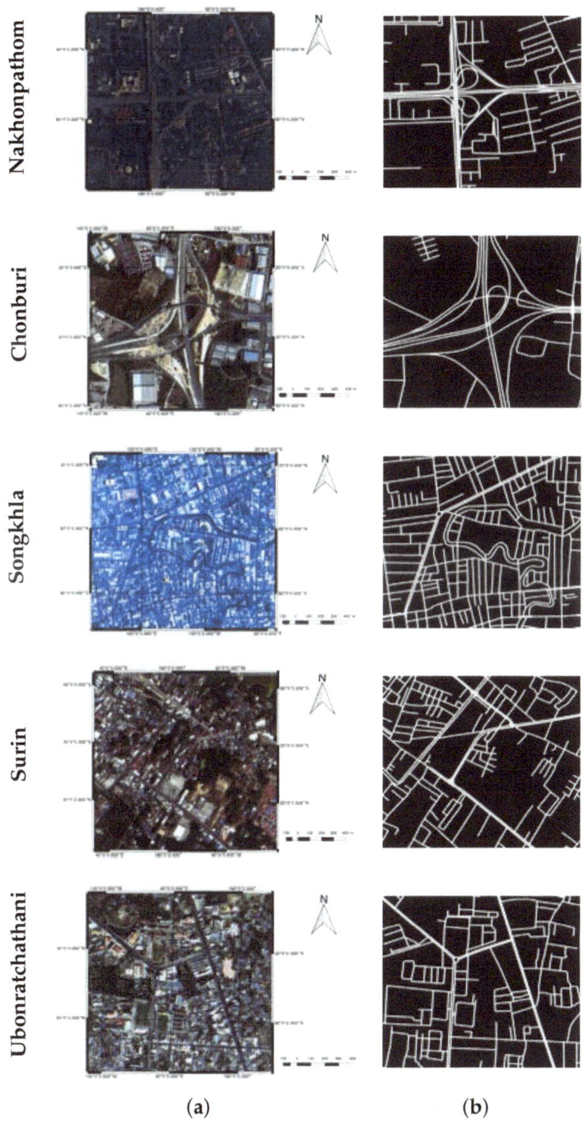

Figure 5. Sample satellite images from five provinces of our data sets; each row refers to a single sample image from one province (Nakhonpathom, Chonburi, Songkhla, Surin, and Ubonratchathani) in a satellite image format (**a**) and in a binary map (**b**), which is served as a ground truth image denoting the location of roads.

4.3. Evaluation

The road extraction task can be considered as binary classification, where road pixels are positives and the remaining non-road pixels are negatives. Let TP denote the number of true positives (the number of correctly classified road pixels), TN denote the number of true negatives (the number of correctly classified non-road pixels), FP denote the number of false positives (the number of mistakenly classified road pixels), and FN denote the number of false negatives (the number of mistakenly classified non-road pixels).

The performance measures used are *precision*, *recall*, and $F1$ as shown in Equations (6)–(8). Precision is the percentage of correctly classified road pixels among all predicted pixels by the classifier. Recall is the percentage of correctly classified road pixels among all actual road pixels. $F1$ is a combination of precision and recall.

$$Precision = \frac{TP}{TP + FP} \tag{6}$$

$$Recall = \frac{TP}{TP + FN} \tag{7}$$

$$F1 = \frac{2 \times Precision \times Recall}{Precison + Recall} \tag{8}$$

5. Experimental Results and Discussions

This section illustrates details of our experiments. The proposed deep learning network is based on SegNet with three improvements: (**1**) it employs the ELU activation function; (**2**) it uses LMs to filter incorrect detected roads; and (**3**) it applies CRFs to sharpen broad roads. Thus, there are three variations of the proposed methods as shown in Table 2.

Table 2. Variations of our proposed deep learning methods. LM: landscape metric; CRF: conditional random field.

Abbreviation	Description
ELU-SegNet	SegNet + **ELU activation**
ELU-SegNet-**LMs**	SegNet + ELU activation + **Landscape Metrics**
ELU-SegNet-LMs-**CRFs**	SegNet + ELU activation + Landscape Metrics + **CRFs**

For the experimental setup, there are three experiments on two remotely-sensed data sets: the Massachusetts road data set and THEOS data sets (details in Section 4). The experiments aim to illustrate that each proposed strategy can really improve the performance. First, ELU-SegNet is compared to SegNet for the ELU strategy. Second, ELU-Segnet-LMs is compared to ELU-SegNet for the LM strategy. Third, the full proposed technique (ELU-Segnet-LMs-CRFs) is compared to existing methods for the CRF technique.

The implementation is based on a deep learning framework, called "Lasagne", which is extended from Theano. All experiments were conducted on a server with Intel Core i5-4590S Processor (6M Cache, up to 3.70 GHz), 32 GB of memory, Nvidia GeForce GTX 960 (4 GB), and Nvidia GeForce GTX 1080 (8 GB). Instead of using the whole image (1500 $\times$ 1500 pixels) to train the network, we randomly cropped all images to be 224 $\times$ 224 as inputs of each epoch.

5.1. Results on Aerial Imagery (Massachusetts Data Set)

In this sub-section, the experiment was conducted on the Massachusetts aerial corpus. To achieve the highest accuracy, the network must be configured and trained many epochs until all parameters in the network are converged. Figure 6a illustrates that the proposed network has been properly set and trained until it really is converged. Furthermore, Figure 6b shows that the higher number of epochs tends to show a better $F1$-score. Thus, the number of chosen epochs based on the validation data is 29 (the best model for this data set).

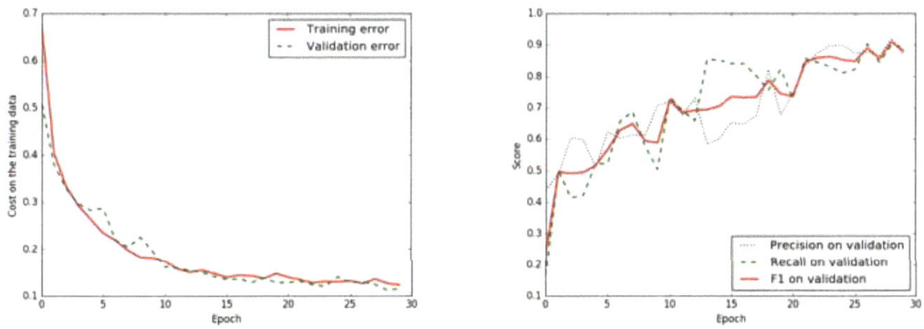

(a) (b)

Figure 6. Iteration plot on Massachusetts aerial corpus of the proposed technique, ELU-SegNet-LMs-CRFs; x refers to epochs and y refers to different measures. (**a**) Plot of model loss (cross entropy) on training and validation data sets, and (**b**) Performance plot on the validation data set.

The result is shown in Table 3 by comparing between baselines and variations of the proposed techniques. It shows that our network with all strategies (ELU-SegNet-LMs-CRFs) outperforms other methods. More details will be discussed to show that each of the proposed techniques can really improve an accuracy. Only in this experiment, there are four baselines, including Basic-model, FCN-no-skip, FCN-8s, and SegNet. Note that SegNet has been implemented and tested on the experimental data set, while the results of other three baselines are carried from the original paper [2].

Table 3. Results on the testing data of Massachusetts aerial corpus between four baselines and three variations of our proposed techniques in terms of *precision*, *recall*, and $F1$. FCN: fully convolutional network.

	Model	Precsion	Recall	F1
Baselines	Basic-model [2]	0.657	0.657	0.657
	FCN-no-skip [2]	0.742	0.742	0.742
	FCN-8s [2]	0.762	0.762	0.762
	SegNet	0.773	0.765	0.768
Proposed Method	**ELU**-SegNet	0.852	0.733	0.788
	ELU-SegNet-**LMs**	0.854	0.861	0.857
	ELU-SegNet-LMs-**CRFs**	**0.858**	**0.894**	**0.876**

5.1.1. Results of Enhanced SegNet (ELU-SegNet)

Our first strategy aims to increase an accuracy of the network by using ELU as an activation function (ELU-SegNet) rather than the traditional one, ReLU (SegNet). Details are shown in Section 3.2. From Table 3, $F1$ of ELU-SegNet (0.788) outperforms that of SegNet (0.768); this yields higher $F1$

at 2.6%. The main reason is due to higher *precision*, but slightly lower *recall*. This can imply that ELU is more robust than ReLU to detect road pixels.

5.1.2. Results of Enhanced SegNet with Landscape Metrics (ELU-SegNet-LMs)

Our second mechanism focuses on applying LMs (details in Section 3.5) on top of ELU-SegNet to filter false road objects. From Table 3, the *F*1 of ELU-SegNet-LMs (0.857) is superior to that of ELU-SegNet (0.788) and SegNet (0.768); this yields higher *F*1 at 6.9% and 8.9%, consecutively. Although LM is specifically designed to increase *precision*, the result shows that it can increase both *precision* (0.854) and *recall* (0.861). It is interesting that *recall* is also improved since all noises in the training images have been removed by the LMs filtering technique resulting in a better quality of the training data set.

5.1.3. Results of All Modules (ELU-SegNet-LMs-CRFs)

Our last strategy aims to sharpen road objects (details in Section 3.6) by integrating CRFs into our deep learning network. From Table 3, *F*1 of ELU-SegNet-LMs-CRFs (0.876) is the winner; it clearly outperforms not only the baselines, but also all previous generations. Its *F*1 is higher than SegNet (0.768) at 10.8%. Also, the result illustrates that CRFs can enhance both *precision* (0.858) and *recall* (0.894).

Figure 7 shows two sample results from the proposed method. By applying all strategies, the images in the last column (Figure 7e) look very close to the ground truths (Figure 7b). Furthermore, *F*1-results are improved for each strategy we added to the network as shown in Figure 7c–e.

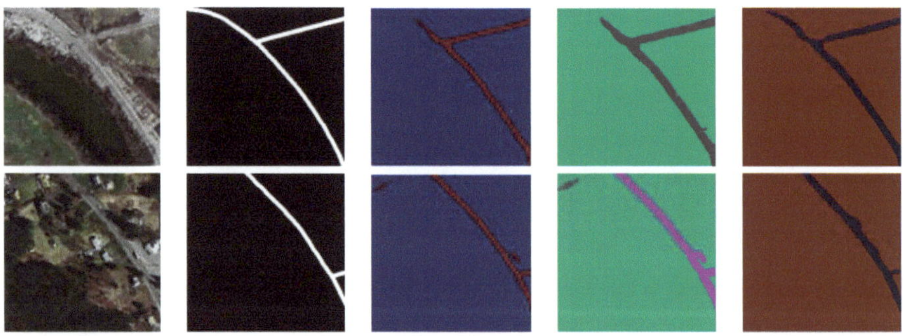

(a) (b) (c) (d) (e)

Figure 7. Two sample input and output aerial images on Massachusetts corpus, where rows refer different images. (**a**) Original input image; (**b**) Target road map (ground truth); (**c**) Output of ELU-SegNet; (**d**) Output of ELU-SegNet-LMs; and (**e**) Output of ELU-SegNet-LMs-CRFs.

5.2. Results for Satellite Imagery (THEOS Data Sets)

In this sub-section, the experiment was conducted on THEOS satellite images. There are five data sets referring to different provinces: Nakhonpathom, Chonburin, Songkla, Surin, and Ubonratchathani; therefore, there are five learning models. Figure 8 shows that each model is properly set up and trained until it is converged and obtained the best *F*1. The best epochs (models) for each province are 25, 15, 30, 21, and 20, respectively.

The results are shown in Tables 4–6 for measures in terms of *F*1, *precision*, and *recall*, respectively. It is interesting that the proposed network with all strategies (ELU-SegNet-LMs-CRFs) is the winner showing the best performance on any measures and provinces. Also, an improvement in the satellite images is higher than that in the aerial images. More details on each proposed strategy will be discussed.

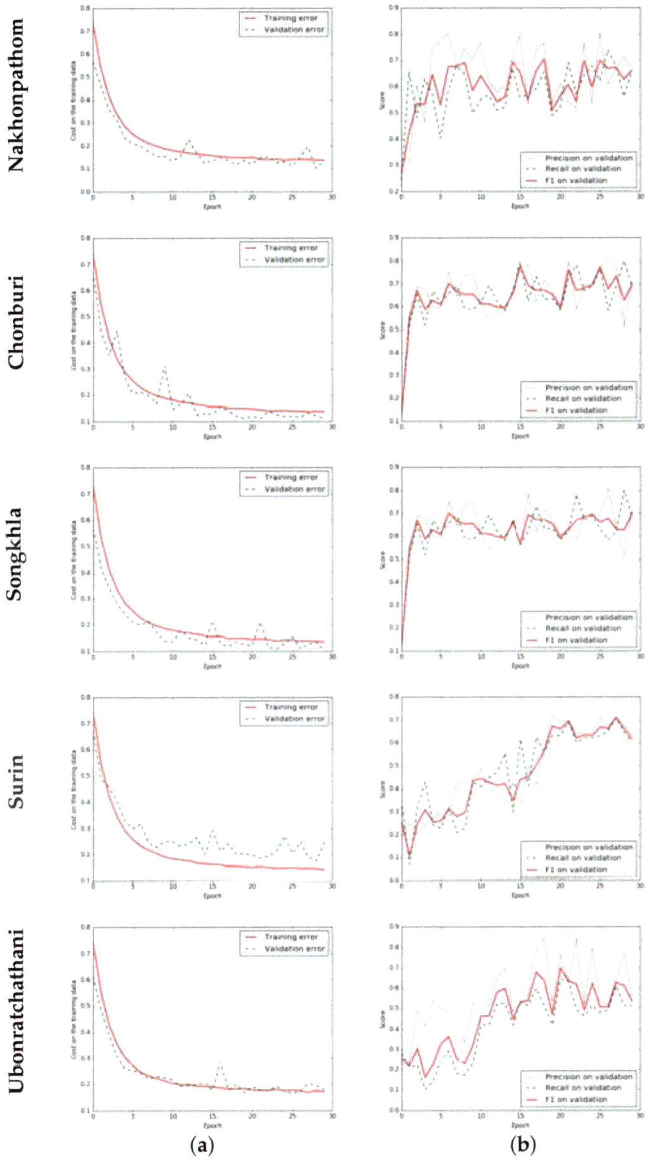

Figure 8. Iteration plot on THEOS satellite data sets of the proposed technique, ELU-SegNet-LMs-CRFs. x refers to epochs and y refers to different measures. Each row refers to different data set (province). (**a**) Plot of model loss (cross entropy) on training and validation data sets; and (**b**) Performance plot on the validation data set.

Table 4. *F*1 on the testing data of the Thailand Earth Observation System (THEOS) satellite data sets between baseline (SegNet) and three variations of our proposed techniques; columns refer to five different provinces (data sets).

	Model	Nakhon.	Chonburi	Songkhla	Surin	Ubon.	Avg.
Baseline	SegNet	0.422	0.572	0.424	0.501	0.406	0.465
Proposed Method	**ELU**-SegNet	0.463	0.690	0.497	0.591	0.534	0.555
	ELU-SegNet-**LMs**	0.488	0.732	0.526	0.625	0.562	0.587
	ELU-SegNet-LMs-**CRFs**	**0.550**	**0.775**	**0.607**	**0.707**	**0.608**	**0.649**

Table 5. *precision* on the testing data of THEOS satellite data sets between baseline (SegNet) and three variations of our proposed techniques; columns refer to five different provinces (data sets).

	Model	Nakhon.	Chonburi	Songkhla	Surin	Ubon.	Avg.
Baseline	SegNet	0.435	0.668	0.456	0.598	0.601	0.552
Proposed Method	**ELU**-SegNet	0.410	0.702	0.478	**0.840**	0.852	0.656
	ELU-SegNet-**LMs**	0.494	0.852	0.557	0.770	0.867	0.708
	ELU-SegNet-LMs-**CRFs**	**0.535**	**0.909**	**0.650**	0.786	**0.871**	**0.751**

Table 6. *recall* on the testing data of THEOS satellite data sets between baseline (SegNet) and three variations of our proposed techniques; columns refer to five different provinces (data sets).

	Model	Nakhon.	Chonburi	Songkhla	Surin	Ubon.	Avg.
Baseline	SegNet	0.410	0.499	0.395	0.431	0.306	0.408
Proposed Method	**ELU**-SegNet	0.532	**0.678**	0.517	0.456	0.389	0.515
	ELU-SegNet-**LMs**	0.483	0.642	0.498	0.526	0.416	0.513
	ELU-SegNet-LMs-**CRFs**	**0.566**	**0.676**	**0.570**	**0.643**	**0.467**	**0.584**

5.2.1. Results of Enhanced SegNet (ELU-SegNet)

The ELU activation function can increase the performance of the network. In terms of *F*1, Table 4 shows that ELU-SegNet outperforms the traditional network (SegNet) for all provinces. It performs better than SegNet by 9.08% on average for all provinces, where Ubonratchathani and Chonburi show the highest *F*1-improvement, at over 10%. For *precision* and *recall*, Tables 5 and 6 illustrate that almost all data sets can be improved employing the ELU function with improvements of 10.48% and 10.68% on average for all provinces, respectively, .

5.2.2. Results of Enhanced SegNet with Landscape Metrics (ELU-SegNet-LMs)

The LMs filtering strategy aims to remove all inaccurately extracted roads (false positives: FP) resulting in higher *precision* and *F*1, but this might imply a slight loss in *recall*. Comparing to the previous generation (ELU-SegNet), there are improvements by LMs on average for all provinces of 5.2% and 3.2% in terms of *precision* (Table 5) and *F*1 (Table 4), respectively, with a slight loss of −0.22% in terms of *recall* (Table 6). Compared to the baseline, LMs outperforms SegNet on all performance measures.

5.2.3. Results of All Modules (ELU-SegNet-LMs-CRFs)

To further improve the performance, CRFs is integrated into the network from the previous section. This is considered to use all proposed modules: ELU, LMs, and CRFs. From Tables 4–6, the results show that ELU-SegNet-LMs-CRFs is the winner compared the previous generations and baseline (SegNet) on any of the measures (*precision*, *recall*, and *F*1). As of *F*1 average of all provinces, it outperforms ELU-SegNet-LMs, ELU-SegNet, and SegNet by 6.28%, 9.44% and 18.44%, respectively.

Figures 9–13 show sample results from the proposed method on five provinces. The results of the last column look closest to the ground truth in the second column.

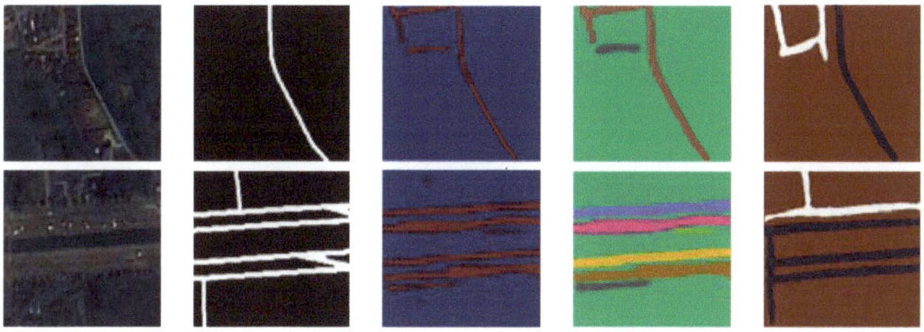

(a) (b) (c) (d) (e)

Figure 9. Two sample input and output THEOS satellite images on the Nakhonpathom data set, where rows refer different images. (**a**) Original input image; (**b**) Target road map (ground truth); (**c**) Output of ELU-SegNet; (**d**) Output of ELU-SegNet-LMs; and (**e**) Output of ELU-SegNet-LMs-CRFs.

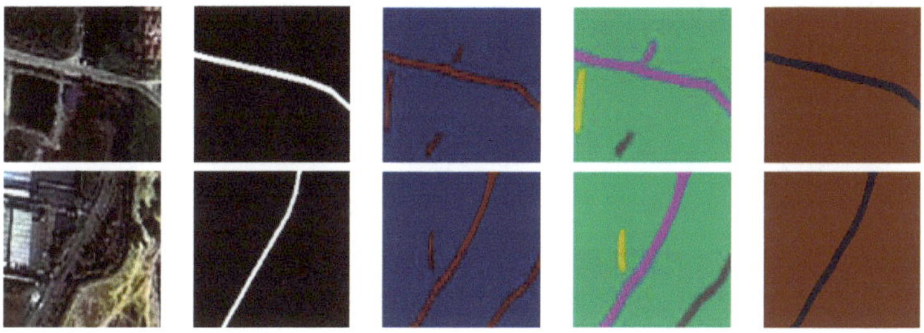

(a) (b) (c) (d) (e)

Figure 10. Two sample input and output THEOS satellite images on the Chonburi data set, where rows refer different images. (**a**) Original input image; (**b**) Target road map (ground truth); (**c**) Output of ELU-SegNet; (**d**) Output of ELU-SegNet-LMs; and (**e**) Output of ELU-SegNet-LMs-CRFs.

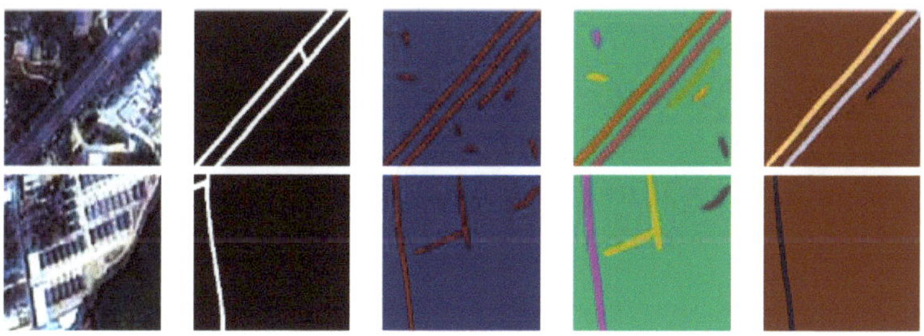

(a) (b) (c) (d) (e)

Figure 11. Two sample input and output THEOS satellite images on the Songkhla data set, where rows refer different images. (**a**) Original input image; (**b**) Target road map (ground truth); (**c**) Output of ELU-SegNet; (**d**) Output of ELU-SegNet-LMs; and (**e**) Output of ELU-SegNet-LMs-CRFs.

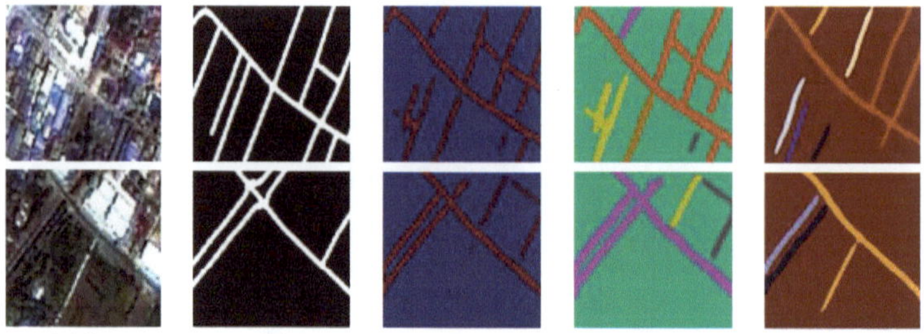

(**a**) (**b**) (**c**) (**d**) (**e**)

Figure 12. Two sample input and output THEOS satellite images on the Surin data set, where rows refer different images. (**a**) Original input image; (**b**) Target road map (ground truth); (**c**) Output of ELU-SegNet; (**d**) output of ELU-SegNet-LMs; and (**e**) Output of ELU-SegNet-LMs-CRFs.

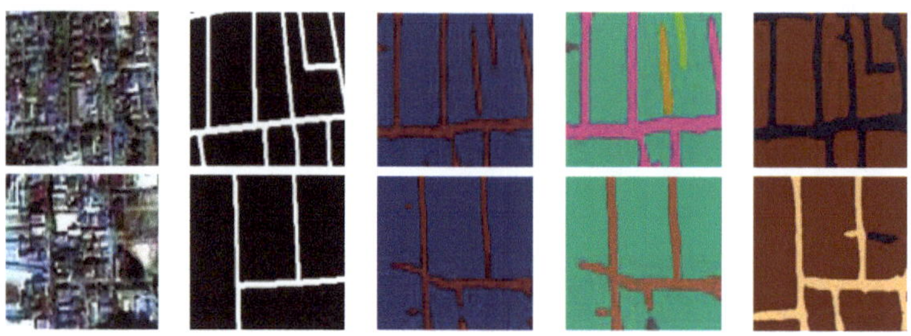

(**a**) (**b**) (**c**) (**d**) (**e**)

Figure 13. Two sample input and output THEOS satellite images on Ubonratchathani data set, where rows refer different images. (**a**) Original input image; (**b**) Target road map (ground truth); (**c**) Output of ELU-SegNet; (**d**) Output of ELU-SegNet-LMs; and (**e**) Output of ELU-SegNet-LMs-CRFs.

5.3. Discussions

In terms of accuracy ($F1$-measure), the results have shown that our proposed framework with all strategies (ELU-SegNet-LMs-CRFs) outperforms the state-of-the-art algorithm, SegNet. On the aerial imagery, our $F1$ (0.876) is greater than SegNet's $F1$ (0.768) by 10.8%. On the satellite imagery, our $F1$ (0.6494) is greater than SegNet's $F1$ (0.465) by 18.44% on average for all five provinces. In terms of the computational cost, our framework requires slightly additional training time compared to the baseline approach, SegNet, by about 6.25% (2–3 h). In our experiment, SegNet's training procedure took approximately 48 h per data set, and finished after 200 epochs with 864 s per epoch. Our framework is built on top of SegNet. There is no additional time required by changing an activation function from ReLU to ELU. The LMs and CRF processes took around 1–2 h and 1 h, consecutively, so there are approximately 2–3 additional hours required on top of SegNet (48 h).

Although our work does not solely rely on the color feature like previous attempts in road extraction, it is recommended for application to high- and very-high resolution remotely-sensed images. It is difficult to identify roads from low- and medium-resolution images, even by humans.

6. Conclusions and Future Work

In this study, we present a novel deep learning network framework to extract road objects from both aerial and satellite images. The network is based on the deep convolutional encoder–decoder network (DCED), called "SegNet". To improve the network's precision, we incorporate the recent activation function, called the exponential linear unit (ELU), into our proposed method. The method is also further improved to detect more road patterns by utilizing landscape metrics and conditional random fields. Excessive detected roads are then eliminated by applying landscape metrics thresholding. Finally, we extend the SegNet network to ELU-SegNet-LMs-CRFs. The experiments were conducted on a Massachusetts road data set as well as THEOS (Thailand) road data sets, and compared to the existing techniques. The results show that our proposed (ELU-SegNet-LMs-CRFs) outperforms the original method on both aerial and satellite imagery for $F1$ as well as for all other baselines.

In future work, more choices of image segmentation, optimization techniques and/or other activation functions will be investigated and compared to obtain the best DCED-based framework for semantic road segmentation.

Acknowledgments: We greatly acknowledge Geo-informatics and Space Technology Development Agency (GISTDA), Thailand, for providing satellite imagery used in this study. T. Panboonyuen thanks the scholarship from Chulalongkorn University to commemorate the 72nd Anniversary of H.M. King Bhumibala Aduladeja.

Author Contributions: The experiment design was carried out by all of the authors. Teerapong Panboonyuen and Peerapon Vateekul performed the experiments and results analysis. Kulsawasd Jitkajornwanich, Siam Lawawirojwong and Panu Srestasathiern supervised research and reviewed results. The article was co-written by the five authors. All authors read and approved the submitted manuscript.

Conflicts of Interest: The authors declare no conflict of interest.

Abbreviations

The following abbreviations are used in this manuscript:

CCL	connected component labeling
CNN	convolutional neural network
CRFs	conditional random fields
DCED	deep convolutional encoder-decoder
DCNN	deep convolutional neural network
DL	deep learning
ELU	exponential linear unit
FCIS	fully convolutional instance-aware semantic segmentation
FCN	fully convolutional network
FN	false negative
FP	false positive
GISTDA	geo-informatics and apace technology development agency
HR	high resolution
LMs	landscape metrics
PASCAL VOC	pascal visual object classes
R-CNN	region-based convolutional neural network
ReLU	rectified linear unit
RGB	red-green-blue
SGD	stochastic gradient descent
TN	true negative
TP	true positive
VGG	visual geometry group
VHR	very-high resolution
VOC	visual object classes

References

1. Poullis, C. Tensor-Cuts: A simultaneous multi-type feature extractor and classifier and its application to road extraction from satellite images. *ISPRS J. Photogramm. Remote Sens.* **2014**, *95*, 93–108.
2. Muruganandham, S. Semantic Segmentation of Satellite Images using Deep Learning. Master Thesis, Lulea University of Technolog, Lulea, Sweden, 2016.
3. Saito, S.; Yamashita, T.; Aoki, Y. Multiple object extraction from aerial imagery with convolutional neural networks. *Electron. Imaging* **2016**, *2016*, 1–9.

4. Long, J.; Shelhamer, E.; Darrell, T. Fully convolutional networks for semantic segmentation. In Proceedings of the IEEE Conference on Computer Vision and Pattern Recognition, Boston, MA, USA, 8–10 June 2015; pp. 3431–3440.
5. Noh, H.; Hong, S.; Han, B. Learning deconvolution network for semantic segmentation. In Proceedings of the IEEE International Conference on Computer Vision, Santiago, Chile, 13–16 December 2015; pp. 1520–1528.
6. Badrinarayanan, V.; Handa, A.; Cipolla, R. Segnet: A deep convolutional encoder-decoder architecture for robust semantic pixel-wise labelling. *arXiv* **2015**, arXiv:1505.07293.
7. Mnih, V. Machine Learning for Aerial Image Labeling. Ph.D. Thesis, University of Toronto, Toronto, ON, Canada, 2013.
8. Badrinarayanan, V.; Kendall, A.; Cipolla, R. Segnet: A deep convolutional encoder-decoder architecture for image segmentation. *arXiv* **2015**, arXiv:1511.00561.
9. Volpi, M.; Ferrari, V. Semantic segmentation of urban scenes by learning local class interactions. In Proceedings of the IEEE Conference on Computer Vision and Pattern Recognition, Boston, MA, USA, 8–10 June 2015; pp. 1–9.
10. Liu, J.; Liu, B.; Lu, H. Detection guided deconvolutional network for hierarchical feature learning. *Pattern Recognit.* **2015**, *48*, 2645–2655.
11. Hong, S.; Noh, H.; Han, B. Decoupled deep neural network for semi-supervised semantic segmentation. *Adv. Neural Inf. Processing Syst.* **2015**, 1495–1503, arXiv:1506.04924.
12. Ronneberger, O.; Fischer, P.; Brox, T. U-net: Convolutional networks for biomedical image segmentation. In Proceedings of the International Conference on Medical Image Computing and Computer-Assisted Intervention, Munich, Germany, 5–9 October 2015; Springer: New York, NY, USA, 2015; pp. 234–241.
13. Andrearczyk, V.; Whelan, P.F. Using filter banks in convolutional neural networks for texture classification. *Pattern Recognit. Lett.* **2016**, *84*, 63–69.
14. Wang, J.; Song, J.; Chen, M.; Yang, Z. Road network extraction: A neural-dynamic framework based on deep learning and a finite state machine. *Int. J. Remote Sens.* **2015**, *36*, 3144–3169.
15. Visin, F.; Ciccone, M.; Romero, A.; Kastner, K.; Cho, K.; Bengio, Y.; Matteucci, M.; Courville, A. Reseg: A recurrent neural network-based model for semantic segmentation. In Proceedings of the IEEE Conference on Computer Vision and Pattern Recognition Workshops, Las Vegas, NV, USA, 26 June–1 July 2016; pp. 41–48.
16. Liu, Z.; Li, X.; Luo, P.; Loy, C.C.; Tang, X. Deep Learning Markov Random Field for Semantic Segmentation. *arXiv* **2016**, arXiv:1606.07230.
17. Koltun, V. Efficient inference in fully connected crfs with gaussian edge potentials. *Adv. Neural Inf. Process. Syst.* **2011**, *2*, 4.
18. Chen, L.C.; Papandreou, G.; Kokkinos, I.; Murphy, K.; Yuille, A.L. Semantic image segmentation with deep convolutional nets and fully connected crfs. *arXiv* **2014**, arXiv:1412.7062.
19. Chen, L.C.; Papandreou, G.; Kokkinos, I.; Murphy, K.; Yuille, A.L. Deeplab: Semantic image segmentation with deep convolutional nets, atrous convolution, and fully connected crfs. *arXiv* **2016**, arXiv:1606.00915.
20. Audebert, N.; Saux, B.L.; Lefèvre, S. Segment-before-Detect: Vehicle Detection and Classification through Semantic Segmentation of Aerial Images. *Remote Sens.* **2017**, *9*, 368.
21. Clevert, D.A.; Unterthiner, T.; Hochreiter, S. Fast and accurate deep network learning by exponential linear units (elus). *arXiv* **2015**, arXiv:1511.07289.
22. Wang, Q.; Fang, J.; Yuan, Y. Adaptive road detection via context-aware label transfer. *Neurocomputing* **2015**, *158*, 174–183.
23. Yuan, Y.; Jiang, Z.; Wang, Q. Video-based road detection via online structural learning. *Neurocomputing* **2015**, *168*, 336–347.
24. Simonyan, K.; Zisserman, A. Very deep convolutional networks for large-scale image recognition. *arXiv preprint* **2014**, arXiv:1409.1556.
25. Li, Y.; Qi, H.; Dai, J.; Ji, X.; Wei, Y. Fully Convolutional Instance-aware Semantic Segmentation. *arXiv* **2016**, arXiv:1611.07709.
26. He, K.; Gkioxari, G.; Dollár, P.; Girshick, R. Mask r-cnn. *arXiv* **2017**, arXiv:1703.06870.
27. Ioffe, S.; Szegedy, C. Batch normalization: Accelerating deep network training by reducing internal covariate shift. *arXiv* **2015**, arXiv:1502.03167.

28. Panboonyuen, T.; Vateekul, P.; Jitkajornwanich, K.; Lawawirojwong, S. An Enhanced Deep Convolutional Encoder-Decoder Network for Road Segmentation on Aerial Imagery. In *Recent Advances in Information and Communication Technology Series*, Proceedings of International Conference on Computing and Information Technology, Tunis, Tunisia, 27–28 April 2017; Volume 566.
29. Kendall, A.; Badrinarayanan, V.; Cipolla, R. Bayesian segnet: Model uncertainty in deep convolutional encoder-decoder architectures for scene understanding. *arXiv* **2015**, arXiv:1511.02680.
30. Kingma, D.; Ba, J. Adam: A method for stochastic optimization. *arXiv* **2014**, arXiv:1412.6980.
31. Gonzalez, R.; Woods, R. *Digital Image Processing*; Prentice Hall: Upper Saddle River, NJ, USA, 2008.
32. McGarigal, K. *Landscape Metrics for Categorical Map Patterns*. Available online: http://studylib.net/doc/7944344/landscape-metrics-for-categorical-map-patterns (accessed on 1 December 2008).
33. Huang, X.; Zhang, L. Road centreline extraction from high-resolution imagery based on multiscale structural features and support vector machines. *Int. J. Remote Sens.* **2009**, *30*, 1977–1987.

remote sensing

MDPI

Article

Hourglass-Shape Network Based Semantic Segmentation for High Resolution Aerial Imagery

Yu Liu [1,2,*], Duc Minh Nguyen [1], Nikos Deligiannis [1], Wenrui Ding [2,3] and Adrian Munteanu [1]

1 ETRO Department, Vrije Universiteit Brussel, Pleinlaan 2, 1050 Brussels, Belgium; mdnguyen@etrovub.be (D.M.N.); ndeligia@etrovub.be (N.D.); acmuntea@etrovub.be (A.M.)
2 School of Electronic and Information Engineering, Beihang University, 37 Xueyuan Rd., Haidian District, Beijing 100191, China; ding@buaa.edu.cn
3 Collaborative Innovation Center of Geospatial Technology, Wuhan 430079, China
* Correspondence: yliub@etrovub.be

Academic Editors: Qi Wang, Nicolas H. Younan, Carlos López-Martínez, Gonzalo Pajares Martinsanz and Prasad S. Thenkabail
Received: 5 April 2017; Accepted: 18 May 2017; Published: 25 May 2017

Abstract: A new convolution neural network (CNN) architecture for semantic segmentation of high resolution aerial imagery is proposed in this paper. The proposed architecture follows an hourglass-shaped network (HSN) design being structured into encoding and decoding stages. By taking advantage of recent advances in CNN designs, we use the composed inception module to replace common convolutional layers, providing the network with multi-scale receptive areas with rich context. Additionally, in order to reduce spatial ambiguities in the up-sampling stage, skip connections with residual units are also employed to feed forward encoding-stage information directly to the decoder. Moreover, overlap inference is employed to alleviate boundary effects occurring when high resolution images are inferred from small-sized patches. Finally, we also propose a post-processing method based on weighted belief propagation to visually enhance the classification results. Extensive experiments based on the Vaihingen and Potsdam datasets demonstrate that the proposed architectures outperform three reference state-of-the-art network designs both numerically and visually.

Keywords: semantic labeling; convolutional neural networks; remote sensing; deep learning; aerial images

1. Introduction

Semantic segmentation in remote sensing aims at accurately labeling each pixel in an aerial image by assigning it to a specific class, such as vegetation, buildings, vehicles or roads. This is a very important task that facilitates a wide set of applications ranging from urban planning to change detection and automated-map making [1]. Semantic segmentation has received much attention for many years, and yet, it remains a difficult problem. One of the major challenges is given by the ever-increasing spatial and spectral resolution of remote sensing images. High spatial resolutions bring the great benefit of being able to capture a large amount of narrow objects and fine details in remote sensing imagery. However, increasing spatial resolutions incurs semantic segmentation ambiguities due to the presence of many small objects within one image and brings along a high imbalance of class distribution, huge intra-class variance and small inter-class differences. For example, a road in the shadows of buildings is similar to buildings with dark roofs, whereas the colors of cars may vary widely, which could cause confusions for the semantic classifiers. High spectral resolutions provide abundant information for Earth observations, but selecting, fusing and classifying hyperspectral images remain significant research challenges in remote sensing [2,3].

Semantic segmentation is often viewed in a supervised learning setting. Like many other supervised learning problems, the general approach for supervised semantic segmentation consists of four main steps: (i) feature extraction; (ii) model design and training; (iii) inference; and (iv) post-processing. In this paper, we focus on the semantic segmentation of high-resolution aerial images and propose a CNN-based solution by following this generic design methodology for supervised-learning.

In the literature, supervised methods have focused much on the feature extraction step and proposed to use a variety of hand-crafted descriptors. Classical methods focus on extracting spatial or spectral features using low-level descriptors, such as GIST [4], ACC [5] or BIC [6]. These descriptors capture both the global color and texture features. In hyperspectral imagery, salient band selection can help feature extraction by reducing the high spectral-resolution redundancy. Lately, mid-level descriptors have became more and more popular in computer vision. One of the most successful descriptors is the bag-of-visual-words (BoVW) descriptor [7,8]. Thanks to its effectiveness, the BoVW descriptor has been widely used in remote sensing in scene recognition and semantic labeling. Sub-space learning techniques were proposed to automatically determine the feature representation of a given dataset by optimizing the feature space [9–11]. By making use of a broad variety of descriptors, an image can be represented by many different features. Each feature has its own advantages and drawbacks; hence, selecting the best features for a specific type of data is particularly important. To achieve this goal, several feature selection frameworks were proposed, such as that of Tokarczyk et al. [12], who designed a boosting-based method to select optimal features in the training process from a vast randomized quasi-exhaustive (RQE) set of feature candidates.

In recent years, the focus was put on feature learning and using learned features for semantic segmentation. Cheriyadat [13] proposed to use sparse coding to guide feature learning. In [14], an improved object detection performance is reached by using a spatial sparse coding bag-of-words model. Recently, the rapid development in deep learning, especially in convolutional neural networks, has brought unified solutions for both feature learning and semantic classification of remote sensing images. Having started as a breakthrough in image classification [15], CNNs have proven to be able to significantly improve state-of-the-art performance in numerous computer vision domains [16]. For example, CNNs with a ResNetarchitecture [17] have won the ILSVRC2015 competition with an error rate of 3.6%, which even surpasses human performance for image classification. For pixel-wise vision tasks like semantic segmentation, CNNs also outperform classical methods [18,19]. In remote sensing, more and more research has been focused on designing and applying CNNs for semantic segmentation. Paisitkriangkrai et al. applied both patch-based CNNs and hand-crafted features to predict the label of each pixel [20]. In addition, conditional random field (CRF) processing follows prediction to provide a smooth final result. Kampffmeyer et al. applied a fully-convolutional network structure to solve pixel-wise labeling of high resolution aerial images in an end-to-end fashion [21]. A weighted loss function was used in their network to address the class imbalance problem. Volpi et al. proposed to apply several learnable transpose convolutional layers to up-sample the scores to the input size, trying to avoid the possible spatial information loss during the up-sampling stage [22]. Nevertheless, existing methods in the literature, especially deep learning-based methods, suffer from two major problems, namely the insufficient spatial information in the inference phase and the lack of contextual information. These problems result in poor segmentations around object boundaries, as well as in other difficult areas, such as shadow regions.

To overcome these problems, in this paper, we introduce a novel hourglass-shaped network architecture for pixel-wise semantic labeling of high-resolution aerial images. Our network is structured into two parts. These parts, namely encoding and decoding, perform down-sampling and up-sampling respectively to infer class maps from input images. Compared to existing designs, our novel contributions are as follows:

- We leverage skip connections with residual units and an inception module in a generic CNN encoder-decoder architecture to improve semantic segmentation of remote sensing data.

This combination benefits multi-scale inference and forwards spatial and contextual information directly to the decoding stage.

- We propose to apply overlapped inference in semantic segmentation, which systematically improves classification performance.
- We propose a weighted belief-propagation post-processing module, which addresses the border effects and smooths the results. This module improves the visual quality, as well as the classification results on segment boundaries.

Extensive experiments on two well-known high resolution remote sensing datasets demonstrate the effectiveness of our proposed architecture compared to state-of-the-art network designs.

The remainder of the paper is organized as follows. A brief review of convolutional neural networks is given in Section 2, followed by an analysis of existing architectures for semantic segmentation in remote sensing. Section 3 presents our proposed hourglass-shaped network architecture and details the training and inference methods. Experimental settings and results are presented in Section 4. Section 5 discusses the proposed approach and experimental results, while Section 6 concludes our work.

2. Convolutional Neural Networks

Convolutional neural networks [15] stem from conventional neural network designs. CNNs consist of layers of neurons, where each neuron has learnable weights and biases. The whole network serves as a complex non-linear function, which transforms the inputs into target variables. The difference with respect to conventional networks is that CNNs comprise specific types of layers and composing elements dedicated to perform specific functions, such as computing convolution, down-sampling or up-sampling operations.

In this section, we first present a short overview of the common layer types employed in CNN architectures. This is subsequently followed by a summary of existing CNN architectures for semantic segmentation.

2.1. Composition Elements

In this section, we present the four basic types of layers that are used in CNNs for semantic segmentation: the convolutional layer, transposed convolutional layer, non-linear function layer and the spatial pooling layer. These are detailed next.

2.1.1. Convolutional Layer

The convolutional layer is the core of CNNs. It can be seen as a bank of simple filters with learnable parameters. As illustrated in Figure 1a, the layer takes the input X of size $W_1 \times H_1 \times C_1$ and convolves it with the filter bank by sliding of stride S and padding the border with P units. The result of this operation is an output volume Y with size $W_2 \times H_2 \times C_2$. Equation (1) formulates the calculation of the output at spatial position (i, j) as:

$$Y_{ij} = W \times N_{ij} + b \tag{1}$$

where (W, b) are the learnable parameters (weights and bias) of the layer, N_{ij} is the corresponding receptive field (or a window surrounding X_{ij}) and $W \times N$ denotes the dot product between W and N.

The spatial dimensions of the output of the convolutional layer are given by $W_2 = (W_1 - F + 2P)/S + 1$, $H_2 = (H_1 - F + 2P)/S + 1$ where F is the size of the receptive field, which also corresponds to the spatial size of the filters. In general, each filter can take different widths and heights, but conventionally, most CNN architectures employ filters with square masks of dimension F. In our work, we consider only filters with square masks.

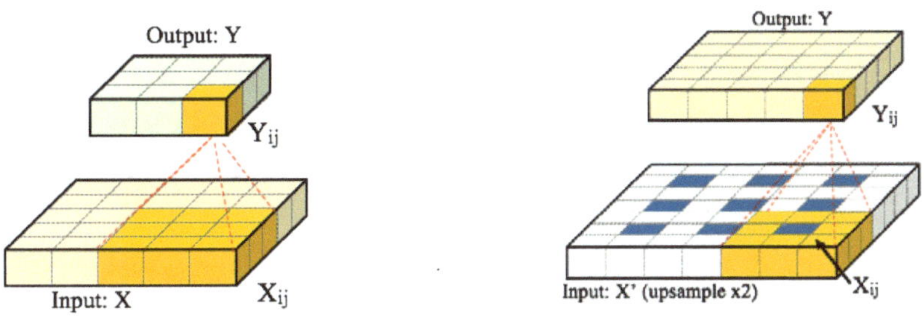

(**a**) Convolutional layer: the input size is $W_1 = H_1 = 5$; the receptive field $F = 3$; the convolution is performed with stride $S = 1$ and no padding ($P = 0$). The output Yis of size $W_2 = H_2 = 3$.

(**b**) Transposed convolutional layer: input size $W_1 = H_1 = 3$; transposed convolution with stride $S = 2$; padding with $P = 1$; and a receptive field of $F = 3$. The output Yis of size $W_2 = H_2 = 5$.

Figure 1. Illustration of elementary modules for the convolutional layer. (**a**) Convolutional layer and (**b**) Transposed convolutional layer.

Neurons in the output volume Y can be considered as filters of size $F \times F \times C_1$. Intuitively, each neuron looks for a specific pattern in the input volume X. Since we want to look for the same pattern across all spatial locations in the input volume, the learnable weights and bias for all neurons in a channel of Y are shared. This is often called parameter sharing, and by doing this, the output volume Y consists of the values obtained when applying C_2 filters on the input volume X. The parameter sharing also reduces the number of weights of a convolutional layer to $C_2 \times F \times F \times C_1$, which is much smaller than that of a fully-connected layer. This helps mitigate the problem of overfitting in neural network training.

2.1.2. Transposed Convolutional Layer

The transposed convolutional layer, also known as the deconvolution layer, was first introduced in [23]. An example of the transposed convolutional layer is shown in Figure 1b. This layer is commonly employed for up-sampling operations in CNNs [18]. As shown in Figure 1b, the input is first up-sampled by a factor of stride S and padded spatially with P units if necessary. After that, convolution is applied to the up-sampled input with a filter bank that has a receptive field of size F. Transposed convolution can be thought of as the inverse operation of convolution. Filter parameters can be set to follow conventional bilinear interpolation [18] or can be set to be learned.

2.1.3. Non-Linear Function Layer

The convolution layer is often followed by a non-linear function layer, also called an activation function. The role of this layer is similar to that of a fully-connected layer in traditional neural networks. This layer introduces non-linearity in the network and enables the network to express a more complex function. Common activation functions include the Sigmoid function, the Tanh function, the rectified linear unit (ReLU) function [24] and the leaky ReLU function [25]. Among these functions, the ReLU function $f(x) = max(0, x)$ is the most commonly used in deep-learning research. In our proposed network design, we also select ReLU as the activation function due to its efficiency and light computational complexity.

2.1.4. Spatial Pooling Layer

The spatial pooling layer is used to spatially reduce the size of the input volume [26]. A small filter (typical size: 2×2 or 3×3) is used to slide through the volume to carry out a simple spatial pooling function. Common pooling functions include max, mean and sum functions. One notes that it is also possible to use the convolutional layer to replace the pooling layer [27]. However, this practice does not necessarily lead to performance benefits and would cost extra memory and training effort [28]. Among the common pooling functions, the max function is most commonly used in the literature. We also employ the max pooling function in our network design.

2.2. CNN Architectures for Semantic Segmentation of Remote Sensing Images

In the literature, there are two basic approaches for semantic segmentation, namely patch-based and pixel-based approaches. In this section, we present an analysis of both categories.

2.2.1. Patch-Based Methods

Patch-based approaches infer the label of each pixel independently based on its small surrounding region. In these approaches, a classifier is designed and trained to predict a single label from a small image patch. In the inference phase, a sliding window is used to extract patches around all pixels in the input image, which are subsequently forwarded through the classifier to get the target labels [29]. Several techniques have been proposed to achieve high performance with patch-based approaches. For instance, replacing the fully-connected layer in the network with convolutional layers can lead to more efficient algorithms by avoiding overlapping computations [22,29]. Multi-scale inference and recurrent refinements can also lead to performance gains [30,31]. Nevertheless, patch-based approaches are often outperformed by pixel-based methods in remote sensing semantic segmentation tasks [21,22]. As a result, in this work, we put more emphasis on the pixel-based approach and follow such a paradigm in our design.

2.2.2. Pixel-Based Methods

Unlike patch-based approaches, pixel-wise methods infer the labels for all of the pixels in the input image at the same time. One of the first CNN architectures for pixel-wise semantic segmentation is the fully-convolutional network (FCN) method introduced by Long et al. in [18]. In this method, a transposed convolutional layer is employed to perform up-sampling. This operation is essential in order to produce outputs of the same spatial dimensions as the inputs.

The FCN architecture was recently employed for semantic segmentation of remote sensing images in [21]. Its architecture, shown in Figure 2, can be divided into two parts, namely encoding and decoding. The latter is depicted within the dotted-line box in the figure. The encoding part follows the same architecture as the VGG-net of [32], which is one of the most powerful architectures for image classification. In Figure 2, the layers A, B, C and D are convolutional layers; their configurations (width, height, depth) are shown in Table 1. Each convolutional layer is followed by a batch normalization layer [33] and ReLU activation function. The final convolutional layer of Type D is followed by a 1×1 convolution, producing an output with scores for each classes. Layer E is a max pooling layer with size $F = 2$ and stride $S = 2$. It performs a down-sampling operation with a factor of two in each dimension. Layer F is a transposed convolutional layer, with filter size $F = 16$ and stride number $S = 8$. It up-samples the scores to original image size. It should be noted that after each pooling layer, the number of filters in the next convolutional layers is doubled to compensate the spatial information loss. To train the network, a median frequency weighted softmax loss layer (Layer G) is appended after the last transposed convolutional layer.

In this FCN design [21], the transposed convolutional layer up-samples the score by a large factor of eight in each dimension. This incurs the risk of introducing classification ambiguities in the up-sampled result. To mitigate this problem, in [22], Volpi et al. proposed to use multiple transposed convolutional layers to progressively up-sample the classification scores. This design is named full patch labeling by learned up-sampling (FPL) [22], its architecture being depicted in Figure 2. Similar to FCN, the FPL network also consists of encoding and decoding modules. However, unlike the FCN design, which incorporate unique layer types in each convolutional module, in FPL, the convolutional modules consist of all four different convolutional layer types, A, B, C and D (see Figure 2). Their configurations are shown in Table 1. Each convolutional module is followed by a max pooling layer, batch normalization layer and leaky ReLU activation. In Figure 2, the pooling and leaky ReLU layers in FPL are grouped together and shown as Layer E. In the decoding stage, three transposed convolutional layers (Type F) are stacked sequentially to spatially up-sample the score to the input image size. They all have an up-sampling factor of two in each spatial dimension. For training, a softmax loss layer (Type G) is appended at the end of the network. The FCL design aims at improving the output classification result by allowing the transpose convolutional layers to learn to recover the fine spatial details. Semantic segmentation results on the Vaihingen dataset reported in [22] show that the FPL network outperforms the FCN design in terms of overall accuracy.

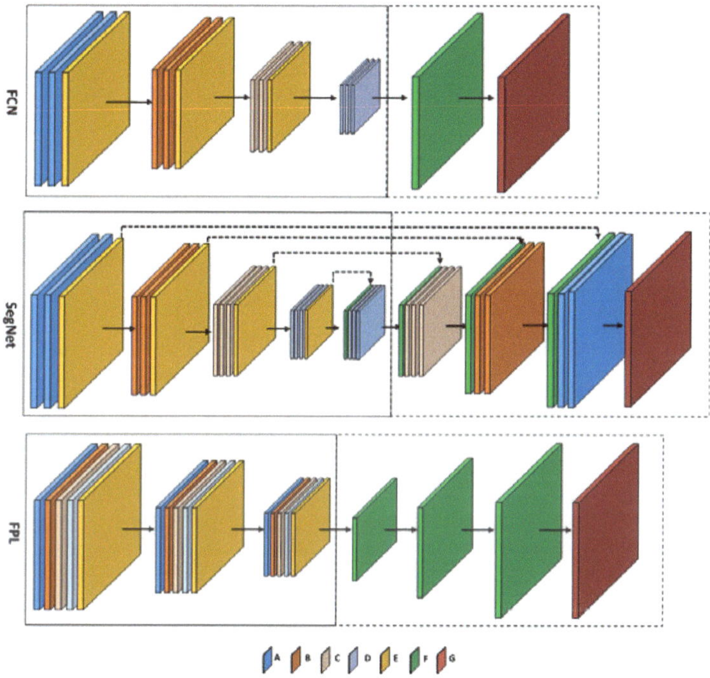

Figure 2. The fully-convolutional network (FCN) [21], SegNet [19] and full patch labeling (FPL) [22] network designs. A, B, C and D are convolutional layers; E is a pooling layer; F is a transposed convolutional layer or unpooling layer (in SegNet); G is a loss layer.

Table 1. Configurations of convolutional and transposed convolutional layer types in the FCN [21], SegNet [19] and FPL [22] architectures.

Layer ID	A	B	C	D	F
FCN	$3 \times 3, 64$	$3 \times 3, 128$	$3 \times 3, 256$	$3 \times 3, 512$	$16 \times 16, 6$
SegNet	$3 \times 3, 64$	$3 \times 3, 128$	$3 \times 3, 256$	$3 \times 3, 512$	Unpooling
FPL	$7 \times 7, 64$	$5 \times 5, 64$	$5 \times 5, 128$	$5 \times 5, 256$	$2 \times 2, 512$

Besides using the transposed convolutional layer for up-sampling in the decoding stage, Vijay et al. proposed to use unpooling in SegNet [19] for pixel-wise segmentation tasks. The encoder part of SegNet (see Figure 2) consists of consecutive convolution layers with uniform 3×3 size filters, followed by ReLU activations and pooling layers. The detailed network parameter settings are given in Table 1. The decoder uses pooling indices computed in the max-pooling step of the corresponding encoder to perform non-linear up-sampling (via an unpooling Layer F), followed by mirror-structured convolution layers to produce the pixel-wise full size label map. Finally, a loss Layer G is attached for network training. The SegNet design aims at preserving the essential spatial information by remembering the pooling indices in the encoding part, which produces state-of-the-art accuracy in generic image segmentation tasks.

Both FCN and FPL architectures suffer from two problems, namely the insufficient spatial information in the decoding stage and the lack of contextual information. Due to the first problem, the FCN and FPL networks often mislabel small objects like cars and produce poor results around object boundaries. Due to the second problem, the lack of contextual information makes it difficult for these architectures to correctly infer classes in difficult areas, such as shadow regions projected by high-altitude buildings and trees.

SegNet effectively mitigates the insufficient spatial information problem by adopting unpooling layers in the decoder part, but it may also suffer from the lack of contextual information. Furthermore, as shown in Table 2, SegNet has three-times more trainable weights than FCN and FPL, making the training phase much more difficult. In this paper, we propose a novel network architecture to address these issues.

Table 2. Trainable weight counts in the FCN [21], SegNet [19], FPL [22] and the proposed HSN architectures.

Network	FCN	SegNet	FPL	HSN
#Trainable weights	7.82M	15.27M	5.66M	5.56M

3. Proposed CNN Architecture for Semantic Segmentation

In this section, we present our novel CNN architecture for semantic segmentation of remote sensing images. The section details first the network design, followed by the training and inference strategies, our post-processing technique and a brief analysis.

3.1. Proposed Hourglass-Shaped Convolutional Neural Network

Our CNN follows a pixel-wise design paradigm, which has been shown to produce state-of-the-art results in semantic segmentation. However, as mentioned in Section 2.2, existing pixel-wise network architectures suffer from the spatial-information loss problem. To overcome this problem, we propose a novel hourglass-shaped network (HSN) architecture. Our HSN design was partially inspired from recent important works in deep learning research [17,34,35].

3.1.1. Network Design

Similar to FCN and FPL, our HSN architecture follows the generic encoder-decoder paradigm, as illustrated in Figure 3. In the figure, the encoder and decoder parts are delimited by continuous and dashed rectangular boxes, respectively. As mentioned in Section 2.2, one key point is to use transposed convolutional layers to progressively up-sample the pixels' class scores to the original spatial resolution of the input image. However, novel components are brought in the network design. Inspired by the hourglass-shaped network introduced for human pose estimation [36] and image depth estimation [35], we propose a network that features (i) multi-scale inference by using inception modules [34] replacing simple convolutional layers and (ii) forwarding information from the encoding layers directly to decoding ones by skip connections.

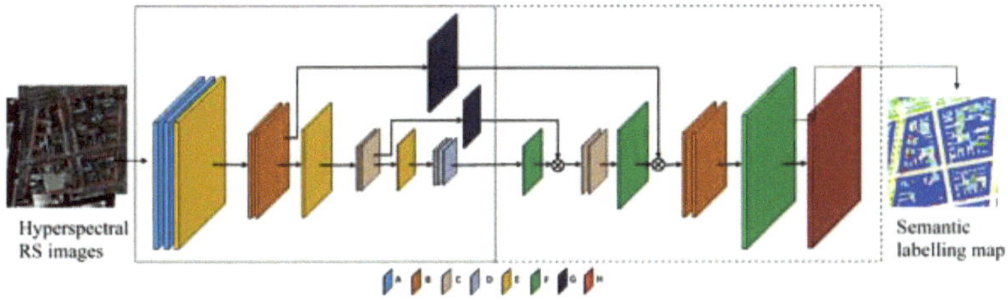

Figure 3. The proposed hourglass-shaped network (HSN) architecture. A and B are convolutional layers; C and D are inception modules; E is the max pooling layer; F is the transposed convolutional layer; G is the residuals modules; H is the loss layer.

The network starts with two layers of A and two layers of B, which are common convolutional layers with filter size $F = 3$. The number of filters are 64 and 128 for Layers A and B, respectively. Each convolution layer is followed by a batch normalization layer and ReLU activation. Layer E is a max pooling layer, with a down-sampling factor of two. Layers C and D are composed of inception modules, as shown in Figure 4a. The configurations of convolutional layers in the inception modules are shown in Table 3. As can be seen from the table, filters of different sizes are assembled in one inception module to enable multi-scale inference through the network.

In the encoding part, after the second Layer B and after Layer C, two skip branches are made with Layer G, forwarding information directly to the corresponding layers in the decoding part. Layer G is a residual module inspired by ResNet [17]. The residual module is shown in Figure 4b, where *conv*1_1 is a bank of 128 filters with size 1×1, and *conv*1_2 is another bank of 128 filters with size 3×3. The input of the module is directly element-wise added to the output of *conv*1_2. It is worth mentioning that, due to the use of filters with size 1×1, the number of trainable weights for the whole network is significantly reduced. As shown in Table 2, the total number of trainable weights of HSN is comparable to that of FPL and nearly three-times less than that of SegNet.

In the decoding part, Layer F serves as the transposed convolutional layer, with the same up-sampling factor of two. After the first and second up-sampling, data directly forwarded from the encoding part are concatenated with the outputs of the transposed convolutional layers. Finally, Layer H, which is a weighted softmax layer, is used in the training phase of the network.

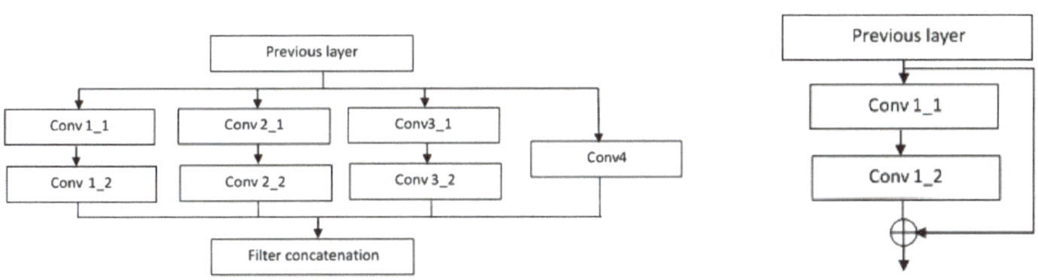

(**a**) Inception module (**b**) Residual module

Figure 4. Composition modules in the proposed HSN architecture. (**a**) Inception module; (**b**) Residual module.

Table 3. Configurations of convolutional layers in the inception modules.

Layer ID	*conv1_1*	*conv1_2*	*conv2_1*	*conv2_2*	*conv3_1*	*conv3_2*	*conv4*
C	$1 \times 1, 128$	$3 \times 3, 128$	$1 \times 1, 64$	$5 \times 5, 32$	$1 \times 1, 32$	$7 \times 7, 32$	$1 \times 1, 64$
D	$1 \times 1, 256$	$3 \times 3, 384$	$1 \times 1, 64$	$5 \times 5, 32$	$1 \times 1, 32$	$7 \times 7, 32$	$1 \times 1, 64$

3.1.2. Median Frequency Balancing

We train our network using the cross-entropy loss function, which is summed over all of the pixels. Nevertheless, the ordinary cross-entropy loss can be heavily affected by the imbalance of the class distribution when applied to high-resolution remote sensing data. To address this problem, the loss for each pixel is weighted based on the median frequency balancing [21,37] technique. The weighted loss for a pixel i is calculated as:

$$L(i) = -\sum_{c=1}^{C} [y_i = c] log(\hat{p}_i^{(c)}) \times w_c \tag{2}$$

where y_i is the ground-truth class of pixel i, w_c is the weight for class c, f_c is the pixel frequency of the class and:

$$w_c = \frac{median(f_c | c \in C)}{f_c} \tag{3}$$

3.2. Training Strategy

We train the network to optimize the weighted cross-entropy loss function using mini-batch stochastic gradient descent (SGD) with momentum [38]. The parameters are initialized following [39]. The learning rate is set to step down 10-times from 1×10^{-5} every 50 epochs, with momentum set to 0.99. The batch size is set to fit the memory. Data augmentation is carried out to mitigate overfitting. The image patches are extracted with size 256 × 256 with 50% of overlap and flipped horizontally and vertically. Each patch is also rotated at 90 degree intervals. In total, this produces eight augmentations for each overlapping patch. We train our network from scratch until the loss converges. Batch normalization is employed, similar to existing network architectures. The training and testing processes are performed on a desktop machine equipped with Nvidia GeForce Titan X (12 Gb vRAM).

3.3. Overlap Inference

In the inference stage, due to the memory limit, the input high-resolution images can be sliced into small non-overlapping patches to feed-in the network. However, this may cause inconsistent segmentation across the patch borders and hence result in degraded accuracy.

To address such boundary effects, overlap inference is employed whereby input images are split into overlapped patches. At the output of the network, the class scores in overlapped areas are averaged. We experimentally justify the benefit of this strategy compared to non-overlapping inference in Section 4.

3.4. Post-Processing with Weighted Belief Propagation

Semantic segmentation for high-resolution remote sensing imagery often requires accurate and visually clear results to serve further automatic processing or manual investigations. However, the raw network output may feature zigzag segment borders and incorrect blobs. Some examples are shown in Figures 5–7. To address this problem, we propose to use weighted belief propagation for post-processing the raw network outputs.

In the proposed HSN architecture, the semantic label for a pixel at an arbitrary position i is determined as $L(i) = \arg\max_c f_i(c)$, where $f_i(c)$ denotes the score of class c for pixel i. This corresponds to the top one class prediction, i.e., the class label with the highest score. Similarly, the top two prediction for any arbitrary pixel is defined as the set of class labels when taking the best two scores for that pixel. We experimentally observed that the top two prediction accuracy for the validation data is around 97% on the Vaihingen dataset. This shows that most of the time, the right labels lie in the top two scores determined by the network.

Let $d_i = f_i(c_1) - f_i(c_2)$, in which $f_i(c_1)$ and $f_i(c_2)$ refer to the top two scores, i.e., the highest and second highest class scores for pixel i, respectively. Intuitively, for a trained network, the higher d_i is, the more confident the network is about its prediction. Therefore, d_i can be thought of as the confidence of the output at position i.

We consider post-processing as a pixel labeling problem and formulate a Markov random field (MRF) model to solve it. A node i in our MRF model corresponds to a pixel in the original image I, which is directly connected to its four spatial neighbors N_i. y_i denotes the class label assigned to node i. We find the optimal labels for the whole image by minimizing the following energy function:

$$E = \sum_{i \in I} E_d(y_i) + \sum_{i,j \in I} E_s(y_i, y_j) \tag{4}$$

where E_d, defined in Equation (5), refers to the data energy term describing how confident the estimated label y_i is; E_s is the smoothness energy defined in Equation (6), which penalizes the inconsistency between node i and its neighbors N_i:

$$E_d(y_i) = \frac{\exp f_i(y_i)}{\Sigma_{j \in C} \exp f_i(j)} \tag{5}$$

$$E_s(y_i, y_j) = v_2 \exp(-\frac{1 - \delta(y_i - y_j)}{T}) \tag{6}$$

where v_2 and T are hyper-parameters, which are set empirically, and $\delta(x)$ is the Dirac delta function. We employ the weighted belief propagation algorithm (WBP) [40,41] to iteratively minimize the energy function E. At each iteration, the update rule of WBP is expressed by Equations (7) and (8) below:

$$m_{ij}(y_j) = w_i \sum_{y_i}^{C} E_s(y_i, y_j) E_d(y_i) \prod_{y_k \in N_i \backslash y_j} m_{ki}(y_i) \tag{7}$$

$$b_i(y_i) = E_d(y_i) \prod_{y_k \in N(y_i)} m_{ki}(y_i) \tag{8}$$

in which $m_{ij}(y_j)$ is the message passed from node i to node j; w_i is the weight for node i, which is set to its confidence value d_i; $b_i(y_i)$ is the belief, which represent how confident the node i is to take label y_i.

The messages are updated until convergence. The final label $\hat{y}_i$ at node i is determined by $\hat{y}_i = \arg\max_{y_i} b_i(y_i)$.

4. Experimental Results

We carried out extensive experiments to assess the effectiveness of our proposed HSN architecture. We employed two well-known datasets in the semantic segmentation literature, namely the Vaihingen and Postdam datasets [42,43]. In this section, we describe our experimental settings and report quantitative and qualitative results. We evaluate the benefits of each of the components in our proposed method and compare our results to those of the FCN [21], SegNet [19] and FPL [22] networks. It should be noted that, as Kampffmeyer et al. [21] do not provide their trained model, we strictly followed their network design and training configuration to reproduce their results. For FPL [22], we have carried out experiments using the original FPL network, which was trained on the Vainhingen and Potsdam datasets and was publicly made available by its authors. Concerning SegNet, it was originally devised and tested on *generic* image datasets; to produce the results, we employed the network provided by the authors and trained it from scratch using the aforementioned remote sensing datasets.

4.1. Datasets

4.1.1. Vaihingen Dataset

The Vaihingen dataset consists of thirty three very high-resolution true orthophoto (TOP) tiles and their corresponding digital surface models (DSMs). Normalized DSMs (nDSMs), which limit the effects of varying ground height, are also provided by Gerke et al. [44]. The tiles have a spatial resolution of 2949 × 2064, with the number of pixels varying from three million to 10 million pixels. Each TOP image is composed of three channels: near-infrared (NIR), red (R) and green (G), with a spatial resolution of 9 cm. Ground-truth labeled images for sixteen out of thirty three tiles were provided by ISPRS. In these images, pixels are labeled as one of the six classes: impervious surfaces, building, low vegetation, tree, car and clutter/background. Examples of the TOP, nDSMs and the corresponding ground truth images are shown in Figure 6.

Following the same training and testing procedures as set by FCN [21] and FPL [22], we used the sixteen annotated tiles in our experiments. Eleven tiles (areas: 1, 3, 5, 7, 13, 17, 21, 23, 26, 32, 37) were selected for training, while the other five tiles (areas: 11, 15, 28, 30, 34) were reserved for testing.

4.1.2. Potsdam Dataset

The Potsdam 2D segmentation dataset includes 38 tiles of high resolution remote sensing images. All of them feature a spatial resolution of 5 cm and have a uniform resolution of 6000 × 6000 pixels. For each tile, five channels are provided, namely near-infrared (NIR), red (R), green (G), blue (B), together with the digital surface models (DSMs). The normalized DSMs (nDSMs) are also made available by Gerke et al. [44]. Twenty four tiles are provided with ground-truth pixel labels, using the same six classes as in the Vaihingen dataset. In our experiments, we employed all five channels, namely NIR-R-G-B and the nDSMs as inputs to the networks. Following the practice in [22], six tiles (02_12, 03_12, 04_12, 05_12, 06_12, 07_12) were selected as testing set, while the other eighteen among the annotated tiles were used for training.

4.2. Evaluation Metrics

To compare our results with the state-of-the-art, we strictly use the same evaluation metrics as in [20–22,42]. Besides the conventional pixel-wise ground truth, in both datasets, border-eroded ground-truth label images are also available. In these images, borders between classes are eroded with a disk radius of three pixels [42,43]. We report our results for both ground-truth versions. All pixels are considered for the conventional pixel-wise ground-truth version, while for the eroded version, border pixels are not accounted for.

We evaluate the performance of the different methods based on three criteria, namely, per-class F-score, overall accuracy and average F-score. The F-score is defined as:

$$\text{F-score} = 2 \times \frac{precision \times recall}{(precision + recall)} \tag{9}$$

The overall accuracy is the total number of correctly-labeled pixels divided by the total number of pixels. In the Vaihingen dataset, the clutter class only accounts for an extremely small number of pixels. As a result, following the common practice [20–22], we neglect the clutter class when reporting the result for this dataset. For the Potsdam dataset, we report the results on all six classes.

Confusion matrices are also provided in the Appendix A for the experiments based on the eroded ground-truth for both datasets. We averaged the values in the confusions matrices across all tested tiles and reported the results for the proposed HSN and the reference techniques.

It is also worth mentioning that ambiguities and mislabeling exist in the provided dataset [20]. There are also some errors for the input normalized DSM [44].

4.3. Overlap Inference Size

Table 4 reports the experimental results obtained on the Vaihingen dataset with four different overlap inference sizes, namely 0%, 25%, 50% and 75%. The results are organized into two groups, corresponding to the two ground-truth versions used in the evaluation: the eroded version (indicated by erGT) and the original version (denoted by GT). It can be observed that the classification performance improves when increasing the overlap size. Overlap inference solves potential border effects at tile boundaries and returns the final classification results by performing a multi-hypothesis prediction of pixel classes instead of single-hypothesis prediction performed in the non-overlapped case. Further increasing the overlap size beyond 75% does not lead to significant improvements in classification performance.

Table 4. Experimental results for different overlap sizes for the Vaihingen dataset.

	Overlap Percent	Imp.Surf	Buildings	Low Veg	Tree	Car	Average F-score	Overall Accuracy
erGT	0%	90.89	94.51	78.83	87.84	81.87	86.79	88.32
	25%	91.18	94.60	79.57	88.19	83.23	87.35	88.67
	50%	91.23	94.64	79.54	88.20	83.74	87.47	88.70
	75%	**91.32**	**94.66**	**79.73**	**88.30**	**83.60**	**87.52**	**88.79**
GT	0%	87.57	92.20	75.03	84.44	75.16	82.88	84.92
	25%	87.88	92.30	75.69	84.76	76.20	83.37	85.27
	50%	87.92	92.34	75.64	84.77	76.61	83.46	85.29
	75%	**88.01**	**92.37**	**75.83**	**84.86**	**76.50**	**83.51**	**85.38**

4.4. Skip Connections and Inception Modules

We further analyze the influence on the performance of our key design components by performing the following experiments: firstly, we remove all skip connections from HSN to study the possible benefit brought by the residual modules; secondly, we keep the residual modules, but replace all of the inception layers with normal convolutional layers to check the influence of inception modules. The results are reported in Table 5 for the first and second set of experiments denoted as HSN-NS (no skip) and HSN-NI (no inception), respectively.

Table 5. Experimental results on the effect of skip connections (Vaihingen dataset). erGT, eroded ground-truth; NS, no skip; NI, no inception.

	Network	Imp. Surf	Buildings	Low Veg	Tree	Car	Average F-Score	Overall Accuracy
erGT	HSN	**90.89**	**94.51**	**78.83**	**87.84**	**81.87**	**86.79**	**88.32**
	HSN-NS	89.40	93.68	78.90	87.57	62.17	82.34	87.48
	HSN-NI	85.63	92.83	74.60	85.74	62.18	80.17	84.89
GT	HSN	**87.57**	**92.20**	**75.03**	**84.44**	**75.16**	**82.88**	**84.92**
	HSN-NS	85.94	91.25	74.78	84.08	56.26	78.46	83.92
	HSN-NI	82.34	90.56	71.05	82.31	55.76	76.41	81.52

From Table 5, it can be observed that both residual and inception modules critically contribute in the HSN design. When removing the residual module, corresponding to the HSN-NS results, a sharp drop in the F-score of the car class is observed. Replacing the inception module with normal convolutional layers leads to a nearly 4% drop in overall accuracy when compared to the eroded ground truth (see NSN-NI results in Table 5).

Visually, from Figure 5, we can observe that the segmentation result of HSN is more coherent compared to the results of HSN-NI and HSN-NS. For instance, when removing the inception layers, there are mislabeled artifacts on the bottom of the image or on the building on the right up corner, the result of HSN being more clean. When removing skip connections, the same effect can also be observed on the road segmentation in the middle bottom of the image.

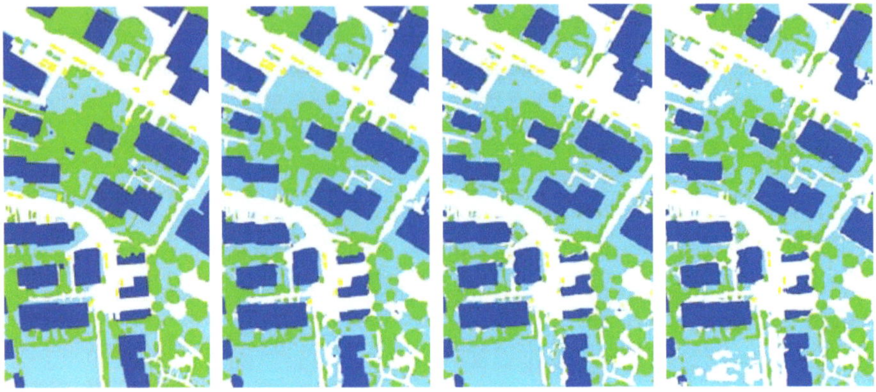

(a) Ground truth **(b)** HSN **(c)** HSN-NS **(d)** HSN-NI

Figure 5. Full tile prediction for tile No. 34. Legend on the Vaihingen dataset: white: impervious surface; blue: buildings; cyan: low vegetation; green: trees; yellow: cars; red: clutter (best viewed in color). (**a**) Ground truth; (**b**) HSN; (**c**) HSN-NS; (**d**) HSN-NI.

4.5. *Performance Evaluations*

In this section, we report extensive experimental results obtained with the proposed HSN and other networks, namely FCN [21], SegNet [19] and FPL [22], which serve as baselines. The HSN applied in this section includes both the inception layers and residual modules. Overlap inference with 75% overlapping size and post-processing with weighted belief propagation are also integrated to demonstrate their effectiveness.

4.5.1. Vaihingen Dataset

Numerical results

Table 6 reports the experimental results obtained in the Vaihingen dataset. The results are organized in the same manner as in Table 4. From the table, it can be observed that the proposed HSN network outperforms the other networks in terms of overall performance. For all classes, except the buildings class, HSN reaches a better performance. Especially in the car class category, HSN significantly outperforms FCN and FPL by more than 10%, and outperforms SegNet by around 5%. Further, by consulting the confusion matrix provided in Table A1, we find that the car class is often mislabeled as impervious surface; trees and low vegetation are also easily confused by the network. It can also be observed that the augmentation in HSN's average F-score is mainly due to the improvement in the car class. Overlap inference (OI) systematically improves the prediction accuracy for each class, bringing up the average F-score to 87.52%. This proves the effectiveness of overlap inference. Post-processing with WBP slightly improves the overall accuracy to 88.82%.

In case border pixels are taken into account (GT), all of the networks perform worse than in the case in which the border pixels are ignored (erGT). This is due to the ambiguities around object boundaries. In case the original GT is used as the reference, post-processing with WBP shows minor performance degradation in some classes, such as car, impervious surface and buildings; yet, the overall accuracy is not affected, and the visual results are improved, as we will see next.

In both cases, all of the networks have high accuracy on the building class thanks to the provided normalized DSM.

Table 6. Experimental results on the Vaihingen dataset [42]. OI, overlap inference.

	Methods	Imp. Surf	Buildings	Low Veg	Tree	Car	Average F-Score	Overall Accuracy
erGT	FCN [21]	89.41	93.80	76.46	86.63	71.32	83.52	86.75
	SegNet [19]	90.15	94.11	77.35	87.40	77.31	85.27	87.59
	FPL [22]	90.43	94.62	78.11	86.81	66.81	83.36	87.70
	HSN	90.89	94.51	78.83	87.84	81.87	86.79	88.32
	HSN + OI	91.32	94.66	79.73	88.30	**83.60**	87.52	88.79
	HSN + OI + WBP	**91.34**	**94.67**	**79.83**	**88.31**	83.59	**87.55**	**88.82**
GT	FCN [21]	85.82	91.27	72.39	83.30	63.10	79.18	83.18
	SegNet [19]	86.68	91.74	73.22	83.99	71.36	81.40	84.07
	FPL [22]	86.62	92.03	73.73	82.73	57.68	78.56	83.69
	HSN	87.57	92.20	75.03	84.44	75.16	82.88	84.92
	HSN + OI	**88.01**	**92.37**	75.83	84.86	**76.50**	**83.51**	85.38
	HSN + OI + WBP	88.00	92.34	**75.92**	**84.86**	75.95	83.41	**85.39**

Table 7 shows the average inference time per image on the test dataset (five images in total). As the proposed HSN employs a more complex architecture, it takes 15.87 s (3.17 s per image) to finish inference on the five test images with an average size of 2563×1810 pixels. While HSN gives the best overall accuracy, it almost doubles the inference time when compared to SegNet, which shows the trade-off between performance and time efficiency. We also note that in [22], the authors of FPL report an average time of 6.2 seconds for inference on the same dataset; this longer inference time for FPL may be caused by the implementation of the network (FCN, SegNet and the proposed HSN are implemented based on the Caffe framework, while FPL is provided in MatConvNet).

Table 7. Average inference time per image tile (on Vaihingen test set) for CNNs.

Network	FCN	SegNet	FPL	HSN
Average inference time (s)	0.78	1.54	6.2	3.17

Qualitative Results

As semantic segmentation often serves other remote sensing applications, visual output quality plays also an important role besides pixel-wise accuracy. For a visual demonstration, Figure 6 shows the labeling results for a complete tile, while Figure 7 zooms into certain areas showing the details of the outputs.

From the figures, it can be seen that the shadows from tall buildings or trees pose great difficulties for semantic labeling. For example, in Figure 7d, we can observe that the road on the left of the building is completely shadowed by the buildings in the middle. In this case, both FCN and FPL methods label this part as the low vegetation class. SegNet managed to detect the road existence, but the segmentation accuracy is quite low. The proposed HSN managed to roughly tag the road. We argue that the inception module design may contribute to this advantage, since using filters of different sizes in one layer allows the network to access multi-scale receptive areas. This aids the network to acquire richer contextual information, which is essential to predict pixels in occluded or shadowed regions.

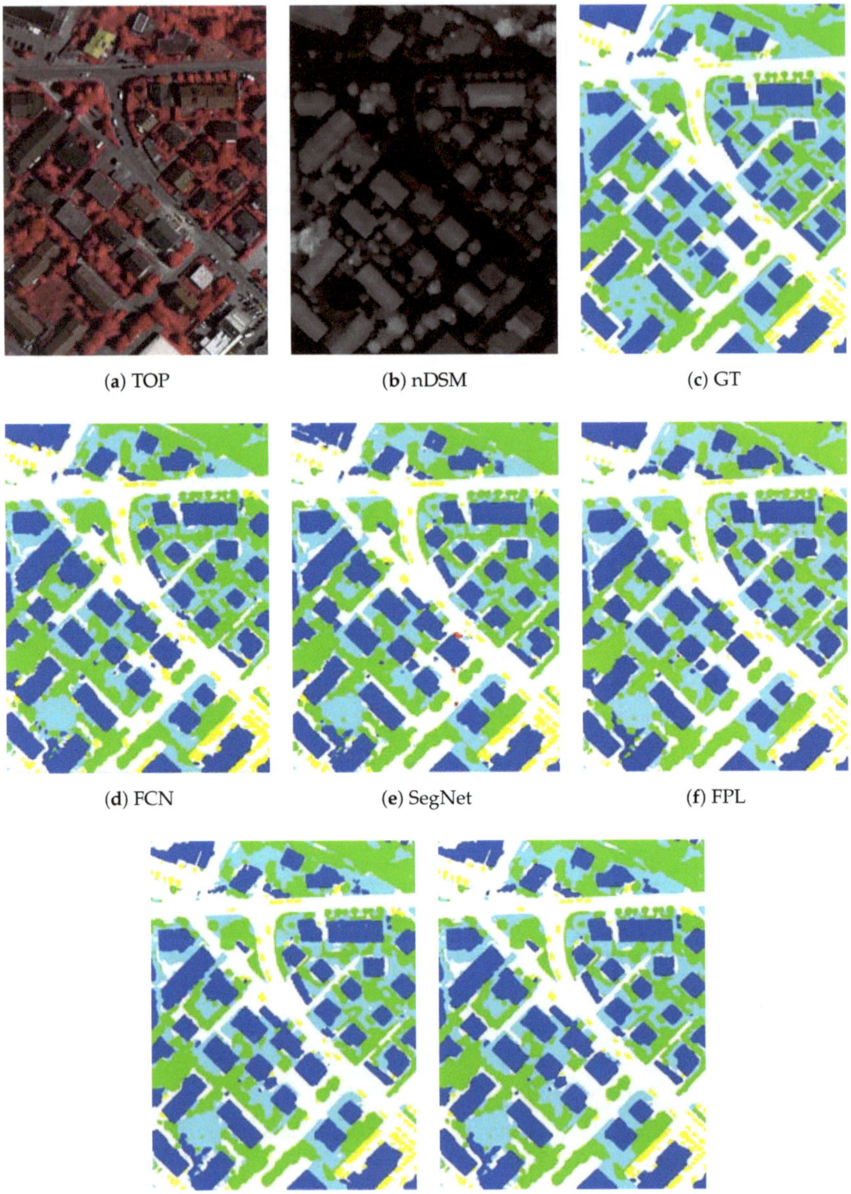

(**a**) TOP (**b**) nDSM (**c**) GT

(**d**) FCN (**e**) SegNet (**f**) FPL

(**g**) HSN (**h**) HSN + WBP

Figure 6. Full tile prediction for No. 30. Legend on the Vaihengen dataset: white: impervious surface; blue: buildings; cyan: low vegetation; green: trees; yellow: cars; red: clutter (best viewed in color). (**a**) TOP, true orthophoto; (**b**) nDSM, normalized DSM; (**c**) GT, Ground truth labeling; (**d**–**g**) the inference result from FCN, SegNet, FPL and HSN respectively; (**h**) HSN + WBP, HSN inference result after WBP post-processing.

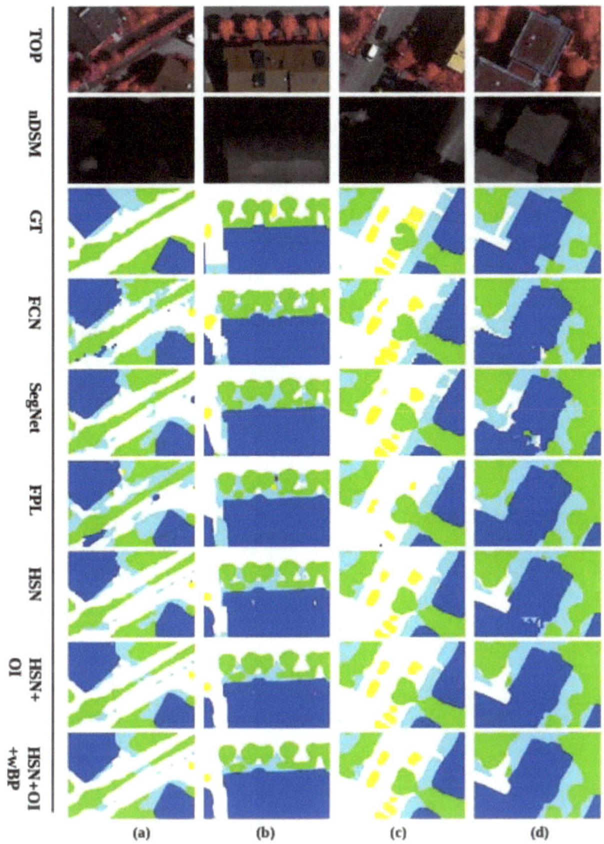

Figure 7. Semantic segmentation results for some patches of Vaihingen dataset. white: impervious surface; blue: buildings; cyan: low vegetation; green: trees; yellow: cars; red: clutter (best viewed in color). Four different tiles from Vaihingen are included: (**a**) a narrow passage; (**b**) shadowed areas from trees and buildings; (**c**) cars in the shadow; and (**d**) building roofs with depth discontinuities.

The car class is quite difficult to deal with, since in the images, cars have various colors leading to a large intra-class difference, whereas dark colored cars are quite similar to the road under shadows (see Figure 7c, for example). FPL fails to label most of the cars, as shown in Figure 7c, due to shadows. HSN successfully detects most of the cars, and the pixel-wise labeling is clear and precise compared to the ground truth. One notes that, since the cars are rather small objects compared to the other classes like buildings, they take fewer pixels in total which in general leads to the class imbalance problem. Median frequency balancing puts a larger weight on the loss for the car class, compensating for its lower occurrence rate in the training phase.

Due to limitations in GPU memory, the high resolution remote sensing image is often split into small-sized patches to perform network inference. As explained in Section 3.3, this practice may possibly introduce erroneous artifacts in the result. For example, in Figure 7d, in the center of the building, both the raw results of HSN and SegNet show artifacts by mislabeling part of the building as low vegetation. However, overlap inference effectively solves this problem by performing multi-hypothesis prediction, whereby the class for each pixel is identified in several overlapping patches. This always leads to more robust results compared to single-hypothesis prediction performed when using non-overlapping inference. Moreover, each patch provides different contextual information for classification, which again contributes to improved classification accuracy compared to the raw HSN.

The provided normalized DSMs help the segmentation of the buildings and trees, as for all of the results, the building segmentation is coherent with the ground-truth. FCN results show obvious zigzags on the class boundaries, while HSN produces sharper and more accurate boundaries (for example, see in Figure 7d the building segment boundaries). Both the hourglass design and post-processing with WBP contribute to this improvement. Thanks to the skip connections with residual modules, information from the encoding stage can be passed directly to the decoding stage. In the early layers of encoding, the data maintain high spatial resolution. Hence, when being fed forward directly to the decoding stage, this information helps with reducing the spatial ambiguities. The WBP in the post-processing stage encourages continuity by propagating the class confidences across pixels throughout the output, hence making the results smoother and correcting small erroneous blobs.

4.5.2. Potsdam Dataset

Numerical Results

Table 8 shows experimental results for the Potsdam dataset. The results are organized similar to those reported in Table 6 for the Vaihengen dataset. The F-score for each class and overall performance are shown respectively for erGT and GT.

Table 8. Experimental results on the Potsdam dataset [42].

	Methods	Imp. Surf	Buildings	Low Veg	Tree	Car	Clutter	Average F-Score	Overall Accuracy
erGT	FCN [21]	89.73	94.87	84.24	76.67	81.64	**28.39**	75.92	87.40
	SegNet [19]	90.44	95.34	83.48	78.49	84.84	25.81	**76.41**	88.37
	FPL [22]	90.59	95.34	83.54	75.58	85.62	17.59	74.71	88.12
	HSN	91.39	95.49	83.91	78.86	86.28	17.77	75.62	88.97
	HSN + OI	91.63	95.65	84.28	79.42	87.47	17.95	76.07	89.29
	HSN + OI + WBP	**91.77**	**95.71**	**84.40**	**79.56**	**88.25**	17.76	76.24	**89.42**
GT	FCN [21]	87.36	93.83	81.73	74.06	76.63	**29.01**	73.77	85.04
	SegNet [19]	88.10	94.37	81.05	75.76	79.40	24.72	**73.90**	86.02
	FPL [22]	88.55	94.31	81.13	72.90	80.52	16.30	72.29	85.93
	HSN	89.01	94.42	81.18	76.09	81.05	15.35	72.85	86.56
	HSN + OI	89.26	94.60	81.54	76.63	82.08	15.36	73.25	86.89
	HSN + OI + WBP	**89.45**	**94.66**	**81.67**	**76.78**	**82.97**	15.12	73.44	**87.05**

From the table, it can be seen that the raw HSN outperforms FCN [21], SegNet [19] and FPL [22] in terms of accuracy for all but the clutter class. In terms of overall accuracy, the proposed HSN outperforms the reference techniques, but SegNet outperform HSN in terms of average F-score and F-score in the clutter class. Overlap inference and WBP help further improve the accuracy, leading to higher overall performance compared to the other three network architectures.

In the Potsdam dataset, the clutter class accounts for a higher percentage of pixels than in the Vaihingen dataset, making it non-negligible. Nevertheless, various types of objects like pedestrian, fence, playground, constructions sets, etc., are all labeled as clutter. This high intra-class variance makes it challenging for the networks to correctly classify clutter pixels (see Figures 8 and 9). As can been seen from Table 8, all of the networks, except FCN and SegNet, give a F-score with values below 20 in the clutter class; Table A2 also shows that the clutter class is often mislabeled as impervious surface and buildings. In contrast, for the building class, all networks reach a high accuracy of more than 95%. We claim that this saturation is due to the provided nDSM channel as the height of the surface gives a strong indication of buildings when combined with other channels' information.

Generally, all of the networks perform better in the Potsdam dataset compared to the Vaihingen dataset, since the images in the Potsdam dataset have a higher spatial resolution (of 5 cm) and an extra blue channel is available. In addition, more data are available in the Potsdam dataset, which leads to better training of the networks.

Qualitative Results

Full tile prediction results from different networks are depicted in Figure 8. Certain clips are selected and shown in Figure 9 to illustrate and analyze the performance of the networks.

(**a**) TOP (**b**) nDSM (**c**) GT

(**d**) FCN (**e**) SegNet (**f**) FPL

(**g**) HSN (**h**) HSN + WBP

Figure 8. Full tile prediction for tile No. 04_12. Legend on the Potsdam dataset: white: impervious surface; blue: buildings; cyan: low vegetation; green: trees; yellow: cars; red: clutter (best viewed in color). (**a**) TOP, true orthophoto; (**b**) nDSM, normalized DSM; (**c**) GT, Ground truth labeling; (**d–g**) the inference result from FCN, SegNet, FPL and HSN respectively; (**h**) HSN + WBP, HSN inference result after WBP post-processing.

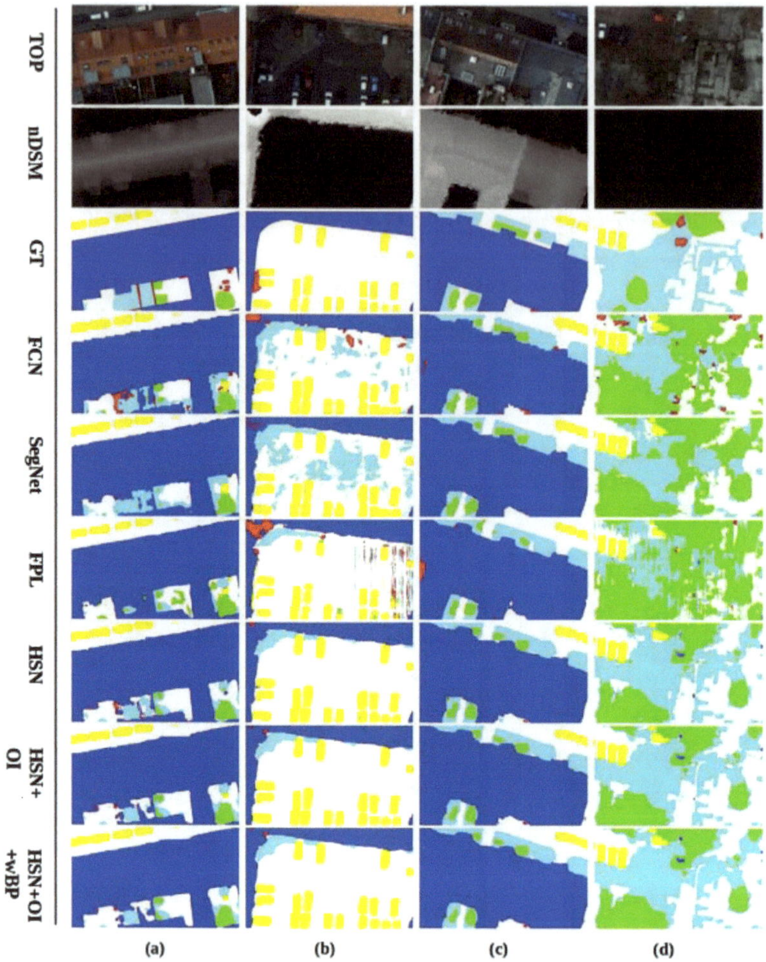

Figure 9. Semantic segmentation results for some patches of Potsdam dataset.white: impervious surface; blue: buildings; cyan: low vegetation; green: trees; yellow: cars; red: clutter (best viewed in color). Four tiles from Potsdam are included: (**a**) buildings with backyards; (**b**) parking lot; (**c**) rooftops; and (**d**) low vegetation areas.

The buildings are always well labeled thanks to the aid provided by the nDSM channel, as shown in Figure 8. However, in Figure 9c, the building roofs show a complex pattern, which leads to partial mislabeling for FCN and FPL. For HSN, the inception module mitigates this problem, as it provides the network with multi-scale contextual information. The same effect can be also observed in Figure 9b. The label maps from both SegNet and FCN are quite noisy, with low vegetation class scattered among the road. FPL provides better results, but still with some mislabeling, like part of the small car in the center is labeled as tree. HSN provides a more accurate and visually improved result.

FPL infers pixel labels using a patch with a smaller size of 64×64 compared to the other networks, which may lead to a restricted receptive area. As shown in Figure 9a, the court yard behind the buildings is mislabeled as buildings, while the other two networks label the yard correctly. As shown in Figure 9, for all three network structures, the clutter areas are hard to accurately label;

from the same figure, we can also observe smoother borders in the class map obtained with the proposed networks.

It is also worth mentioning that, in the Potsdam dataset, most trees are not covered with leaves, which causes difficulties for the networks to detect and segment them accurately. As shown in Figure 9d, trees can be barely distinguished from the surrounding grasses. All reference networks mislabeled nearly half a part of the tree class, but HSN can still correctly distinguish the tree class from the low vegetation.

5. Discussion

The experimental results in Section 4 prove that state-of-the-art performance on well-known remote sensing datasets is achieved with our approach. On the Vaihingen dataset, the proposed approach outperforms reference methods by substantial margins in terms of both average F-score and overall accuracy. On the Potsdam dataset, it is marginally worse than SegNet in term of average F-score, but noticeably better in terms of overall accuracy. Besides, the proposed approach systematically performs better than FCN and FPL on this dataset. In addition, this high performance is achieved with relatively low complexity. The number of trainable parameters in our network is just slightly higher than that of FPL while being far lower than those of FCN and especially SegNet, which has three-times more parameters than the proposed network.

We argue that the effectiveness of the propose approach comes from the highly complementary characteristics of different components in the architecture. Firstly, the use of skip connection with residual modules helps with transferring spatial information from the encoder directly to the decoder, improving the segmentation around object borders. Secondly, the use of inception provides the decoder with richer contextual information. This helps the network to label difficult areas such as roads, which are shadowed and which can be correctly inferred if enough surrounding contexts are available. Richer spatial and contextual information in the decoder also resolves the class ambiguities, especially in high resolution images. Thirdly, the weight balancing employed during training mitigates the class imbalance problem and improves the labeling of classes that account for a small number of pixels, e.g., the car class. This is of particular significance when working with remote sensing data of high resolutions. Fourthly, overlapped inference, which returns the final segmentation making use of multi-hypothesis prediction, diminishes the patch border effects and improves the robustness of the results. Finally, post-processing based on weighted belief propagation corrects the object borders and erroneous small blobs and systematically improves the segmentation results both quantitatively and visually. Combining all of these components, especially the skip connections and inception module in the CNN, mitigates the two problems of existing approaches in the literature, namely insufficient spatial information and lack of contextual information.

Possible directions for future research include: reducing the memory consumption while keeping efficiency and enough spatial and contextual information for high quality segmentation; improving the generalizability of the network by employing more data augmentation. This will be highly relevant in some applications in which large datasets are impossible or expensive to obtain.

6. Conclusions

In this paper, we propose a novel hourglass-shape network architecture for semantic segmentation of high-resolution aerial remote sensing images. Our architecture adopts the generic encoder-decoder paradigm and integrates two powerful modules in state-of-the-art CNNs, namely the inception and residual modules. The former assembles differently-sized filters into one layer, allowing the network to extract information from multi-scale receptive areas. The latter is employed together with the skip connection, feeding forward information from the encoder directly to the decoder, making use more effectively of the spatial information. Furthermore, our solution for remote sensing semantic segmentation employs (i) weighted cross-entropy loss to address the class imbalance problem in the training phase, (ii) overlap processing in inference phase and (iii) weighted belief propagation for post-processing.

Extensive experiments on well-known high-resolution remote sensing datasets demonstrate the effectiveness of our proposed approach. Our hourglass-shaped network outperforms state-of-the-art networks on these datasets in terms of overall accuracy and average F-score while being relatively simpler in terms of the number of trainable parameters.

Acknowledgments: The research has been supported by Fonds Wetenschappelijk Onderzoek (project no. G084117), Brussels Institute for Research and Innovation (project 3DLicornea) and Vrije Universiteit Brussel (PhD bursary Duc Minh Nguyen).

Author Contributions: Yu Liu, Duc Minh Nguyen and Adrian Munteanu proposed the network architecture design. Yu Liu performed the experiments and analyzed the data. Yu Liu, Duc Minh Nguyen wrote the paper. Adrian Munteanu, Wenrui Ding, Nikos Deligiannis revised the paper and provided valuable advices for the experiments.

Conflicts of Interest: The authors declare no conflict of interest.

Appendix A. Confusion Matrices for Vaihingen and Potsdam Datasets

In this section, we report the confusion matrices for both the proposed HSN and the reference techniques tested on the Vaihingen and Potsdam datasets. The values are given in percentages, and the diagonal elements are highlighted in bold.

Appendix A.1. Vaihingen Dataset

Table A1. Confusion matrix on the Vaihingen dataset.

	Reference→ Predictions↓	Imp. Surf	Buildings	Low Veg	Tree	Car
FCN	**Imp. Surf**	**88.99**	3.14	5.39	1.09	1.38
	Buildings	3.89	**93.21**	2.22	0.57	0.11
	Low Veg	5.88	2.47	**74.11**	17.32	0.22
	Tree	0.92	0.37	9.36	**89.35**	0.01
	Car	15.60	1.71	1.00	0.57	**81.11**
SegNet	**Imp. Surf**	**91.68**	2.46	3.87	1.18	0.81
	Buildings	4.16	**93.22**	2.02	0.55	0.55
	Low Veg	6.62	2.44	**73.63**	17.22	0.09
	Tree	1.09	0.34	0.90	**97.66**	0.01
	Car	17.31	0.80	0.90	0.72	**80.27**
FPL	**Imp. Surf**	**91.66**	2.24	4.47	3.96	1.24
	Buildings	3.24	**93.46**	2.74	4.14	1.40
	Low Veg	6.47	1.75	**76.21**	15.50	0.07
	Tree	1.32	0.51	9.87	**88.28**	0.03
	Car	10.06	1.54	2.69	0.3	**85.67**
HSN	**Imp. Surf**	**92.64**	2.54	3.71	0.65	0.46
	Buildings	3.50	**94.11**	2.18	0.18	0.03
	Low Veg	6.73	2.44	**78.09**	12.67	0.08
	Tree	1.24	0.35	10.96	**87.44**	0.01
	Car	15.91	1.96	1.22	0.32	**85.59**

Appendix A.2. Potsdam Dataset

Table A2. Confusion matrix on the Potsdam dataset.

	Reference→ Predictions↓	Imp. Surf	Buildings	Low Veg	Tree	Car	Clutter
FCN	**Imp. Surf**	**85.52**	2.36	4.84	1.80	1.09	4.39
	Buildings	1.69	**93.79**	1.59	1.55	0.30	1.08
	Low Veg	2.24	0.74	**87.19**	8.30	0.12	1.41
	Tree	4.01	0.88	15.06	**78.54**	0.87	0.64
	Car	0.65	0.87	0.13	0.20	**96.74**	1.41
	Clutter	16.87	17.99	12.83	2.87	8.27	**41.17**
SegNet	**Imp. Surf**	**87.42**	1.81	6.72	1.81	0.80	1.43
	Buildings	2.33	**94.19**	1.96	0.84	0.14	0.54
	Low Veg	2.22	0.57	**89.44**	7.34	0.04	0.38
	Tree	2.97	0.94	16.04	**79.07**	0.81	0.17
	Car	1.39	1.07	1.30	0.32	**95.73**	1.37
	Clutter	27.13	17.68	2.18	2.87	7.54	**22.60**
FPL	**Imp. Surf**	**92.08**	2.54	2.42	1.00	0.29	1.66
	Buildings	2.56	**95.21**	0.71	0.42	0.15	0.95
	Low Veg	5.79	0.91	**85.45**	6.84	0.01	1.00
	Tree	7.05	2.36	16.33	**73.12**	0.21	0.92
	Car	4.74	2.56	0.34	2.36	**83.14**	6.85
	Clutter	44.03	13.67	8.42	1.93	1.59	**30.37**
HSN	**Imp. Surf**	**90.69**	2.05	4.92	0.60	0.59	1.14
	Buildings	2.45	**95.06**	1.13	0.66	0.20	0.51
	Low Veg	4.26	0.70	**87.17**	7.56	0.06	0.25
	Tree	3.31	0.95	17.70	**77.05**	0.92	0.07
	Car	1.04	1.87	0.08	0.11	**96.81**	0.08
	Clutter	37.60	26.33	12.72	2.22	7.36	**13.75**

References

1. Rees, W.G. *Physical Principles of Remote Sensing*; Cambridge University Press: Cambridge, UK, 2013.
2. Wang, Q.; Lin, J.; Yuan, Y. Salient band selection for hyperspectral image classification via manifold ranking. *IEEE Trans. Neural Netw. Learn. Syst.* **2016**, *27*, 1279–1289.
3. Yuan, Y.; Ma, D.; Wang, Q. Hyperspectral anomaly detection by graph pixel selection. *IEEE Trans. Cybern.* **2016**, *46*, 3123–3134.
4. Oliva, A.; Torralba, A. Modeling the shape of the scene: A holistic representation of the spatial envelope. *Int. J. Comput. Vis.* **2001**, *42*, 145–175.
5. Huang, J.; Kumar, S.R.; Mitra, M.; Zhu, W.J.; Zabih, R. Image indexing using color correlograms. In Proceedings of the IEEE Conference on Computer Vision and Pattern Recognition (CVPR), 17–19 June 1997; pp. 762–768.
6. Stehling, R.O.; Nascimento, M.A.; Falcão, A.X. A Compact and Efficient Image Retrieval Approach Based on Border/Interior Pixel Classification. In Proceedings of the International Conference on Information and Knowledge Management (CIKM), McLean, VA, USA, 4–9 November 2002; pp. 102–109.
7. Avila, S.; Thome, N.; Cord, M.; Valle, E.; AraúJo, A.D.A. Pooling in image representation: The visual codeword point of view. *Comput. Vis. Image Underst.* **2013**, *117*, 453–465.
8. Lazebnik, S.; Schmid, C.; Ponce, J. Beyond bags of features: Spatial pyramid matching for recognizing natural scene categories. In Proceedings of the IEEE Conference on Computer Vision and Pattern Recognition (CVPR), New York, NY, USA , 17–22 June 2006; pp. 2169–2178.
9. Agrawal, R.; Gehrke, J.; Gunopulos, D.; Raghavan, P. Automatic subspace clustering of high dimensional data. *Data Min. Knowl. Discov.* **2005**, *11*, 5–33.

10. Lu, H.; Plataniotis, K.N.; Venetsanopoulos, A.N. A survey of multilinear subspace learning for tensor data. *Pattern Recognit.* **2011**, *44*, 1540–1551.
11. Peng, X.; Yu, Z.; Yi, Z.; Tang, H. Constructing the L2-graph for robust subspace learning and subspace clustering. *IEEE Trans. Cybern.* **2017**, *47*, 1053–1066.
12. Tokarczyk, P.; Wegner, J.D.; Walk, S.; Schindler, K. Features, Color Spaces, and Boosting: New Insights on Semantic Classification of Remote Sensing Images. *IEEE Trans. Geosci. Remote Sens.* **2015**, *53*, 280–295.
13. Cheriyadat, A.M. Unsupervised feature learning for aerial scene classification. *IEEE Trans. Geosci. Remote Sens.* **2014**, *52*, 439–451.
14. Sun, H.; Sun, X.; Wang, H.; Li, Y.; Li, X. Automatic Target Detection in High-Resolution Remote Sensing Images Using Spatial Sparse Coding Bag-of-Words Model. *IEEE Geosci. Remote Sens. Lett.* **2012**, *9*, 109–113.
15. Krizhevsky, A.; Sutskever, I.; Hinton, G.E. ImageNet Classification with Deep Convolutional Neural Networks. In Proceedings of the Advances in Neural Information Processing Systems (NIPS), Stateline, NV, USA, 3–8 December 2012; pp. 1097–1105.
16. Razavian, A.S.; Azizpour, H.; Sullivan, J.; Carlsson, S. CNN Features Off-the-Shelf: An Astounding Baseline for Recognition. In Proceedings of the IEEE Conference on Computer Vision and Pattern Recognition Workshops (CVPRW), Columbus, OH, USA, 23–28 June 2014; pp. 512–519.
17. He, K.; Zhang, X.; Ren, S.; Sun, J. Deep residual learning for image recognition. In Proceedings of the IEEE Conference on Computer Vision and Pattern Recognition (CVPR), Las Vegas, NV, USA, 27–30 June 2016; pp. 770–778.
18. Long, J.; Shelhamer, E.; Darrell, T. Fully convolutional networks for semantic segmentation. In Proceedings of the IEEE Conference on Computer Vision and Pattern Recognition (CVPR), Boston, MA, USA, 7–12 June 2015; pp. 3431–3440.
19. Badrinarayanan, V.; Kendall, A.; Cipolla, R. Segnet: A deep convolutional encoder-decoder architecture for image segmentation. *Comput. Vis. Pattern Recognit.* **2015**, arXiv:1511.00561.
20. Paisitkriangkrai, S.; Sherrah, J.; Janney, P.; Hengel, V.D.; others. Effective semantic pixel labeling with convolutional networks and conditional random fields. In Proceedings of the IEEE Conference on Computer Vision and Pattern Recognition Workshops (CVPRW), Boston, MA, USA, 7–12 June 2015; pp. 36–43.
21. Kampffmeyer, M.; Salberg, A.B.; Jenssen, R. Semantic Segmentation of Small Objects and Modeling of Uncertainty in Urban Remote Sensing Images Using Deep Convolutional Neural Networks. In Proceedings of the IEEE Conference on Computer Vision and Pattern Recognition Workshops (CVPRW), Las Vegas, NV, USA, 26 June–1 July 2016; pp. 1–9.
22. Volpi, M.; Tuia, D. Dense Semantic Labeling of Subdecimeter Resolution Images With Convolutional Neural Networks. *IEEE Trans. Geosci. Remote Sens.* **2017**, *55*, 1–13.
23. Zeiler, M.D.; Taylor, G.W.; Fergus, R. Adaptive deconvolutional networks for mid and high level feature learning. In Proceedings of the IEEE International Conference on Computer Vision (ICCV), Barcelona, Spain, 6–13 November 2011; pp. 2018–2025.
24. Glorot, X.; Bordes, A.; Bengio, Y. Deep Sparse Rectifier Neural Networks. In Proceedings of the International Conference on Artificial Intelligence and Statistics (AISTATS), Fort Lauderdale, FL, USA, 11–13 April 2011; pp. 315–323.
25. Maas, A.L.; Hannun, A.Y.; Ng, A.Y. Rectifier nonlinearities improve neural network acoustic models. In Proceedings of the International Conference on Machine Learning (ICML), Atlanta, GA, USA, 16–21 June 2013.
26. Saxe, A.; Koh, P.W.; Chen, Z.; Bhand, M.; Suresh, B.; Ng, A.Y. On random weights and unsupervised feature learning. In Proceedings of the International conference on machine learning (ICML), Bellevue, WA, USA, 28 June–2 July 2011; pp. 1089–1096.
27. Springenberg, J.T.; Dosovitskiy, A.; Brox, T.; Riedmiller, M. Striving for simplicity: The all convolutional net. In Proceedings of the International Conference on Learning Representations (ICLR), San Diego, CA, USA, 7–9 May 2015.
28. Lee, C.Y.; Gallagher, P.W.; Tu, Z. Generalizing pooling functions in convolutional neural networks: Mixed, gated, and tree. In Proceedings of the International Conference on Artificial Intelligence and Statistics (AISTATS), Cadiz, Spain, 9–11 May 2016.

29. Sermanet, P.; Eigen, D.; Zhang, X.; Mathieu, M.; Fergus, R.; LeCun, Y. Overfeat: Integrated recognition, localization and detection using convolutional networks. In Proceedings of the International Conference on Learning Representations (ICLR), Banff, AB, Canada, 14–16 April 2014.
30. Farabet, C.; Couprie, C.; Najman, L.; LeCun, Y. Learning hierarchical features for scene labeling. *IEEE Trans. Pattern Anal. Mach. Intell.* **2013**, *35*, 1915–1929.
31. Pinheiro, P.; Collobert, R. Recurrent convolutional neural networks for scene parsing. In Proceedings of the International Conference on Machine Learning (ICML), Beijing, China, 21–26 June 2014; pp. 82–90.
32. Simonyan, K.; Zisserman, A. Very deep convolutional networks for large-scale image recognition. In Proceedings of the International Conference on Learning Representations (ICLR), San Diego, CA, USA, 7–9 May 2015.
33. Ioffe, S.; Szegedy, C. Batch Normalization: Accelerating Deep Network Training by Reducing Internal Covariate Shift. In Proceedings of the International Conference on Machine Learning (ICML), San Diego, CA, USA, 7–9 May 2015; pp. 448–456.
34. Szegedy, C.; Liu, W.; Jia, Y.; Sermanet, P.; Reed, S.; Anguelov, D.; Erhan, D.; Vanhoucke, V.; Rabinovich, A. Going deeper with convolutions. In Proceedings of the IEEE Conference on Computer Vision and Pattern Recognition (CVPR), Boston, MA, USA, 7–12 June 2015; pp. 1–9.
35. Chen, W.; Fu, Z.; Yang, D.; Deng, J. Single-image depth perception in the wild. In Proceedings of the Advances in Neural Information Processing Systems (NIPS), Barcelona, Spain, 5–10 December 2016; pp. 730–738.
36. Newell, A.; Yang, K.; Deng, J. Stacked Hourglass Networks for Human Pose Estimation. In Proceedings of the European Conference on Computer Vision (ECCV), Amsterdam, The Netherlands, 8–16 October 2016; pp. 483–499.
37. Eigen, D.; Fergus, R. Predicting depth, surface normals and semantic labels with a common multi-scale convolutional architecture. In Proceedings of the IEEE International Conference on Computer Vision (ICCV), Los Alamitos, CA, USA, 7–13 December 2015; pp. 2650–2658.
38. Tieleman, T.; Hinton, G. Lecture 6.5-rmsprop: Divide the gradient by a running average of its recent magnitude. *Neural Netw. Mach. Learn.* **2012**, *4*, 26–30.
39. He, K.; Zhang, X.; Ren, S.; Sun, J. Delving Deep into Rectifiers: Surpassing Human-Level Performance on ImageNet Classification. In Proceedings of the IEEE International Conference on Computer Vision (ICCV), Los Alamitos, CA, USA, 7–13 December 2015; pp. 1026–1034.
40. Murphy, K.P.; Weiss, Y.; Jordan, M.I. Loopy belief propagation for approximate inference: An empirical study. In Proceedings of the Conference on Uncertainty in artificial intelligence (UAI), Stockholm, Sweden, 30 July–1 August 1999; pp. 467–475.
41. Kschischang, F.R.; Frey, B.J.; Loeliger, H.A. Factor graphs and the sum-product algorithm. *IEEE Trans. Inf. Theory* **2001**, *47*, 498–519.
42. ISPRS Vaihingen 2D Semantic Labeling Dataset. Available online: http://www2.isprs.org/commissions/comm3/wg4/2d-sem-label-vaihingen.html (accessed on 24 May 2017).
43. ISPRS Potsdam 2D Semantic Labeling Dataset. Available online: http://www2.isprs.org/commissions/comm3/wg4/2d-sem-label-potsdam.html (accessed on 24 May 2017).
44. Gerke, M. *Use of the Stair Vision Library within the ISPRS 2D Semantic Labeling Benchmark (Vaihingen)*; Technical Report; University of Twente: Enschede, The Netherlands, 2015.

remote sensing

MDPI

Article

Learning Dual Multi-Scale Manifold Ranking for Semantic Segmentation of High-Resolution Images

Mi Zhang [1], Xiangyun Hu [1,2,*], Like Zhao [1], Ye Lv [1], Min Luo [1] and Shiyan Pang [2,3]

1 School of Remote Sensing and Information Engineering, 129 Luoyu Road, Wuhan University, Wuhan 430079, China; mizhang@whu.edu.cn (M.Z.); lenci_zhao@whu.edu.cn (L.Z.); ye.lv@whu.edu.cn (Y.L.); luo_min@whu.edu.cn (M.L.)

2 Collaborative Innovation Center of Geospatial Technology, Wuhan University, Wuhan 430079, China; psy@whu.edu.cn

3 School of Resource and Environmental Sciences, 129 Luoyu Road, Wuhan University, Wuhan 430079, China

* Correspondence: huxy@whu.edu.cn; Tel.: +86-27-6877-1528

Academic Editors: Qi Wang, Nicolas H. Younan, Carlos López-Martínez and Prasad S. Thenkabail
Received: 09 March 2017; Accepted: 12 May 2017; Published: 19 May 2017

Abstract: Semantic image segmentation has recently witnessed considerable progress by training deep convolutional neural networks (CNNs). The core issue of this technique is the limited capacity of CNNs to depict visual objects. Existing approaches tend to utilize approximate inference in a discrete domain or additional aides and do not have a global optimum guarantee. We propose the use of the multi-label manifold ranking (MR) method in solving the linear objective energy function in a continuous domain to delineate visual objects and solve these problems. We present a novel embedded single stream optimization method based on the MR model to avoid approximations without sacrificing expressive power. In addition, we propose a novel network, which we refer to as dual multi-scale manifold ranking (**DMSMR**) network, that combines the dilated, multi-scale strategies with the single stream MR optimization method in the deep learning architecture to further improve the performance. Experiments on high resolution images, including close-range and remote sensing datasets, demonstrate that the proposed approach can achieve competitive accuracy without additional aides in an end-to-end manner.

Keywords: semantic segmentation; deep convolutional neural networks; manifold ranking; single stream optimization; high resolution image

1. Introduction

Semantic image segmentation, which aims to classify each pixel into one of the given categories, is an important task for understanding [1–3] and inferring objects [4–6] and their observed relations in a scene. As a bridge towards high-level tasks, semantic segmentation is adopted in various applications in computer vision and remote sensing areas, such as autonomous vehicle driving [2,7,8], human pose estimation [9–11], remote sensing image interpretation [12–16], and 3D reconstruction [17–19]. Over the last five years, remarkable success in the semantic scene labeling area has been gained through the usage of convolutional neural networks (CNNs) [20–26] in dense prediction. Naturally, the ability to express the complex input–output relationships and the efficiency of integrated into the end-to-end learning framework are attributed to fully convolutional neural networks (FCNs).

Generally, recent semantic segmentation methods have often been formulated to convert the architecture of existing CNNs to FCNs [22,23,27–29]. Coarse pixel-wise labeling is obtained by multi-scale and dilation strategies, whereas the fine segmentation is conducted by optionally integrating contextual information into the output map. Although active research has been conducted on these aspects, semantic image segmentation remains a challenging issue because of the complexity of balancing contextual information and pixel-level accuracy [24,26,29–31]. Contextual relationships model the interactions between predicted labels and provide structured cues for dense prediction. In addition, various approaches in formulating compatible relations within contextual information have been proposed for performance improvement. A dominant paradigm for modeling contextual relationships advocates the use of the conditional random field (CRF), which computes unary and pairwise potentials for further refinement, on top of CNNs [25,26,32]. By combining CRF and FCNs, the interactions between the predicted labels and the contextual information are well counterpoised. A few of these approaches utilize the pairwise or higher order CRF [33,34] as a post-process on FCN output to preserve sharp boundaries, while others formulate pixel-wise labeling problems with the CRF in conjunction with FCNs [26,35] in a unified framework and train in an end-to-end manner.

These leading approaches perform dense prediction in a discrete domain, and hence end with learning approximate mean-filed inference or graph model optimization in a fixed number of iterations. However, these methods require additional aides and do not guarantee the convergence of the inference process to the global or even local optimum [26,35]. Therefore, the efficiency of the expressive power might be lost if the uncertainty of the predicted label increases in each iteration.

In this paper, we propose a novel approach to address the issues mentioned. In contrast to the approaches optimized in the discrete domain, we formulate the pixel-wised labeling issue as a special case of manifold ranking (MR) problem in a continuous domain on top of CNNs. Motivated by [36–39], we observe that the MR model has a unique global optimal solution and is guaranteed to converge as a type of graphical model. Moreover, global optimum can be efficiently obtained by solving a linear equation. Unlike the Gaussian graphical models [26,35] that are performed in unary and pairwise streams in the sub-networks, we use the embedded manifold ranking optimization method only on a single stream by constructing the Laplacian matrix generated from possible pairs of vertices.

Numerous strategies without CRF optimization have been established to improve the semantic segmentation accuracy in the FCN or deconvolution manner, and each of them has its own superiorities [25–27,29,35,40]. In order to take these advantages, we propose a framework called dual multi-scale manifold ranking (**DMSMR**) network to estimate the predicted labels in an end-to-end fashion. In each scale, the dilated and non-dilated convolution layers are jointly optimized by MR. With the dual multi-scale contextual information, the combined results achieve competitive accuracy without any additional aides. An overview of our proposed approach is illustrated in Figure 1.

We conduct experiments on high spatial resolution remote sensing and close-range images to validate the effectiveness of the proposed approach. Both high spatial resolution remote sensing and close-range images are rich in details, such as texture and color information. The close-range images can be viewed as a special kind of high-resolution images and can guide us to find better CNN architectures to deal with high-resolution remote sensing images. In summary, the main contributions of our work are as follows:

(1) **Multi-label MR graphical model for semantic segmentation.** Unlike existing approaches that utilize the CRF as the post-processing or approximate inference in the discrete domain, we propose to model the MR method for semantic segmentation in a continuous domain. Our model is end-to-end optimization that can be linearly solved and guarantee a global optimal solution.

(2) **Embedded feedforward single stream optimization method.** In contrast to Gaussian graphical models, we propose an embedded single stream technique that requires only the Laplacian matrix obtained from pairs of vertices, which makes the gathering of the low-level cues as the contextual information more efficient.

(3) **Dual multi-scale manifold ranking network.** We adopt the multi-scale strategy to construct the dual-dilated and non-dilated networks and jointly optimize them with MR in a unified framework for semantic image segmentation. Our model is the first work to back propagate through manifold ranking and integrate it to deep learning architecture in the area of remote sensing.

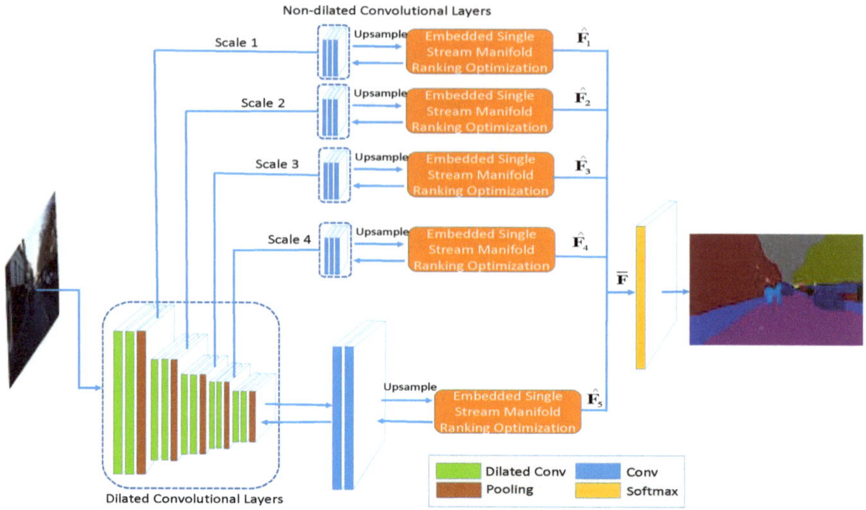

Figure 1. Dual multi-scale manifold ranking (**DMSMR**) network overview. For each dilated convolutional layer, a non-dilated convolution layer is applied following the pooling layer in each scale. The dilated and non-dilated convolution layers form a dual layer, in which the corresponding layers are optimized with the embedded feedforward single stream manifold ranking network. The scale factor is implicitly represented by the pooling layer in each block. Figure 2 illustrates how to embed the manifold ranking optimization method into the single stream network (marked with orange color in this figure). The optimized outputs of each scale, that is, $\widehat{\mathbf{F}}_l$ generated in each scale, are combined by Equation (17).

2. Related Work

In the past decade, convolutional networks have been driving advances in object recognition. Therefore numerous semantic segmentation tasks have preferred to conduct dense prediction based on CNNs in both computer vision and remote sensing areas.

In [21,41,42], each semantic object is refined from region proposals by CNN features. In contrast to these instance-awarded methods, Mostajabi et al. [20] and Dai et al. [43] sought to preserve the shape information for dense labeling from superpixel-wise proposal segments. Unlike these approaches, Farabet et al. [44] trained on the entire image with a multi-scale strategy and labeled each pixel with the category of the object to which it belongs. A remarkable breakthrough was recently made by Shelhamer et al. [22]. In their approach, the contemporary classification networks are converted into fully convolutional networks (FCNs) and the fully connected layers in standard CNNs are viewed as convolutional layers with large receptive filed. Yu et al. [23] presented a dilated module to the FCNs to further broaden the receptive filed on the convolution layer. Instead of adopting the "convolution by pooling" schema in the classification task, they used a dilated rectangular prism on the convolution layer to preserve the receptive field. Similar strategies were proposed by Chen et al. [24,45] in the DeepLab framework. With the "hole" algorithm, a fast dense prediction is allowed on modern GPUs. More recently, Bearman et al. [46] exploited a point-wise annotation for semantic segmentation, which creatively makes a better trade-off between training annotation cost and accuracy. In the area of remote sensing, Camps-Valls and Romero et al. [47,48] proposed the use of greedy layer-wise unsupervised pre-training that learns sparse features for remote sensing image classification. Tschannen et al. [49] introduced a structured CNNs that employed Haar wavelet-based trees for identifying the semantic category of every pixel of remote sensing image. Piramanayagam et al. [50] further exploited a multi-path CNNs that support both true ortho photo and digital surface model (DSM) for land cover classification. Marcu et al. [51] presented a dual path, that is VGG-Net path and

AlexNet path, to learn local and global representations of aerial images. Yuan et al. [52] also conducted a dual clustering approach to select optimal bands for hyperspectral remote sensing images. A few of these approaches are derived from basic FCNs model and utilize different strategies, such as multi-scale pyramid pooling, dilated convolution, dual-path representations and symmetric structures, to improve the inner stability of CNNs. Nevertheless, these networks still need to be properly initialized from pre-trained model or additional aides and may lack of contextual information.

As special extensions to basic FCNs, the symmetric encoder/decoder structures are further exploited by numerous recent approaches. The symmetric structures are able to delineate finer details of the upsampled output. In [27,53], Kendall and Badrinarayanan et al.presented a novel semantic pixel-wise segmentation architecture called SegNet. The architecture comprises an encoder that corresponds to the 13 convolutional layers in the VGG-16 [54] model and a decoder that maps the final features up to the full original image resolution. A similar schema was proposed by Hong and Hyeonwoo et al. [28,55]. The deconvolution network is composed of convolution and unpooling layers, thereby mitigating the limitations of the existing methods based on FCNs and handling the object in multi-scale space. Such symmetric structures were also applied to remote sensing image processing. Audebert et al. [56] exploited the symmetric encoder-decoder structure to detect, segment and classify different varieties of wheeled vehicles from aerial images. Huang et al. [57] further presented two symmetric encoder-decoder structures to fine-tune the networks from RGB and NRG bands. Audebert et al. [58] combined the SegNet with SVM to generate the geometrically corrected orthophoto. These symmetric structures reduce possible loss in the uppooling procedure of CNNs. However, these approaches may suffer from the bottleneck of GPU memory and contextual information embedding in terms of training remote sensing images.

To overcome the above issues, various recent approaches use discrete CRF models on top of CNNs. The CRF is an effective optimization method that can further boost the performance of semantic segmentation. By exploiting more contextual information, the rough segments are able to infer the relationship with their surround pixels. In [32], dense CRF [33,40] was proposed for the first time to improve accuracy by utilizing CRF as a post-process with more contextual information for fine predictions on top of CNNs. To make better use of contextual cues, Lin et al. [29] exploited an efficient "patch-patch" and "patch-background" schema to improve the performance by the CRF optimization framework. Unlike [24], Zheng et al. [25] introduced a mean-filed approximate inference for CRF that has the advantages of CNNs and CRF and is easily incorporated to the CNNs. Furthermore, Vemulapalli et al. [35] and Chandra et al. [26] proposed the use of simple Gaussian conditional random field (G-CRF) for the task of structured prediction. In [59], CNN features and hand-crafted features were combined to parse remote sensing images. Alam et al. [60] further introduced a framework that combined with mean-field CRF inference and performed superpixel-level labellings on remote sensing images. Sherrah [60] exploited the effectiveness of CRF post-processing approaches on top of CNNs and analyzed the major differences between close-range and remote sensing images in terms of contextual information. However, these methods either serve as a post-process or end up with mean-filed approximation and do not guarantee a global optimum.

Hence, we combine CNNs with the MR method, which guarantees a global optimum in a unified framework without additional aides. The multi-scale, dilated convolution strategies are also incorporated on top of CNNs to better delineate visual objects in remote sensing images. The MR method presented in [36,37,39] is an effective graph-based ranking method that aims to find the underlying cluster or manifold structure from the given datasets. For a query data, MR seeks to rank the neighborhood relevance to the query. Unlike the CRF, the optimal ranking solution is linearly solved by constructing the Laplacian matrix [61] from the neighbor contextual information, guaranteeing a global optimal solution in the continuous domain. Quan [62] et al. exploited such characteristics and utilized the MR based co-segmentation strategy to find the common objects contained in a set of relevant images. Wang et al. [63] presented an effective approach for salient band selection for hyperspectral image classification via MR. They put the band vectors in a more accurate manifold space and treats the salient band selection problem from a ranking perspective. Moreover, the MR method has been applied to estimate the status of many other complex low-level vision tasks, such as saliency detection [38,64], image retrieval [65,66] and visual tracking [67]. Considering that the semantic segmentation task also has a manifold structure, in which each pixel is first assigned several probabilities (ranking) that belong to the given categories (underlying clusters) and then the maximum probability is obtained from them, we apply the MR method embedded in CNNs to exploit the efficient global optimal solution to semantic segmentation. Combined with dilated, multi-scale strategies, the MR method, which can further establish the foundation of the dense prediction task in an end-to-end manner, is introduced into this field.

3. Manifold Ranking Formulation

The goal of graph based manifold ranking is to find the rank of a neighborhood relevance to the query node. Learning the objective function, which defines the relevance of neighbor nodes and query, is necessary to achieve this goal. In this section, we briefly describe the manifold ranking algorithm in a binary case and further extend it to multi-label situations that can be applied to the semantic segmentation task.

3.1. Binary Manifold Ranking

In [65], a binary ranking method was presented to exploit the manifold structure of the dataset. Given a set of data $\chi = \{x_1, x_2, \cdots x_i \cdots x_n\} \subset \Re^n$, a graph $G = (V, E)$ with vertices $v \subset V$ and edges $e \subset E$ can be built on the dataset. The weight between two vertices $v_i \in V$ and $v_j \in V$ connected by the edge $e_{ij} \in E$ is denoted by w_{ij}, which is commonly obtained by the Gaussian weighting function, that is $w_{ij} = \exp\left(-\gamma\|x_i - x_j\|^2\right)$. In addition, the degree of a vertex v_i is given by $d_i = \sum_j w_{ij}$. If we let $\mathbf{f} : \mathbb{R}^2 \to \mathbb{R}^n$ as a ranking function that assigns each point x_i two ranking scores $f_0(x_i)$, $f_1(x_i)$, and $\mathbf{y} = [y_1, y_2, \cdots y_n]^T$ as a binary indication vector in which $y_i = 1$ if $f_1(x_i) > f_0(x_i)$ and $y_i = 0$ otherwise, then the normalized Laplacian matrix $\overline{\mathbf{L}}$ is computed as follows:

$$\overline{\mathbf{L}} = \mathbf{D}^{-\frac{1}{2}}\mathbf{W}\mathbf{D}^{-\frac{1}{2}}, \tag{1}$$

where $\mathbf{D} = diag\{d_1, d_2, ..., d_n\}$, $\mathbf{W} = [\mathrm{w_{ij}}]$ and each element $\overline{L}_{ij}$ in the normalized Laplacian matrix $\overline{\mathbf{L}}$ is given by

$$\overline{L}_{ij} = \begin{cases} -w_{ij} & \text{if } i \text{ and } j \text{ are connected} \\ d_i & \text{if } i = j \\ 0 & \text{otherwise} \end{cases}. \tag{2}$$

And the optimal ranking score vector is obtained by solving the following manifold ranking energy function associated with $\mathbf{f}$:

$$E(\mathbf{f}) = \arg\min_{\mathbf{f}} \sum_{v_i \in V} \|\mathbf{f}(x_i) - \mathbf{f}^*(x_i)\|^2 + \lambda \sum_{e_{ij} \in E} w_{ij}\|\mathbf{f}(x_i) - \mathbf{f}(x_j)\|^2, \tag{3}$$

where $\mathbf{f} = \begin{bmatrix} \mathbf{f}(x_1) & \mathbf{f}(x_2) & \ldots & \mathbf{f}(x_i) & \ldots & \mathbf{f}(x_j) & \ldots & \mathbf{f}(x_n) \end{bmatrix}^T$, $\mathbf{f}(x_i) = \begin{bmatrix} f_0(x_i) & f_1(x_i) \end{bmatrix}^T$ and $\mathbf{f}^*(x_i) = \begin{bmatrix} f_0{}^*(x_i) & f_1{}^*(x_i) \end{bmatrix}^T$ is the corresponding posterior probability for each point x_i. The first term in the energy function is a data term that encodes the intrinsic structure of the given dataset, and the second term is a smoothness term that demonstrates the compatibility of the query data with its neighbors. By minimizing the energy function, we obtain the optimal ranking scores $\hat{\mathbf{f}}$ through the following close form

$$\begin{aligned}\hat{\mathbf{f}} &= (\mathbf{I}+2\lambda\mathbf{D}-2\lambda\mathbf{W})^{-1}\mathbf{f}^{*} \\ &= (\mathbf{I}+2\lambda\mathbf{L})^{-1}\mathbf{f}^{*}\end{aligned}, \tag{4}$$

where $\hat{\mathbf{f}} = \left[\begin{array}{cccccccc}\hat{\mathbf{f}}(x_1) & \hat{\mathbf{f}}(x_2) & \dots & \hat{\mathbf{f}}(x_i) & \dots & \hat{\mathbf{f}}(x_j) & \dots & \hat{\mathbf{f}}(x_n)\end{array}\right]^T$, $\hat{\mathbf{f}}(x_i) = \left[\begin{array}{cc}\hat{f}_0(x_i) & \hat{f}_1(x_i)\end{array}\right]^T$, $\mathbf{f}^* = \left[\begin{array}{cccccc}\mathbf{f}^*(x_1) & \mathbf{f}^*(x_2) & \dots & \mathbf{f}^*(x_i) & \dots & \mathbf{f}^*(x_n)\end{array}\right]^T$, $\mathbf{D}$ is the degree of the vertices, $\mathbf{W}$ is the compatibility matrix as mentioned in Equation (1), $\mathbf{L}$ is the unnormalized Laplacian matrix which is calculated as $\mathbf{L} = \mathbf{D} - \mathbf{W}$, λ is the regulation coefficient, and $\mathbf{I}$ is the identity matrix. Given the optimal ranking score, the corresponding optimal indicator $\hat{y}_i$ for each query point x_i can be achieved by:

$$\hat{y}_i = \begin{cases}1 & \text{if } \hat{f}_1(x_i) > \hat{f}_0(x_i) \\ 0 & \text{otherwise}\end{cases}. \tag{5}$$

3.2. Multi-Label Manifold Ranking

In the previous subsection, we introduced the basic optimal manifold ranking solution to a binary label case in which each data has a unique binary indicator. In this section, we extend the binary MR solution to a multi-label situation and apply it to the semantic image segmentation task. As previously mentioned, given a set of pixels $\{p_i\}_{i=1}^{M\times N} \in \mathcal{P}$ in an image $\mathcal{I}^{M\times N}$, the semantic segmentation task aims to classify each pixel p_i to one of the K possible classes. In other words, each pixel p_i is assigned to the index of the K variables that has the highest ranking score. If we let $f_k(p_i)$ denote the ranking score of the kth class, then the assigned label for pixel p_i is

$$y_i^*(f) = \underset{k\in\{1,2....K\}}{\arg\max}\ f_k(p_i), \tag{6}$$

where k also stands for the index corresponding to the ranking score $f_k(p_i)$ in each pixel.

Although our objective is to assign each pixel p_i an optimal discrete label y_i^*, we first find the optimal ranking vector $\hat{\mathbf{f}}(p_i) = \left[\begin{array}{cccccccc}\hat{f}_1(p_i) & \hat{f}_2(p_i) & \dots & \hat{f}_i(p_i) & \dots & \hat{f}_j(p_i) & \dots & \hat{f}_K(p_i)\end{array}\right]^T$ and then obtain the optimal ranking score $f_k^{\max}(p_i) = \max\{\hat{f}_1(p_i), \hat{f}_2(p_i), ..., \hat{f}_K(p_i)\}$ of each pixel p_i in the continuous domain. Once we find the maximum ranking score for each pixel , we can easily assign each pixel p_i a discrete label using Equation (6).

In order to compute the optimal ranking score vector $\hat{\mathbf{f}}(p_i)$ for the multi-label situation, we extend the Equation (3) to the generalized energy function as follows:

$$E\left(\tilde{\mathbf{f}}\right) = \underset{\tilde{\mathbf{f}}}{\arg\min} \sum_{v_i\in V} \mu_i \left\|\tilde{\mathbf{f}}(p_i) - \tilde{\mathbf{f}}^*(p_i)\right\|^2 + \lambda \sum_{e_{ij}\in E} w_{ij} \left\|\tilde{\mathbf{f}}(p_i) - \tilde{\mathbf{f}}(p_j)\right\|^2, \tag{7}$$

where $\tilde{\mathbf{f}} = \left[\tilde{\mathbf{f}}(p_1)\, \tilde{\mathbf{f}}(p_2) \dots \tilde{\mathbf{f}}(p_i) \dots \tilde{\mathbf{f}}(p_j) \dots \tilde{\mathbf{f}}(p_n)\right]^T$, $\tilde{\mathbf{f}}(p_i) = \left[\tilde{f}_1(p_i)\, \tilde{f}_2(p_i) \dots \tilde{f}_K(p_i)\right]^T$ and $\tilde{\mathbf{f}}^*(p_i) = \left[\tilde{f}_1^*(p_i)\, \tilde{f}_2^*(p_i) \dots \tilde{f}_K^*(p_i)\right]^T$ is the posterior probability vector for each pixel p_i. The corresponding cost function in matrix form is

$$\begin{aligned}\mathcal{L}\left(\tilde{\mathbf{F}}\right) &= 2\lambda\mathrm{Trace}\left(\tilde{\mathbf{F}}^T\left(\tilde{\mathbf{D}} - \tilde{\mathbf{W}}\right)\tilde{\mathbf{F}}\right) + \mathrm{Trace}\left(\left(\tilde{\mathbf{F}} - \tilde{\mathbf{F}}^*\right)^T \mathbf{D}_\mu \left(\tilde{\mathbf{F}} - \tilde{\mathbf{F}}^*\right)\right) \\ &= 2\lambda\mathrm{Trace}\left(\tilde{\mathbf{F}}^T\tilde{\mathbf{L}}\tilde{\mathbf{F}}\right) + \mathrm{Trace}\left(\left(\tilde{\mathbf{F}} - \tilde{\mathbf{F}}^*\right)^T \mathbf{D}_\mu \left(\tilde{\mathbf{F}} - \tilde{\mathbf{F}}^*\right)\right),\end{aligned} \tag{8}$$

where $\tilde{\mathbf{D}}$ and $\tilde{\mathbf{W}}$ are the matrices accounting for the degree of the vertices and the compatibility for the multi-label case, $\tilde{\mathbf{L}} = \tilde{\mathbf{D}} - \tilde{\mathbf{W}}$ denotes the unnormalized Laplacian matrix in a multi-label situation, $\mathbf{D}_\mu = diag\{\mu_1, \mu_2, ..., \mu_n\}$ is a diagonal matrix containing the regulation coefficients μ_i for the data term, and $\widetilde{\mathbf{F}} \in \Re^{(M\times N)\times K}$ and $\widetilde{\mathbf{F}}^* \in \Re^{(M\times N)\times K}$ are built from the ranking score vectors $\widetilde{\mathbf{f}}(p_i) \in \Re^K$ and $\widetilde{\mathbf{f}}^*(p_i) \in \Re^K$, respectively.

The solution is optimal if the derivative of $\widetilde{\mathbf{F}} \in \Re^{(M\times N)\times K}$ yields zero in the Equation (8). Specifically,

$$\frac{d\mathcal{L}\left(\widetilde{\mathbf{F}}\right)}{d\widetilde{\mathbf{F}}} = 4\lambda\widetilde{\mathbf{F}}^T\widetilde{\mathbf{L}} + 2\left(\widetilde{\mathbf{F}} - \widetilde{\mathbf{F}}^*\right)^T\mathbf{D}_\mu = 0. \tag{9}$$

Therefore, the optimal solution to Equation (8) is

$$\begin{aligned}\widehat{\mathbf{F}} &= \left(2\lambda\left(\widetilde{\mathbf{D}} - \widetilde{\mathbf{W}}\right) + \mathbf{D}_\mu\right)^{-1}\mathbf{D}_\mu\widetilde{\mathbf{F}}^* \\ &= \left(2\lambda\widetilde{\mathbf{L}} + \mathbf{D}_\mu\right)^{-1}\mathbf{D}_\mu\widetilde{\mathbf{F}}^*.\end{aligned} \tag{10}$$

4. Deep Multi-Scale Manifold Ranking Network

In order to incorporate the proposed multi-label manifold ranking algorithm into CNNs, we first embed the single stream manifold ranking method in a feedforward schema [20] into the network. Figure 2 shows how the MR optimization method is embedded to the single stream network. By exploiting the derivative of the learned parameters with respect to the loss function in the feedforward network, the required parameters can be trained in an end-to-end manner. Then, a **DMSMR** network is constructed, in which the dilated [23] and non-dilated networks are jointly optimized through the multi-scale feedforward manifold ranking method.

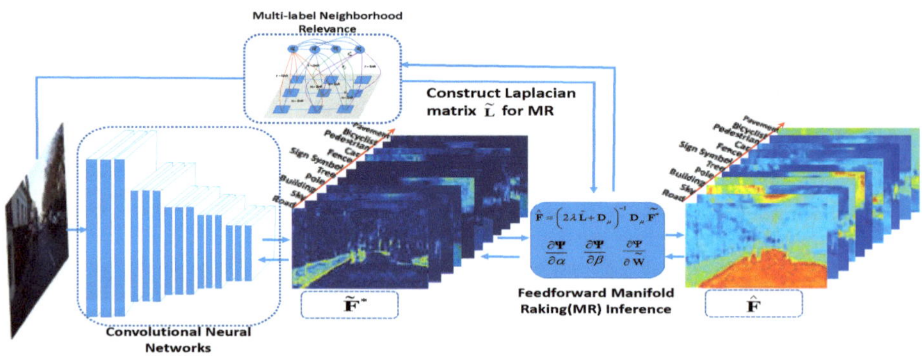

Figure 2. The embedded feedforward single stream manifold ranking optimization network. The output of the convolutional features that upsample to full image resolution for each class, such as road, sky and building, within the CamVid dataset [68,69] depicted in the figure, serves as the initial manifold ranking score $\widetilde{\mathbf{F}}^*$ to be optimized. By applying the feedforward MR inference with the contextual information extracted from the input image, the optimal MR score $\widehat{\mathbf{F}}$ of each class can be obtained by Equation (10). The only requirement for the proposed network is the multi-label neighborhood relationship, which is designed for constructing the Laplacian matrix $\widetilde{\mathbf{L}}$ in a single stream rather than the unary and pairwise streams presented in [26,29].

4.1. Embedded Feedforward Single Stream Manifold Ranking Optimization

Calculating the derivative of the learned parameters with respect to the loss is necessary to train the embedded multi-label MR network. In the following subsection, we describe the inference procedure for the manifold ranking algorithm in detail and describe the mathematical form of the derivatives.

4.1.1. Manifold Ranking Inference

As previously mentioned, the key to manifold ranking is seeking the neighborhood relevance to the query. For the semantic segmentation task, we model the neighborhood relevance, that is, the smoothness term in Equation (7), as follows:

$$\begin{aligned}\mathbf{k}\left(f_i, f_j\right) &= w_{ij}\left\|\widetilde{\mathbf{f}}(p_i) - \widetilde{\mathbf{f}}\left(p_j\right)\right\|^2 = \alpha\mathbf{k}_1\left(f_i, f_j\right) + \beta\mathbf{k}_2\left(f_i, f_j\right) \\ &= \alpha \exp\left(-\frac{\left\|p_i - p_j\right\|^2 + \left\|I_i - I_j\right\|^2}{2\sigma_1^2}\right) + \beta\exp\left(-\frac{\left\|p_i - p_j\right\|^2}{2\sigma_2^2}\right),\end{aligned} \tag{11}$$

where the first kernel (Here the notation "kernel" refers to Potts model.) $\mathbf{k}_1\left(f_i, f_j\right)$ measures the color likelihood nearby and the second term $\mathbf{k}_2\left(f_i, f_j\right)$ weights the spatial position correlation. α and β are the smoothness coefficients. I_i and I_j are the image intensities, p_i and p_j denote the position of neighbor pixels, σ_1 and σ_2 are the degrees of nearness and similarity, respectively.

Our formulation is based on the energy hypothesis proposed in Equation (7), and the inference to this energy function for semantic image segmentation is provided by Equation (10). Given the smoothness relationship in Equation (11), we can easily setup a single stream manifold ranking neuron from the compatibility matrix $\widetilde{\mathbf{W}}$. We only need to learn the smoothness coefficients α, β and the compatibility matrix $\widetilde{\mathbf{W}}$ in a single stream rather than two streams in the network, that is, the unary and pairwise streams presented in [26,29].

In our work, the preceding parameters are determined by the stochastic gradient descent (SGD) algorithm [70]. The loss between the predicted label y_l^* in Equation (6) and the ground truth y is indicated by $\mathbf{\Psi}\left(y_l^*, y\right)$. Therefore, the derivative of y_l^* with respect to $\mathbf{\Psi}\left(y_l^*, y\right)$ can be represented as

$$\nabla\mathbf{\Psi} = \frac{\partial\mathbf{\Psi}}{\partial y_l^*}. \tag{12}$$

In our experiment, we use softmax loss as the loss function. In order to learn the smoothness coefficients α, β and compatibility matrix $\widetilde{\mathbf{W}}$ via SGD, the derivatives of these parameters, that is, $\frac{\partial\mathbf{\Psi}}{\partial\alpha}$, $\frac{\partial\mathbf{\Psi}}{\partial\beta}$, $\frac{\partial\mathbf{\Psi}}{\partial\widetilde{\mathbf{W}}}$, for loss function are necessary.

4.1.2. Derivative to Smoothness Coefficients

The derivative of loss function in terms of smoothness coefficients α, β can be obtained by the chain rule shown below:

$$\frac{\partial\mathbf{\Psi}}{\partial\alpha} = \nabla\mathbf{\Psi}\cdot\frac{\partial y_l^*}{\partial\alpha} = \nabla\mathbf{\Psi}\cdot\delta\cdot\mathbf{k}_1\left(f_i, f_j\right) \tag{13}$$

$$\frac{\partial\mathbf{\Psi}}{\partial\beta} = \nabla\mathbf{\Psi}\cdot\frac{\partial y_l^*}{\partial\beta} = \nabla\mathbf{\Psi}\cdot\delta\cdot\mathbf{k}_2\left(f_i, f_j\right) \tag{14}$$

where δ is the delta function for the derivative result of $\widetilde{\mathbf{F}}$ with respect to y_l^*, $\mathbf{k}_1\left(f_i, f_j\right)$ and $\mathbf{k}_2\left(f_i, f_j\right)$ are the smoothness kernels.

4.1.3. Derivative to Compatibility Matrix

Similar to the derivative to smoothness coefficients, the derivative of the compatibility matrix $\widetilde{\mathbf{W}}$ with respect to the loss function can be represented as

$$\frac{\partial\mathbf{\Psi}}{\partial\widetilde{\mathbf{W}}} = \nabla\mathbf{\Psi}\cdot\frac{\partial y_l^*}{\partial\widetilde{\mathbf{W}}} = \nabla\mathbf{\Psi}\cdot\delta\cdot\nabla\mathbf{\Psi}\otimes\widetilde{\mathbf{F}}, \tag{15}$$

where $\widetilde{\mathbf{F}}$ is the linear solution to manifold ranking energy function in Equation (8), $\otimes$ denotes the Kronecker product, and δ and $\nabla\mathbf{\Psi}$ represent the same as those in Equation (14).

4.2. Dual Multi-Scale Manifold Ranking Network

The recent works [23,26–28] shows that the CNNs have a remarkable capacity to implicitly represent a feature in a multi-scale space. The capacity of CNNs to find objects is dramatically improved by training the dataset with varying kernel sizes or pooling rates (i.e., in an atrous spatial pyramid pooling (ASPP) [24] schema). Meanwhile, the dilated rectangular prism of convolution layers [23] is a natural choice for boosting the performance and broadening the receptive field in each layer.

In our proposed network, we use a dual approach to handle the scale variability for the semantic image segmentation task. On the basis of the work presented in [71], the dual approach aims to minimize the residual produced by dilated and non-dilated networks in each scale. Let $\widehat{\mathbf{F}}_l : \mathbb{R} \to \mathbb{R}$ be a discrete function that denotes the optimized ranking score with scale factor of l in a given convolutional layer and $s : \Omega_s \to \mathbb{R}$ be the dilation filter in this layer. The objective function for the **DMSMR** network can be represented as follows:

$$\begin{aligned}\mathbf{\Delta} &= \mathbf{\Theta}\left(\left(\widehat{\mathbf{F}}_l * s\right)(\overline{x}), \widehat{\mathbf{F}}_l(\overline{x})\right) \\ &= \frac{1}{2}\left\|\theta_1\left(\widehat{\mathbf{F}}_l * s\right)(\overline{x}) - \theta_2\widehat{\mathbf{F}}_l(\overline{x})\right\|^2,\end{aligned} \tag{16}$$

where $\mathbf{\Theta}(\cdot)$ denotes the objective function that measures the output difference between the dilated and non-dilated layers, $\overline{x}$ is the input obtained from the non-dilated convolutional layer with a scale factor of $l-1$, $*$ is the dilated convolution operator, and θ_1 and θ_2 represent the weights for the dual outputs, that is, the dilated output $\left(\widehat{\mathbf{F}}_l * s\right)(\overline{x})$ and the non-dilated output $\widehat{\mathbf{F}}_l(\overline{x})$, respectively. The objective function in Equation (16) models how to combine the dilated and non-dilated layers in the l scale. The final results from all the scales are fused by the following equation:

$$\overline{\mathbf{F}} = \frac{1}{N}\sum_{l=1}^{N}\widehat{\mathbf{F}}_l, \tag{17}$$

where $\overline{\mathbf{F}}$ is the fusion result for the multi-scale space, and N is the total number of scales. Figure 1 illustrates the corresponding relation.

5. Experiments

We have devised two groups of experiments on high resolution datasets, including close-range images (PASCAL VOC dataset and CamVid dataset) and remote sensing images (ISPRS Vaihingen dataset and EvLab-SS dataset), to validate the effectiveness of our model and find the approach that can be potentially applied to remote sensing image processing. For fair evaluation, the first group, which includes the PASCAL VOC dataset [72] and ISPRS Vaihingen dataset [73], is designed for comparison with a few recent state-of-the-art methods whose results are publicly available online. In this group, we evaluate our model by submitting the results to the server, wherein the ground truth of testing images are not available to all researchers. The second group, which includes the CamVid dataset [68,69] and the EvLab-SS dataset (See Section 5.2.2), is used to evaluate the capacity of the proposed **DMSMR** approach by comparing the methods that employ only one of the three strategies, namely, multi-scale convolution (**MS**), broader receptive field (**Dilated**) and MR optimization (**MR-opti**) approaches. The detailed structures of the network with different strategies are explained in the Appendix (See Figure A1 and Table A1).

In our **DMSMR** model, the first five blocks are developed from the standard VGG-16 [54] structures, which comprise convolutional and non-dilated convolutional layers. The dilation kernel sizes are 6, 4, 2, 2, and 1 pixels. For each scale, the pooling layer is followed by the non-dilated layers, which comprise three convolutional layers. The parameters of our implementation are shown in detail in Table 1. The dilated and non-dilated layers are optimized with single stream manifold ranking algorithm and fused by Equation (17). The structure is illustrated in Figure 1. In the table and figure, the "ReLU" active function [74] is implicitly employed in each convolutional layer. In our model, all layers are randomly initialized without using the pre-trained VGG-16 model. The hyper-parameters, such as learning rate, momentum and weight decay, are confirmed via cross validation. The entire net is trained in an end-to-end manner using SGD algorithm. σ_1 and σ_2 in Equation (11) are both set to 3.0 as in [32] in our experiments.

The proposed architectures are implemented using Caffe [75] in a Win7 x64 platform running on an Intel I7-4790 CPU @ 3.6 GHz with a single GeForce GTX 1070 (8 GB RAM). Our model requires only 5523 MB of GPU memory. The source code is implemented with C++ and the model is publicly available at http://earthvisionlab.whu.edu.cn/zm/SemanticSegmentation/index.html.

Table 1. Detailed implementation of the **DMSMR** networks.

(a) Dilated Convolutional Layers						
Scale (Block)	**Name**	**Kernel Size**	**Pad**	**Dilation**	**Stride**	**Number of Output**
0	input	-	-	-	-	3
1	conv1-1	3 × 3	6	6	1	64
	conv1-2	3 × 3	6	6	1	64
	pool1	3 × 3	1	0	2	64
2	conv2-1	3 × 3	4	4	1	128
	conv2-2	3 × 3	4	4	1	128
	pool2	3 × 3	1	0	2	128
3	conv3-1	3 × 3	2	2	1	256
	conv3-2	3 × 3	2	2	1	256
	pool3	3 × 3	1	0	2	256
4	conv4-1	3 × 3	2	2	1	512
	conv4-2	3 × 3	2	2	1	512
	pool4	3 × 3	1	0	1	512
5	conv5-1	3 × 3	2	2	1	512
	conv5-2	3 × 3	2	2	1	512
	pool5	3 × 3	1	0	1	512
-	fc6	3 × 3	1	1	1	1024
	fc7	1 × 1	0	1	1	1024
*	fc8	1 × 1	0	1	1	12
-	Manifold Ranking Optimization					12
(b) Non-Dilated Convolutional Layers						
Scale (Block)	**Name**	**Kernel Size**	**Pad**	**Dilation**	**Stride**	**Output Size**
1	pool1-conv-1	3 × 3	1	1	4	128
	pool1-conv-2	1 × 1	0	1	1	128
	pool1-conv-3	1 × 1	0	1	1	12
-	Manifold Ranking Optimization					12
2	pool2-conv-1	3 × 3	1	1	2	128
	pool2-conv-2	1 × 1	0	1	1	128
	pool2-conv-3	1 × 1	0	1	1	12
-	Manifold Ranking Optimization					12
3	pool3-conv-1	3 × 3	1	1	1	128
	pool3-conv-2	1 × 1	0	1	1	128
	pool3-conv-3	1 × 1	0	1	1	12
-	Manifold Ranking Optimization					12
4	pool4-conv-1	3 × 3	1	1	1	128
	pool4-conv-2	1 × 1	0	1	1	128
	pool4-conv-3	1 × 1	0	1	1	12
-	Manifold Ranking Optimization					12

5.1. Experiment on Close-Range Dataset

As a special kind of high resolution image, close-range imagery is rich in details. Many of the recent breakthroughs [12–14,49,50,76] in the remote sensing area used pre-trained models on this kind of high resolution images. We adopt the PASCAL VOC dataset [72] and the CamVid dataset [68,69] for training and testing and to evaluate the proposed approach on close-range images. The PASCAL VOC dataset is a golden standard measurement for semantic segmentation evaluation. Meanwhile the CamVid dataset comprises a small number of training images, and is a reasonable choice for evaluating the intrinsic capacity of the network that employs different strategies.

5.1.1. Evaluation on PASCAL VOC

The PASCAL VOC 2012 segmentation dataset comprises 20 object classes and one background class with 1464, 1449 and 1456 images for training, validation and testing, respectively. In our experiment, we use the extra annotations provided by [77], thus obtaining a total of 10582 augmented training images [77,78]. For our model, we resize the images to 321 × 321 pixels as in DeepLab model [24] and evaluate the model by remotely submitting the predictions to the test server (Our result on PASCAL VOC dataset is available at http://host.robots.ox.ac.uk:8080/leaderboard). The evaluation metric is the standard Intersection-over-Union (IoU) averaged across the 21 classes. In our experiment, we train the model with the initial learning rate, momentum and weight decay 1e-9, 0.9 and 0.0005, respectively. The momentum and weight decay terms are utilized as suggested in FCNs framework [22]. In addition, the learning rate is confirmed via cross validation. The initial parameters for smoothness coefficients α and β are set to 3 and 5, respectively. The drop-out layers are removed in our proposed approach. Our network converges after 60,000 iterations with a mini-batch size of 8.

Numerous methods have been applied to the PASCAL VOC 2102 dataset and achieve the high accuracy. However, the complexity has been increasing due to the gradual addition of aides, which unfortunately does not reveal the true performance of the deep architecture as stated by Kendall et al. [27]. Our work in this benchmark do not aim to obtain the top score using additional aides, such as CRF post-processing [24], region proposal [28], multi-stage inference [25], and pre-trained model from other dataset (e.g., Microsoft COCO [79]). Instead, we seek to improve the performance by applying three main strategies, which include multi-scale convolution, a broader receptive field, and a single stream MR optimization method, to jointly upgrade the intrinsic structure of the network. The multi-scale strategy has the advantage of deep architecture because the potential scale is implicitly expressed by a pooling layer in the CNN. The broader receptive filed is captured by a dilated operation [28], thus preventing the loss of resolution. By contrast, the feedforward single stream MR optimization method allows obtaining the optimal solution without the complicated inference procedure and can be trained in an end-to-end manner. Though we embed the feedforward MR optimization algorithm into the network, the optimal solution can be solved linearly rather than in a multi-stage inference schema.

Table 2 presents the results of the comparison to recent methods, and a few of the corresponding intuitive results are depicted in Figure 3. In the table, we compare our method with several models that can be potentially applied to remote sensing area. We choose the listed models rather than all top scored approaches for the following reasons. First, the model should utilize as less additional aides as possible. Additional aides can hide the true performance of a network and are not easily transplanted to remote sensing application. Several models on the table, such as FCN-8s [22], DeconvNet [28] and SegNet [27], have been applied to process remote sensing images. Second, the selected model needs to be tested on PASCAL VOC 2012 server and does not repeat with previous methods. Algorithms, such as DeepLab [24], CRF-RNN [25], DilatedConv [28], and G-CRF [35], are milestones on PASCAL VOC 2012 benchmark and satisfy such requirements. Third, training the model is not too much time consuming, especially when dealing with remote sensing images, which are usually bigger than close range indoor/outdoor images. The recent state-of-the-art approach, such as RefineNet [80], employs ResNet-101 structures that may suffer from high GPU consumption and need MS-COCO dataset support. In the area of remote sensing, however, we do not have the large number extensions of labeled samples for training.

Table 2. PASCAL VOC12 dataset [72] results. We compare our proposed network with recent methods that support inference techniques. Additional aides, such as region proposal, multi-stage inference, and extra unary initialized model, are unnecessary in our approach. Some of the methods use the CRF as a post optimization procedure. In contrast, our proposed approach achieves competitive accuracy without post-processing in an end-to-end manner.

	Aeroplane	Bicycle	Bird	Boat	Bottle	Bus	Car	Cat	Chair	Cow	Diningtable	Dog	Horse	Motorbike	Person	Pottedplant	Sheep	Sofa	Train	Tvmonitor	Mean IoU
SegNet [27]	73.6	37.6	62.0	46.8	58.6	79.1	70.1	65.4	23.6	60.4	45.6	61.8	63.5	75.3	74.9	42.6	63.7	42.5	67.8	52.7	59.9
FCN-8s [22] (Multi-stage training)	76.8	34.2	68.9	49.4	60.3	75.3	74.7	77.6	21.4	62.5	46.8	71.8	63.9	76.5	73.9	45.2	72.4	37.4	70.9	55.1	62.2
DeepLab-Msc [24] (VGG-16 initialization)	74.9	34.1	72.6	52.9	61.0	77.9	73.0	73.7	26.4	62.2	49.3	68.4	64.1	74.0	75.0	51.7	72.7	42.5	67.2	55.7	62.9
DilatedConv Front end [28] (VGG-16 initialization)	82.2	37.4	72.7	57.1	62.7	82.8	77.8	78.9	28	70	51.6	73.1	72.8	81.5	79.1	56.6	77.1	49.9	75.3	60.9	67.6
DeconvNet + CRF [28] (Region Proposals)	**87.8**	41.9	80.6	63.9	67.3	88.1	78.4	81.3	25.9	73.7	61.2	72.0	77.0	79.9	78.7	59.5	78.3	**55.0**	75.2	61.5	70.5
CRF-RNN [25] (Multi-stage training)	87.5	39.0	79.7	64.2	68.3	87.6	80.8	84.4	30.4	78.2	60.4	**80.5**	77.8	83.1	80.6	59.5	82.8	47.8	78.3	**67.1**	72.0
DMSMR	87.6	40.3	80.6	62.9	**71.3**	88.1	**84.4**	84.7	29.6	77.8	58.5	79.9	**80.9**	85.4	**82.1**	54.9	83.8	48.2	**80.2**	65.3	72.4
G-CRF [35] (Unary Initialized with DeepLab CNN)	85.2	**43.9**	**83.3**	**65.2**	68.3	**89.0**	82.7	**85.3**	**31.1**	**79.5**	**63.3**	**80.5**	79.3	**85.5**	81.0	**60.5**	**85.5**	52.0	77.3	65.1	**73.2**

In the Table 2, the proposed **DMSMR** performs significantly (averaged approximately eight points) better than the similar methods without additional aides (methods without qualifying comments in Table 2). This is because our method is composed of the dilated, multi-scale strategies and has characteristics that complement to a few basic networks, such as SegNet [27], dilated convolutional network [28] and DeepLab-Msc [24]. Compared to recent methods, such as CRF-RNN [25] and G-CRF [35], our method achieves a similar score by optimizing with a single stream MR algorithm in an end-to-end manner. However, our approach does not require multi-stage inference or training two streams (i.e., unary term and pairwise stream, with unary initialized by other networks). Furthermore, some approaches, such as DeepLab [24], have a worse result when they do not use all of the additional aides with a pre-trained model. However, our model yields superior results without these pre-trained weights.

(**a**) Input (**b**) SegNet (**c**) FCN-8s (**d**) DeepLab-Msc (**e**) DilatedConv (**f**) DeconvNet (**g**) **DMSMR** (**h**) **GT**

Figure 3. Several semantic segmentation results on PASCAL VOC 2012 validation images. **DMSMR**: Semantic segmentation result predicted by dual multi-scale manifold ranking network. **GT**: Ground Truth.

5.1.2. Evaluation on CamVid

CamVid dataset [68,69], which is captured from high-definition (HD) video sequences with high quality, is designed for the road scene understanding. However, a relatively few number of images exist for training purpose. The dataset comprises 367 training images, 101 validation images and 233 testing images. The challenge data contains 11 semantic object classes which are downsampled to 640×480 pixels.

The overall training parameter settings for this dataset are as follows. The learning rate, momentum and weight decay are set to 1e-3, 0.9 and 0.0005, respectively. The momentum and weight decay terms are utilized as suggested in FCNs framework [22]. In addition, the learning rate is confirmed via cross validation. The proposed network is trained at the default resolution of 640×480 with a mini-batch size of 2. The initial values for α and β are set to 3 and 5, respectively, through cross validation. Our network converges after 40,000 iterations.

We employ the pixel mean intersection over union (mIoU) measurement with respect to the band width around the object boundaries as in [24] on the CamVid benchmark to analyze the expressive power of the proposed **DMSMR** network. The experimental results are illustrated in Figure 4. The comparisons between the **DMSMR** approach and the networks employing different strategies are reported in Table 3. We also analyze the accuracy change with respect to boundary in Figure 5. As shown in Figure 5a, we consider a narrow band, that is, trimap [81] boundary, on CamVid dataset. A trimap divides an image into three regions of foreground, background and unknown. Figure 5b shows boundary accuracy as the trimap width is varied. In this experiment, we set the same parameters as those in the **DMSMR** model but with different strategies as previously stated. The three strategies, namely, multi-scale convolution (**MS**), broader receptive field (**Dilated**) and manifold ranking optimization (**MR-Opti**) approaches, are utilized for comparison. Obviously, different strategies yield different performance for each of the classes. The **MS** and **Dilated** approaches help boost the performance in the situation where color and texture are uniformly distributed. In addition, the **MR-Opti** achieves a score that is approximately 2.5% better than those of the **MS** and **Dilated** methods because more contextual information are considered. The results demonstrate that the combination of **MS**, **Dilated** and **MR-Opti** approaches is possibly a better approach for semantic segmentation task on close-range images. Figure 5 shows that improving the recognition of pixels around the boundary helps delineate the object because the smoothness potentials of the correctly detected pixels increase. Additionally, as can be seen from Table 3, the **DMSMR** method outperforms the approaches that employ only one strategy, indicating that the **DMSMR** approach can improve the semantic segmentation result further by combing these strategies in close-range situations.

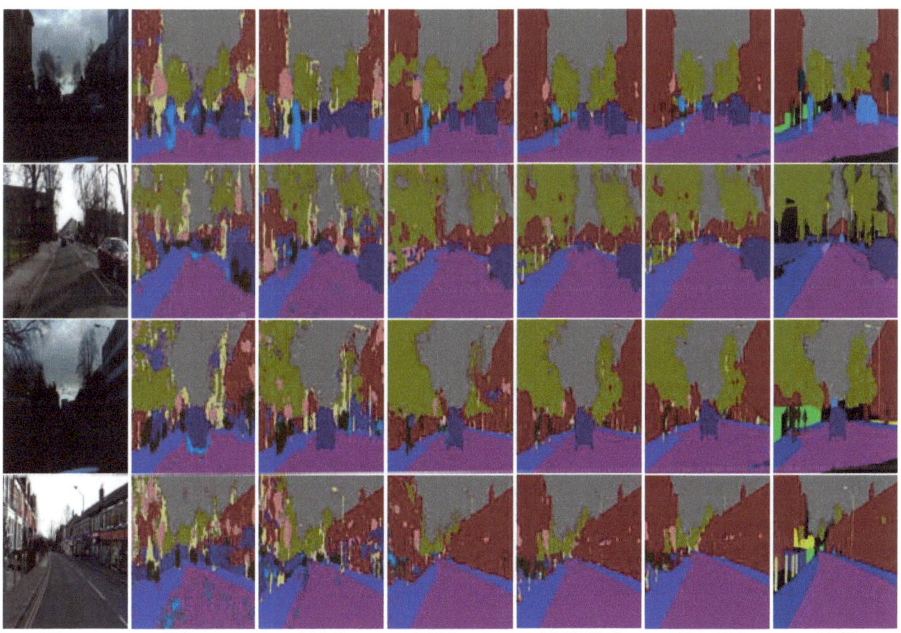

(**a**) Input (**b**) Before (**c**) MS (**d**) Dilated (**e**) MR-Opti (**f**) DMSMR (**g**) GT

Figure 4. Semantic segmentation results on CamVid images. **DMSMR**: Semantic segmentation result predicted by dual multi-scale manifold ranking network (**DMSMR**). **GT**: Ground Truth.

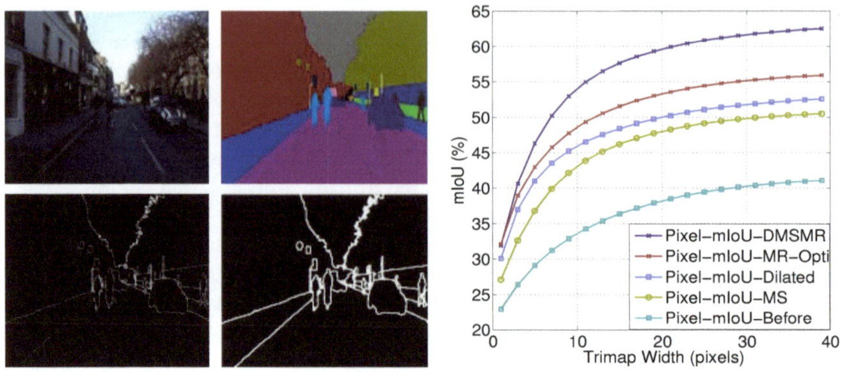

(a) (b)

Figure 5. Accuracy analysis with respect to boundary on CamVid dataset. (**a**) Trimap visualization on CamVid dataset. Top-left: source image. Top-right: ground truth. Bottom-left: trimap with one pixel band width. Bottom-right: trimap with three pixels band width. (**b**) Pixel mIoU with respect to band width around object boundaries. We measure the relationship of our model before and after employing the multi-scale (**MS**), dilated convolution (**Dilated**), single stream Manifold Ranking (**MR-Opti**) and joint strategies (**DMSMR**).

Table 3. Quantitative evaluation of the semantic segmentation results on CamVid dataset [68,69]. The proposed **DMSMR** approach outperforms the methods employing only one strategy.

	Building	Tree	Sky	Car	Sign	Road	Pedestrian	Fence	Pole	Sidewalk	Bicyclist	Mean IoU
Before	45.5	73.5	78.0	23.7	14.5	87.2	11.3	36.9	2.5	74.3	13.1	41.9
MS	81.4	88.1	80.3	40.1	16.3	95.6	26.2	40.0	3.7	82.0	37.4	53.7
Dilated	59.8	82.8	79.5	29.0	19.4	91.0	17.5	48.0	6.7	81.2	44.7	50.9
MR-Opti	90.6	**95.1**	74.6	**94.6**	21.9	**98.2**	53.1	64.3	**9.8**	92.6	42.1	54.8
DMSMR	**93.1**	94.5	**82.9**	92.7	**45.5**	97.4	**72.5**	**77.2**	7.2	**94.5**	**68.9**	**63.6**

5.2. Experiment on High Resolution Remote Sensing Dataset

Compare to the close-range imagery, high resolution remote sensing images have a few special features, which are different from that of commonly encountered indoor/outdoor close-range images in the area of computer vision. High resolution remote sensing images are large and contain a potentially-unlimited scene context (i.e., the road could possibly pass through the entire image). In addition, the object scale on high resolution images dramatically varies when employing the training dataset captured from different satellites (i.e., GF-1 with spatial resolution 2.1 m, QuickBird with spatial resolution of 0.6 m), whereas the close-range images do not. In the following experiments, we adopt two kinds of benchmarks: the ISPRS 2D Vaihingen dataset and EVLab-SS dataset. The ISPRS 2D Vaihingen benchmark is a well-known high resolution aerial imagery semantic labeling database, whose spatial resolution is 0.9 cm with uniform color and texture distributions. The EVLab-SS benchmark, which is designed for evaluating the semantic segmentation results on remote sensing imagery, contains the images captured from different platforms (both aerial and satellite images are included) with different types of spatial resolutions (ranging from 0.1 m to 2 m). In addition, the images vary in color, gradient, and texture.

5.2.1. Evaluation on Vaihingen Dataset

The Vaihingen dataset comprises 6 classes with 33 image tiles, out of which 16 are fully annotated (tile numbers 1, 3, 5, 7, 11, 13, 15, 17, 21, 23, 26, 28, 30, 32, 34 and 37). The dataset is cropped from an aerial orthophoto mosaic (GSD 9 cm) with three spectral bands (i.e., red, green and near-infrared bands) that are rich in detail. The categories to be classified for each pixel are *impervious surfaces, buildings, low vegetation, trees, and cars*. In our experiment, we randomly sample 2932 patches of 480×360 pixels from annotated images by sliding window. All patches are reserved for training. For the objective evaluation of the proposed approach, we submit the predicted results to the organizers who keep the ground truth.

The training procedure is performed with the SGD algorithm. The mini-batch size is set to 8, and each batch contains the cropped images that are randomly selected from training patches. These patches are resized to 321×321 pixels. We employ the "poly" learning policy, and the base learning rate is 1e-7 with the power of 0.9. The momentum and weight decay are set to 0.9 and 0.0005, respectively, as recommended by Krizhevsky et al. [82]. Smoothness coefficients α and β are set to 3 and 5, respectively. Our network converges after 50,000 iterations on this benchmark.

The experimental results on the Vaihingen testing images are available online (Our result on Vaihingen dataset is available at http://ftp.ipi.uni-hannover.de/ISPRS_WGIII_website/ISPRSIII_4_Test_results/2D_labeling_vaih/2D_labeling_Vaih_details_Ano2/index.html). Figure 6 visualizes the comparative results on a few testing images (tile numbers 2, 4, 6 and 8) with different methods. The quantitative evaluations of the corresponding state-of-the-art methods and our proposed network architecture are reported in Table 4. In this experiment, we employ the averaged F1 score and the overall pixel-wise accuracy as the evaluation metrics.

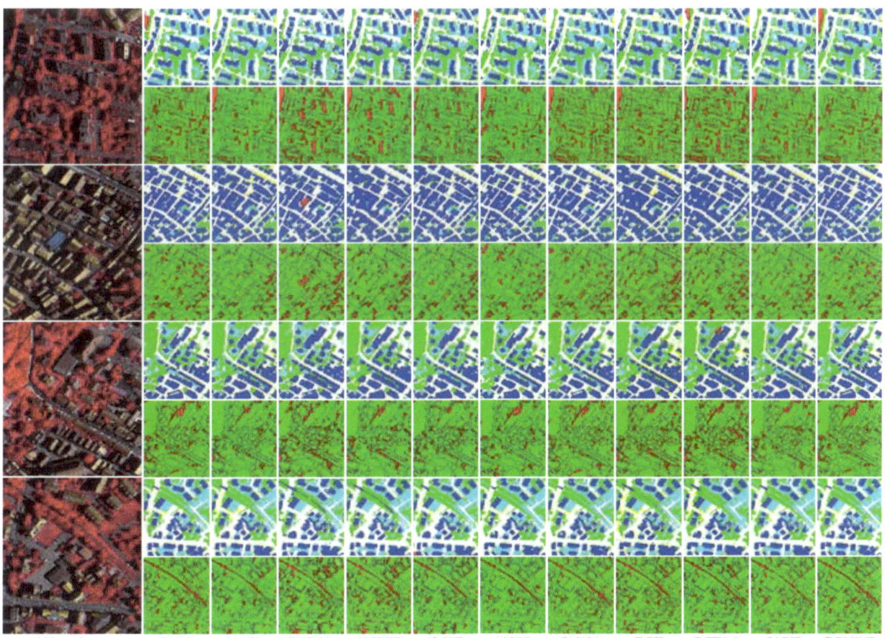

Figure 6. Visualization of the comparative results on a few Vaihingen testing imagery (tile numbers 2, 4, 6 and 8). For each image, we generate the dense prediction results and corresponding error maps (red/green image) with different approaches.

Figure 6 presents the visual comparison of these approaches. It can be seen from the error map that the CRF post-processing method (**ADL** [59] and **HUST** [83]) indeed helps improve the performance. Nevertheless, the upper left corner of the error map in the first row shows that even if the CRF post-processing method is employed, more incorrectly classified pixels will exist if the initial predictions are poorly provided. In Table 4, we compare our approach with the methods using additional aides, such as the VGG-16 pre-trained model [29,76,84], digital surface model (DSM) [49,85,86], and the CRF post-processing [59,83]. We also compare our approach with traditional feature based methods [87]. Recent advances in the area of computer vision have shown that very deep networks can improve the semantic segmentation accuracy [27,54]. Therefore, our **DMSMR** approach reasonably outperforms the "**SVL**" method by approximately 4% in overall pixel-wise accuracy and 6% on global F1 score. Although additional aides help improve accuracy, they are not the *core to segmentation engine* [53]. Our networks do not need these aides but achieve competitive scores compared with these approaches. For the fine-tuned networks from the pre-trained VGG-16 model (**ONE** [84], **DLR** [76], **UOA** [29], **RIT** [50]), their performances are not always steady compared to that of the proposed **DMSMR** approach. Our overall accuracy varies approximately 0.1% (see **Ano** (**Ano** is available at http://ftp.ipi.uni-hannover.de/ISPRS_WGIII_website/ISPRSIII_4_Test_results/2D_labeling_vaih/2D_labeling_Vaih_details_Ano/index.html) and **Ano2** in the ISPRS leader board. **Ano** and **Ano2** are initialized with the same hyper-parameters, but the weights and biases terms are randomly initialized.) when tested on this benchmark. This is mainly caused by uncertainty of weights when trying to transfer the VGG-16 classification networks into semantic segmentation task. The dense prediction problem, such as semantic segmentation, is structurally different from image classification [23]. Thus these performances are not as stable as expected. Our approach somehow utilizes the dual-dilated and non-dilated convolutional layers to prevent such instability.

Table 4. Vaihingen dataset [88] results. We compare our proposed approach with a few recent state-of-the-art methods listed on the ISPRS Vaihingen 2D contest leader board. Traditional approaches and methods that employ additional aides (methods with qualifying comments) are referenced for comparison.

	Imp.surf.	Building	Low veg.	Tree	Car	Overall F1	Overall Acc.
SVL [87] (Feature based)	86.1	90.9	77.6	84.9	59.9	79.88	84.7
ADL [59] (CRF post-processing)	89.0	93.0	81.0	87.8	59.5	82.06	87.3
UT_Mev [85] (DSM supported)	84.3	88.7	74.5	82.0	9.9	67.88	81.8
HUST [83] (CRF post-processing)	86.9	92.0	78.3	86.9	29.0	74.62	85.9
ONE [84] (VGG-16 pre-trained model)	87.8	92.0	77.8	86.2	50.7	78.90	85.9
DLR [76] (VGG-16 pre-trained model)	90.3	92.3	**82.5**	**89.5**	**76.3**	86.18	88.5
UOA [29] (VGG-16 pre-trained model)	89.8	92.1	80.4	88.2	82.0	**86.50**	87.6
RIT [50] (DSM supported, VGG-16 pre-trained model)	88.1	93.0	80.5	87.2	41.9	78.14	86.3
ETH_C [86] (DSM supported)	87.2	92.0	77.5	87.1	54.5	79.66	85.9
DST [49] (DSM supported)	90.3	**93.5**	**82.5**	88.8	73.9	85.80	**88.7**
DMSMR	**90.4**	93.0	81.4	88.6	74.5	85.58	88.4

5.2.2. Evaluation on EvLab-SS Dataset

The EvLab-SS benchmark (EvLab-SS dataset can be downloaded from our website http://earthvisionlab.whu.edu.cn/zm/SemanticSegmentation/index.html.) is designed for the evaluation of the semantic segmentation algorithms on real engineered scenes, which aims to find a good deep learning architecture for the high resolution pixel-wise classification task in remote sensing area. The dataset is originally obtained from the Chinese Geographic Condition Survey and Mapping Project, and each image is fully annotated by the Geographic Conditions Survey (NO.GDPJ 01—2013) [89] standards. The average resolution of the dataset is approximately 4500 × 4500 pixels. The EvLab-SS dataset contains 11 major classes, namely, *background, farmland, garden, woodland, grassland, building, road, structures, digging pile, desert and waters*, and currently includes 60 frames of images captured by different platforms and sensors. The dataset comprises 35 satellite images, 19 frames of which are captured by the World-View-2 satellite [90] (re-sample GSD 0.2 m), 5 frames are captured by the GeoEye satellite [91] (re-sample GSD 0.5 m), 5 frames are captured by the QuickBird satellite [92] (re-sample GSD 2 m), 6 frames are captured by the GF-2 satellite [93] (re-sample GSD 1 m). The dataset also has 25 aerial images, 10 images of which with spatial resolution of 0.25 m and 15 images have a spatial resolution of 0.1 m. In our experiment, we divide the dataset into 37 frames for training, 8 frames for validation, and 15 frames for testing. We produce the training dataset by applying the sliding window with a stride of 128 pixels to the training images, thereby resulting in 48,622 patches with a resolution of 640 × 480 pixels. Similar methods are utilized on validation images, thus generating 13,539 patches for validation. The ***Garden*** class, which is reserved for validating the expressive power of CNNs in real scenes, is absent in our validation images.

In the training procedure, each iteration comprises a feed-forward pass in which the model weights are adjusted by the SGD algorithm. Each training patch image in a batch is resized to 321 × 321 pixels. The mini-batch size is set to 12 and the corresponding training patches are randomly selected. We employ the "poly" learning policy and start with a learning rate 1e-7 with the power of 0.9. Smoothness coefficients α and β are set to 3 and 5 in our experiments, respectively. The momentum and weight decay are set to 0.9 and 0.0005, respectively, as recommended by Krizhevsky et al. [82]. Our network converges after 70,000 iterations on this dataset. In the following experiments, we set the same learning parameters for the methods employing only one strategy (**MS**, **Dilated** or **MR-Opti**) as the **DMSMR** approach.

Figure 7 is the visualization of the results on the validation patches with different methods. Figure 8 illustrates the comparative results of employing different strategies with respect to the varying trimap band width. Quantitative results are shown in Table 5. In our experiments, we adopt the overall pixel-wise accuracy and mean intersection over union (mIoU) measurements to evaluate the effectiveness of different approaches.

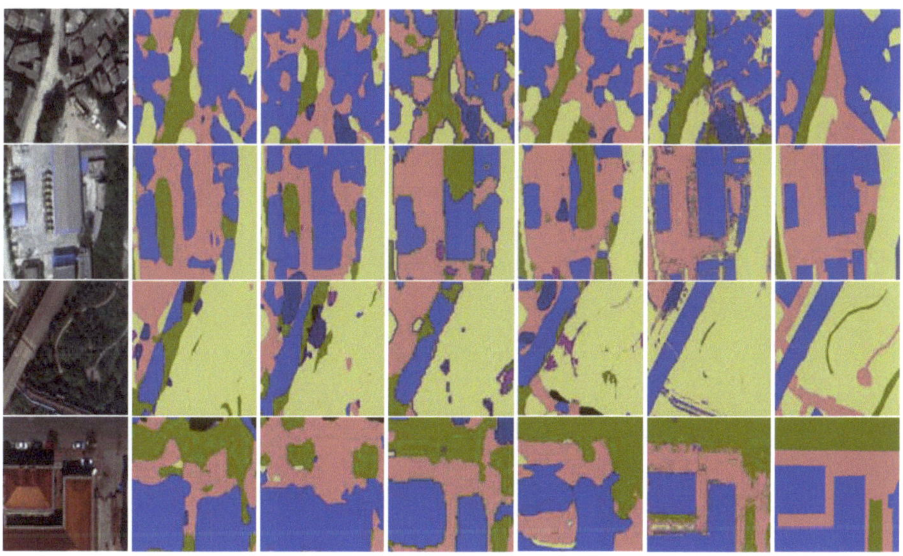

Figure 7. Semantic segmentation results with different strategies on the EvLab-SS validation patches. Four kinds of image patches with different spatial resolutions and illuminations are depicted in the figure. The first and second rows are the GeoEye and World-View 2 satellite images with resample GSD of 0.5 m and 0.2 m. The third and the last rows are the aerial images with resample GSD of 0.25 m and 0.1 m, respectively. **MS**: Predictions with multi-scale approach. **MR-Opti**: Semantic segmentation results using manifold ranking optimization method. **DMSMR**: Segmentation result predicted by dual multi-scale manifold ranking network. **GT**: Ground Truth.

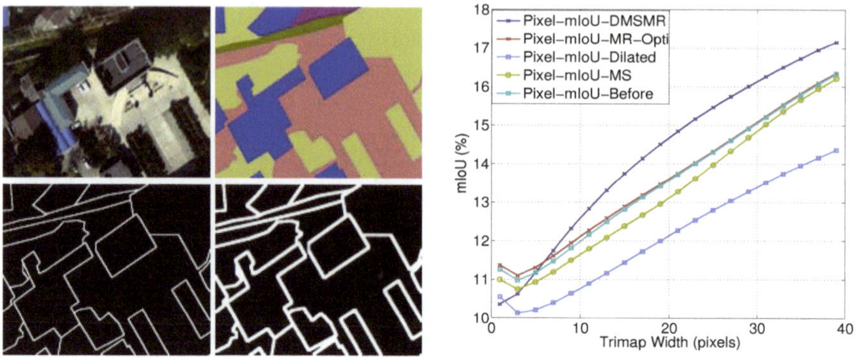

(a) **(b)**

Figure 8. Accuracy analysis with respect to boundary on EvLab-SS dataset. (**a**) Visualization of trimap for EvLab-SS dataset. Top-left: source patch. Top-right: ground truth. Bottom-left: trimap with one pixel band width. Bottom-right: trimap with three pixels band width. (**b**) Pixel mIoU with respect to band width around object boundaries. We measure the relationship for our model before and after employing the multi-scale (**MS**), dilated convolution (**Dilated**), single stream Manifold Ranking (**MR-Opti**) and joint strategies (**DMSMR**) on the EvLab-SS dataset.

Table 5. Quantitative evaluation of the semantic segmentation results on the EvLab-SS dataset. The proposed **DMSMR** approach outperforms the methods that employ only one strategy.

	Background	**Farmland**	**Garden**	**Woodland**	**Grassland**	**Building**	**Road**	**Structures**	**Digging Pile**	**Desert**	**Waters**	**Overall Accuracy**	**Mean IoU**
Before	75.16	35.73	0.0	51.65	8.99	66.59	35.12	46.19	**19.05**	3.56	3.13	49.76	21.35
MS	75.73	**39.36**	0.0	49.33	**11.89**	65.85	32.80	46.94	12.91	**16.69**	5.87	48.93	21.42
Dilated	40.59	29.18	0.0	46.48	11.36	61.74	**40.46**	42.54	18.10	11.57	**19.84**	46.8	19.03
MR-Opti	**79.44**	20.52	0.0	57.84	2.95	**74.29**	28.96	49.60	17.55	0.10	0.99	53.51	21.85
DMSMR	40.59	22.14	0.0	**62.47**	8.11	68.84	39.80	**51.06**	14.56	16.52	19.45	**54.15**	**22.17**

Compare to the 2D Vaihingen dataset provided by the ISPRS organization, the EvLab-SS dataset is inconsistently distributed in terms shape, color, and texture. The resolutions of the images captured from different sensors are dramatically varying. The buildings, roads and other classes are not obtained in the same scale. Therefore, the EvLab-SS dataset poses more challenge to researchers. It intuitively can be seen from Figure 7 that the **DMSMR** method can better delineate the boundary of an object. The results demonstrate the superiority of the combination of multi-scale (**MS**), broader receptive field (**Dilated**), and manifold ranking optimization (**MR-Opti**) strategies, which can more accurately classify each pixel with varying spatial resolutions. Figure 8 shows that although the mIoU score of the proposed DMSMR approach is relatively low with a small trimap width, it has become increasingly stable and competitive. By contrast, the mIoU scores of the MS, dilated, and MR-Opti approaches are unstable, even decreasing with a few small trimap widths. The main reason attribute to this phenomena is that the spatial resolution is different in the training patches, which may be ignored by only employing one strategy. In Table 5, the special class (***Garden***) is detected as 0.0% in all approaches, indicating that these methods can preserve the intrinsic nature of CNNs well. For the real engineered remote sensing data, the **Dilated** approach does not appear to boost performance and decreases in overall accuracy and mean IoU by approximately 2.96%, 2.32%, respectively. This can be attributed to the numerous inhomogeneous objects in the training patches. For example, the road and buildings may not be completely covered in a single patch, which renders training with dilation operations in some layer meaningless. Although the **MR-Opti** approach improves the overall accuracy by approximately 4%, this approach may disregard a few classes, such as the *Desert and Waters*, due to insufficient contextual information with varying illumination and color. However, the **MS** approach retains more contextual information in each scale space but still suffers from the optimization problem in each scale, resulting in 0.8% decrease in overall accuracy. Notably, the proposed **DMSMR** approach can take the superior features of these strategies and overcome the drawbacks, achieving approximately 5% and 1% improvements in overall accuracy and mIoU score under the condition of limited training images and varying spatial resolutions.

6. Conclusions

In this paper, we present a **DMSMR** network for semantic image segmentation in a continuous domain. By extending the binary manifold ranking (MR) algorithm to a multi-label case, the assignment of a discrete label to each pixel can be linearly solved and a unique global optimum can be guaranteed. In addition, with the single stream MR method embedded into CNNs in a feedforward schema, the required parameters can be trained in an end-to-end manner. Furthermore, we propose to utilize dilated and non-dilated networks, which form dual layers to jointly optimize the results from the single stream manifold ranking network rather than on two separate streams, that is, unary and pairwise streams. Combined with multi-scale (**MS**), broader receptive field (**Dilated**) and manifold ranking optimization (**MR-Opti**) strategies, the proposed **DMSMR** network enables training without additional aides, such as multi-stage inference, region proposals, VGG-16 initialization, digital surface model (DSM) and CRF post-processing. Two groups of experiments on close-range and remote sensing high resolution datasets are designed to evaluate the performance. When discriminatively trained by submitting the results to the server on PASCAL VOC and ISPRS Vaihingen benchmarks, the proposed **DMSMR** network can achieve competitive results without additional aides compared to recent methods. Our experiments on publicly available datasets, including CamVid and EvLab-SS datasets, demonstrate the superior capacity of the proposed **DMSMR** approach over the methods that employ only one strategy. For the real world application in remote sensing, the combined strategy steadily boosts the performance even under limited training images and the varying spatial resolutions.

Nevertheless, the proposed approach may be further improved in the following ways. First, more prior information, such as orientation and texture, is expected to be integrated into the smoothness term in the multi-label manifold ranking objective function to delineate the visual objects with varying illumination and spatial resolution. Second, the generative adversarial nets [94–96] (GAN) can be introduced to boost the performance by combining the adversarial term in the loss function with the limited number of training images. Third, model parallelism should be investigated when incorporating more prior knowledge to our model. For example, buildings and roads are the salient objects in remote sensing images that can guide the semantic contextual information. The prior information might be parallel-trained in a distributed system. Finally, the superpixel segmentation can be applied as a pre-processing step to reduce the number of optimization elements in the proposed multi-label MR graphical model.

Acknowledgments: This work was partially supported by the National Key Research and Development Program of China (Project No. 2016YFB0501403) and National Natural Science Foundation of China (Project No. 41371401 and 91438203). The authors would like to thank Jiasi Yi and Yun He from Earth Vision Laboratory, School of Remote Sensing and Information Engineering, Wuhan University for designing the work flow to produce the EvLab-SS benchmark. The author would also like to thank the group of undergraduates from School of Remote Sensing and Information Engineering, Wuhan University for producing the EvLab-SS dataset.

Author Contributions: Mi Zhang designed the **DMSMR** network and performed the experimental analysis. He also wrote the paper. Xiangyun Hu guided the algorithm design, initiated the EvLab-SS dataset production and revised the paper. Shiyan Pang help organize the paper. Like Zhao, Ye Lv and Min Luo contributed to the design of project homepage and edited the manuscript.

Conflicts of Interest: The authors declare no conflict of interest.

Appendix A

In this section, we derive the rule of weights updating for learning the parameters mentioned in the paper and detailedly depict the implementation structures of the networks, which include the networks before and after employing the multi-scale (**MS**), dilated convolution (**Dilated**), and manifold ranking optimization (**MR-Opti**) approaches.

Appendix A.1. Learning Parameter α and β

To compute the term $\frac{\partial y_l^*}{\partial \alpha}$ in Equation (13), we apply the chain rule through the following equation:

$$\frac{\partial y_l^*}{\partial \alpha} = \frac{\partial y_l^*}{\partial f_k^{\max}} \cdot \frac{\partial f_k^{\max}}{\partial \alpha} = \frac{\partial y_l^*}{\partial f_k^{\max}} \cdot \frac{\partial \mathcal{L}\left(\widetilde{\mathbf{F}}\right)}{\partial \mathbf{S}} \cdot \frac{\partial \mathbf{S}}{\partial \alpha}, \tag{A1}$$

where y_l^*, $f_k^{\max}$, $\mathcal{L}\left(\widetilde{\mathbf{F}}\right)$, and α are the symbols that have the same meaning as previously mentioned. $\mathbf{S}$ is the simplified representation of the smoothness term in Equation (7), which is specifically denoted by

$$\begin{aligned} \mathbf{S} &= \sum_{e_{ij}\in E} w_{ij} \left\| \widetilde{\mathbf{f}}(p_i) - \widetilde{\mathbf{f}}(p_j) \right\|^2 = \alpha \mathbf{k}_1 (f_i, f_j) + \beta \mathbf{k}_2 (f_i, f_j) \\ &= \alpha \exp\left(-\frac{\|p_i - p_j\|^2 + \|I_i - I_j\|^2}{2\sigma_1^2} \right) + \beta \exp\left(-\frac{\|p_i - p_j\|^2}{2\sigma_2^2} \right). \end{aligned} \tag{A2}$$

Since the term $\frac{\partial y_l^*}{\partial f_k^{\max}}$ is equal to delta function δ, the term $\frac{\mathcal{L}(\widetilde{\mathbf{F}})}{\partial S}$ is equal to the identity matrix, the term $\frac{\partial \mathcal{L}(\widetilde{\mathbf{F}})}{\partial \alpha}$ and $\frac{\partial \mathcal{L}(\widetilde{\mathbf{F}})}{\partial \beta}$ are obviously represented by $\mathbf{k}_1 (f_i, f_j)$ and $\mathbf{k}_2 (f_i, f_j)$. The derivative of α, β with respect to y_l^* are obtained by the following form:

$$\frac{\partial y_l^*}{\partial \alpha} = \delta \cdot \mathbf{k}_1 (f_i, f_j). \tag{A3}$$

$$\frac{\partial y_l^*}{\partial \beta} = \delta \cdot \mathbf{k}_2 (f_i, f_j). \tag{A4}$$

Appendix A.2. Learning Compatibility Matrix $\widetilde{\mathbf{W}}$

Similar to the derivative of smoothness parameters α and β, the derivative of compatibility matrix $\widetilde{\mathbf{W}}$ with respect to y_l^* can be denoted by:

$$\begin{aligned} \frac{\partial y_l^*}{\partial \widetilde{\mathbf{W}}} &= \frac{\partial y_l^*}{\partial f_k^{\max}} \cdot \frac{\partial f_k^{\max}}{\partial \widetilde{\mathbf{W}}} \\ &= \delta \cdot \frac{\partial f_k^{\max}}{\partial \widetilde{\mathbf{W}}} = \delta \cdot \frac{\partial \mathcal{L}\left(\widetilde{\mathbf{F}}\right)}{\partial \widetilde{\mathbf{W}}}. \end{aligned} \tag{A5}$$

As discussed in the main paper, the optimal solution to the multi-label manifold ranking method is achieved by the following matrix form:

$$\begin{aligned} \widehat{\mathbf{F}} &= \left(2\lambda \left(\widetilde{\mathbf{D}} - \widetilde{\mathbf{W}}\right) + \mathbf{D}_\mu\right)^{-1} \mathbf{D}_\mu \widetilde{\mathbf{F}}^* \\ &= \left(2\lambda \widetilde{\mathbf{L}} + \mathbf{D}_\mu\right)^{-1} \mathbf{D}_\mu \widetilde{\mathbf{F}}^*. \end{aligned} \tag{A6}$$

From Petersen et al. [88], we recall that the derivative of the inverse of matrix **A** with respect to **A** is

$$\frac{\partial \mathbf{A}^{-1}}{\partial \mathbf{A}} = -\mathbf{A}^{-\mathrm{T}} \otimes \mathbf{A}^{-1}. \tag{A7}$$

For the preceding term $\frac{\partial \mathcal{L}(\widetilde{\mathbf{F}})}{\partial \widetilde{\mathbf{W}}}$, the corresponding matrix form can be represented by:

$$\begin{aligned}\frac{\partial \mathcal{L}\left(\widetilde{\mathbf{F}}\right)}{\partial \widetilde{\mathbf{W}}} &= \left(\mathbf{D}_\mu + 2\lambda\left(\widetilde{\mathbf{D}} - \widetilde{\mathbf{W}}\right)\right)^{-T} \otimes \left(\mathbf{D}_\mu + 2\lambda\left(\widetilde{\mathbf{D}} - \widetilde{\mathbf{W}}\right)\right)^{-1} \mathbf{D}_\mu \widetilde{\mathbf{F}}^* \\ &= \left(\mathbf{D}_\mu + 2\lambda\left(\widetilde{\mathbf{D}} - \widetilde{\mathbf{W}}\right)\right)^{-T} \otimes \widetilde{\mathbf{F}} \\ &= \left(\left(\mathbf{D}_\mu \widetilde{\mathbf{F}}^*\right)^{-1} \widetilde{\mathbf{F}}\right)^{T} \otimes \widetilde{\mathbf{F}} \\ &= \nabla \mathbf{\Psi} \otimes \widetilde{\mathbf{F}}.\end{aligned} \tag{A8}$$

Therefore, the derivative of $\widetilde{\mathbf{W}}$ with respect to y_l^* is

$$\frac{\partial y_l^*}{\partial \widetilde{\mathbf{W}}} = \delta \cdot \nabla \mathbf{\Psi} \otimes \widetilde{\mathbf{F}}. \tag{A9}$$

Appendix A.3. Network with Different Strategies

In this part, we explain in detail for the methods that employ only one of the three strategies, namely, multi-scale convolution (**MS**), broader receptive field (**Dilated**) and MR optimization (**MR-opti**) approaches. Figure A1 shows the general structures of these approaches and Table A1 presents the corresponding implementation parameters in each convolutional layer. In the table and figure, the "ReLU" active function [74] is implicitly employed in each convolutional layer. The network depicted in Figure A1a serves as the baseline convolutional network for comparison. Figure A1c,d are the networks that use only the dilated convolutional kernel [23] and manifold ranking optimization methods, respectively. The only difference between network in Figure A1a,c is the dilation kernel. In our experiment, we set the kernel sizes in each block as 6, 4, 2, 2 and 1, as illustrated in Table A1a. For the MR optimization layer embedded in the baseline network shown in Figure A1d, initial parameters of α and β are set to 3 and 5, respectively. Figure A1b presents the network with multi-scale strategy on the baseline network. After applying the pooling layer in each block, a convlutional block is adopted with three convolutional layers (named as poolx-conv-y in Table A1b. The scale is implicitly expressed in the pooling layer by factor 2.0.

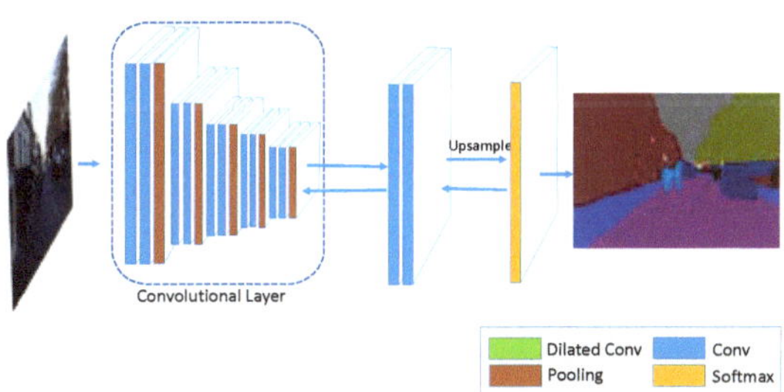

(a)

Figure A1. *Cont.*

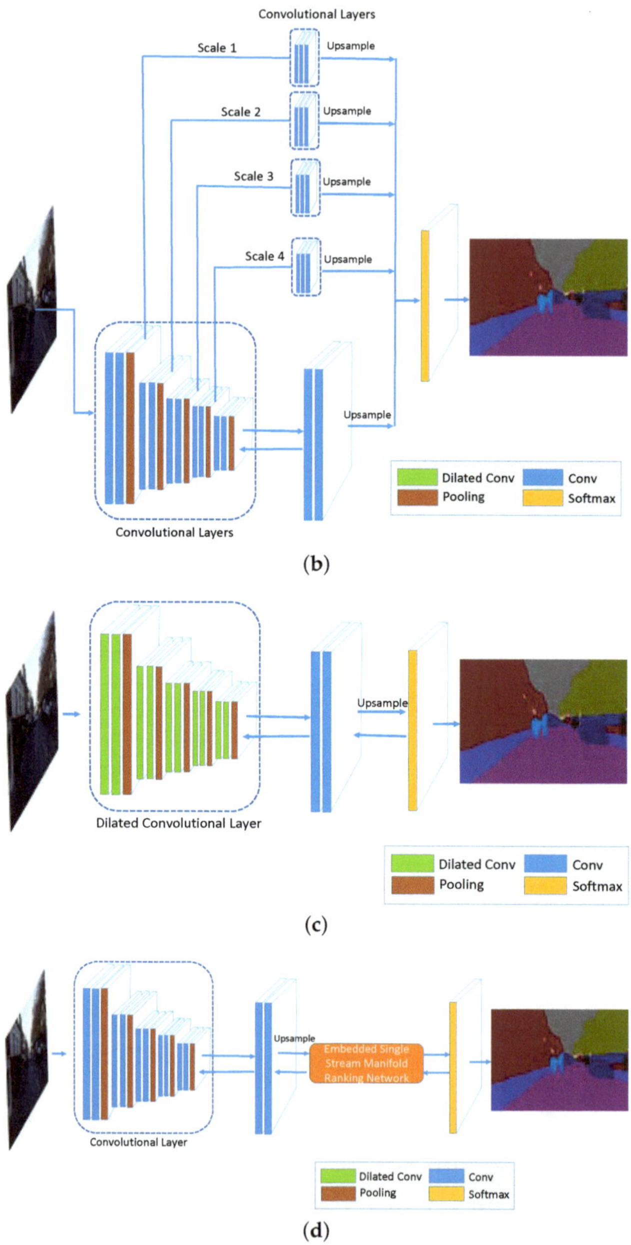

Figure A1. The architectures of the networks with different strategies: (**a**) Convolutional networks before employing the strategies (**Before**); (**b**) Networks using multi-scale strategy (**MS**); (**c**) Networks using dilated method (**Dilated**); (**d**) Networks using manifold ranking optimization (**MR-Opti**).

Table A1. Implementation details of the networks with different strategies.

(a) Networks before Employing the Strategies (Before)						
Block	**Name**	**Kernel Size**	**Pad**	**Dilation**	**Stride**	**Number of Output**
0	input	-	-	-	-	3
1	conv1-1	3 × 3	1	1	1	64
	conv1-2	3 × 3	1	1	1	64
	pool1	3 × 3	1	0	1	64
2	conv2-1	3 × 3	1	1	1	128
	conv2-2	3 × 3	1	1	1	128
	pool2	3 × 3	1	0	2	128
3	conv3-1	3 × 3	1	1	1	256
	conv3-2	3 × 3	1	1	1	256
	pool3	3 × 3	1	0	2	256
4	conv4-1	3 × 3	1	1	1	512
	conv4-2	3 × 3	1	1	1	512
	pool4	3 × 3	1	0	1	512
5	conv5-1	5 × 5	2	1	1	512
	conv5-2	5 × 5	2	1	1	512
	pool5	3 × 3	1	0	1	512
-	fc6	3 × 3	1	1	1	1024
	fc7	1 × 1	0	1	1	1024
*	fc8	1 × 1	0	1	1	12
-	output	1 × 1	0	1	1	12

(b) Networks Using Multi-Scale Strategy (MS)						
Scale (Block)	**Name**	**Kernel Size**	**Pad**	**Dilation**	**Stride**	**Number of Output**
0	input	-	-	-	-	3
1	conv1-1	3 × 3	1	1	1	64
	conv1-2	3 × 3	1	1	1	64
	pool1	3 × 3	1	0	2	64
2	conv2-1	3 × 3	1	1	1	128
	conv2-2	3 × 3	1	1	1	128
	pool2	3 × 3	1	0	2	128
3	conv3-1	3 × 3	1	1	1	256
	conv3-2	3 × 3	1	1	1	256
	pool3	3 × 3	1	0	2	256
4	conv4-1	3 × 3	1	1	1	512
	conv4-2	3 × 3	1	1	1	512
	pool4	3 × 3	1	0	1	512
5	conv5-1	5 × 5	2	1	1	512
	conv5-2	5 × 5	2	1	1	512
	pool5	3 × 3	1	0	1	512
-	fc6	3 × 3	1	1	1	1024
	fc7	1 × 1	0	1	1	1024
*	fc8	1 × 1	0	1	1	12
1	pool1-conv-1	3 × 3	1	1	4	128
	pool1-conv-2	1 × 1	0	1	1	128
	pool1-conv-3	1 × 1	0	1	1	12
2	pool2-conv-1	3 × 3	1	1	2	128
	pool2-conv-2	1 × 1	0	1	1	128
	pool2-conv-3	1 × 1	0	1	1	12
3	pool3-conv-1	3 × 3	1	1	1	128
	pool3-conv-2	1 × 1	0	1	1	128
	pool3-conv-3	1 × 1	0	1	1	12
4	pool4-conv-1	3 × 3	1	1	1	128
	pool4-conv-2	1 × 1	0	1	1	128
	pool4-conv-3	1 × 1	0	1	1	12
-	output	1 × 1	0	1	1	12

Table A1. *Cont.*

(c) Networks Using Dilated Method (Dilated)						
Block	**Name**	**Kernel Size**	**Pad**	**Dilation**	**Stride**	**Number of Output**
0	input	-	-	-	-	3
1	conv1-1	3 × 3	6	6	1	64
	conv1-2	3 × 3	6	6	1	64
	pool1	3 × 3	1	0	2	64
2	conv2-1	3 × 3	4	4	1	128
	conv2-2	3 × 3	4	4	1	128
	pool2	3 × 3	1	0	2	128
3	conv3-1	3 × 3	2	2	1	256
	conv3-2	3 × 3	2	2	1	256
	pool3	3 × 3	1	0	2	256
4	conv4-1	3 × 3	2	2	1	512
	conv4-2	3 × 3	2	2	1	512
	pool4	3 × 3	1	0	1	512
5	conv5-1	3 × 3	2	2	1	512
	conv5-2	3 × 3	2	2	1	512
	pool5	3 × 3	1	0	1	512
-	fc6	3 × 3	1	1	1	1024
	fc7	1 × 1	0	1	1	1024
*	fc8	1 × 1	0	1	1	12
-	output	1 × 1	0	1	1	12
(d) Networks Using Manifold Ranking Optimization (MR-Opti)						
Block	**Name**	**Kernel Size**	**Pad**	**Dilation**	**Stride**	**Number of Output**
0	input	-	-	-	-	3
1	conv1-1	3 × 3	1	1	1	64
	conv1-2	3 × 3	1	1	1	64
	pool1	3 × 3	1	0	1	64
2	conv2-1	3 × 3	1	1	1	128
	conv2-2	3 × 3	1	1	1	128
	pool2	3 × 3	1	0	2	128
3	conv3-1	3 × 3	1	1	1	256
	conv3-2	3 × 3	1	1	1	256
	pool3	3 × 3	1	0	2	256
4	conv4-1	3 × 3	1	1	1	512
	conv4-2	3 × 3	1	1	1	512
	pool4	3 × 3	1	0	1	512
5	conv5-1	5 × 5	2	1	1	512
	conv5-2	5 × 5	2	1	1	512
	pool5	3 × 3	1	0	1	512
-	fc6	3 × 3	1	1	1	1024
	fc7	1 × 1	0	1	1	1024
*	fc8	1 × 1	0	1	1	12
-	Manifold Ranking Optimization					12
-	output	1 × 1	0	1	1	12

References

1. Ladicky, L.; Torr, P.; Zisserman, A. Human Pose Estimation using a Joint Pixel-wise and Part-wise Formulation. In Proceedings of the 2013 IEEE Conference on Computer Vision and Pattern Recognition, Portland, OR, USA, 23–28 June 2013.
2. Romera, E.; Bergasa, L.; Arroyo, R. Can we unify monocular detectors for autonomous driving by using the pixel-wise semantic segmentation of CNNs? *arXiv* **2016**, arXiv:1607.00971
3. Barrnes, D.; Maddern, W.; Posner, I. Find Your Own Way: Weakly-Supervised Segmentation of Path Proposals for Urban Autonomy. *arXiv* **2016**, arXiv:1610.01238.
4. Kendall, A.; Cipolla, R. Modelling Uncertainty in Deep Learning for Camera Relocalization. *arXiv* **2015**, arXiv:1509.05909.
5. Xiao, J.; Quan, L. Multiple View Semantic Segmentation for Street View Images. In Proceedings of the 2009 IEEE 12th International Conference on Computer Vision, Kyoto, Japan, 29 September–2 October 2009.
6. Floros, G.; Leibe, B. Joint 2D-3D Temporally Consistent Semantic Segmentation of Street Scenes. In Proceedings of the 2012 IEEE Conference on Computer Vision and Pattern Recognition, Providence, RI, USA, 16–21 June 2012.
7. Huval, B.; Wang, T.; Tandon, S.; Kiske, J.; Song, W.; Pazhayampallil, J.; Mujica, F. An empirical evaluation of deep learning on highway driving. *arXiv* **2015**, arXiv:1504.01716.
8. Chen, C.; Seff, A.; Kornhauser, A.; Xiao, J. Deepdriving: Learning affordance for direct perception in autonomous driving. In Proceedings of the 2015 IEEE International Conference on Computer Vision, Santiago, Chile, 7–13 December 2015.
9. Toshev, A.; Szegedy, C. DeepPose: Human Pose Estimation via Deep Neural Networks. *arXiv* **2014**, arXiv:1312.4659
10. Tompson, J.J.; Jain, A.; LeCun, Y.; Bregler, C. Joint training of a convolutional network and a graphical model for human pose estimation. *arXiv* **2014**, arXiv:1406.2984.
11. Jackson, A.; Valstar, M.; Tzimiropoulos, G. A CNN Cascade for Landmark Guided Semantic Part Segmentation. *arXiv* **2016**, arXiv:1609.09642.
12. Maggiori, E.; Tarabalka, Y.; Charpiat, G.; Alliez, P. High-Resolution Semantic Labeling with Convolutional Neural Networks. *arXiv* **2016**, arXiv:1611.01962.
13. Kampffmeyer, M.; Salberg, A.B.; Jenssen, R. Semantic segmentation of small objects and modeling of uncertainty in urban remote sensing images using deep convolutional neural networks. In Proceedings of the 2016 IEEE Conference on Computer Vision and Pattern Recognition Workshops (CVPRW), Las Vegas, NV, USA, 26 June–1 July 2016.
14. Audebert, N.; Saux, B.L.; Lefèvre, S. Semantic Segmentation of Earth Observation Data Using Multimodal and Multi-scale Deep Networks. *arXiv* **2016**, arXiv:1609.06846.
15. Längkvist, M.; Kiselev, A.; Alirezaie, M.; Loutfi, A. Classification and segmentation of satellite orthoimagery using convolutional neural networks. *Remote Sens.* **2016**, *8*, 329.
16. Muruganandham, S. Semantic Segmentation of Satellite Images Using Deep Learning. Master's Thesis, Department of Computer Science, Electrical and Space Engineering, Luleå University of Technology, Luleå, Sweden, 2016.
17. Wu, Z.; Song, S.; Khosla, A.; Yu, F.; Zhang, L.; Tang, X.; Xiao, J. *3d Shapenets: A Deep Representation for Volumetric Shapes*; Princeton University: Princeton, NJ, USA, 2015.
18. Kendall, A.; Grimes, M.; Cipolla, R. PoseNet: A Convolutional Network for Real-Time 6-DOF Camera Relocalization. *arXiv* **2016**, arXiv:1505.07427.
19. Barron, J.T.; Poole, B. The fast bilateral solver. *arXiv* **2016**, arXiv:1511.03296.
20. Mostajabi, M.; Yadollahpour, P.; Shakhnarovich, G. Feedforward semantic segmentation with zoom-out features. *arXiv* **2015**, arXiv:1412.0774v1.
21. Dai, J.; He, K.; Sun, J. Instance-aware Semantic Segmentation via Multi-task Network Cascades. *arXiv* **2015**, arXiv:1512.04412.
22. Shelhamer, E.; Long, J.; Darrell, T. Fully Convolutional Networks for Semantic Segmentation. *arXiv* **2015**, arXiv:1605.06211.

23. Yu, F.; Koltun, V. Multi-Scale Context Aggregation by Dilated Convolutions. *arXiv* **2016**, arXiv:1511.07122.
24. Chen, L.; Papandreou, G.; Kokkinos, I.; Murphy, K.; Yuille, A. DeepLab: Semantic Image Segmentation with Deep Convolutional Nets, Atrous Convolution, and Fully Connected CRFs. *arXiv* **2015**, arXiv:1606.00915.
25. Zheng, S.; Jayasumana, S.; Paredes, B.; Vineet, V.; Su, Z.; Du, D.; Huang, C.; Torr, P. Conditional Random Fields as Recurrent Neural Networks. *arXiv* **2015**, arXiv:1502.03240.
26. Chandra, S.; Kokkinos, I. Fast, Exact and Multi-Scale Inference for Semantic Image Segmentation with Deep Gaussian CRFs. *arXiv* **2016**, arXiv:1603.08358.
27. Badrinarayanan, V.; Handa, A.; Cipolla, R. SegNet: A Deep Convolutional Encoder-Decoder Architecture for Image Segmentation. *arXiv* **2015**, arXiv:1511.00561.
28. Hyeonwoo, N.; Hong, S.; Han, B. Learning deconvolution network for semantic segmentation. *arXiv* **2015**, arXiv:1505.04366.
29. Lin, G.; Shen, C.; Hengel, A.; Reid, I. Efficient piecewise training of deep structured models for semantic segmentation. *arXiv* **2016**, arXiv:1504.01013.
30. Eigen, D.; Fergus, R. Predicting Depth, Surface Normals and Semantic Labels with a Common Multi-Scale Convolutional Architecture. *arXiv* **2015**, arXiv:1411.4734.
31. Chen, L.; Schwing, A.; Yuille, A.; Urtasun, R. Learning Deep Structured Models. *arXiv* **2015**, arXiv:1407.2538.
32. Chen, L.; Papandreou, G.; Kokkinos, I.; Murphy, K.; Yuille, A. Semantic Image Segmentation with Deep Convolutional Nets and Fully Connected CRFs. *arXiv* **2015**, arXiv:1412.7062.
33. Krähenbühl, P.; Koltun, V. Efficient Inference in Fully Connected CRFs with Gaussian Edge Potentials. *arXiv* **2012**, arXiv:1210.5644.
34. Arnab, A.; Jayasumana, S.; Zheng, S.; Torr, P. Higher Order Conditional Random Fields in Deep Neural Networks. *arXiv* **2016**, arXiv:1511.08119.
35. Vemulapalli, R.; Tuzel, O.; Liu, M.; Chellappa, R. Gaussian Conditional Random Field Network for Semantic Segmentation. In Proceedings of the 2016 IEEE Conference on Computer Vision and Pattern Recognition (CVPR), Las Vegas, NV, USA, 27–30 June 2016.
36. Zhou, D.; Weston, J.; Gretton, A.; Bousquent, O.; Scholkopf, B. Ranking on data manifolds. In Proceedings of the 16th International Conference on Neural Information Processing Systems, Whistler, BC, Canada, 9–11 December 2003.
37. Zhou, D.; Bousquent, O.; Lal, T.; Weston, J.; Scholkopf, B. Learning with Local and Global Consistency. In Proceedings of the 16th International Conference on Neural Information Processing Systems, Whistler, BC, Canada, 9–11 December 2003.
38. Yang, C.; Zhang, L.; Lu, H.; Ruan, X.; Yang, M. Saliency Detection via Graph-Based Manifold Ranking. In Proceedings of the 2013 IEEE Conference on Computer Vision and Pattern Recognition, Portland, OR, USA, 23–28 June 2013.
39. Bencherif, M.A.; Bazi, Y.; Guessoum, A.; Alajlan, N.; Melgani, F.; AlHichri, H. Fusion of Extreme Learning Machine and Graph-Based Optimization Methods for Active Classification of Remote Sensing Images. *IEEE Geosci. Remote Sens. Lett.* **2015**, *12*, 527–531.
40. Krähenbühl, P.; Koltun, V. Parameter Learning and Convergent Inference for Dense Random Fields. In Proceedings of the 30th International Conference on Machine Learning, Atlanta, GA, USA, 16–21 June 2013.
41. Gupta, S.; Girshick, R.; Arbeláez, P.; Malik, J. Learning Rich Features from RGB-D Images for Object Detection and Segmentation. *arXiv* **2014**, arXiv:1407.5736.
42. Hariharan, B.; Arbeláez, P.; Girshick, R.; Malik, J. Simultaneous Detection and Segmentation. *arXiv* **2014**, arXiv:1407.1808.
43. Dai, J.; He, K.; Sun, J. Convolutional Feature Masking for Joint Object and Stuff Segmentation. *arXiv* **2015**, arXiv:1412.1283.
44. Farabet, C.; Couprie, C.; Najman, L.; LeCun, Y. Learning Hierarchical Features for Scene Labeling. *IEEE Trans. Pattern Anal. Mach. Intell.* **2013**, *35*, 1915–1929.
45. Chen, L.; Yang, Y.; Wang, J.; Xu, W.; Yuille, A. Attention to Scale: Scale-aware Semantic Image Segmentation. *arXiv* **2016**, arXiv:1511.03339.
46. Bearman, A.; Russakovsky, O.; Ferrari, V.; Li, F.F. What's the Point: Semantic Segmentation with Point Supervision. *arXiv* **2016**, arXiv:1506.02106.
47. Romero, A.; Gatta, C.; Camps-Valls, G. Unsupervised Deep Feature Extraction for Remote Sensing Image Classification. *IEEE Trans. Geosci. Remote Sens.* **2015**, *54*, 1349–1362.

48. Campos-Taberner, M.; Romero-Soriano, A.; Gatta, C.; Camps-Valls, G.; Lagrange, A.; Le Saux, B.; Randrianarivo, H. Processing of Extremely High-Resolution LiDAR and RGB Data: Outcome of the 2015 IEEE GRSS Data Fusion Contest–Part A: 2-D Contest. *IEEE J. Sel. Top. Appl. Earth Obs. Remote Sens.* **2016**, *9*, 5547–5559.
49. Tschannen, M.; Cavigelli, L.; Mentzer, F.; Wiatowski, T.; Benini, L. Deep Structured Features for Semantic Segmentation. *arXiv* **2016**, arXiv:1609.07916.
50. Piramanayagam, S.; Schwartzkopf, W.; Koehler, F.W.; Saber, E. Classification of remote sensed images using random forests and deep learning framework. *SPIE Remote Sens. Int. Soc. Opt. Photonics* **2016**, doi:10.1117/12.2243169.
51. Marcu, A.; Leordeanu, M. Dual Local-Global Contextual Pathways for Recognition in Aerial Imagery. *arXiv* **2016**, arXiv:1605.05462.
52. Yuan, Y.; Lin, J.; Wang, Q. Dual-clustering-based hyperspectral band selection by contextual analysis. *IEEE Trans. Geosci. Remote Sens.* **2016**, *54*, 1431–1445.
53. Kendall, A.; Badrinarayanan, V.; Cipolla, R. Bayesian SegNet: Model Uncertainty in Deep Convolutional Encoder-Decoder Architectures for Scene Understanding. *arXiv* **2015**, arXiv:1511.02680.
54. Simonyan, K.; Zisserman, A. Very Deep Convolutional Networks for Large-Scale Image Recognition. *arXiv* **2015**, arXiv:1409.1556.
55. Hong, S.; Noh, H.; Han, B. Decoupled Deep Neural Network for Semi-supervised Semantic Segmentation. *arXiv* **2015**, arXiv:1506.04924.
56. Audebert, N.; Saux, B.L.; Lefèvre, S. Segment-before-Detect: Vehicle Detection and Classification through Semantic Segmentation of Aerial Images. *Remote Sens.* **2017**, *9*, 368.
57. Huang, Z.; Cheng, G.; Wang, H.; Li, H.; Shi, L.; Pan, C. Building extraction from multi-source remote sensing images via deep deconvolution neural networks. In Proceedings of the IEEE International Geoscience and Remote Sensing Symposium (IGARSS), Beijing, China, 10–15 July 2016.
58. Audebert, N.; Boulch, A.; Lagrange, A.; Le Saux, B.; Lefevre, S. *Deep Learning for Remote Sensing*; Technical Report; ONERA The French Aerospace Lab, DTIM & Univ. Bretagne-Sud & ENSTA ParisTech: Palaiseau, France, 2016.
59. Paisitkriangkrai, S.; Sherrah, J.; Janney, P.; Hengel, V.D. Effective semantic pixel labelling with convolutional networks and conditional random fields. In Proceedings of the 2015 IEEE Conference on Computer Vision and Pattern Recognition Workshops, Boston, MA, USA, 7–12 June 2015.
60. Alam, F.I.; Zhou, J.; Liew, A.W.C.; Jia, X. CRF learning with CNN features for hyperspectral image segmentation. In Proceedings of the IEEE International Geoscience and Remote Sensing Symposium, Beijing, China, 10–15 July 2016.
61. He, X.; Cai, D.; Niyogi, P. Laplacian Score for Feature Selection. In Proceedings of the 18th International Conference on Neural Information Processing Systems, Vancouver, BC, Canada, 5–8 December 2005.
62. Quan, R.; Han, J.; Zhang, D.; Nie, F. Object co-segmentation via graph optimized-flexible manifold ranking. In Proceedings of the 2016 IEEE Conference on Computer Vision and Pattern Recognition (CVPR), Las Vegas, NV, USA, 27–30 June 2016.
63. Wang, Q.; Lin, J.; Yuan, Y. Salient band selection for hyperspectral image classification via manifold ranking. *IEEE Trans. Neural Netw. Learn. Syst.* **2016**, *27*, 1279–1289.
64. Yang, C.; Zhang, L.; Lu, H. Graph-regularized saliency detection with convex-hull-based center prior. *IEEE Signal Process. Lett.* **2013**, *20*, 637–640.
65. Xu, B.; Bu, J.; Chen, C.; Cai, D.; He, X.; Liu, W.; Luo, J. Efficient Manifold Ranking for Image Retrieval. In Proceedings of the 34th international ACM SIGIR Conference on Research and Development in Information Retrieval, Beijing, China, 24–28 July 2011.
66. Hsieh, C.; Han, C.; Shih, J.; Lee, C.; Fan, K. 3D Model Retrieval Using Multiple Features and Manifold Ranking. In Proceedings of the 2015 8th International Conference on Ubi-Media Computing (UMEDIA), Colombo, Sri Lanka, 24–26 August 2015.
67. Zhou, T.; He, X.; Xie, K.; Fu, K.; Zhang, J.; Yang, J. Robust visual tracking via efficient manifold ranking with low-dimensional compressive features. *Pattern Recognit.* **2015**, *48*, 2459–2473.
68. Brostow, G.; Shotton, J.; Fauqueur, J.; Cipolla, R. Segmentation and Recognition Using Structure from Motion Point Clouds. In Proceedings of the 10th European Conference on Computer Vision, Marseille, France, 12–18 October 2008.

69. Brostow, G.; Fauqueur, J.; Cipolla, R. Semantic object classes in video: A high-definition ground truth database. *Pattern Recognit. Lett.* **2009**, *30*, 88–97.
70. Ruder, S. An overview of gradient descent optimization algorithms. *arXiv* **2016**, arXiv:1609.04747.
71. He, K.; Zhang, X.; Ren, S.; Sun, J. Deep Residual Learning for Image Recognition. *arXiv* **2015**, arXiv:1512.03385.
72. Everingham, M.; Gool, L.; Williams, C.K.I.; Winn, J.; Zisserman, A. The Pascal Visual Object Classes (VOC) Challenge. *Int. J. Comput. Vis.* **2010**, *88*, 303–338.
73. Rottensteiner, F.; Sohn, G.; Jung, J.; Gerke, M.; Baillard, C.; Benitez, S.; Breitkopf, U. The ISPRS benchmark on urban object classification and 3D building reconstruction. *ISPRS Ann. Photogramm. Remote Sens. Spat. Inf. Sci.* **2012**, *I-3*, 293–298.
74. Glorot, X.; Bordes, A.; Bengio, Y. Deep Sparse Rectifier Neural Networks. In Proceedings of the 14th International Conference on Artificial Intelligence and Statistics, Fort Lauderdale, FL, USA, 11–13 April 2011; Volume 15, pp. 315–323.
75. Jia, Y.; Shelhamer, E.; Donahue, J.; Karayev, S.; Long, J.; Girshick, R.; Guadarrama, S.; Darrell, T. Caffe: Convolutional Architecture for Fast Feature Embedding. In Proceedings of the 22nd ACM International Conference on Multimedia, Orlando, FL, USA, 3–7 November 2014.
76. Marmanis, D.; Wegner, J.D.; Galliani, S.; Schindler, K.; Datcu, M.; Stilla, U. Semantic Segmentation of Aerial Images with an Ensemble of CNSS. *ISPRS Ann. Photogramm. Remote Sens. Spat. Inf. Sci.* **2016**, *3*, 473–480.
77. Hariharan, B.; Arbeláez, P.; Bourdev, L.; Maji, S.; Malik, J. Semantic Contours from Inverse Detectors. In Proceedings of the 2011 IEEE International Conference on Computer Vision (ICCV), Barcelona, Spain, 6–13 November 2011.
78. Zoran, D.; Weiss, Y. From Learning Models of Natural Image Patches to Whole Image Restoration. In Proceedings of the 2011 IEEE International Conference on Computer Vision (ICCV), Barcelona, Spain, 6–13 November 2011.
79. Lin, T.; Maire, M.; Belongie, S.; Bourdev, L.; Girshick, R.; Hays, J.; Perona, P.; Ramanan, D.; Zitnick, C.; Dollár, P. Microsoft coco: Common objects in context. *arXiv* **2014**, arXiv:1405.0312.
80. Lin, G.; Milan, A.; Shen, C.; Reid, I. RefineNet: Multi-Path Refinement Networks with Identity Mappings for High-Resolution Semantic Segmentation. *arXiv* **2016**, arXiv:1611.06612.
81. Kohli, P.; Torr, P.H. Robust higher order potentials for enforcing label consistency. *Int. J. Comput. Vis.* **2009**, *82*, 302–324.
82. Krizhevsky, A.; Sutskever, I.; Hinton, G.E. Imagenet classification with deep convolutional neural networks. In Proceedings of the 25th International Conference on Neural Information Processing Systems, Lake Tahoe, NV, USA, 3–6 December 2012.
83. Quang, N.T.; Thuy, N.T.; Sang, D.V.; Binh, H.T.T. An efficient framework for pixel-wise building segmentation from aerial images. In Proceedings of the Sixth International Symposium on Information and Communication Technology, Hue City, Vietnam, 3–4 December 2015.
84. Boulch, A. *DAG of Convolutional Networks for Semantic Labeling*; Technical Report; Office National d'études et de Recherches Aérospatiales: Palaiseau, France, 2015.
85. Gerke, M.; Speldekamp, T.; Fries, C.; Gevaert, C. Automatic semantic labelling of urban areas using a rule-based approach and realized with mevislab. *Unpublished* **2015**, doi:10.13140/RG.2.1.3345.0408.
86. Sherrah, J. Fully convolutional networks for dense semantic labelling of high-resolution aerial imagery. *arXiv* **2016**, arXiv:1606.02585.
87. Gerke, M. *Use of the Stair Vision Library within the ISPRS 2D Semantic Labeling Benchmark (Vaihingen)*; Technical Report; University of Twente: Enschede, The Netherlands, 2015.
88. Petersen, K.; Pedersen, M. *The Matrix Cookbook*; Technical University of Denmark: Kongens Lyngby, Denmark, 2008.
89. The National Survey of Geographical Conditions Leading Group Office, Sate Council, P.R.C. *General Situation and Index of Geographical Conditions (Chinese Manual, GDPJ 01-2013)*; The National Survey of Geographical Conditions Leading Group Office, Sate Council, P.R.C.: Beijing, China, 2013.
90. Immitzer, M.; Atzberger, C.; Koukal, T. Tree species classification with random forest using very high spatial resolution 8-band WorldView-2 satellite data. *Remote Sens.* **2012**, *4*, 2661–2693.
91. Dribault, Y.; Chokmani, K.; Bernier, M. Monitoring seasonal hydrological dynamics of minerotrophic peatlands using multi-date GeoEye-1 very high resolution imagery and object-based classification. *Remote Sens.* **2012**, *4*, 1887–1912.

92. Onojeghuo, A.O.; Blackburn, G.A. Mapping reedbed habitats using texture-based classification of QuickBird imagery. *Int. J. Remote Sens.* **2011**, *32*, 8121–8138.
93. Junwei, S.; Youjing, Z.; Xinchuan, L.; Wenzhi, Y. Comparison between GF-1 and Landsat-8 images in land cover classification. *Prog. Geogr.* **2016**, *35*, 255–263.
94. Goodfellow, I.; Pouget-Abadie, J.; Mirza, M.; Xu, B.; Warde-Farley, D.; Ozair, S.; Courville, A.; Bengio, Y. Generative adversarial nets. In *Advances in Neural Information Processing Systems*; The MIT Press: Cambridge, MA, USA, 2014; pp. 2672–2680.
95. Mirza, M.; Osindero, S. Conditional generative adversarial nets. *arXiv* **2014**. arXiv:1411.1784.
96. Luc, P.; Couprie, C.; Chintala, S.; Verbeek, J. Semantic Segmentation using Adversarial Networks. *arXiv* **2016**, arXiv:1611.08408.

remote sensing

MDPI

Article

Multi-Scale Analysis of Very High Resolution Satellite Images Using Unsupervised Techniques

Jérémie Sublime [1,2,*], Andrés Troya-Galvis [3] and Anne Puissant [4]

[1] LISITE Laboratory, RDI Team–Institut Supérieur d'Électronique de Paris, 10 rue de Vanves, 92130 Issy Les Moulineaux, France

[2] CNRS UMR 7030 LIPN–Université Paris 13, Sorbonne Paris Cité, 99 av. J-B Clément, 93430 Villetaneuse, France

[3] CNRS UMR 7357 ICube–Université de Strasbourg, 300 bd Sébastien Brant-CS 10413, F-67412 Illkirch CEDEX, France; troyagalvis@unistra.fr

[4] CNRS UMR 7362 LIVE–Université de Strasbourg, 3 rue de l'Argonne, 67000 Strasbourg, France; anne.puissant@live-cnrs.unistra.fr

* Correspondence: jeremie.sublime@isep.fr or sublime@lipn.univ-paris13.fr; Tel.: +33-149-545-219

Academic Editors: Qi Wang, Nicolas H. Younan, Carlos López-Martínez, Sangram Ganguly and Prasad S. Thenkabail

Received: 15 March 2017; Accepted: 12 May 2017; Published: 18 May 2017

Abstract: This article is concerned with the use of unsupervised methods to process very high resolution satellite images with minimal or little human intervention. In a context where more and more complex and very high resolution satellite images are available, it has become increasingly difficult to propose learning sets for supervised algorithms to process such data and even more complicated to process them manually. Within this context, in this article we propose a fully unsupervised step by step method to process very high resolution images, making it possible to link clusters to the land cover classes of interest. For each step, we discuss the various challenges and state of the art algorithms to make the full process as efficient as possible. In particular, one of the main contributions of this article comes in the form of a multi-scale analysis clustering algorithm that we use during the processing of the image segments. Our proposed methods are tested on a very high resolution image (Pléiades) of the urban area around the French city of Strasbourg and show relevant results at each step of the process.

Keywords: very high resolution images; segmentation; multi-scale clustering

1. Introduction

The recent advances of remote sensing technologies for Earth observation have led to a surge in the number of large and complex available data to process. For example, very high spatial resolution (VHR) satellite images covering large areas are nowadays commonly delivered by remote sensors (Pléiades, Worldview, Quickbird, Ikonos). The manual analysis of such images by experts to extract useful information would be overwhelming, and the use of machine learning techniques is more than ever necessary to obtain satisfactory results in a fair amount of time. However, the majority of popular machine learning techniques for classification purposes (known as supervised learning) also require human intervention in the sense that the computer can only learn to recognize things that have already been learned and identified by humans based on similar data. In the case of VHR images, since they have a high level of detail and deal with a wide variety of landscapes, such knowledge to feed the machine learning algorithm is quite often unavailable or incomplete.

Within this context, in this article, we propose a complete methodology for an almost fully-unsupervised analysis of VHR images requiring only minimal knowledge on the data and little human intervention. We will discuss the different steps and challenges of going from the raw satellite image to the final segmented and clustered image where the different elements of interest can be linked to expert classes. In particular, the main novelty of this work lies in the proposition of a clustering algorithm that can process image segments and find multi-scale clusters matching the different scales of interest that can be found on VHR images.

Furthermore, our algorithm also provides minimal semantic information that can be used to link the clusters to land cover classes. Unlike the majority of methods in the literature, our proposed model focuses on object-based image analysis (OBIA) rather than pixel-based analysis. Indeed, it makes more sense to focus on objects rather than pixels that have little semantic value when using very high resolution [1,2].

Works closely related to this article include other unsupervised algorithms that have been proposed recently to process datasets built from the segments of non-hyperspectral VHR images:

- In [3], the authors propose an unsupervised algorithm that provides some low level semantic information on the clusters. This algorithm is the base that we used for the multi-scale method proposed in the learning step of this article. The improvements that we bring include that our proposed method covers the segmentation step, while the original algorithm does not. Furthermore, this algorithm was designed to produce a non-hierarchical hard partition, whereas our method can find the object at several scales of interest and produces multi-scale hierarchical clusters.
- In [4], the authors also tackle image data acquired from image segments. The method they used is based on the self-organizing map (SOM) algorithm, a known unsupervised neural network used for dimension reduction. While this methods considers dimension reduction aspects that our proposed algorithm does not handle, it is also limited to the learning step and can only provide hard partitions computed at a single scale of interest.

The remainder of this article is organized as follows: In Section 2, we present the different steps involved in VHR image processing and discuss the various challenges and state of the art methods for each step. In Section 3, we introduce the material and methods that we use in our experiments. In particular, we give the details of the multi-scale clustering algorithm that we use to process our data. Section 4 shows our experimental results and features various discussions on the results. Finally, in Section 5, we give our conclusions on this work, as well as some perspective on future extensions.

2. State of the Art on Unsupervised VHR Images Processing

The fully-automated analysis of a satellite image can usually be decomposed into three steps: (1) a pre-processing step during which the image is prepared from raw sources (merging pictures, orthorectification, etc.); (2) a segmentation step that consists of grouping together adjacent pixels that are similar given a certain homogeneity criterion; these groups, called segments, should ideally be a good estimation of the geographical objects in the image [5,6]; (3) the segments created during Step 2 can then be fed to a supervised or unsupervised machine learning algorithm in order to recognize the elements in the image.

This succession of steps, all dependent on the previous ones, is summed up in Figure 1. As one can see, errors are quite likely to accumulate through the process.

In the next subsections, we will discuss the state of the art methods used during the segmentation and clustering step: we will go into detail on explaining which difficulties are encountered during each step and which techniques can be used to reduce the risk of error accumulation in order to ensure the best possible final results.

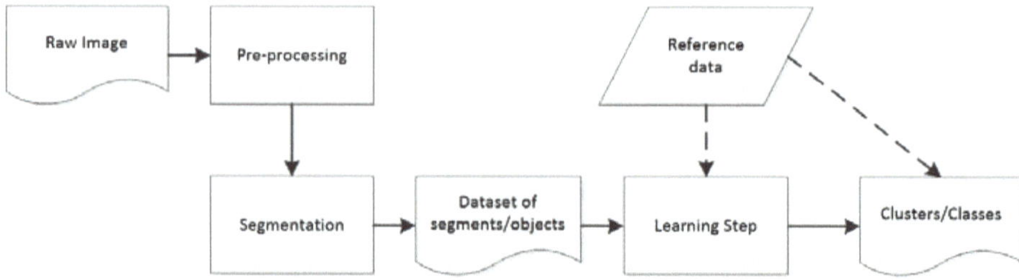

Figure 1. Step by step approach to image processing.

2.1. Image Segmentation

Image segmentation is the first critical step within the OBIA workflow and aims at finding segments that will correspond to the objects of interest in the image. Indeed, poor quality segments would likely lead to the computation of inaccurate and irrelevant descriptors, making the dataset difficult or even impossible to exploit by machine learning techniques in further steps.

A wide range of segmentation approaches and their ad hoc variants devoted to specific applications can be found in the literature. The reader interested in general segmentation approaches may refer to [7] for a complete survey on this topic.

In the context of remote sensing imaging, the most popular approaches are mainly these relying on region-based and spectral homogeneity paradigms. For instance, the mean-shift [8,9] applies a technique for estimating local modes in a multivariate distribution [10] to a joint spatial and spectral domain. For each pixel, local modes are computed with respect to a spectral and spatial similarity ranges so that in the end, each pixel is associated with the local mode's spectral signature and the spatial location of its density probability distribution. Finally, pixels sharing the same local mode are merged together to generate the segments. Region-growing approaches, such as [11,12], are also commonly employed. They usually start by considering each pixel as a segment and then iteratively merge similar pixels based on a given homogeneity criterion. Other constraints such as a minimum or maximum segment size are often considered as well during the merging procedure. Other popular segmentation algorithms are based on the watershed transformation [13,14]. The main idea consists of considering the gradient image as a topographic surface. This surface is then flooded starting from the local minima of the image gradient. When two different flooding basins are about to merge, the process stops, and a watershed (segment boundary) is drawn. Finally, hierarchical strategies [15] are based on graph theory and consider the image (and the segments being created) as a tree structure in which lower level objects are close to the leafs and more abstract objects are at higher hierarchical levels. This structure allows focusing on objects at different levels of resolution or semantics.

While few efforts have been made in this area, evaluating the quality of a segmentation remains a key issue: image segmentation is an ill-posed problem, so almost any partition of the image can be considered as a correct segmentation given the general definition of image segmentation (i.e., partitioning the image by grouping similar pixels given a certain criterion). Thus, the definition of segmentation quality is usually dependent on a given application. In a remote sensing context, a perfect segmentation should map each segment to an object of interest in the image. Given this definition of quality, it is possible to distinguish mainly two kinds of segmentation errors: over-segmentation where objects are split into several segments; and under-segmentation where a single segment may contain several objects. There exist mainly three families of quality criteria:

- Subjective criteria, which basically rely on a visual examination of segmentation results. This task is long, tedious and does not provide an objective and quantitative evaluation.
- Supervised criteria [16,17], which consist of measuring the distance between one segmentation and a gold-standard segmentation. However, such a ground-truth generally has to be manually

generated. Thus, it is very rare to dispose of complete reference datasets in remote sensing applications, making supervised metrics less reliable.

- Finally, unsupervised criteria [18], which consist of exploiting intrinsic segment and image properties. It is then necessary to accurately define and model the notion of quality without any external information. Many of these metrics rely on the number and size of segments [19], as well as statistics, such as band mean values or the standard deviation [17,20], or on local (per segment) quality estimation based on some homogeneity criterion in order to compute global quality metrics by aggregation of the local scores [21].

In short, segmentation algorithms should be used along with different quality metrics so that the produced segmentation has as few segmentation errors as possible. In practice, over-segmentation errors are usually tolerated as they can be easily corrected by further analysis; however, under-segmentation has to be avoided as much as possible, see Figure 2.

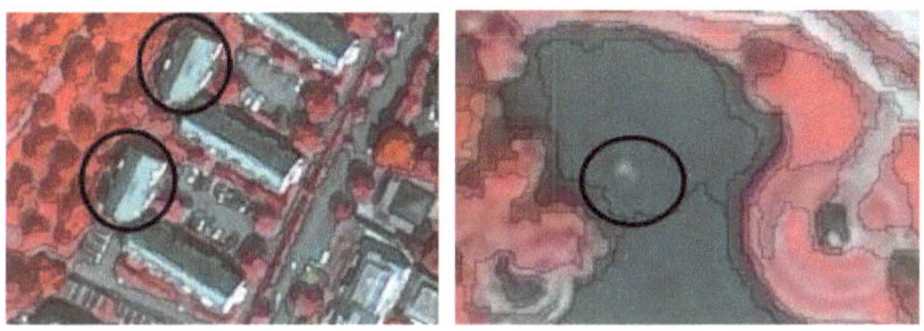

(a) (b)

Figure 2. Examples of over-segmentation and under-segmentation. (**a**) Example of an over-segmentation on two houses that could be fixed during the clustering step: the algorithm may still detect that these two segments are part of the same cluster; (**b**) example of an under-segmentation where the white object in the middle of the lake was not detected during the segmentation step and will never be since it is now merged with a lake segment.

2.2. Unsupervised Analysis of the Segments

The objects extracted from an image during the segmentation can be seen as regular data, where each segment is described by several features from the original image, such as color attributes, as well as new features created during the segmentation process: surface of the segments, perimeter and elongation, shape information, color extrema, variance and average value of the attributes in the pixels of a given segment, texture information, contrast with the neighboring segments, etc.

Because the segments and their attributes can vary greatly depending on the image or the algorithm used for the segmentation process, it is very difficult to find similar data using the same attributes that could be used to train a supervised classifier to process such a segment-based dataset. Unsupervised methods are therefore most convenient to process such data acquired from a segmentation. In particular, clustering techniques that consist of finding groups of similar data in a dataset are usually a good choice since the clusters can be built without external knowledge and can usually be easily linked to expert-defined classes once they have been built. These methods are therefore popular for both object-based and pixel-based image analysis [22–25].

The main known weakness of unsupervised approaches for object identification in images is that there is no warranty that the clusters found by the algorithms will end up being pertinent classes. A first possible solution consists of using semi-supervised approaches instead of fully-unsupervised ones: In the case of pixel-based analysis of VHR images, a solution proposed in the literature is to guide the clustering process using ontologies [26,27], a tool commonly used in supervised process. The results achieved using these methods are promising, but seem limited to a very low number of clusters/classes. The second solution that is usually preferred in the context of OBIA is to use a mixed clustering and Markov random field (MRF) approach [28–30] with the goal of using all of the extra attributes from the segmentation in the clustering process (shapes, texture, contrast), but also to use the information from the neighborhood dependencies in order to influence the clustering of each segment based on both its characteristics and the cluster to which the neighboring segments belong. Other approaches have been attempted using topological clustering instead of MRF-based techniques [4] for OBIA.

One advantage of MRF-based approaches is that these methods are used for both segmentation, classification and clustering. In our case, we are particularly interested in the segmentation and clustering uses. Using MRF-based methods has the advantage that it can deal with over-segmented data just fine, thus reducing of error accumulation from the segmentation step during the clustering step. In the remainder of this section, we will focus on the MRF-based approach, as it will be the basis of our proposed method in the experiments.

The clustering task using MRF models can be seen as a graph partitioning problem, where each segment is a node of the graph and the edges are represented by the neighborhood dependencies between the segments. Assigning each segment to a cluster based on its features and its neighbors (see Figure 3) is indeed equivalent to finding the optimal cuts in the graph to separate dissimilar neighbor segments. This process will provide both the clusters and a new segmentation as a by-product.

Figure 3. Illustration of the MRF clustering problem with very few features: in this example, we try to guess the cluster of the central segment based on five features and the clusters of its neighbor segments (identified using the colors).

There are many methods in the literature to solve this kind of problem: the graph-cut algorithm [31], the integer projected fixed point method [32], the graduated non-convexity and concavity procedure [33], the iterated conditional modes (ICM) [34] and hybrid algorithms mixing the principle of expectation-maximization and the ICM algorithm [35].

In the case of segments from VHR images, approaches with the lowest complexity are usually preferred due to the expected large size of the graph. To this end, an adaptation of the hybrid EM-ICM approach capable of assessing the affinities between neighbor segments of different pixel was proposed in the form of a semantic-rich ICM [3] (SR-ICM). This algorithm is similar to what already existed for semantic-rich pixel-based MRF models [36], but adapted to the case of segments that have an irregular number of neighbors, instead of always four neighbors for pixel-based models.

To better explain this idea of adding semantics to the MRF model, in the case of Figure 3, using a regular ICM approach, the neighborhood dependencies would encourage putting the central segment in the light green cluster (which is the majority neighbor). However, using a semantic-rich ICM algorithm, it may be possible to put this very same segment in any cluster having a good neighborhood compatibility with the light green segment.

We will now give the details of this algorithm. Let us consider a dataset that contains N segments: $X = \{x_1, \cdots, x_N\}, x_i \in \mathbb{R}^d$ where each x_i represents a segment having d real-valued features. We will denote $V_{x_i} \subset X$ the set containing all of the neighbor segments of any segment x_i. We suppose for now that we are looking for K hard clusters and that K is known in advance:

we denote $S = \{s_1, \cdots, s_N\}, s_i \in [1 \cdots K]$ the clustering solution that links each of the N segments to a cluster. We make the hypothesis that each cluster C_k can be represented as following a Gaussian distribution of parameters $\theta = \{\pi_k, \mu_k, \Sigma_k\}$ where the π_k are the mixing probabilities, the μ_k are the mean of each cluster and the Σ_k the variance-covariance matrices of each cluster. Finally, we define $A = (a_{ij})_{(K\times K)}, a_{ij} \in [0,1], \forall i \sum_j a_{ij} = 1$, the affinity matrix between neighbor segments [3,30], where each a_{ij} denotes the probability for a segment of cluster C_i of having a neighbor segment belonging to cluster C_j. Using these notations, the goal of the SR-ICM algorithm is to optimize the following function:

$$\{S, \Theta, A\} = \underset{S,\Theta,A}{\text{Argmax}} \prod_{n=1}^{N} \left(\pi_{s_n} \mathcal{N}(\mu_{s_n}, \Sigma_{s_n}, x_n) \times \prod_{v \in V_{x_n}} (a_{s_v, s_n})^{\tau_{x,v}} \right) \tag{1}$$

where $\tau_{x,v}$ is the percentage of the border shared between neighbor segments (replaced by one when this information is not available).

The optimization of Equation (1), where S, μ, π, Σ and A are unknown, is usually done in two steps: the first step using the regular EM algorithm [37] for the Gaussian mixture model on the data without the neighborhood dependencies. This step will be used to determine π, μ and Σ and to initialize S and A. The second step using a maximization-maximization process analog with the EM algorithm is then used to refine S and A with Θ fixed.

$$s_n = \underset{k}{\text{Argmax}} \left[\pi_k \times \mathcal{N}(\mu_k, \Sigma_k, x_n) \times \prod_{v \in V_{x_n}} a_{s_v,k}^{\tau_{x,v}} \right] \tag{2}$$

$$a_{ij} = \frac{\sum_{x_n \in C_i} \sum_{v \in V_{x_n}} \delta_{s_v, j}}{\sum_{x_n \in C_i} \sum_{v \in V_{x_n}} 1} \tag{3}$$

As one can see from Algorithm 1, the optimization is quite simple and has a linear complexity, which makes it convenient to use with large datasets. The stopping criterion of this algorithm is the trace of the affinity matrix A. This criterion comes from the idea that the original ICM algorithm is a segmentation algorithm and tries to create large and homogeneous areas of elements in the same cluster. Since the diagonal elements of the matrix contain the self-transition probabilities, the trace of the matrix assesses the overall compactness of the newly-created area using the SR-ICM algorithm.

Algorithm 1: Semantic-rich ICM algorithm.

```
Find Θ and initialize S with the EM algorithm
Initialize A using Equation (3)
while Tr(A) is increasing do
    Update S using Equation (2) over all of the data
    Update A from the new distribution S using Equation (3)
end
return S and A
```

As stated in the Introduction, one of the main issues with the unsupervised analysis of VHR data is that the lack of supervision sometimes makes it difficult to map the clusters to the expert classes. One advantage of the SR-ICM algorithm is that in addition to providing a partition of the data, it returns the affinity matrix A, which gives useful information on the relationship between the clusters. The affinity matrix therefore serves a dual purpose: first it helps improve the clustering by enriching the data with neighborhood compatibility information; second, it contains low semantic level information on how the clusters relate to each other in the image. This information can either be used to help identify the expert classes or simply be translated into a description of the image once the clusters have been mapped to land cover classes.

Figure 4 shows an example of a simple affinity matrix with four clusters. In this figure, we can see how each value can be interpreted. It is easy to see how such a matrix can then be translated into a description of the image. This would lead to sentences, such as: "urban areas are surrounded by area of vegetation" or "urban areas are rarely in direct contact with water areas", "water areas have a low compactness", and so forth. While this may not seem like much, even this low level of description is not possible with other unsupervised algorithms.

$$A = \begin{matrix} 0{,}12 & 0{,}01 & 0{,}17 & 0{,}70 \\ 0{,}02 & 0{,}81 & 0{,}01 & 0{,}16 \\ 0{,}13 & 0{,}03 & 0{,}50 & 0{,}24 \\ 0{,}36 & 0{,}25 & 0{,}14 & 0{,}25 \end{matrix}$$

- Areas of clusters 1 are surrounded by areas of cluster 4
- Areas of clusters 1 and 4 have a low compactness
- Clusters 1 and 2, as well as 2 and 3 are incompatible neighbors

Figure 4. Example of an affinity matrix: Diagonal values indicate whether or not the clusters are forming compact areas (high value) or are scattered elements in the image (low value). Non-diagonal elements indicate which clusters are often neighbors on the image (high value) or incompatible neighbors (low value).

3. Material and Methods

3.1. Presentation of the Strasbourg Dataset

In this section, we present the data used in our experiments. The original set is an extract of a multispectral VHR pan-sharpened image with 0.5-m spatial resolution and four spectral bands (red, green, blue and near-infrared) from the Pléiades satellite Airbus, ©CNES, orthorectified and geo-referenced in Lambert93, acquired on 14 August 2012 covering the metropolitan area of Strasbourg, see Figure 5. In this article, we use only a subset of this image (9211×11,275 pixels), which is multispectral and not hyperspectral.

The data were later enriched with a hierarchical land cover/use database featuring 15 classes at the finest level (Level 4) from the metropolitan area of Strasbourg (Figure 6a). This database is a combination of existing vector databases (buildings, roads, railways, bare soil, crops, water) and a semi-automatic extraction of vegetation classes from several Pléiades images.

However, this hierarchical land cover/use database had to be modified because some classes such as 'grass' and 'urban grass' or 'bare soils' and 'winter crops' cannot be distinguished from the sky. Therefore, in order to propose a nomenclature adapted to an extraction from a VHR image, we have proposed the modified hierarchical typology detailed in Figure 6b. This modified database can be considered as the reference data for our research.

Nevertheless, some pre-processing was necessary in order to reduce the bias due to the misalignments between the land cover polygons and the Pléiades image (Figure 7): the reference data provide accurate labels, as well as very regular polygons (Figure 7a). However, when inspecting them in detail, one realizes that the polygons are not well aligned with the represented objects (Figure 7b). The misalignments are possibly due to orthorectification procedures during the pre-processing of the image or because of a date difference between the geographic information system (GIS) data and the image acquisition. Therefore, any comparison against these data would result in a difficult to quantify, yet certain bias. In order to make these data more reliable to evaluate our results, it is necessary to find a solution to improve their quality, especially in terms of segment alignment.

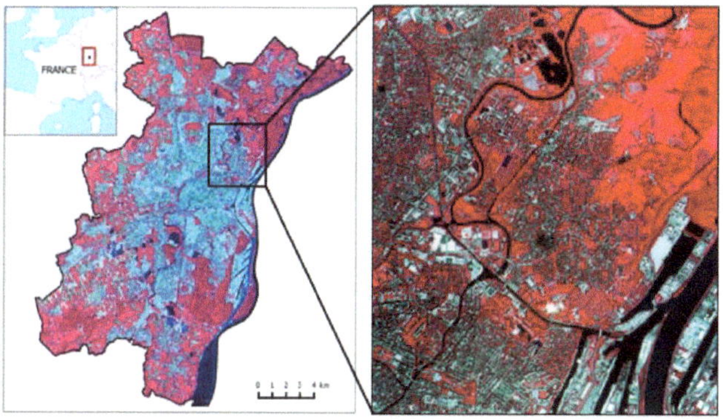

Figure 5. (**Left**) the metropolitan area of Strasbourg (Spotimage ©CNES, 2012); (**right**) extract of the Pan-sharpened Pléiades image (Airbus ©CNES, 2012).

(**a**) Hierarchical land cover/use thematic classes proposed by experts

(**b**) Hierarchical nomenclature retained for the experiments (+ expert classes in grey)

Figure 6. Expert classes (**a**) and hierarchical classes retained for the experiments (**b**).

To cope with this issue of misaligned reference data, we propose hereafter a refining procedure that consists of superposing an over-segmentation of the image with the GIS polygons in order to propagate the GIS labels into the segments. The procedure goes as follows: for every segment s_i in the over-segmentation, we find the set of GIS polygons intersecting s_i. Then, it is possible to assign a single label to s_i by taking the label of the GIS polygon with the largest intersection area with respect to s_i. By proceeding this way, we ensure that the new reference database and the segments are actually aligned with the objects of interest, since the boundaries of the segments tend to align well with actual object boundaries. It is also possible to reinforce the quality of the produced labels by adding a threshold over the intersection area. Thus, one would only consider GIS polygons intersecting more than 50% of the area of s_i, for example. Another possibility is to consider the labels of all intersecting polygons and to construct a fuzzy reference dataset in which each class c is weighted by the intersection area of GIS polygons labeled with class c.

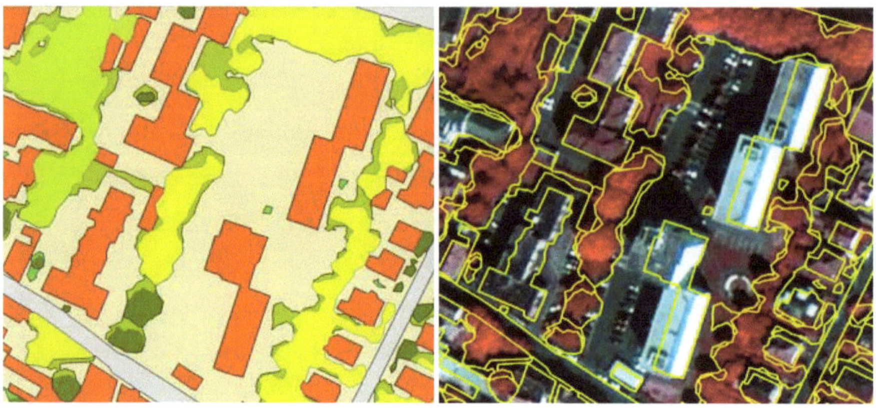

(**a**) (**b**)

Figure 7. Example of reference data from geographic information systems (GIS). (**a**) GIS labeled data; (**b**) contours of the GIS polygons.

In this paper, we opted to build a hard reference hybrid reference dataset using a simple majority vote.

3.2. Segmentation and Feature Computation

For our experiments, we ran the multi-resolution image segmentation (MRIS) implemented in the eCognition software ©Definens (2014) on the raw image. We chose this algorithm because it gives good performance for the retrieval of land cover/use classes [38]. MRIS is an algorithm of segmentation by "region growing", where a scale parameter is used as the maximum heterogeneity threshold during the fusion process [11]. This heterogeneity parameter includes a spectral criterion and a shape one. Then, a level of segmentation with a scale parameter of 160 was chosen after several runs based on a statistical method developed in [39]: this method relies on the potential of the local variance to detect scale transitions in geospatial data. The tool detects the number of layers added to a project and segments them iteratively with a multi-resolution segmentation algorithm in a bottom-up approach, where the scale factor in the segmentation, namely the scale parameter, increases with a constant increment. The average local variance value of the objects in all of the layers is computed and serves as a condition for stopping the iterations: when a scale level records a local variance value that is equal to or lower than the previous value, the iteration ends, and the objects segmented in the previous level are retained.

A wide range of features available in eCognition has been computed for each segment, including spectral, textural and shape features that were exported in a CSV file. A total of 27 attributes have been calculated for 187,057 segments and are shown in Table 1, where *XS1* stands for blue, *XS2* green, *XS3* red, and *XS4* near infra-red.

Table 1. The 27 attributes computed for the 187,057 segments.

Attribute	Type	Comments
Brightness	Spectral	
Max. difference	Spectral	
Mean XS1	Spectral	Blue
Mean XS2	Spectral	Green
Mean XS3	Spectral	Red
Mean XS4	Spectral	near-infrared
Standard deviation XS1	Spectral	Blue
Standard deviation XS2	Spectral	Green
Standard deviation XS3	Spectral	Red
Standard deviation XS4	Spectral	Near-infrared
Ratio XS1	Spectral	Blue
Ratio XS2	Spectral	Green
Ratio XS3	Spectral	Red
Ratio XS4	Spectral	Near-infrared
Mean Diff. to neighbors XS1	Spectral	Blue
Mean Difference to neighbors XS2	Spectral	Green
Mean Difference to neighbors XS3	Spectral	Red
Mean Difference to neighbors XS4	Spectral	Near-infrared
Area	Shape	in pixels
Elliptic fit	Shape	
Density	Shape	
Rectangular Fit	Shape	
Shape index	Shape	
Asymmetry	Shape	
Gray level co-occurrence matrix contrast (all dir.)	Textural	
Gray level co-occurrence matrix entropy (all dir.)	Textural	
Gray level co-occurrence matrix correlation (all dir.)	Textural	

3.3. Adaptation of MRF-Based Methods to a Multi-Scale Context

As we explained in the previous section, the clusters form a hierarchical structure depending on the desired level of detail. It is obvious that exploiting these hierarchical relationship between the clusters could lead to improved results and that hierarchical clustering would have the advantage of directly providing several scales of interest [2]. However, most hierarchical clustering algorithms in the literature do not handle neighborhood relationships between data and have an algorithmic complexity that is between $O(N^2 logN)$ and $O(N^3)$. Such high complexity does not scale for large datasets typically used in VHR image analysis.

To solve this problem, in our experimental section, we propose to use a modified version of the SR-ICM algorithm presented in Section 2.2. This modified version allows the user to search for different number of clusters (different scales of interest) and then runs several SR-ICM in parallel with a modified optimization function that encourages each algorithm to build hierarchical clusters depending on the other algorithms' partitions. To this end, let us consider J scales of interest, and let us define $\Omega_{i,j}$ the confusion matrix between any scales i with K_i cluster and j with K_j clusters so that:

$$\Omega^{i,j} = \begin{pmatrix} \omega^{i,j}_{1,1} & \cdots & \omega^{i,j}_{1,K_j} \\ \vdots & \ddots & \vdots \\ \omega^{i,j}_{K_i,1} & \cdots & \omega^{i,j}_{K_i,K_j} \end{pmatrix} \text{ where } \omega^{i,j}_{a,b} = \frac{|C^i_a \cap C^j_b|}{|C^i_a|} \quad (4)$$

The confusion matrix from Equation (4) defines how each cluster of the SR-ICM algorithm at scale i maps into the clusters of the SR-ICM algorithm at scale j. This matrix is in fact very similar to the affinity matrix from the SR-ICM model and plays the same role as a multi-scale level instead of a geographic one. From there, favoring the construction of hierarchical clusters is done by minimizing the following entropy function:

$$\mathcal{H} = \sum_{i=1}^{J} \sum_{j \neq i}^{J} \frac{-1}{K_i \times \ln(K_j)} \sum_{l=1}^{K_i} \sum_{m=1}^{K_j} \omega_{l,m}^{i,j} \ln(\omega_{l,m}^{i,j}) \tag{5}$$

To optimize Equation (5) while ensuring that the solutions remain coherent, we modify Algorithm 1 as follows:

$$s_n^i = \underset{k \in [1..K_i]}{\text{Argmax}} \left[\left(\pi_k^i \times \mathcal{N}(\mu_k^i, \Sigma_k^i, x_n) \times \prod_{v \in V_{x_n}} a_{s_v,k}^{\tau_{x,v}} \right) \times \prod_{j \neq i}^{J} \omega_{s_n^j,k}^{j,i} \right] \tag{6}$$

$$\begin{cases} \mu_k^i = \frac{1}{|C_k^i|} \sum_{n=1}^{N} s_n^i(k) \cdot x_n \\ \Sigma_k^i = \frac{1}{|C_k^i|} \sum_{n=1}^{N} s_n^i(k) \cdot (x_n - \mu_k^i)(x_n - \mu_k^i)^T \\ \pi_k^i = \frac{|C_k^i|}{N} \end{cases} \tag{7}$$

As one can see, Algorithm 2 is a simple parallelization of the SR-ICM algorithm presented in Algorithm 1, with a slightly modified likelihood function to which an extra prior has been added to account for the decisions made at the other scales of interest. The stopping criterion is also slightly modified, and the new criterion is that the parallel solutions found by the algorithms must be as compatible as possible. The main difference is that unlike in the original SR-ICM algorithms, the parameter Θ is not fixed in our proposed method. As we will show bellow, this does not affect the convergence properties and has the advantages of keeping up to date clusters when using the hierarchical dependencies.

Algorithm 2: Parallel SR-ICM for hierarchical clusters.

Initialize all S^i, Θ^i and A^i using Algorithm 1, and compute the confusion matrices $\Omega^{i,j}$
while $\mathcal{H}$ *is decreasing* **do**
  for $i \in [1..J]$ **do**
    Update S using Equation (6) over all of the data
    Update A using Equation (3)
    Update Θ^i using the regular GMM rules from Equation (7)
  end
  Update the $\Omega^{i,j}$ using Equation (5)
end
return all S^i

This algorithm has a complexity of $O(NJ)$ for a dataset of size N and J different scales of interest. The convergence of the process is ensured because the algorithm optimizes the global log-likelihood function of the whole system, whose form is shown in Equation (8). In this equation, $\mathcal{L}^i(X, \Theta^i, S^i)$ is a local log-likelihood for an algorithm at scale i, and $H(S^i, S^j)$ denotes the joint entropy between the solutions at scales i and j.

$$L(\mathbf{S}, \mathbf{\Theta}) = \sum_{i=1}^{J} \left(\mathcal{L}^i(X, \Theta^i, S^i) - \sum_{j \neq i} H(S^i, S^j) \right) \tag{8}$$

Equation (8) can be transformed into Equation (9) by summing over all local likelihoods and entropies to get a global likelihood over all local models and an entropy over the whole system. Please note that $\boldsymbol{H}(\boldsymbol{S})$ is equivalent to the entropy in Equation (5).

$$L(\boldsymbol{S},\boldsymbol{\Theta}) = \boldsymbol{\mathcal{L}}(X,\boldsymbol{\Theta},S) - \boldsymbol{H}(S) \tag{9}$$

Since we optimize Equation (9) using a maximization-maximization process over all algorithms, this is equivalent to the variational EM algorithm proposed by Neal et al. [40] and has the same convergence properties: we know that the system will converge in a finite time toward an optimum. However, we have no warranty that it will be the global optimum.

4. Experimental Results

In this section, we present the results of the clustering done from the CSV files containing the segments information, as well as the subsequent mapping to the expert classes. The experiments were therefore done on the 187,057 segments acquired from the previous steps. Each segment is described by its id, 27 geometric and radiometric attributes and its neighborhood dependencies.

Using the hierarchy established in Figure 6b, we ran three SR-ICM algorithms in parallel using Algorithm 2 searching for 4, 6 and 10 clusters. The results are shown in the next subsection.

4.1. Numerical Results

We first propose an experimental setting in which we compare our proposed method with three others from the literature: the EM algorithm using a diagonal variance-covariance matrix [37], the ICM algorithm using the Gaussian mixture model and a regular prior [35], the regular non multiscale SR-ICM algorithm [3] and the SOM algorithm for VHR images [4].

We ran a dozen simulations with each algorithm for the three scales of interest with 4, 6 and 10 clusters. In Table 2, we show the results of these simulation with the average values for four different indexes:

- The Davies–Bouldin index [41]: It is a clustering index assessing that the clusters are compact and well separated. Its value is better when it is lower and tends to be biased towards a lower number of clusters.
- The silhouette index [42]: It is another clustering index assessing that each datum is closer to its clusters centroid than from the other clusters'. It takes its values between -1 and one and is better when closer to one.
- The Rand index [43]: It is an external index assessing the degree of similitude between two vectors. In the case of this experiment, we compared our solution vectors with our GIS hybrid reference data. It takes its values between zero and one, with one being a 100% match.
- An entropy measure assessing the entropy between each algorithm solutions and the GIS hybrid reference data using the confusion matrix as shown in Equation (10). It takes its values between zero and one, with zero being a 100% match and achievable only if the solution and the reference data have the same number of classes/clusters. This measure is therefore better when close to zero and is biased toward a greater number of clusters.

$$H = \frac{-1}{K\ln(15)} \sum_{l=1}^{K} \sum_{k=1}^{15} \omega_{l,m}^{S,GT} \ln(\omega_{l,m}^{S,GT}) \tag{10}$$

Note that in Equation (10), we use the value of 15, because there are 15 classes in the expert reference data.

Table 2. Comparative results of our proposed "multi-scale semantic-rich iterated conditional modes (ms-SR-ICM)" approach with other methods of the literature using 2 internal indexes and 2 external indexes. SOM, self-organizing map.

Algorithm	Davies–Bouldin Index	Silhouette Index	Rand Index	Entropy
EM (4 clusters)	**2.09**	**0.23**	0.69	0.64
EM (6 clusters)	**2.10**	0.19	0.72	0.62
EM (10 clusters)	**2.59**	**0.11**	0.70	0.61
GMM-ICM (4 clusters)	3.65	0.14	**0.72**	0.59
GMM-ICM (6 clusters)	2.52	0.16	0.73	0.58
GMM-ICM (10 clusters)	3.92	0.07	0.75	0.58
SR-ICM (4 clusters)	4.16	0.15	**0.72**	0.58
SR-ICM (6 clusters)	2.49	0.19	0.75	0.58
SR-ICM (10 clusters)	3.50	0.10	0.78	0.57
ms-SR-ICM (4 clusters)	4.27	0.14	**0.72**	**0.57**
ms-SR-ICM (6 clusters)	2.47	**0.20**	**0.77**	**0.57**
ms-SR-ICM (10 clusters)	3.33	0.10	**0.80**	**0.55**
SOM (6 clusters)	2.23	0.17	0.75	0.60
SOM (10 clusters)	4.04	0.05	0.75	0.63

From Table 2, we can draw several conclusions: First, if we look at the unsupervised indexes (Davies–Bouldin and silhouette), we can see that the expectation-maximization algorithm mostly outperforms all algorithms. This result was to be expected in the sense that both indexes assess the quality of clusters and that the EM algorithm is the only "pure" clustering method that we used here. All three variations of the ICM use spatial dependencies to bend the original clusters toward more realistic classes, hence the degradation that we observe in the unsupervised indexes. It is therefore logical that the EM algorithm has the best results for unsupervised indexes. It is followed by the GMM-ICM and SR-ICM with their modified priors. Then comes the SOM algorithm. Finally, our proposed multi-scale SR-ICM is lagging behind because it has two priors that further bend the partitions away from the usual spherical and well-separated clusters.

This leads us to the interpretation of the supervised indexes (Rand index and entropy). Given the final goal of our application, which is the automatic classification (and not the clustering) of objects in very high resolution images, it is these two indexes that matter most for real applications. As one can see, the results are reversed: our proposed ms-SR-ICM algorithm slightly outperforms both other ICM algorithms; the SOM algorithm still has average performances; and the EM algorithm scores last.

In terms of performances, our proposed parallelized multi-scale version of the semantic-rich ICM algorithm achieves the lowest entropies on the three scales of interest and up to an 80% match with the reference data, which is approximately 2% ahead of the second best algorithm. We also note that on the four clusters' scale, there is no difference between the results of the three ICM algorithms. There are two possible explanations for these results: First, with only four clusters compared with the 15 reference classes, the Rand Index may not be able to discriminate between the algorithms. Second, multi-scale approaches are known to favor scales with more clusters: it is easier to check that clusters have been properly divided from a scale with less clusters because there is less information dispersion than checking that they have been properly merged from a scale with more clusters. Therefore, our proposed method is mostly beneficial for the six clusters and 10 clusters scales.

Beyond the efficiency of our proposed method, this experiment highlights that there is a strong disconnection between clustering indexes that are used by most unsupervised methods and the supervised indexes that are used in real applications. This difficulty that we have been discussing since the Introduction is a real challenge for the conception of future automated detection systems.

In Figure 8, we show the typical hierarchical clusters found by our proposed method.

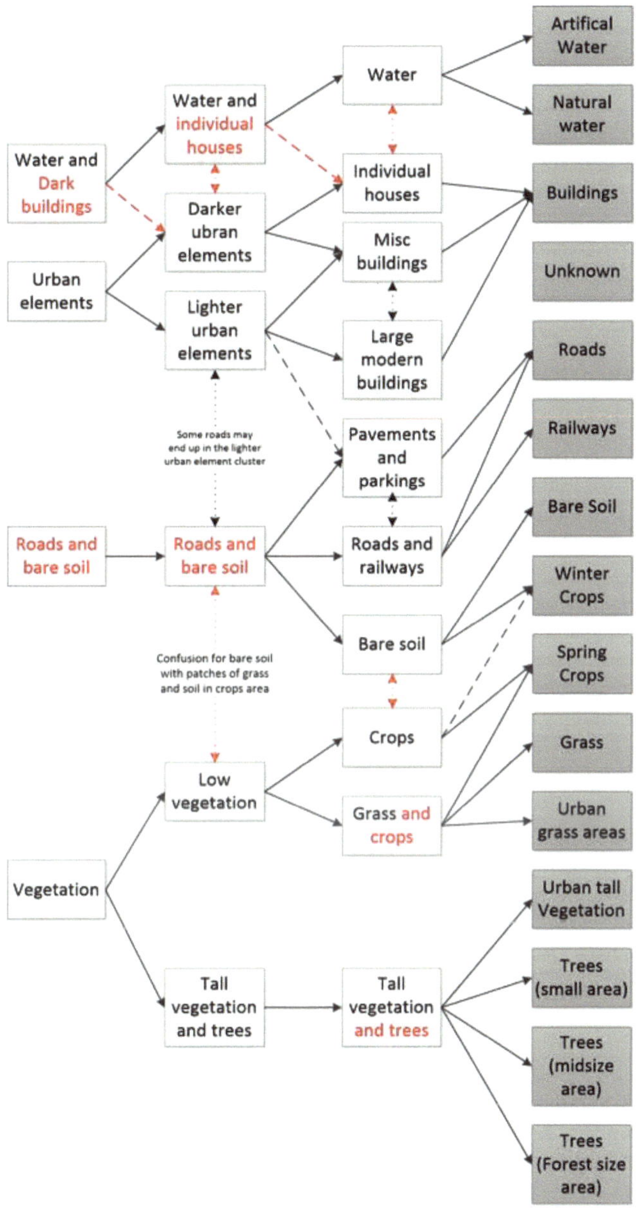

Figure 8. Expert classes in grey (**right**) and hierarchical clusters extracted from the confusion matrices Ω found by our proposed method (**left**): The plain arrows highlight strong links, dashed arrows mild links and dotted arrows weak links. The arrows and characters in red highlight potentially armful errors in the clusters or their hierarchy when compared with the expected classes.

As one can see, the two main differences with Figure 6 come from the inclusion of road elements grouped with bare soil areas at scales of four and six clusters and from the difficulty to properly separate water from a dark building and then individual houses at the same scales. Our explanation for the difference in the hierarchy is the following: Unlike in pixel-based clustering where only color attributes are considered, leading to easily separating water from the other classes, our OBIA approach uses a larger number of non-color based attributes, which delay this separation. While color-based attributes are still the most influential, since we use the Euclidean distance, in which all features have the same weight, their discrimination power is significantly reduced. Consequently, our four first clusters regroup elements that have close enough colors and shapes, thus regrouping the water (dark blue) and some dark urban elements (dark grey and black) in the first cluster, brighter (light grey or white) urban elements in another, roads and bare soil (brown and grey) in a third one and vegetation (green) in the final cluster. Then, at the six clusters' scale, the shape of the segments seems to become significant enough to separate most large darker buildings from the water cluster.

On the other hand, small individual houses with blue tiled roofs or shadow areas have segments whose shape is very similar to water areas. Furthermore, all three tend to be surrounded by a similar vegetation environment, thus making the differentiation difficult even using the neighborhood semantic matrix. Therefore, a decent separation of the water from the other elements is only achieved at the 10 clusters' scale. While this may be problematic in the sense that the supervised algorithm usually learns first to detect water, in the case of unsupervised learning, this was to be expected, since there is no supervision at all. Other unsupervised algorithms applied to OBIA suffer from the same issue as satellite images [3,4], but our method still handles this problem when there are enough clusters.

Other minor flaws when comparing the clustering to what could have been expected from a supervised algorithm include: The regrouping of roads and bare soil in the same cluster; the different types of tree areas generally grouped in a single cluster. However, this matched with the reference data and therefore is not really a problem; the confusions that occurs between some bare soil and vegetation areas due to the fact that there may be patches of grass or crops in bare soil areas and patches of bare soil in crops and low vegetation areas. This problem is in our opinion impossible to solve without changing the segmentation.

Our proposed method also created some unexpected clusters, such as one containing large modern buildings (mostly industrial buildings) and another one differentiating roads from parking and pavements (based on the cluster's shape and semantic surrounding). In fact, our method gives three types of buildings where the expert found only one and where we expected to find only two. Furthermore, except for the minor confusions between crops and low vegetation, the hierarchical tree found by our method globally matches the one given in Figure 6.

4.2. Visual Results

In this section, we show some visual extracts of the results obtained by our method and the algorithms used in the previous section. As such, the explanations that follow are purely based on our interpretation of these visual results. To get the exact accuracy of the clusters displayed in Figures 9 and 10, you can refer to the "Rand index" column of Table 2.

In Figure 9, we show the visual result of our method looking for six clusters in the center area of the city of Strasbourg. Our results are compared with these of two others algorithms from the literature. We tried to use similar color codes for all figures despite the variety of classes and clusters: blue is used for water, different scales of green and yellow for vegetation areas, grey for roads, pink and violet for buildings.

If we first look at Figure 9b with the raw polygons and Figure 9c with the hybrid reference data, we can see that the original GIS reference data of this area in Figure 9b have much less and more linear objects than the segmentation Figure 9c. For this reason, the hybrid ground-truth shown in Figure 9c features large homogeneous areas of the same class that clearly should be separated when we look at the original image in Figure 9a. This is a visual confirmation that our ground-truth used for Table 2 is not perfect and further explains why this hybrid ground-truth cannot be used for supervised learning.

Moving to Figure 9d–f, we can see a comparison between one of our SR-ICM results at the scale with six clusters and the visualization of a result from the SOM algorithm [4] and EM algorithm using the Gaussian mixture model for the same area. First, we can see in Figure 9d that our algorithm correctly detects the river, whereas the SOM algorithm (Figure 9e) only partially does so, and the EM algorithm Figure 9f fails to do so. The same can be said for the stadium on the top right of the image. We can see that all three algorithms mostly correctly identify vegetation areas, with the EM algorithm making slightly more mistakes. Finally, all three algorithms make several confusions between individual houses and water areas due to the roofs' color as we had already mentioned when commenting on Figure 8. This proves that this issue is not isolated to our method.

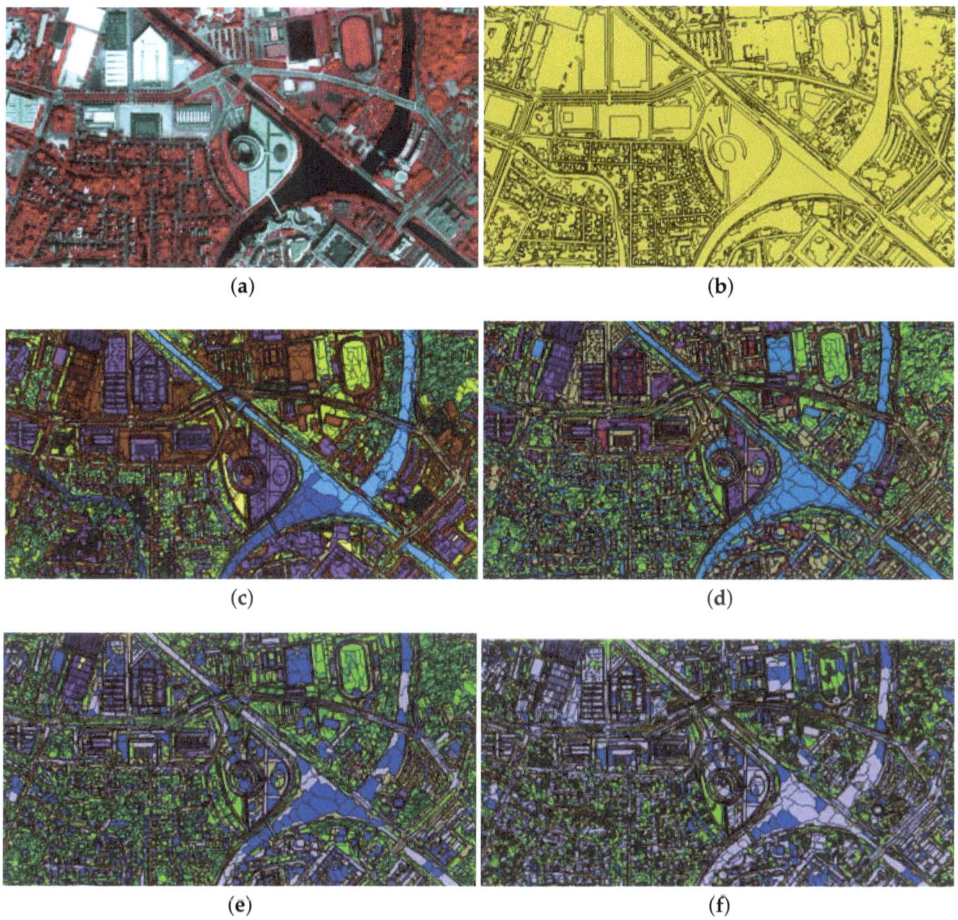

Figure 9. Original image (extract), reference data images and results using different algorithms looking for six clusters. (**a**) Original image, Pléiades©Airbus, CNES 2012; (**b**) reference data©EMS 2012: raw polygons; (**c**) hybrid reference data; (**d**) multi-scale SR-ICM at the six clusters' scale; (**e**) SOM algorithm [4] with six clusters; (**f**) EM algorithm with six clusters.

In Figure 10, we show the result of our algorithms at scales of six and 10 clusters when applied to a non-urban area of our satellite image. As we can see, while the confusion between water and individual houses is less frequent at the 10 clusters' scales, it remains present for several segments. Nevertheless, several areas are correctly classified: roads, rivers and several types of vegetation areas.

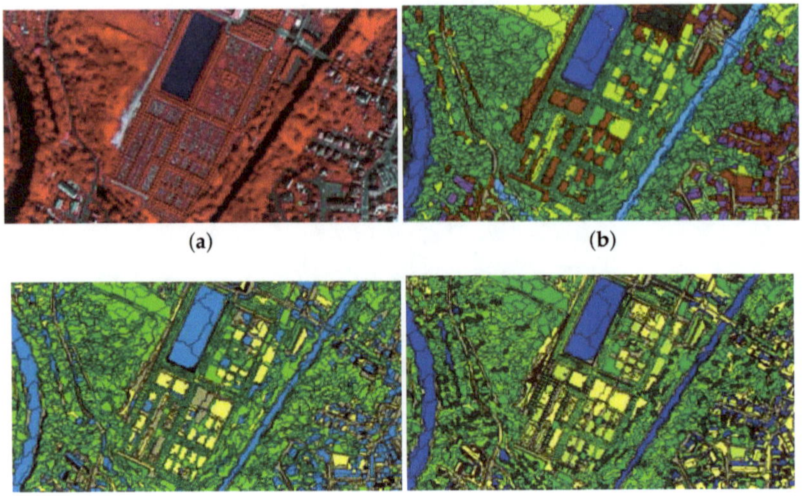

Figure 10. Original image (extract), reference data and our algorithm at scales of six and 10 clusters. (**a**) Original image, Pléiades ©Airbus, CNES 2012; (**b**) hybrid reference data; (**c**) multi-scale SR-ICM at the six clusters' scale; (**d**) multi-scale SR-ICM at the 10 clusters' scale.

4.3. Discussion

We now would like to conclude this experimental section. For the clustering step, our proposed method uses a multi-scale analysis that is both adapted to this type of images, but also helps achieve better results. We have compared our method with three other methods available from the literature, and while we have seen that our method still has flaws (also found in other unsupervised methods), our algorithm achieves better results in terms of supervised indexes, unsupervised indexes and also visual results.

It is true that the results in terms of supervised and unsupervised indexes are not overwhelming when compared to those of other methods, but the projection of our results on the original images makes it clear that our method gives the best results. Furthermore, our algorithm has the advantage of keeping both the semantic analysis aspect of the original SR-ICM algorithm and to add the description of the cluster hierarchy at different scales. This latter addition is extremely valuable to interpret the strengths and weaknesses of our method and helps to adjust the algorithms' parameters to achieve the best possible results.

Possible future works to improve the results of our method, both during the segmentation step and the clustering step, could include a pre-selection of the attributes of interest based on saliency criteria at the considered scale. To this end, several inspiring works exist in hyperspectral image analysis [44,45] to select the optimal bands. These works could be adapted to weight attributes instead of bands and may lead to improved results.

5. Conclusions

In this article, we have been concerned with the challenges and issues that lie with the unsupervised analysis of very high resolution satellite images. After an overview of the different steps to achieve this goal and a short summary of the methods available in the literature with their strengths and weaknesses, we have proposed our own contribution in the form of a multi-scale version of the semantic-rich ICM algorithm that covers the need for multi-scale algorithms to analyze very high resolution images.

In order to demonstrate the efficiency of our method, we have detailed the step by step processing of a satellite image of the French city of Strasbourg, using methods available from the literature for the cleaning and segmentation steps and then comparing our proposed method to others during the unsupervised analysis of the images segments with the goal of finding the final classes of interests at several scales. During these steps, we have highlighted the difficulties encountered by all methods including ours.

In addition to its low computational complexity and the ease to choose the scales of interest to which to apply a clustering process, our method has shown competitive performances when compared to other state of the art algorithms. Furthermore, our proposed algorithm retains low level semantic information that can be easily used to map the clusters to the expert classes of interest.

In our future work, we look forward to proposing similar multi-scale implementations during the segmentation step of a satellite image with the goal of producing better segments, thus reducing the accumulation of errors during the different steps of the image processing. It would also be interesting to use feature selection criteria in order to better detect objects of interest at the different scales, but also to avoid using redundant attributes.

Acknowledgments: This work has been supported by the ANR project COCLICO , ANR-12-MONU-0001. The authors would like to thank the Pléiades Users Thematic Commissioning, Program ORFEO (Optical and Radar Federated Earth Observation) Accompaniment Program (CNES), for Pléiades images availability. Thanks to the metropolitan area of Strasbourg for the land cover/use database used as reference data for our experiments.

Author Contributions: Anne Puissant is one of the main contributors in creating most of the source material for the ANR project COCLICO and this paper. She has also worked in collaboration with Simon Rougier (LIVE laboratory) on the segmentation step that produced the CSV file with feature data used in this paper. Andrés Troya-Galvis has participated in the creation of the CSV files describing the segments used for the unsupervised learning part of this article, as well as the creation of the hybrid reference data. Jérémie Sublime designed the unsupervised algorithms introduced in this article (several of them are from his PhD thesis) and ran all clustering experiments proposed in this work. The paper was co-written by all three authors.

Conflicts of Interest: The authors declare no conflict of interest. The ANR funding sponsor had no role in designing the study or the experiments, nor in the decision to write and publish this paper.

Abbreviations

The following abbreviations are used in this manuscript:

ANR	"Agence Nationale de la Recherche" (French National Agency for Research)
EM	Expectation maximization
GIS	Geographic information systems
ICM	Iterated conditional modes
MRF	Markov random fields
MRIS	Multi-resolution image segmentation
OBIA	Object-based image analysis
VHR	Very high resolution

References

1. Blaschke, T. Object based image analysis for remote sensing. *ISPRS J. Photogramm. Remote Sens.* **2010**, *65*, 2–16.
2. d'Oleire-Oltmanns, S.; Eisank, C.; Dragut, L.; Blaschke, T. An Object-Based Workflow to Extract Landforms at Multiple Scales from Two Distinct Data Types. *IEEE Geosci. Remote Sens. Lett.* **2013**, *10*, 947–951.
3. Sublime, J.; Troya-Galvis, A.; Bennani, Y.; Gancarski, P.; Cornuéjols, A. Semantic Rich ICM Algorithm for VHR Satellite Image Segmentation. In Proceedings of the IAPR International Conference on Machine Vision Applications (MVA 2015), Tokyo, Japan, 18–22 May 2015.
4. Grozavu, N.; Rogovschi, N.; Cabanes, G.; Troya-Galvis, A.; Gançarski, P. VHR satellite image segmentation based on topological unsupervised learning. In Proceedings of the 14th IAPR International Conference on Machine Vision Applications, Tokyo, Japan, 18–22 May 2015; pp. 543–546.
5. Pal, N.R.; Pal, S.K. A review on image segmentation techniques. *Pattern Recognit.* **1993**, *26*, 1277–1294.

6. Gonzalez, R.C.; Woods, R.E. *Digital Image Processing*, 3rd ed.; Prentice-Hall, Inc.: Upper Saddle River, NJ, USA, 2006.
7. Cheng, H.; Jiang, X.; Sun, Y.; Wang, J. Color image segmentation: Advances and prospects. *Pattern Recognit.* **2001**, *34*, 2259–2281.
8. Comaniciu, D.; Meer, P. Mean shift: A robust approach toward feature space analysis. *IEEE Trans. Pattern Anal. Mach. Intell.* **2002**, *24*, 603–619.
9. Michel, J.; Youssefi, D.; Grizonnet, M. Stable mean-shift algorithm and its application to the segmentation of arbitrarily large remote sensing images. *IEEE Trans. Geosci. Remote Sens.* **2015**, *53*, 952–964.
10. Fukunaga, K.; Hostetler, L. The estimation of the gradient of a density function, with applications in pattern recognition. *IEEE Trans. Inf. Theory* **1975**, *21*, 32–40.
11. Baatz, M.; Schäpe, A. Multiresolution Segmentation: An optimization approach for high quality multi-scale image segmentation. In *Angewandte Geographische Informationsverarbeitung XII. Beiträge zum AGIT-Symposium Salzburg 2000*; Strobl, J., Ed.; Herbert Wichmann Verlag: Karlsruhe, Germany, 2000; pp. 12–23.
12. Wang, Z.; Jensen, J.R.; Im, J. An automatic region-based image segmentation algorithm for remote sensing applications. *Environ. Model. Softw.* **2010**, *25*, 1149–1165.
13. Derivaux, S.; Forestier, G.; Wemmert, C.; Lefèvre, S. Supervised image segmentation using watershed transform, fuzzy classification and evolutionary computation. *Pattern Recognit. Lett.* **2010**, *31*, 2364–2374.
14. Li, D.; Zhang, G.; Wu, Z.; Yi, L. An edge embedded marker-based watershed algorithm for high spatial resolution remote sensing image segmentation. *IEEE Trans. Image Process.* **2010**, *19*, 2781–2787.
15. Peng, B.; Zhang, L.; Zhang, D. A survey of graph theoretical approaches to image segmentation. *Pattern Recognit.* **2013**, *46*, 1020–1038.
16. Paglieroni, D.W. Design considerations for image segmentation quality assessment measures. *Pattern Recognit.* **2004**, *37*, 1607–1617.
17. Corcoran, P.; Winstanley, A.; Mooney, P. Segmentation performance evaluation for object-based remotely sensed image analysis. *Int. J. Remote Sens.* **2010**, *31*, 617–645.
18. Srubar, S. Quality Measurement of Image Segmentation Evaluation Methods. In Proceedings of the 2012 Eighth International Conference on Signal Image Technology and Internet Based Systems (SITIS), Naples, Italy, 25–29 November 2012; pp. 254–258.
19. Zhang, X.; Xiao, P.; Feng, X. An Unsupervised Evaluation Method for Remotely Sensed Imagery Segmentation. *IEEE Geosci. Remote Lett.* **2012**, *9*, 156–160.
20. Johnson, B.; Xie, Z. Unsupervised image segmentation evaluation and refinement using a multi-scale approach. *ISPRS J. Photogramm.* **2011**, *66*, 473–483.
21. Troya-Galvis, A.; Gançarski, P.; Passat, N.; Berti-Équille, L. Unsupervised quantification of under and over segmentation for object based remote sensing image analysis. *IEEE J. Sel. Top. Appl. Earth Obs. Remote Sens.* **2015**, *8*, 1936–1945.
22. Liu, N.; Li, J.; Li, N. A Graph-segment-based Unsupervised Classification for Multispectral Remote Sensing Images. *WSEAS Trans. Inf. Sci. Appl.* **2008**, *5*, 929–938.
23. Asmus, V.V.; Buchnev, A.A.; Pyatkin, V.P. Cluster analysis of earth remote sensing data. *Optoelectron. Instrum. Data Process.* **2010**, *46*, 149–155.
24. He, H.; Liang, T.; Hu, D.; Yu, X. Remote sensing clustering analysis based on object-based interval modeling. *Comput. Geosci.* **2016**, *94*, 131–139.
25. Li, H.; Zhang, S.; Ding, X.; Zhang, C.; Dale, P. Performance Evaluation of Cluster Validity Indices (CVIs) on Multi/Hyperspectral Remote Sensing Datasets. *Remote Sens.* **2016**, *8*, 295.
26. Chahdi, H.; Grozavu, N.; Mougenot, I.; Berti-Equille, L.; Bennani, Y. On the Use of Ontology as a priori Knowledge into Constrained Clustering. In Proceedings of the IEEE International Conference on Data Science and Advanced Analytics (DSAA), Montreal, QC, Canada, 17–19 October 2016.
27. Chahdi, H.; Grozavu, N.; Mougenot, I.; Bennani, Y.; Berti-Equille, L. Towards Ontology Reasoning for Topological Cluster Labeling. ICONIP (3). In *Lecture Notes in Computer Science;* Hirose, A., Ozawa, S., Doya, K., Ikeda, K., Lee, M., Liu, D., Eds.; Springer: Cham, Switzerland, 2016; Volume 9949, pp. 156–164.
28. Roth, S.; Black, M.J. Fields of experts. In *Markov Random Fields for Vision and Image Processing;* Blake, A., Kohli, P., Rother, C., Eds.; MIT Press: Cambridge, MA, USA, 2011; pp. 297–310.

29. Ardila, J.; Tolpekin, V.; Bijker, W. Markov random field based super-resolution mapping for identification of urban trees in VHR images. In Proceedings of the IEEE International Geoscience & Remote Sensing Symposium, Honolulu, HI, USA, 25–30 July 2010; pp. 1402–1405.
30. Sublime, J.; Cornuéjols, A.; Bennani, Y. *A New Energy Model for the Hidden Markov Random Fields*; Lecture Notes in Computer Science; Springer: Cham, Switzerland, 2014; Volume 8835, pp. 60–67.
31. Boykov, Y.; Veksler, O.; Zabih, R. Fast Approximate Energy Minimization via Graph Cuts. *IEEE Trans. Pattern Anal. Mach. Intell.* **2001**, *23*, 1222–1239.
32. Leordeanu, M.; Hebert, M.; Sukthankar, R. An Integer Projected Fixed Point Method for Graph Matching and MAP Inference. In Proceedings of the Advances in Neural Information Processing Systems 22: 23rd Annual Conference on Neural Information Processing Systems, Vancouver, BC, Canada, 7–10 December 2009; Bengio, Y., Schuurmans, D., Lafferty, J.D., Williams, C.K.I., Culotta, A., Eds.; Curran Associates, Inc.: Red Hook, NY, USA, 2009; pp. 1114–1122.
33. Liu, Z.Y.; Qiao, H.; Su, J.H. *Neural Information Processing: 21st International Conference, ICONIP 2014, Kuching, Malaysia, November 3–6, 2014. Proceedings, Part II*; Springer: Cham, Switzerland, 2014; pp. 404–412.
34. Besag, J. On the statistical analysis of dirty pictures. *J. R. Stat. Soc. Ser. B* **1986**, *48*, 259–302.
35. Zhang, Y.; Brady, M.; Smith, S.M. Segmentation of Brain MR Images through a Hidden Markov Random Field Model and the Expectation Maximization Algorithm. *IEEE Trans. Med. Imaging* **2001**, *20*, 45–57.
36. Xu, K.; Yang, W.; Liu, G.; Sun, H. Unsupervised Satellite Image Classification Using Markov Field Topic Model. *IEEE Geosci. Remote Sens. Lett.* **2013**, *10*, 130–134.
37. Dempster, A.P.; Laird, N.M.; Rubin, D.B. Maximum likelihood from incomplete data via the EM algorithm. *J. R. Stat. Soc. Ser. B* **1977**, *39*, 1–38.
38. Puissant, A.; Rougier, S.; Stumpf, A. Object-oriented mapping of urban trees using Random Forest classifiers. *Int. J. Appl. Earth Obs. Geoinf.* **2014**, *26*, 235–245.
39. Dragut, L.; Csillik, O.; Eisank, C.; Tiede, D. Automated parameterisation for multi-scale image segmentation on multiple layers. *ISPRS J. Photogramm. Remote Sens.* **2014**, *88*, 119–127.
40. Neal, R.M.; Hinton, G.E. A view of the EM algorithm that justifies incremental, sparse, and other variants. In *Learning in Graphical Models*; Springer: Cham, The Netherlands, 1998; pp. 355–368.
41. Davies, D.L.; Bouldin, D.W. A Cluster Separation Measure. *IEEE Trans. Pattern Anal. Mach. Intell.* **1979**, *1*, 224–227.
42. Rousseeuw, P.J.; Leroy, A.M. *Robust Regression and Outlier Detection*; John Wiley & Sons, Inc.: New York, NY, USA, 1987.
43. Rand, W. Objective criteria for the evaluation of clustering methods. *J. Am. Stat. Assoc.* **1971**, *66*, 846–850.
44. Yuan, Y.; Lin, J.; Wang, Q. Dual-Clustering-Based Hyperspectral Band Selection by Contextual Analysis. *IEEE Trans. Geosci. Remote Sens.* **2016**, *54*, 1431–1445.
45. Wang, Q.; Lin, J.; Yuan, Y. Salient Band Selection for Hyperspectral Image Classification via Manifold Ranking. *IEEE Trans. Neural Netw. Learn. Syst.* **2016**, *27*, 1279–1289.

remote sensing

MDPI

Article

Gated Convolutional Neural Network for Semantic Segmentation in High-Resolution Images

Hongzhen Wang [1,2], Ying Wang [1], Qian Zhang [3], Shiming Xiang [1,*] and Chunhong Pan [1]

[1] National Laboratory of Pattern Recognition, Institute of Automation, Chinese Academy of Sciences, 95 Zhongguancun East Road, Beijing 100190, China; hongzhen.wang@nlpr.ia.ac.cn (H.W.); ywang@nlpr.ia.ac.cn (Y.W.); chpan@nlpr.ia.ac.cn (C.P.)

[2] University of Chinese Academy of Sciences, Beijing 101408, China

[3] Alibaba Group, Beijing 100102, China; zhangqiancsuia@163.com

* Correspondence: smxiang@nlpr.ia.ac.cn; Tel.: +86-136-7118-9070

Academic Editors: Qi Wang, Nicolas H. Younan, Carlos López-Martínez, Xiaofeng Li and Prasad S. Thenkabail
Received: 2 April 2017; Accepted: 1 May 2017; Published: 5 May 2017

Abstract: Semantic segmentation is a fundamental task in remote sensing image processing. The large appearance variations of ground objects make this task quite challenging. Recently, deep convolutional neural networks (DCNNs) have shown outstanding performance in this task. A common strategy of these methods (e.g., SegNet) for performance improvement is to combine the feature maps learned at different DCNN layers. However, such a combination is usually implemented via feature map summation or concatenation, indicating that the features are considered indiscriminately. In fact, features at different positions contribute differently to the final performance. It is advantageous to automatically select adaptive features when merging different-layer feature maps. To achieve this goal, we propose a gated convolutional neural network to fulfill this task. Specifically, we explore the relationship between the information entropy of the feature maps and the label-error map, and then a gate mechanism is embedded to integrate the feature maps more effectively. The gate is implemented by the entropy maps, which are generated to assign adaptive weights to different feature maps as their relative importance. Generally, the entropy maps, i.e., the gates, guide the network to focus on the highly-uncertain pixels, where detailed information from lower layers is required to improve the separability of these pixels. The selected features are finally combined to feed into the classifier layer, which predicts the semantic label of each pixel. The proposed method achieves competitive segmentation accuracy on the public ISPRS 2D Semantic Labeling benchmark, which is challenging for segmentation by only using the RGB images.

Keywords: semantic segmentation; CNN; deep learning; ISPRS; remote sensing; gate

1. Introduction

With the recent advances of remote sensing technologies for Earth observation, large number of high-resolution remote sensing images are being generated every day. However, it is overwhelming to manually analyze such massive and complex images. Therefore, automatic understanding of the remote sensing images has become an urgent demand [1–3]. Automatic semantic segmentation is one of the key technologies for understanding remote images and has many important real-world applications, such as land cover mapping, change detection, urban planning and environmental monitoring [4–6]. In this paper, we mainly focus on the task of semantic segmentation in very high-resolution images acquired by the airborne sensors. The target of this problem is to assign an object class label to each pixel in a given image, as shown in Figure 1a,b.

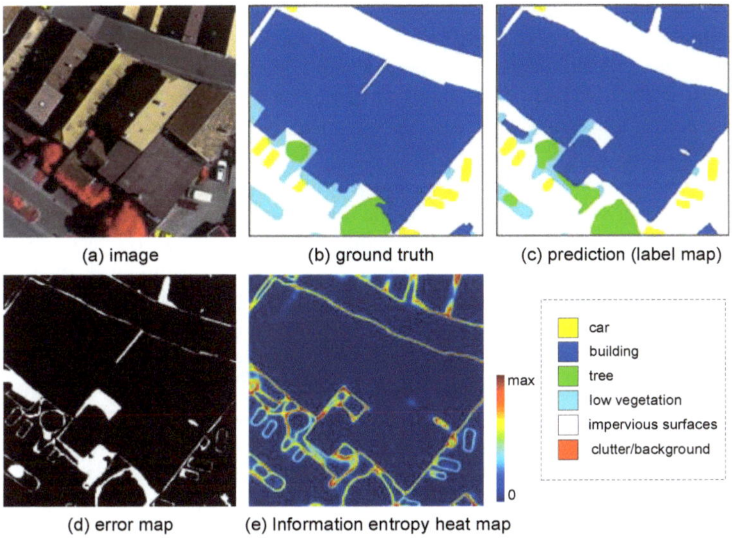

Figure 1. The strong relationship between segmentation error label map with entropy heat map. (**a**) Input image; (**b**) Segmentation reference map; (**c**) Predicted label map; (**d**) Error map with white pixels indicating wrongly classified pixels; (**e**) Corresponding entropy heat map.

Semantic segmentation in remote sensing images is a tough task due to several challenges. First of all, one characteristic of these images is that they often contain a lot of complex objects with various sizes. For example, there are huge buildings and blocks, as well as tiny cars and trees. This factor makes it challenging to simultaneously segment all the objects of various sizes. Another difficulty lies in that resolution improvement can make redundant object details (e.g., building shadow or branches of tree) more clear, which increases the difficulty for semantic segmentation. In addition, high-resolution images contain many objects with high intra-class variance and low inter-class variance [7,8]. Taking the building for example, their roofs look very similar to the roads in term of the appearance. The fact is also true for low vegetation vs. tree. Therefore, features at different levels need to be extracted and jointly combined to fulfill the segmentation task. For one thing, high-level and abstract features are more suitable for the semantic segmentation of large and confused objects, while small objects benefit from low-level and raw features. For another, the ensemble of different level features will provide richer information for semantic segmentation.

Deep convolutional neural network (DCNN) is a well-known model for feature learning. It can automatically learn features of different levels and abstractions from raw images by multiple hierarchically stacking convolutional and pooling layers. In the last few years, DCNN has been extensively studied and demonstrated remarkable learning capability in many applications [9–11]. In the literature, it has also been utilized in the task of image segmentation. Typically, Long et al. [12] adapted the typical DCNN into a fully convolutional network (FCN) for semantic segmentation. FCN achieves pixel-wise classification and now becomes the basic framework for most of the recent state-of-the-art approaches. However, FCN only uses the high-level feature maps (output of the upper convolutional layer) to perform pixel-classification; the low-level feature maps (output of the lower convolutional layer) with rich detailed information are discarded. Although the high-level feature maps are more abstract, they lose a lot of details due to the pooling operation. As a result, FCN has very limited capacity in dealing with small and complex objects. In order to address this issue, reusing low-level feature maps becomes a popular solution as these maps possess rich spatial information and fine-grained details. For example, U-Net [13] modifies and extends the FCN by introducing concatenation structures between the corresponding encoder and decoder layers. The concatenation structure enables the decoder layers to reuse low-level feature maps with more details to achieve a more precise pixel-wise classification. Compared with U-Net, SegNet [14] also records the pooling indices in encoder and reuses them in decoder to enable precise segmentation. RefineNet [15], a recent framework, also adopts this strategy, but uses sum operation and introduces many residual convolution units both in the encoder and decoder path.

Basically, these successful models concatenate or sum feature maps without feature map selection. In this study, we notice that only using subsequent convolutional layers for feature fusion might make the network difficult to train. On the one hand, without feature map selection may introduce redundant information into the network and result in over-segmentation when the model tends to receive more information from lower layers. This is because low-level feature maps contain rich detailed information (e.g., branches in trees). On the other hand, this may lose fine-grained details and lead to under-segmentation when the network tends to receive more information from upper layers. Therefore, it is a critical problem to automatically select adaptive features when merging low- and high-level features.

To tackle the above problems, we propose a gated convolutional neural network for the semantic segmentation in high-resolution images, called gated segmentation network (GSN). When combining two feature maps, we introduce an input gate to adaptively decide whether to keep the corresponding information. Generally speaking, our goal is to import extra low-level information at the positions where the pixel labels are difficult to infer by only using the upper layer feature maps. Meanwhile, we prevent low-level information from being imported into the combined features if the pixel labels have already been determined. This is because over-segmentation may arise if we bring overmuch details. The gate mechanism is implemented by calculating the information entropy of the feature maps before the softmax layer (classifier). The generated entropy heat map has strong relationship with the label-error map, as shown in Figure 1d,e. We summarize our contributions as follows:

- A gated network architecture is proposed for adaptive information propagation among feature maps with different level. With this architecture, convolution layers propagate the selected information into the final features. In this way, local and contextual features work with each other for improving the segmentation accuracy.
- An entropy control layer is introduced to implement the gate. It is based on the observation that the information entropy of the feature maps before the classifier are closely related to the label-error map of the segmentation, as shown in Figure 1.
- A new deep learning pipeline for semantic segmentation is proposed. It effectively integrates local details and contextual information and can be trained via an end-to-end manner.
- The proposed method achieves state-of-the-art performance among all the published papers on the ISPRS 2D semantic labeling benchmark. Specifically, our method achieves a mean F_1 score of 88.7% on five categories (ranking 1st) and overall accuracy 90.3% (ranking 1st). It should be noted that these results are obtained using only RGB images with a single model, without Digital Surface Model (DSM) and model ensemble strategy.

The remainder of this paper is organized as follows: Section 2 presents the related work. In Section 3.2, we introduce the proposed GSN architecture. Section 4 validates our approach experimentally, followed the conclusions in Section 5.

2. Related Work

2.1. Deep Learning

In 2012, the AlexNet [16] won the ILSVRC contest, which is a key milestone in deep learning. Since then, DCNNs have got an explosive development. VGG [17], GoogLeNet [18], ResNet [19] have been proposed one after another. These frameworks are usually treated as feature extractor and play an import role in a wide range of computer vision tasks, such as object detection [20], semantic segmentation [21] and scene understanding [22], etc.

2.2. Semantic Segmentation in Remote Sensing

Semantic segmentation is a significant branch in computer vision. There are a considerable number of works focusing on the remote sensing imagery. Full reviews can be found in [23–25]. Generally, these methods can be roughly classified into the pixel-to-pixel and image-to-image segmentation. The pixel-to-pixel method determines a pixel's label based on an image patch enclosing the target pixel. Then other pixels are classified using a sliding window approach [26,27]. With the development of deep learning on remote sensing images, image-to-image segmentation becomes the mainstream. Sherrah and Jamie [8] proposed a deep FCN with no down-sampling to infer a full-resolution label map. Their method employs the strategy of the dilated convolution in DeepLab [21], which uses dilated kernel to enlarge the size of convolution output at the expense of storage cost. Marmanis et al. [28] embedded boundary detection to the SegNet encoder-decoder architecture. The boundary detection significantly improves semantic segmentation performance with extra model complexity. Kampffmeyer et al. [29] focused on small object segmentation through measuring the uncertainty for DCNNs. This approach achieves high overall accuracy as well as good accuracy for small objects. For all the above methods, further improvements can be achieved by using Conditional Random Fields (CRF) [30,31] or additional data (e.g., Digital Surface Model).

2.3. Gate in Neural Networks

Long short-term memory (LSTM) [32] is a famous framework in the natural language and speech processing. Its success largely owes to the design of gate to control the message propagation. Recently, Dauphin et al. [33] introduced the gated convolutional networks to substitute LSTM for language modeling. A convolution layer followed by a sigmoid layer is treated as a gate unit. Similar to [33], GBD-Net [34] also uses convolution layers with the sigmoid non-linearity as gate unit. GBD-Net is designed for object detection. The gate units are used for passing information among features from different RoIs (region of interest). Through analysis of related literature, embedding gate in neural networks is a simple, yet effective way for both feature learning and feature fusion.

3. Method

This section starts with an important observation of DCNNs for semantic segmentation, which motivates us to design the gated segmentation network (GSN). Then we introduce the GSN architecture in detail, which largely improves the performance of semantic segmentation in remote sensing images.

3.1. Important Observation

When applying DCNNs for the semantic segmentation, the softmax (cross entropy) is usually used as the classifier for the given feature maps. The output of the softmax represents a probability distribution of each pixel over K different categories. With the estimated probabilities of pixel x, we can calculate the corresponding entropy $H(x)$ with

$$H(x) = E[-\log_2(p_i(x))] = -\sum_{i=1}^{K} p_i(x)\log_2(p_i(x)), \tag{1}$$

where $E[\cdot]$ denotes expectation over all the K categories, and $p_i(x)$ is the probability of pixel x belonging to category i.

We observe that the entropy heat map has strong relationship with the label-error map. As shown in Figure 1d,e, there is a strong possibility that the pixels of high entropy are wrong classified. Generally, entropy is a measure of the unpredictability of states [35]. When the entropy of pixel x is maximized, $p(x)$ approximates an uniform probability distribution, indicating that the network is unable to classify this pixel by using only existing information. At these positions, extra information is needed to help the network to classify the pixels. On the contrary, when the network has a high confidence in the pixel label, the entropy will become lower. According to this consideration, when we combine low-level feature maps with high-level ones, the entropy heat map can be treated as a weight map of the low-level feature maps.

3.2. Gated Segmentation Network

Based on the above observation, we propose a gated convolutional neural network for the semantic segmentation in high-resolution images. An overview of the GSN architecture is shown in Figure 2. Our architecture can be divided into two parts: encoder and decoder. In the encoder part, ResNet-101 is applied for feature extraction. In this process, we can get low-level feature maps containing detailed information from lower layers, as well as high-level feature maps containing high-level contextual information from upper layers. In the decoder part, we first use the high-level feature maps for semantic segmentation and get the entropy heat map. Then the generated entropy heat map is treated as the input weight (pixel-to-pixel) of the low-level feature maps when merged with high-level feature maps. A larger entropy value indicates higher uncertainty about the label of the pixel. Consequently, a higher adoption of the low-level feature maps is necessary. We repeat this operation until all the available low-level feature maps are combined. Additionally, residual convolution module is introduced as the basic processing unit before and after the merging process for better training the network. Finally, the combined feature maps containing both high- and low-level information are fed into the softmax layer to obtain the segmentation result. The details are described in the subsequent subsections.

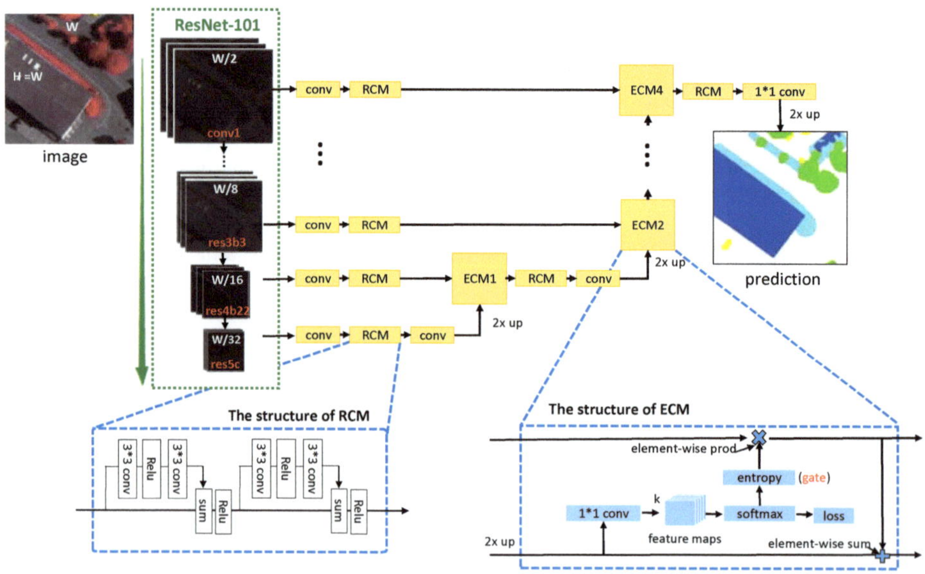

Figure 2. The overview of our gated segmentation network. In the encoder part, we use ResNet-101 as the feature extractor. Then the Entropy Control Module (ECM) are proposed for feature fusion in decoder. In addition, we design the Residual Convolution Module (RCM) as a basic processing unit. The details of RCM and ECM are shown in the dashed boxes.

3.2.1. Entropy Control Module

The bottom-right corner of Figure 2 shows the structure of the proposed entropy control module (ECM). It takes the feature maps f^{upper} (already up-sampled) and f^{lower} as input. The output is represented by F^{fusion}, which combines contextual information and details from f^{upper} and f^{lower} respectively. This feature fusion process is implemented by a gate function, which can be summarized as follows:

$$F^{fusion} = (H[f^{upper} \otimes w_{1*1}] \odot f^{lower}) \oplus f^{upper}, \tag{2}$$

where $\otimes$, $\odot$ and $\oplus$ stands for the convolution operator, the element-wise product operator, and the element-wise sum operator respectively, and w_{1*1} represents the $1*1$ convolutional kernel. As there are K categories in our work setting, the output of the $1*1$ convolutional layer will contain K channels, and each channel records the probabilities of pixels belonging to one of the K categories. In Equation (2), $H[\cdot]$ stands for the entropy calculator, which yields the entropy heat map by Equation (1).

Based on Equation (2), one can see that the designed gate is a binary function, which takes the entropy heat map and the low-level feature map f^{lower} as its inputs. Functionally, it is actually a feature selector on f^{lower} , which is guided by the entropy heat map that is originated from the high-level feature map f^{upper}. Beyond simply fusing the f^{lower}, in this way we build up a mechanism to select the features with their importance for classification. In practice, an entropy control layer is introduced to implement the gate. This layer is only used for calculating the entropy, thus it does not participate in the process of back-propagation.

For clarity, we take Figure 1e as an example to explain our design. Actually, the entropy heat map generated by $H[\cdot]$ offers very helpful information for classifying those pixels that are hard to be classified. As can be witnessed in Figure 1e, most of the high-entropy pixels appear on the object boundaries. Thus, with the gate operation, the information from lower layer will be passed and highly weighted into the final F^{fusion} (see Equation (2)). In contrast, the entropy inside the objects is usually low. Sequentially, the information from lower layer at these positions (e.g., the chimney in the roof in Figure 1e) will be blocked. As a result, over-segmentation can be avoided.

3.2.2. Residual Convolution Module

Inspired by ResNet, residual convolution module (RCM) is introduced as the basic processing unit to ease the training of the network. As shown in the bottom-left corner of Figure 2, there is an identity mapping between the input and output of the module. In the forward propagation, input message can be delivered without loss, and network only needs to learn the residual mapping. In the backward propagation, gradient can be directly propagated from top to bottom, which can settle the problem of gradient vanishing. Compared with the residual blocks in ResNet, the RCM has two differences. First, we removed the $1*1$ convolutional layer. Compute reduction layers have been added at the begin of encoder. Numbers of feature channels are small in the decoder and compute reduction becomes unnecessary. Second, batch normalization layer [36] is removed. Given that the model size is large, we are limited to use small batch size to stay within the GPU memory capacity.

3.2.3. Model Optimization

In the field of neural networks, model optimization is driven by a loss function (also known as objective function). Once the loss function is defined, we can train the network by back-propagation errors [37] in conjunction with gradient descent. To train the proposed architecture, softmax loss function, i.e., cross entropy loss, is adopted. We have a main loss at the end of network and four auxiliary losses in four ECMs. For clarity, we only consider the main loss in the following analysis. Specifically, the softmax function is defined as:

$$L(y, x, \theta) = -\frac{1}{B \cdot P} \sum_{b=1}^{B} \sum_{p=1}^{P} \sum_{k=1}^{K} 1\{y_b^p = k\} \log p_k(x_b^p), \tag{3}$$

where θ represents the parameters of the proposed GSN, B and P are the mini-batch size and number of pixels in each image respectively, $1\{\cdot\}$ is the indicator function, which takes 1 when $1\{true\}$ and 0 otherwise, x_b^p is the p-th pixel in the b-th batch and y_b^p is the corresponding label, and the probability of pixel x_b^p belonging to the k-th class is denoted by $p_k(x_b^p)$, which can be calculated by:

$$p_k(x) = \frac{\exp(W_k^T f(\theta^C, x))}{\sum_{i=1}^{K} \exp(W_i^T f(\theta^C, x))}, \tag{4}$$

where $W_k \in \mathbb{R}^d$ is the j-th filter of the last $1 * 1$ conv layer, d is the feature dimension, θ^C are the rest parameters except the $1 * 1$ conv layer, and $f(\theta^C, x) \in \mathbb{R}^d$ denotes the learned deep features.

To train the GSN in an end-to-end manner, the stochastic gradient descent (SGD) is adopted for the optimization. Thus, the derivatives of the loss to different convolutional layers need to be calculated with chain rule. Taking the $1 * 1$ conv layer as an example, the partial derivative of the loss with respect to W_k is acquired by

$$\frac{\partial L}{\partial W_k} = -\frac{1}{B \cdot P} \sum_{b=1}^{B} \sum_{p=1}^{P} f(\theta^C, x_b^p)(1\{y_b^p = k\} - p_k(x_b^p)). \tag{5}$$

We can get the partial derivative of loss with respect to the parameters in other layers by chain rule. In Algorithm 1, we summarize the learning steps with SGD.

Algorithm 1 The training algorithm for the proposed GSN.

Input: Training data x, maximum iteration T.
Initialize the parameters θ in convolutional layers, learning rate α^t, learning rate policy *ploy*.
Set the initialized iteration $t \leftarrow 0$.
Output: The leanred parameter θ.

```
while t < T do
    t ← t + 1.
    Call network forward to compute the output and loss L.
    Call network backward to compute the gradients ∂L/∂θ.
    Update the parameters θ by θ^{t+1} = θ^t − α^t · ∂L/∂θ.
    Updates the α^{t+1} according to learning rate policy.
end while
```

3.3. Implementation Details

We fine-tune the model weights of ResNet-101 pre-trained on Imagenet [38] to our GSN model. Five kinds of feature maps with different sizes (acquired from the outputs of branches in ["res5c", "res4b22", "res3b3", "res2c", "conv1"]) are prepared to be merged in the decoder part. The spatial sizes of these feature maps are $[W/32 \times W/32, W/16 \times W/16, W/8 \times W/8, W/4 \times W/4, W/2 \times W/2]$ respectively, with input image $I^{W \times W}$. Dropout is applied after these feature maps with ratio 0.5 to avoid overfitting [39]. Moreover, we further add a convolutional (conv) layer after the dropout layer mainly to reduce the channels. The channels of the five branches are set to [256, 128, 128, 64, 64] respectively. Intuitively, similar conv layers should be applied before the up-sampled layers (2× up), since the channels are different between these branches.

The proposed GSN is implemented with Caffe [40] on GPU (TITAN X). Our loss function is the sum of softmax loss, which comes from the final classification and four ECMs. Initial learning rate is 0.0004. We employ the "ploy" learning rate policy. Momentum and weight decay are set to 0.9 and 0.0005 respectively. The bath size is set to 1. The maximum iteration is 30 k. The total training time is about 24 h, and the average testing time of one image (600 × 600) is about 100 ms.

4. Experiments

4.1. Dataset

We evaluate the proposed method on the ISPRS 2D semantic labeling contest [41], which is an open benchmark dataset. The dataset contains 33 very high-resolution true orthophoto (TOP) tiles extracted from a large TOP mosaic as shown in Figure 3. Each tile contains around 2500 × 2000 pixels with a resolution of 9 cm. The dataset has been manually classified into six most common land cover classes, as shown in Figure 1. The clutter class includes water bodies and other objects that look very different from other objects (e.g., containers, tennis courts, swimming pools). As previously done in other methods, the class of *clutter* is not included in the experiments, as the pixels of the clutter class only account for 0.88% of the total image pixels. ISPRS only provides 16 labeled images for training, while the remaining 17 tiles are unreleased and used for the evaluation of submitted results by the benchmark organizers. Following other methods, 4 tiles (image numbers 5, 7, 23, 30) are removed from the training set as a validation set. Experimental results are reported on the validation set if not specified.

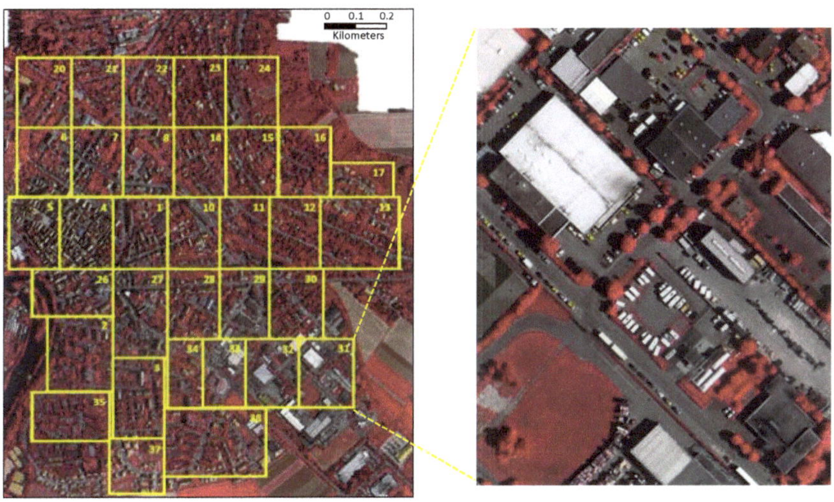

Figure 3. Overview of the ISPRS 2D Vaihingen Labeling dataset. There are 33 tiles. Numbers in the figure refer to the individual tile flag.

Dataset augmentation: The 16 training tiles are first rotated 90 and 180 degrees. Then, we sample 600 × 600 patches from original images with stride (300 pixels) to avoid the insufficiency of GPU memory. Moreover, we also randomly process the input images at the training stage with the following one or combined operations: mirror, rotated between −10 and 10 degrees, resize by a factor between 0.5 and 1.5, and Gaussian blur.

Evaluation: According to the benchmark rules, F_1 score and overall accuracy are used to assess the quantitative performance. F_1 score is calculated by:

$$F_1 = 2 \times \frac{precision \times recall}{precision + recall} \tag{6}$$

where

$$precision = \frac{tp}{tp + fp}, \quad recall = \frac{tp}{tp + fn}, \tag{7}$$

where tp, fp and fn are true positive, false positive and false negative respectively. These values can be calculated by pixel-based confusion matrices per tile, or an accumulated confusion matrix. Overall accuracy is the normalization of the trace from the confusion matrix.

4.2. Model Analysis

For the sake of convenient comparison, we use the result of GSN without entropy control module (ECM) as our baseline, which uses the sum operation to merge the feature maps. As shown in Table 1, the model with ECM outperforms the baseline by a significant margin. This proves that the ECM can effectively control information propagation and integrate features of different level effectively. One can also see that the auxiliary loss in ECM is helpful for model optimization (GSN vs. GSN_noL). The auxiliary loss forces the network to learn accurate contextual feature before merging lower feature maps with high-spatial. Moreover, we notice from the confusion matrix that the *low_veg* and *car* are more likely to be classified into tree and imp_suf respectively. This motivates us to slightly increase the weights of *low_veg* to 1.1 and car to 1.2 in the loss function without accurate selection (GSN vs. GSN_w). Finally, we have reported the result with sliding window overlap and multi-scale input, i.e., GSN_w_mc. Averaging predictions on the overlap regions reduce the risk of error classification, since the borders of one patch is difficult to predict due to the lack of context.

Table 1. The F_1 scores of 5 categories on the validation set. GSN_noL represents that the auxiliary loss in ECM does not participate in the back propagation of the network. GSN_w is the version that assigns different weights to different classes in the loss function. GSN_w_mc represents we test GSN with sliding window overlap and multi-scale input.

Method	Imp Surf	Building	Low_veg	Tree	Car	Overall Accuracy	Mean F_1 Score
baseline	87.6%	93.2%	73.3%	86.9%	54.1%	86.1%	79.0%
GSN	89.2%	94.5%	74.9%	87.5%	79.8%	87.9%	85.2%
GSN_noL	89.1%	94.3%	74.7%	87.4%	78.7%	87.8%	84.8%
GSN_w	89.5%	94.4%	75.9%	87.8%	80.9%	88.3%	85.7%
GSN_w_mc	**90.2%**	**94.8%**	**76.9%**	**88.3%**	**82.3%**	**88.9%**	**86.5%**

4.3. Comparisons with Related Methods

To show the effectiveness of the proposed method, we have performed comparisons against a number of state-of-the-art semantic segmentation methods, as listed in Table 2. Deeplab-v2 [21] and RefineNet [15] are the versions with ResNet-101 as their encoder. In particular, we re-implement the RefineNet with Caffe, since the released code is built on MatConvNet [42]. We can see that GSN significantly outperforms other methods on both overall accuracy and mean F_1 score. Notably, our approach outperforms the RefineNet, within which the feature map merging is implemented by the sum operation. The comparison indicates that the promising performance of GSN can be ascribed to the ECM, which selects low-level information in feature fusion.

Table 2. Comparisons between our proposed GSN with mainstream models.

Method	Imp Surf	Building	Low_veg	Tree	Car	Overall Accuracy	Mean F_1 Score
FCN-8s [12]	87.1%	91.8%	**75.2%**	86.1%	63.8%	85.9%	80.8%
SegNet [14]	82.7%	89.1%	66.3%	83.9%	55.7%	82.1%	75.5%
Deeplab-v2 [21]	88.5%	93.5%	73.9%	86.9%	**84.7%**	86.9%	83.5%
RefineNet [15]	88.1%	93.3%	74.0%	87.1%	65.1%	86.7%	81.5%
GSN	**89.2%**	**94.5%**	74.9%	**87.5%**	79.8%	**87.9%**	**85.2%**

4.4. Model Visualization

To understand GSN better, we have also carried out feature map visualization to examine how entropy gate affects the final performance. Four entropy control modules are embedded in GSN to merge the five kinds of feature maps with different resolutions. In this section, we visualize the entropy heat map, error map and prediction in each ECM.

At each iteration , the prediction will be more fine-grained by merging larger resolution feature maps (ECM 1 $\rightarrow$ ECM 4). An illustration is provided in Figure 4. In ECM 1, we only get a coarse label map, since only the smallest resolution maps are available. Successively merging features from lower layers, we can refine the coarse label map. This is consistent with the analysis of upper-layer feature maps containing more contextual information, and lower-layer feature maps containing more details.

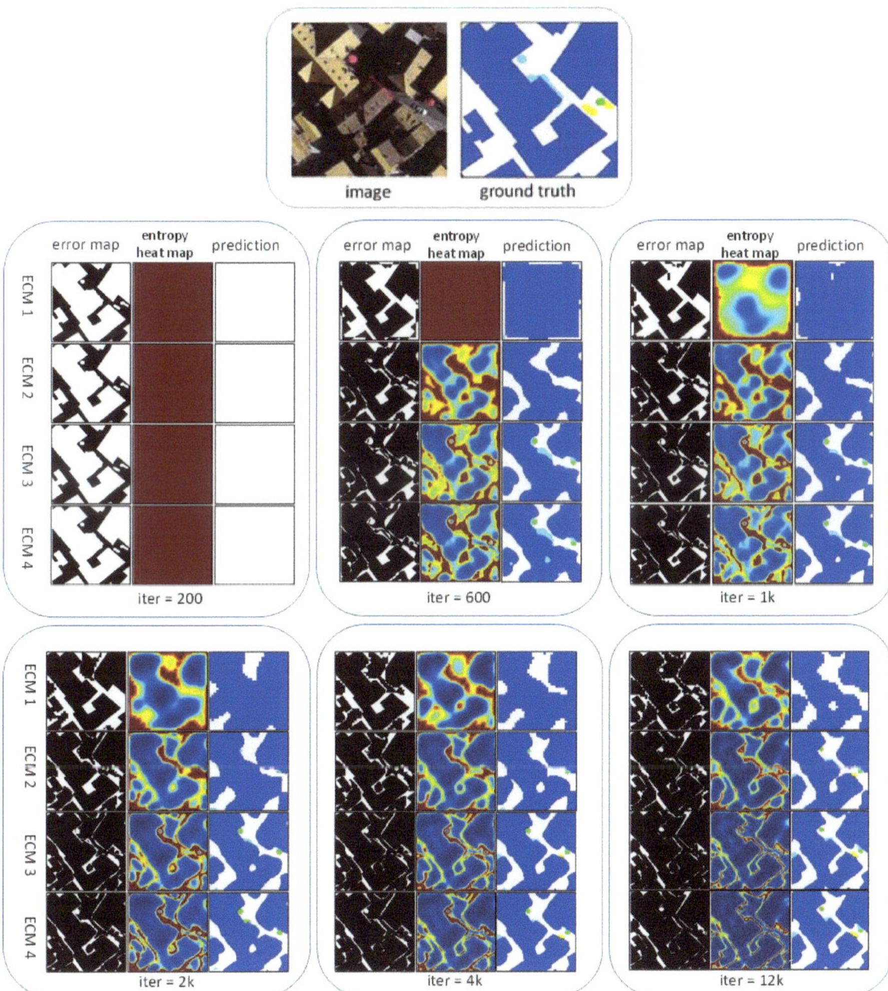

Figure 4. Model visualization. We show the error maps, entropy heat maps, and predictions at different iterations in the training procedure. Four rows at each iteration block correspond to four ECMs, which are used to merge five kinds of feature maps with different resolutions.

In addition, we also visualize the three kinds of maps at different iterations while training the model. At the beginning, the entropy heat maps of four ECMs are almost the same, i.e., red images. It shows that the value of entropy is very high at the beginning, and thus all the gates are at the fully opened state. At this moment, the network has not learned the discriminative features and needs additional information to determine the pixels' labels. As the training proceeds, GSN learns more discriminative features and starts to close the gates at some positions, as shown in 600 or 1 k iterations. Towards the end of the training, we acquire a more satisfying prediction. As can be seen in Figure 4, the positions of high entropy values (similar to error map) almost appear on the boundaries, whose width is very thin. All the above observations once again demonstrate the effectiveness of the proposed ECM.

4.5. ISPRS Benchmark Testing Results

We submitted the results on the unlabelled test images to ISPRS organizers for evaluation. As shown in Table 3, GSN ranks 1st both in mean F_1 score and overall accuracy, compared with all the other published works. Visual performance among related methods is shown in Figure 5. It should be noted that we only use the RGB source images. Neither the additional DSM images offered by ISPRS nor the CRF for post-processing is used in the proposed method, both of which can further improve the performance as described in these compared methods. This is based on the following two considerations. First, we want to sufficiently mine the information contained in RGB images, which will eliminate the need to acquire DSM data. Second, the operation of CRF is time-consuming. Therefore, we manage to build a fast and simple architecture for sematic segmentation in high-resolution remote sensing images. In addition, according to the evaluation of ISPRS, the boundaries of objects in testing labeled images are eroded by a circular disc of 3 pixel radius. Those eroded areas are ignored during evaluation in order to reduce the impact of uncertain border definitions. Thus the performance on testing set is slightly better than that on validation set.

Table 3. Quantitative comparisons between our method and other related methods (already published) on ISPRS test set.

Method	Imp Surf	Building	Low_veg	Tree	Car	Overall Accuracy	Mean F_1 Score
UPB [43]	87.5%	89.3%	77.3%	85.8%	77.1%	85.1%	83.4%
ETH_C [44]	87.2%	92.0%	77.5%	87.1%	54.5%	85.9%	79.7%
UOA [45]	89.8%	92.1%	80.4%	88.2%	82.0%	87.6%	86.5%
ADL_3 [26]	89.5%	93.2%	82.3%	88.2%	63.3%	88.0%	83.3%
RIT_2 [46]	90.0%	92.6%	81.4%	88.4%	61.1%	88.0%	82.7%
DST_2 [8]	90.5%	93.7%	83.4%	89.2%	72.6%	89.1%	85.9%
ONE_7 [47]	91.0%	94.5%	84.4%	89.9%	77.8%	89.8%	87.5%
DLR_9 [28]	**92.4%**	**95.2%**	**83.9%**	**89.9%**	81.2%	**90.3%**	88.5%
GSN	92.2%	95.1%	83.7%	**89.9%**	**82.4%**	**90.3%**	**88.7%**

4.6. Failed Attempts

Before creating entropy control module, many failed attempts have been made to find an effective way for feature fusion. Motivated by [33,34], we once tried to create the gate by using convolutional layer followed by sigmoid non-linearity, which make the information propagation rate in the range of (0, 1). Three modules have been designed based on this idea. As shown in Figure 6, we have attempted to add the gate in the output of the lower or upper layer. In the third module, gate on the output of lower layer is created by the combination of lower and upper layers output. However, as shown in Table 4, these modules are less effective than we expected. It is because they can not learn the right open (or closed) state due to the lack of supervised information. One may consider adding auxiliary losses in these modules to guide learning. However it is not feasible. Sigmoid is just an activation layer that has nothing to do with the label-error map. There is no supervised information to guide the network training. Thus we cannot get the right gate states. In contrast, entropy has a strong relationship with the label-error map, which is the supervised information for controlling the gate states. This is the reason why ECM can effectively select features and improve the segmentation performance.

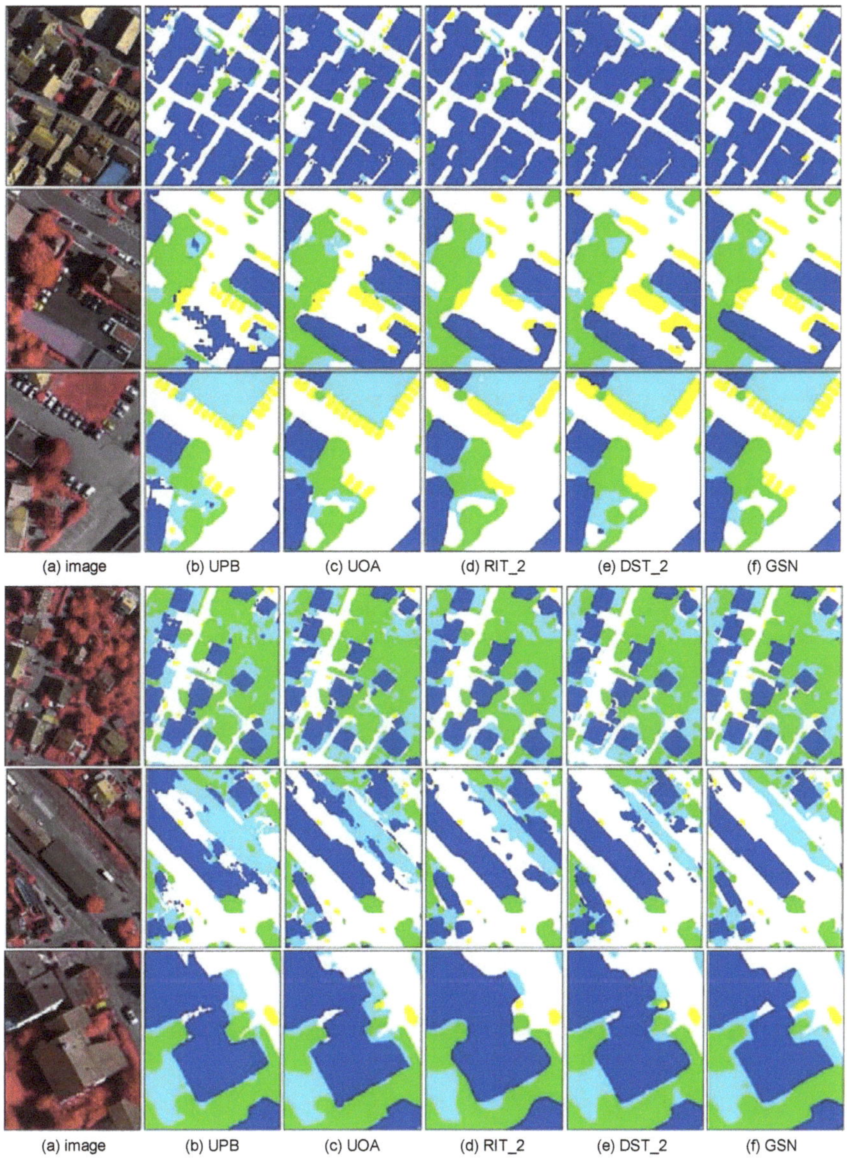

Figure 5. Visual comparisons between GSN and other related methods on ISPRS test set. Images come from the website of ISPRS 2D Semantic Labeling Contest.

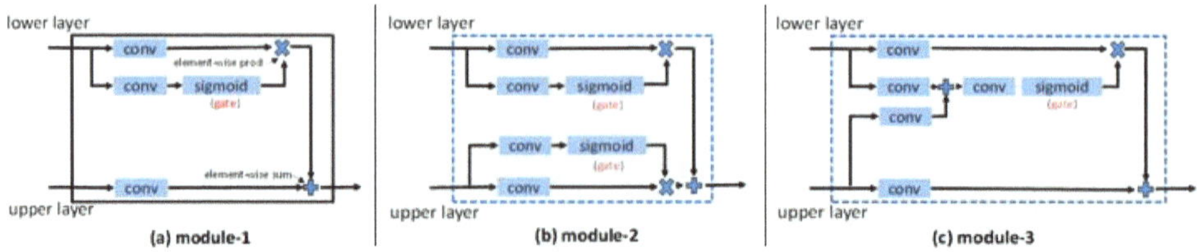

Figure 6. Three failure modules. (**a**) Placing gate on the output of lower layer; (**b**) Placing gate both on the output of lower layer and upper layers; (**c**) Gate on the output of lower layer is created by the combination of lower and upper layers output.

Table 4. Performance of the failure models.

	Model_1	Model_2	Model_3	GSN
overall accuracy	83.4%	60.0%	82.2%	86.1%
mean F_1 score	75.3%	57.3%	74.8%	79.0%

5. Conclusions

In this paper, a gated convolutional neural network was proposed for the semantic segmentation in high-resolution aerial images. We introduced entropy control module (ECM) to guide the message passing between feature maps with different resolutions. The ECM can effectively help for integrating contextual information from the upper layers and details from the lower layers. Extensive experiments on the ISPRS dataset demonstrate that the proposed method achieve clear promising gains compared with the state-of-the art methods. Our approach has the potential to perform better. Actually, the pixels in a certain region are interrelated. However, we calculate the entropy map (gate) pixel-to-pixel, which ignores the relationships between surrounding pixels. In the future work, we will try to incorporate gaussian smoothing into the entropy map to further improve the performance. In addition, we will also try to apply GSN to other fine-grained semantic segmentation tasks.

Acknowledgments: This work was supported in part by the National Natural Science Foundation of China under Grants 91646207, 91338202, 91438105, and the Beijing Natural Science Foundation under Grant 4162064.

Author Contributions: Hongzhen Wang and Shiming Xiang designed the deep learning model; Hongzhen Wang performed the experiments; Ying Wang analyzed the solution to the model; Shiming Xiang and Chunhong Pan analyzed the data; Qian Zhang contributed the analysis tools and comparative methods; Hongzhen Wang and Chunhong Pan wrote the paper.

Conflicts of Interest: The authors declare no conflict of interest.

References

1. Wang, Q.; Lin, J.; Yuan, Y. Salient band selection for hyperspectral image classification via manifold ranking. *IEEE Trans. Neural Netw. Learn. Syst.* **2016**, *27*, 1279–1289.
2. Cheng, G.; Zhu, F.; Xiang, S.; Wang, Y.; Pan, C. Accurate urban road centerline extraction from VHR imagery via multiscale segmentation and tensor voting. *Neurocomputing* **2016**, *205*, 407–420.
3. Yuan, Y.; Lin, J.; Wang, Q. Dual-clustering-based hyperspectral band selection by contextual analysis. *IEEE Trans. Geosci. Remote Sens.* **2016**, *54*, 1431–1445.
4. Matikainen, L.; Karila, K. egment-based land cover mapping of a suburban area—Comparison of high-resolution remotely sensed datasets using classification trees and test field points. *Remote Sens.* **2011**, *3*, 1777–1804.
5. Tang, Y.; Zhang, L. Urban change analysis with multi-sensor multispectral imagery. *Remote Sens.* **2017**, *9*, 252.
6. Yuan, Y.; Lin, J.; Wang, Q. Hyperspectral image classification via multitask joint sparse representation and stepwise MRF optimization. *IEEE Trans. Cybern.* **2016**, *46*, 2966–2977.

7. Zhang, Q.; Seto, K.C. Mapping urbanization dynamics at regional and global scales using multi-temporal DMSP/OLS nighttime light data. *Remote Sens. Environ.* **2011**, *115*, 2320–2329.
8. Sherrah, J. Fully convolutional networks for dense semantic labelling of highresolution aerial imagery. *arXiv* **2016**, arXiv:1606.02585.
9. Girshick, R. Fast r-cnn. In Proceedings of the IEEE International Conference on Computer Vision, Santiago, Chile, 13–16 December 2015; pp. 1440–1448.
10. Goodfellow, I.; Pouget-Abadie, J.; Mirza, M.; Xu, B.; Warde-Farley, D.; Ozair, S.; Courville, A.; Bengio, Y. Generative adversarial nets. In Proceedings of the Advances in Neural Information Processing Systems, Montreal, QC, Canada, 8–13 December 2014; pp. 2672–2680.
11. Yang, S.; Luo, P.; Loy, C.C.; Tang, X. From facial parts responses to face detection: A deep learning approach. In Proceedings of the IEEE International Conference on Computer Vision, Santiago, Chile, 13–16 December 2015; pp. 3676–3684.
12. Long, J.; Shelhamer, E.; Darrell, T. Fully convolutional networks for semantic segmentation. *IEEE Trans. Pattern Anal. Mach. Intell.* **2015**, *79*, 1337–1342.
13. Ronneberger, O.; Fischer, P.; Brox, T. U-net: Convolutional networks for biomedical image segmentation. In Proceedings of the International Conference on Medical Image Computing and Computer-Assisted Intervention, Munich, Germany, 5–9 October 2015; Springer: Cham, Switzerland, 2015; pp. 234–241.
14. Badrinarayanan, V.; Kendall, A.; Cipolla, R. Segnet: A deep convolutional encoder-decoder architecture for image segmentation. *arXiv* **2015**, arXiv:1511.00561.
15. Lin, G.; Milan, A.; Shen, C.; Reid, I. RefineNet: Multi-path refinement networks with identity mappings for high-resolution semantic segmentation. *arXiv* **2016**, arXiv:1611.06612.
16. Krizhevsky, A.; Sutskever, I.; Hinton, G.E. Imagenet classification with deep convolutional neural networks. In Proceedings of the Advances in Neural Information Processing Systems, Lake Tahoe, NV, USA, 3–6 December 2012; pp. 1097–1105.
17. Simonyan, K.; Zisserman, A. Very deep convolutional networks for large-scale image recognition. *arXiv* **2014**, arXiv:1409.1556.
18. Szegedy, C.; Liu, W.; Jia, Y.; Sermanet, P.; Reed, S.; Anguelov, D.; Erhan, D.; Vanhoucke, V.; Rabinovich, A. Going deeper with convolutions. In Proceedings of the IEEE Conference on Computer Vision and Pattern Recognition, Boston, MA, USA, 7–12 June 2015; pp. 1–9.
19. He, K.; Zhang, X.; Ren, S.; Sun, J. Deep residual learning for image recognition. In Proceedings of the IEEE Conference on Computer Vision and Pattern Recognition, Seattle, WA, USA, 27–30 June 2016; pp. 770–778.
20. Ren, S.; He, K.; Girshick, R.; Sun, J. Faster r-cnn: Towards real-time object detection with region proposal networks. In Proceedings of the Advances in Neural Information Processing Systems, Montreal, QC, Canada, 7–12 December 2015; pp. 91–99.
21. Chen, L.C.; Papandreou, G.; Kokkinos, I.; Murphy, K.; Yuille, A.L. Deeplab: Semantic image segmentation with deep convolutional nets, atrous convolution, and fully connected crfs. *arXiv* **2016**, arXiv:1606.00915.
22. Zhou, B.; Khosla, A.; Lapedriza, A.; Oliva, A.; Torralba, A. Learning deep features for discriminative localization. In Proceedings of the IEEE Conference on Computer Vision and Pattern Recognition, Seattle, WA, USA, 27–30 June 2016; pp. 2921–2929.
23. Ghamisi, P.; Dalla Mura, M.; Benediktsson, J.A. A survey on spectral–spatial classification techniques based on attribute profiles. *IEEE Trans. Geosci. Remote Sens.* **2015**, *53*, 2335–2353.
24. Bruzzone, L.; Demir, B. A review of modern approaches to classification of remote sensing data. In *Land Use and Land Cover Mapping in Europe*; Springer: Dordrecht, The Netherlands, 2014; pp. 127–143.
25. Zhang, L.; Zhang, L.; Du, B. Deep learning for remote sensing data: A technical tutorial on the state of the art. *IEEE Geosci. Remote Sens. Mag.* **2016**, *4*, 22–40.
26. Paisitkriangkrai, S.; Sherrah, J.; Janney, P.; van-Den Hengel, A. Effective semantic pixel labelling with convolutional networks and Conditional Random Fields. In Proceedings of the IEEE Conference on Computer Vision and Pattern Recognition Workshops, Boston, MA, USA, 7–12 June 2015; pp. 36–43.
27. Audebert, N.; Le Saux, B.; Lefevre, S. How useful is region-based classification of remote sensing images in a deep learning framework? In Proceedings of the IEEE Conference on Geoscience and Remote Sensing Symposium, Beijing, China, 10–15 July 2016; pp. 5091–5094.
28. Marmanis, D.; Schindler, K.; Wegner, J.D.; Galliani, S.; Datcu, M.; Stilla, U. Classification with an edge: Improving semantic image segmentation with boundary detection. *arXiv* **2016**, arXiv:1612.01337.

29. Kampffmeyer, M.; Salberg, A.B.; Jenssen, R. Semantic segmentation of small objects and modeling of uncertainty in urban remote sensing images using deep convolutional neural networks. In Proceedings of the IEEE Conference on Computer Vision and Pattern Recognition Workshops, Seattle, WA, USA, 27–30 June 2016; pp. 1–9.
30. Arnab, A.; Jayasumana, S.; Zheng, S.; Torr, P.H. Higher order conditional random fields in deep neural networks. In Proceedings of the European Conference on Computer Vision, Amsterdam, The Netherlands, 8–16 October 2016; Springer: Cham, Switzerland, 2016, pp. 524–540.
31. Zheng, S.; Jayasumana, S.; Romera-Paredes, B.; Vineet, V.; Su, Z.; Du, D.; Huang, C.; Torr, P.H. Conditional random fields as recurrent neural networks. In Proceedings of the IEEE Conference on International Conference on Computer Vision, Los Alamitos, CA, USA, 7–13 December 2015; pp. 1529–1537.
32. Hochreiter, S.; Schmidhuber, J. Long short-term memory. *Neural Comput.* **1997**, *9*, 1735–1780.
33. Dauphin, Y.N.; Fan, A.; Auli, M.; Grangier, D. Language modeling with gated convolutional networks. *arXiv* **2016**, arXiv:1612.08083.
34. Zeng, X.; Ouyang, W.; Yan, J.; Li, H.; Xiao, T.; Wang, K.; Liu, Y.; Zhou, Y.; Yang, B.; Wang, Z.; et al. Crafting GBD-Net for Object Detection. *arXiv* **2016**, arXiv:1610.02579.
35. Shannon, C.E. A mathematical theory of communication. *Bell Syst. Tech. J.* **1948**, *5*, 3–55.
36. Ioffe, S.; Szegedy, C. Batch normalization: Accelerating deep network training by reducing internal covariate shift. *arXiv* **2015**, arXiv:1502.03167.
37. Rumelhart, D.E.; Hinton, G.E.; Williams, R.J. Learning representations by back-propagating errors. *Nature* **1986**, *323*, 533–536.
38. Russakovsky, O.; Deng, J.; Su, H.; Krause, J.; Satheesh, S.; Ma, S.; Huang, Z.; Karpathy, A.; Khosla, A.; Bernstein, M. ImageNet large scale visual recognition challenge. *Int. J. Comput. Vis.* **2015**, *115*, 211–252.
39. Hinton, G.E.; Srivastava, N.; Krizhevsky, A.; Sutskever, I.; Salakhutdinov, R.R. Improving neural networks by preventing co-adaptation of feature detectors. *Comput. Sci.* **2012**, *3*, 212–223.
40. Jia, Y.; Shelhamer, E.; Donahue, J.; Karayev, S.; Long, J.; Girshick, R.; Guadarrama, S.; Darrell, T. Caffe: Convolutional architecture for fast feature embedding. *arXiv* **2014**, 675–678, arXiv:1408.5093 .
41. International Society for Photogrammetry and Remote Sensing (ISPRS). 2D Semantic Labeling Contest. Available online: http://www2.isprs.org/commissions/comm3/wg4/semantic-labeling.html (accessed on 1 April 2015).
42. Vedaldi, A.; Lenc, K. Matconvnet: Convolutional neural networks for matlab. In Proceedings of the 23rd ACM international conference on Multimedia, Brisbane, Australia, 26–30 October 2015, 2015; pp. 689–692.
43. Marcu, A.; Leordeanu, M. Dual local-global contextual pathways for recognition in aerial imagery. *arXiv* **2016**, arXiv:1605.05462.
44. Tschannen, M.; Cavigelli, L.; Mentzer, F.; Wiatowski, T.; Benini, L. Deep structured features for semantic segmentation. *arXiv* **2016**, arXiv:1609.07916.
45. Lin, G.; Shen, C.; van den Hengel, A.; Reid, I. Efficient piecewise training of deep structured models for semantic segmentation. In Proceedings of the IEEE Conference on Computer Vision and Pattern Recognition, Seattle, WA, USA, 27–30 June 2016; pp. 3194–3203.
46. Piramanayagam, S.; Schwartzkopf, W.; Koehler, F.; Saber, E. Classification of remote sensed images using random forests and deep learning framework. In *Proceedings of the SPIE Remote Sensing*; International Society for Optics and Photonics: Edinburgh, UK, 2016; p. 100040L.
47. Audebert, N.; Saux, B.L.; Lefèvre, S. Semantic segmentation of earth observation data using multimodal and multi-scale deep networks. *arXiv* **2016**, arXiv:1609.06846.

remote sensing

MDPI

Article

A Novel Affine and Contrast Invariant Descriptor for Infrared and Visible Image Registration

Xiangzeng Liu [1], Yunfeng Ai [2,*], Juli Zhang [1] and Zhuping Wang [1]

1. Xi'an Microelectronics Technology Institute, Xi'an 710068, China; lxzccy20062008@126.com (X.L.); juli2320@sina.com (J.Z.); zxjwl@126.com (Z.W.)
2. University of Chinese Academy of Sciences, Beijing 100049, China

* Correspondence: aiyunfeng@ucas.ac.cn; Tel.: +86-10-8825-6564

Received: 2 April 2018; Accepted: 19 April 2018; Published: 23 April 2018

Abstract: Infrared and visible image registration is a very challenging task due to the large geometric changes and the significant contrast differences caused by the inconsistent capture conditions. To address this problem, this paper proposes a novel affine and contrast invariant descriptor called maximally stable phase congruency (MSPC), which integrates the affine invariant region extraction with the structural features of images organically. First, to achieve the contrast invariance and ensure the significance of features, we detect feature points using moment ranking analysis and extract structural features via merging phase congruency images in multiple orientations. Then, coarse neighborhoods centered on the feature points are obtained based on Log-Gabor filter responses over scales and orientations. Subsequently, the affine invariant regions of feature points are determined by using maximally stable extremal regions. Finally, structural descriptors are constructed from those regions and the registration can be implemented according to the correspondence of the descriptors. The proposed method has been tested on various infrared and visible pairs acquired by different platforms. Experimental results demonstrate that our method outperforms several state-of-the-art methods in terms of robustness and precision with different image data and also show its effectiveness in the application of trajectory tracking.

Keywords: infrared image; image registration; MSER; phase congruency

1. Introduction

In recent years, the rapid development of sensor technology has made it possible to fully perceive an object in complicated scenes. As the two most common visual sensors, infrared and visible sensors are widely applied in various kinds of optoelectronic systems [1]. To make use of both sensors simultaneously, a prerequisite is to achieve the image registration, which is a process of aligning two or more images of a same scene captured by different sensors, at different times, or from distinct viewpoints [2]. The accuracy of image registration has a significant impact on many computer vision tasks, such as image fusion [3], image mosaic, visual-based navigation, and object recognition. In the registration field, infrared and visible image registration is very challenging work mainly due to two reasons. First, as a result of the differences in imaging mechanisms, the same scene's content may be represented by different intensity values, which means that images from two different sources have poor consistency in contrast. This makes it difficult to find the correspondence based on their intensity or gradient values directly, which can be seen from Figure 1. Second, he various intrinsic and extrinsic sensing conditions may lead to large geometric deformations that exist between the images, which further increase the difficulty of registration. A number of related methods have been proposed and applied successfully in the situation where the geometric changes are small [4–8] or can be greatly alleviated according to the capture information [9,10]. However, automatic infrared and visible image registration has not been solved effectively in complicated environments with large geometric changes and significant differences in contrast.

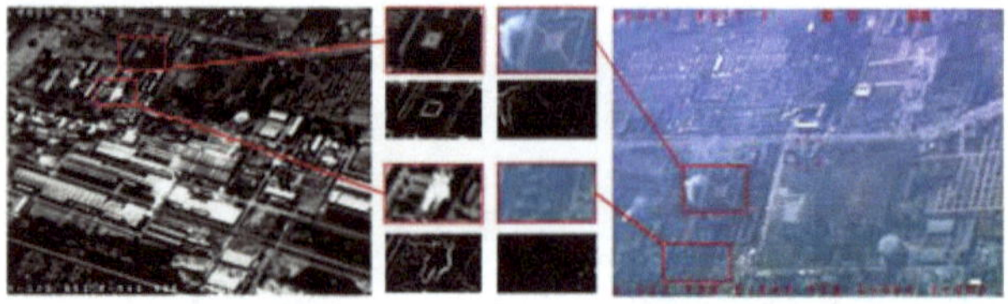

Figure 1. Differences of contrast and viewpoints in input images. (**a**) Infrared image; (**b**) Corresponding regions and their gradient images; and (**c**) Visible image.

This paper presents a novel affine and contrast invariant descriptor for the automatic registration of infrared and visible images. The proposed method detects the significant feature points based on moment ranking analysis and constructs structural features via merging phase congruency images in multiple orientations. This embodies the significance of feature points maximally and makes structural features to be contrast invariant. Descriptors of orientated phase congruency centered on the feature points are constructed in the affine invariant regions detected by maximally stable extremal regions (MSER), which ensures that the descriptors are affine invariant. This paper is organized as follows. Related works in registration for infrared and visible images are described in Section 2. The proposed registration method using a novel affine and contrast invariant descriptor is described in detail in Section 3. Comparative and applied experimental results are discussed in Section 4. Finally, conclusions are drawn, and future work is discussed in Section 5.

2. Related Works

At present, the registration methods for infrared and visible image can be classified into two categories: global region-based methods and local features-based methods. Global region-based methods obtain correspondence by using the whole image content in spatial domain or transform domain, which mainly include mutual information (MI) [8,11,12], phase correlation (PC) [4], Fourier transform [6,13], particle swarm optimization (PSO) [7], gradient information [5,14], and template correlation matching [15,16]. Those methods can get remarkable performance for images with small geometric changes or medical images with high correlation in global intensity. However, contrast reversal, occlusion, uneven heated, and clutters occur frequently in some regions of input images, which result in the global region-based methods being unable to achieve an accurate registration.

In contrast to global region-based methods, local feature-based methods utilize the extracted features to establish correspondence, and they are generally divided into two groups: typical features-based methods and structural features-based methods. In the first group, extracted typical features include edges [17], lines [18–22], contours [23], gradient distribution [15,24], and their variants [25–28]. Those methods above are robust in response to geometrical changes, occlusion, background clutter, and noise. However, they treat all content equally, such that they are highly sensitive to structural disparities caused by insignificant structures. This results in serious degradation in matching performance when large differences in contrast appeared in input images. Two images obtained from the same scene using different modalities may have significantly different intensity characteristics but should have very similar structural features. Therefore, the structural features of the disparate images can be compared in a direct fashion.

Compared with the typical features-based methods, structural features-based methods can extract more robust common features from different modalities and are less sensitive to the contrast differences. Due to these advantages, they have been successfully applied to multimodal image registration [10,21,28–33]. As a valid structural feature extraction method, phase congruency was proposed by Morrone et al. [34], which is the ratio of local energy to the overall path length taken by the local Fourier components in reaching the endpoint. To improve the insensitivity of phase congruency to noise and provide good localization, Kovesi proposed a new sensitivity measure and noise compensation method for phase congruency, which can locate the features that remain constant over scales [35]. Subsequently, Kovesi presented a highly localized feature detector whose responses are invariant to image contrast [36]. These properties make local phase congruency an effective method for creating a structural representation of the images. Wong and Orchard [29] constructed local phase-coherent representations of images and applied their method to multimodal medical image registration successfully. Xia et al. [30] combined phase congruency representations of images with scale-invariant feature transform (SIFT) to achieve multimodal medical image registration. Recently, Liu et al. [31] proposed mean local phase angle (MLPA) and frequency spread phase congruency (FSPC) by using local frequency information on Log-Gabor wavelet transformation space, which improved the robustness compared with traditional multimodal matching. Based on the structural properties of images, Ye et al. [10] developed the histogram of orientated phase congruency (HOPC) descriptor, which outperforms several methods in matching performance. These registration methods that relate to phase congruency are robust against complex nonlinear radiometric differences and have good performance on image pairs with slight geometric changes. However, they cannot obtain satisfactory results for image pairs with large geometric deformations. Zhao et al. [21] proposed a novel multimodality robust line segment descriptor (MRLSD) and developed a MRLSD matching method, which can deal with large-scale and rotation changes in image pairs, while the registration results are poor when line segments or edges are deficient in some regions.

Motivated by the phase congruency-related methods [10,21,31], this paper develops an affine and contrast invariant descriptor and presents a robust registration method based on that descriptor. Firstly, feature points are extracted based on the moment analysis over orientations. Then, the coarse description regions are estimated by Log-Gabor response over scales and orientations centered on the feature points, and the descriptors are constructed by the orientations on the fine regions detected by MSER. Finally, the registration is achieved according to the correspondence of descriptors between image pairs. The whole process of the proposed method is shown in Figure 2.

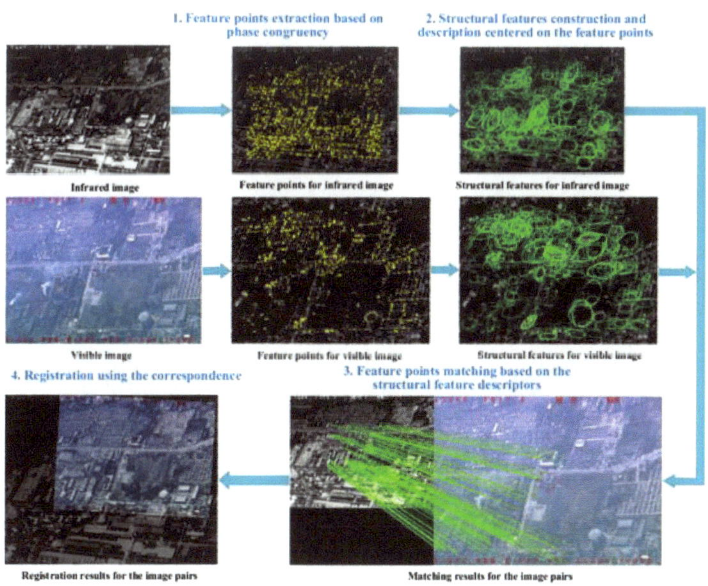

Figure 2. Illustration of registration by using the proposed method.

3. Methodology

The key issues in infrared and visible registration are what type of features should be detected and how to extract the feature form input images. With the idea that feature points of high perceptual significance coincide with points of high structural significance within an image, the salient feature points (SFP) detection method based on the moment analysis in phase congruency images is presented in Section 3.1. Then, the approach of construction for the maximally stable phase congruency (MSPC) descriptor, using orientated phase congruency and MSER [37], is developed in detail in Section 3.2. Finally, the algorithm of registration for infrared and visible image is described in Section 3.3.

3.1. Salient Feature Points Detection

The measure of phase congruency developed by Morrone et al. [34] is follows:

$$\mathrm{PC_1(x)} = \frac{|\mathrm{E(x)}|}{\sum_{\mathrm{n}} \mathrm{A_n(x)}}, \tag{1}$$

where $\mathrm{A_n(x)}$ is an amplitude of Fourier components at a location x in a signal, and $|\mathrm{E(x)}|$ is the magnitude of the vector from the origin to the endpoint. From the definition above, if all the Fourier components are in phase, all the complex vectors would be aligned, and $\mathrm{PC_1(x)}$ would be 1. If there is no coherence of phase, $\mathrm{PC_1(x)}$ falls to a minimum of 0. Phase congruency provides a measure that is independent of the overall magnitude of the signal, making it invariant to variations in image contrast. Subsequently, Kovesi proposed an improved measure [35] as follows:

$$PC(x) = \frac{\sum_{\mathrm{n}} \mathrm{W(x)} \left\lfloor \mathrm{A_n(x)(\cos(f_n(x) - \bar{f}(x)) - \left|\sin(f_n(x) - \bar{f}(x))\right|) - T} \right\rfloor}{\sum_{\mathrm{n}} \mathrm{A_n(x)} + \varepsilon}, \tag{2}$$

where $\mathrm{W(x)}$ is a factor that weights for frequency spread, and $\mathrm{A_n(x)}$ is an amplitude of Fourier components at a location x. $\mathrm{f_n(x)}$ and $\mathrm{\bar{f}(x)}$ are phase angle and weighted mean phase angle, respectively. ε is a small constant, and T is a threshold that eliminates noise influence. The symbol $\lfloor\ \rfloor$ denotes that the enclosed quantity is equal to itself when its value is positive and zero otherwise. Based on the measure, Kovesi presented a highly localized feature detector whose responses are invariant to image contrast [36], which consists of the following steps:

(1) Compute the moment analysis equations at each point in the image as follows:

$$\mathrm{A} = \sum (\mathrm{PC}(\theta)\cos(\theta))^2, \tag{3}$$

$$\mathrm{B} = 2\sum (\mathrm{PC}(\theta)\cos(\theta)) \cdot (\mathrm{PC}(\theta)\sin(\theta)), \tag{4}$$

$$\mathrm{C} = \sum (\mathrm{PC}(\theta)\sin(\theta))^2, \tag{5}$$

where $\mathrm{PC}(\theta)$ refers to the phase congruency value determined at orientation θ.

(2) The minimum moment matrix m and principal axis matrix Φ are given by

$$\mathrm{m} = (\mathrm{C} + \mathrm{A} - \sqrt{\mathrm{B}^2 - (\mathrm{A} - \mathrm{C})^2})/2, \tag{6}$$

$$\Phi = \mathrm{atan}(\mathrm{B}, \mathrm{A} - \mathrm{C})/2. \tag{7}$$

If the minimum moment of phase congruency is still large, then it means that the point should be marked as a 'corner'. The principal axis, corresponding to the axis about which the moment is minimized, provides an indication of the orientation of the feature. Thus, the minimum moment is used for detecting the feature points, and the principal axis matrix is used to guide the construct of the structural feature image in Section 3.2.

Therefore, the SFP extraction (MSFPE) based on salient ranking can be expressed as follows:

(1) Compute the minimum moment matrix m at each point in the input image using (2)–(6).

(2) To ensure the significance of feature points, candidate feature points FP are obtained by filtering m:

$$\mathrm{FP} = \{(x, y) | m(x, y) >\}, \tag{8}$$

where $\mathrm{Th} = \mathrm{mean}(m > 0.1)$ is the mean of values that are larger than 0.1 and adaptive to matrix m.

(3) To make the feature points distributed uniformly, we extract MFP from FP by using non-maximum suppress in the neighborhood of (x, y):

$$\mathrm{MFP} = \left\{ (x + \hat{p}, y + \hat{q}) | \underset{p,q \in [-2,2]}{\mathrm{argmax}} (m(x + p, y + q)) \right\}. \tag{9}$$

(4) The significance ranking space is built by sorting the positions in MFP according to corresponding value in m from maximum to minimum.

(5) The top N of significance ranking space are selected as SFP.

In the above algorithm, the non-maximal suppression over a 5×5 neighborhood of a candidate feature point is adopted to ensure the uniform distribution of feature points. An example for feature points extraction using MSFPE is shown in Figure 3. It can be seen that the feature points are not only significant, but also distributed uniformly in the whole image.

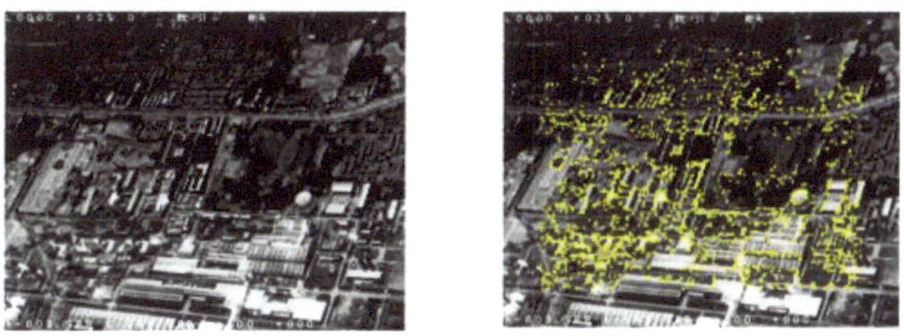

Figure 3. Feature points detection by the method of salient feature points extraction (MSFPE).

3.2. Maximally Stable Phase Congruency Descriptor

Salient feature points indicate that there are significant features around them. Hence, to improve the robustness of feature matching, the description for structural features centered on the feature points in an image is necessary. Consequently, a method of construction for structural features using multi-orientation phase congruency is proposed, and the generation of the MSPC descriptor based on the structural features is developed in this section.

3.2.1. Structural Features Extraction

The calculation model of phase congruency was improved by Kovesi [35] using Log-Gabor wavelets over multiple scales and orientations. To make full use of multi-orientation phase congruency, we construct the structural features from multiple phase congruency images over orientations according to the principal axis information. The detailed calculation steps of the structural features extraction (SFE) are shown as follows:

(1) Compute n different phase congruency images PC_θ with $\theta \in \mathrm{OTS}$ and the principal axis matrix Φ from the input image using (2)–(7).

$$\mathrm{OTS} = \{(i - 1) * \pi/n, i = 1, \ldots, n\}. \tag{10}$$

(2) To embody the significance of structural features over the image maximumly, structural features image (SFI) is constructed from different PC_θ according to the principal axis matrix Φ. The value at (x, y) in SFI can be expressed as follows:

$$SFI(x, y) = PC_{\tilde{\theta}}(x, y), \tag{11}$$

where

$$\tilde{\theta} = \underset{\theta \in OTS}{\operatorname{argmin}} |\Phi(x, y) - \theta|, \tag{12}$$

where $PC_{\tilde{\theta}}$ is the phase congruency image corresponding to $\tilde{\theta}$.

In the algorithm above, each value of SFI is from a special matrix PC_θ, and θ is the closest orientation to the corresponding value in Φ, which ensures that each point of SFI has a maximum response in all orientations. The construction of structural features can be seen in Figure 4.

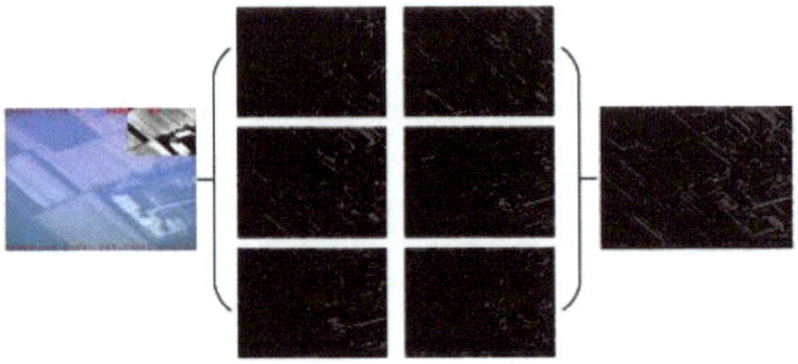

Figure 4. Structural features extraction using multi-orientation phase congruency.

3.2.2. Affine Invariant Structural Descriptor

In order to produce an affine invariant descriptor for a feature point, the coarse shape of the region to be described centered on the feature point should be estimated first. Similar to SIFT [24], the coarse shape can be determined by the feature point's scale and orientation, which can be computed by the responses of Log-Gabor wavelets over multiple scales and orientations.

In frequency domain, the Log-Gabor function is defined as

$$g(\omega) = \exp\left(\frac{-(\log(\omega/\omega_0))^2}{2(\log(\sigma_\omega/\omega_0))}\right), \tag{13}$$

where ω_0 is the central frequency, and σ_w is the related width parameter. Let I denote the image, $LG^e_{n,\theta}$ and $LG^o_{n,\theta}$ denote the even-symmetric and odd-symmetric component of Log-Gabor function at the scale n and orientation θ, respectively. The responses of each quadrature pair of filters can be expressed as

$$[e_{n,\theta}(x), o_{n,\theta}(x)] = [I(x) * LG^e_{n,\theta}, I(x) * LG^o_{n,\theta}]. \tag{14}$$

The values $e_{n,\theta}(x)$ and $o_{n,\theta}(x)$ can be regarded as real and imaginary parts of a complex valued frequency component. The amplitude and phase of the responses at the scale n and orientation θ are given by

$$A_{n,\theta}(x) = \sqrt{e_{n,\theta}(x)^2 + o_{n,\theta}(x)^2}, \tag{15}$$

$$f_{n,\theta}(x) = \operatorname{atan}(e_{n,\theta}(x), o_{n,\theta}(x)). \tag{16}$$

The orientation for a point x in phase congruency is defined as

$$F(x) = \sum_\theta \sum_n e_{n,\theta}(x), \tag{17}$$

$$H(x) = \sum_{\theta}\sum_{n} o_{n,\theta}(x), \tag{18}$$

$$\Phi(x) = \operatorname{atan}(F(x), H(x)). \tag{19}$$

We can see that the results computed by (7) and (19) are the same. The coarse scale of a point x can be obtained based on the responses of Log-Gabor filters, along with its orientation over scales in phase congruency, which can be computed as follows:

$$\widetilde{\sigma}(x) = \underset{n\in\{1,2,\ldots N\}}{\operatorname{argmax}} A_{n,\widetilde{\theta}}(x), \tag{20}$$

where $\widetilde{\theta}$ can be computed by (12) and is the closest orientation θ to the corresponding value in $\Phi(x)$. Based on the coarse scale and orientation of a feature point x, the coarse rectangle shape of its neighborhood can be estimated by

$$[R\_size(x), R\_ang] = [Initial\_size * Mul\_factor^{\widetilde{\sigma}(x)}, \Phi(x)], \tag{21}$$

where $R\_size(x)$ is a two-dimensional (2D) vector that contains the length and width of the rectangle, R_ang is the rotation angle, Initial_size is a given minimum size, and Mul_factor is the scaling factor between successive Log-Gabor filters.

Because the scale of the feature point is approximate, the rectangle neighborhood is also imprecise. Consequently, the fine ellipse region of a feature point is further obtained by MSER on the estimated coarse rectangle neighborhood from SFI according to (21), which is the definitive description area for the point and affine invariant in image content. Structural features computed by (11) indicate the degree of phase congruency in some orientations; however, they cannot represent the significant directions of feature variation [9]. Thus, it is insufficient to use only the amplitude of phase congruency to construct robust feature descriptors. Therefore, we use orientated phase congruency that is weighted by the amplitude of structural features to compute the descriptors. The construction process of the maximally stable phase congruency (MSPC) descriptor can be expressed as follows.

(1) Compute the scale and orientation by using (14)–(20) for each feature point extracted by MSFPE.
(2) Estimate the coarse rectangle shape of the feature point's neighborhood by (21).
(3) Get the fine ellipse region E for the feature point by applying MSER to the coarse rectangle region on SFI obtained by (11).
(4) Normalize the ellipse region E to a circle region C according to the long axis to ensure the affine invariance of the descriptor.
(5) Calculate the weighted statistical histogram with four orientations distributed in $(0^0 - 180^0)$ by structural feature values in the circle region C, in which, the weight of a certain orientation θ can be computed as follows:

$$C(\theta) = \{(x, y) | abs(\Phi(x, y) - \theta) \in [0, \pi/4)\}, \tag{22}$$

$$W(\theta) = \sum_{(x,y)\in C(\theta)} SFI(x, y). \tag{23}$$

(6) The orientation histogram is normalized as a descriptor by

$$Des = h_i \Big/ \sqrt{\sum_{i=1}^{64} h_i}. \tag{24}$$

In the algorithm above, a circle region is divided into 4 × 4 small regions, and each small region is computed in four directions. Therefore, a circle region can be described as a vector of 64 dimensions.

In the process of description, we use both the orientation and amplitude of the phase congruency to compute the descriptor in the ellipse region detected by MSER, which can effectively describe the feature distribution in the orientation and strength of phase congruency and make the descriptors to be affine invariant. The construction example of the descriptor is shown in Figure 5. From that, we can see the descriptor is robust against contrast and geometrical distortion.

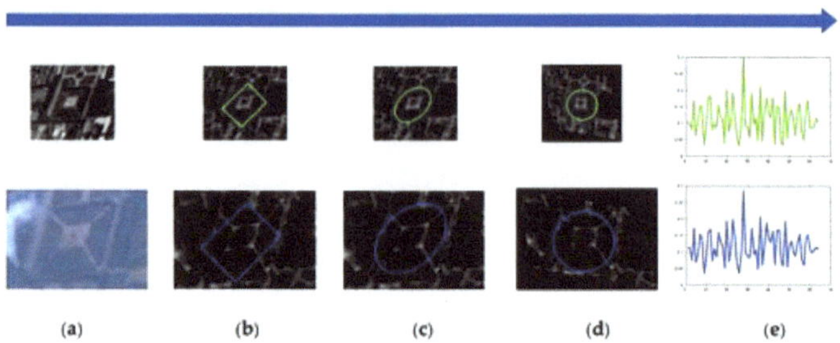

Figure 5. The construction of the maximally stable phase congruency (MSPC) descriptor from input images. (**a**) Original patches around the feature points; (**b**) Rectangle regions from structural features image (SFI) according to the scales and orientations of the feature points; (**c**) Fine ellipse regions detected by maximally stable extremal regions (MSER) based on the rectangle regions; (**d**) Normalized circle regions relate to the ellipse regions; (**e**) MSPC descriptors constructed in the circle regions.

3.3. Registration Using the MSPC Descriptor

After the extraction of salient feature points and the construction of the MSPC descriptors were presented in Sections 3.1 and 3.2, the method of registration for infrared and visible images based on those feature points and descriptors is proposed in this section.

The flow chart of the registration algorithm is shown in Figure 6, and the details are described as follows.

(1) Compute the phase congruency images using Log-Gabor filters over the scales and orientations from infrared and visible images, respectively.
(2) Extract the salient feature points based on the moment analysis of the phase congruency images by the MSFPE algorithm proposed in Section 3.1.
(3) Construct the structural features using the multi-orientation phase congruency by the SFE algorithm presented in Section 3.2.
(4) Generate the descriptors for the salient feature points using the construction algorithm of the MSPC designed in Section 3.2.
(5) Find the matching points via the minimization of the Euclidean distances between the descriptors and refine the matching with random sample consensus (RANSAC).
(6) Obtain the transformation from the matching and achieve the image registration.

In the registration algorithm above, the affine transformation model is used for describing the geometric distortion between the input images, which can be expressed as follows:

$$\begin{bmatrix} \mathrm{X} \\ \mathrm{Y} \\ 1 \end{bmatrix} = \begin{bmatrix} \mathrm{a} & \mathrm{b} & \mathrm{e} \\ \mathrm{c} & \mathrm{d} & \mathrm{f} \\ 0 & 0 & 1 \end{bmatrix} \cdot \begin{bmatrix} \mathrm{x} \\ \mathrm{y} \\ 1 \end{bmatrix}, \quad (25)$$

where $\mathrm{a}, \mathrm{b}, \mathrm{c}$ and d are the combination of scale, rotation, stretch, and twist, and e and f are the translation in the horizonal direction and vertical direction, respectively. (x, y) and (X, Y) are the coordinates of the corresponding points in the input images. The transformation parameters are estimated by applying the least squares on the corresponding point pairs in our algorithm. In addition, the significance of the feature points is ensured by minimum moment analysis and significance ranking. Affine and contrast invariance of the descriptors is guaranteed by the scale and orientation of the feature points and MSER detection. Therefore, the proposed algorithm can achieve good performance for infrared and visible images with significant contrast changes and large geometric deformation, which will be seen in Section 4.

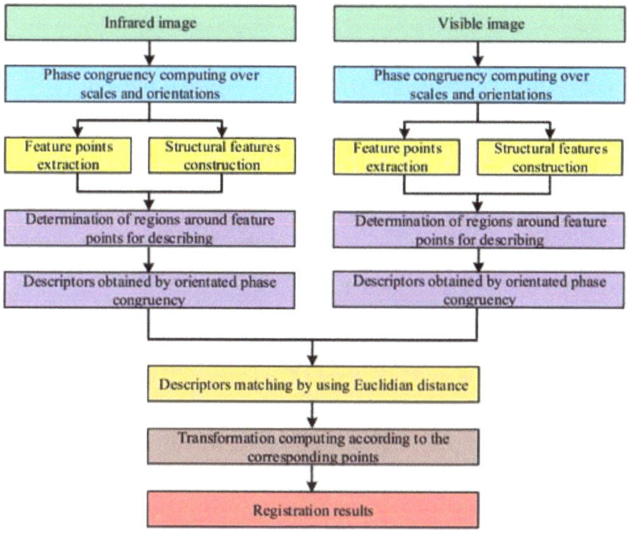

Figure 6. Flow chart of the proposed registration.

4. Experimental Results and Analysis

To test our method in terms of validity and efficiency, three different sets of images were used in comparative and evaluative experiments in this section. There were four infrared and visible pairs from computer vision center (CVC) datasets in the first set, which are used to evaluate the matching performance of the proposed method via a comparison with multimodal-SURF (MM-SURF) [27], fast visual salient and descriptor-rearranging (FVS-DR) [28], local frequency information (LFI) [31], MRSLD [21], and HOPC [10]. The second image set contained 300 image pairs captured from electro-optical pod (EOP) on unmanned aerial vehicle (UAV) with discontinuous focus length change from 25 to 300 mm in a mid-wavelength infrared camera and from 6.5 to 130.2 mm in a visible camera. Those remote sensing images were used to test the validity of our method for registration with significant contrast change and large geometric distortion. Several registration results of our method have been given, and the corresponding registration errors have been compared with those of the related methods. The third image set contained one large Google image and 40 infrared images captured from EOP on UAV, which are used to confirm the practicability of the proposed method in trajectory tracking.

For evaluating the matching performance, precision and repeatability are employed, which can be expressed as follows:

$$\text{Precision} = \frac{\text{NCM}}{\text{NTM}}, \tag{26}$$

$$\text{Repeatability} = \frac{\text{NCM}}{\min(\text{NFP}_{\text{ref}}, \text{NFP}_{\text{sen}})}, \tag{27}$$

where NCM and NTM are the number of correct matched and total correct matched point pairs, respectively, and NFP_{ref} and NFP_{sen} are the number of feature points extracted from the reference and sensed image, respectively. For each feature point in the reference image, we compare its mapped point with the corresponding point in the sensed image. If the Euclidean distance is less than 3 pixels, the match is considered to be correct.

To assess the registration results, root-mean-square error (RMSE) is used in the overlapped area between the reference image and the transformed sensed image, which is calculated as follows:

$$\mathrm{RMSE} = \sqrt{(X_i^r - X_i^{ts})^2 + (Y_i^r - Y_i^{ts})^2 / N}, i = 1, \ldots, N, \tag{28}$$

where (X_i^r, Y_i^r), (X_i^{ts}, Y_i^{ts}) are the coordinates of pixels in the reference image and the transformed sensed image, respectively, and N is the number of pixels in their overlapped area.

4.1. Comparative Experiments

To evaluate the matching performance of the proposed method, four multimodal stereo image pairs from CVC datasets were used to compare with the related methods presented in [10,21,27,28,31] in terms of precision and repeatability. The set of image pairs with size of 506×408 are shown in Figure 7, which have large difference in contrast and small viewpoint changes. Matching results using the proposed method for the image pairs in Figure 7 are shown in Figure 8. It can be seen that our method obtained a good matching when significant difference contrast occurs in the image pairs. In addition to LFI, the other five methods belong to local feature matching and contain the feature points detection steps. To compare the proposed method with LFI conveniently, the feature points are extracted by the Harris corner detector first, and then, the matching of regions is computed by LFI. The precision and repeatability of the matching results of different methods are shown in Table 1. From that, we can see that the proposed method has better performance than the other five related methods. The average precision of the proposed method for the four image pairs is 93.32%, which is 5.79%, 10.43%, and 14.30% higher than that of HOPC, MRLSD, and LFI, respectively. This is mainly due to the affine and contrast invariance of the MSPC constructed by the proposed method. The average precision of both MM-SURF and FVS-DR is less than 75%, which is due to the fact that simple intensity symmetry or reversal cannot eliminate the difference in contrast completely. The average repeatability of our method for the four image pairs is 33.30%, which is 5.88%, 6.02%, and 10.64% higher than that of HOPC, MRLSD, and LFI respectively. This advantage is attributed to the great significance of the extracted feature points in sequence and the high communization of the constructed structural features in the proposed method.

Table 1. Matching performance of the related methods in Figure 8.

Image Pairs		MM-SURF	FVS-DR	LFI	MRLSD	HOPC	Our Method
Precision	(a)	40.72	75.36	80.22	85.58	87.13	91.85
	(b)	35.14	77.81	82.56	88.72	93.37	97.78
	(c)	22.31	73.30	77.28	82.15	91.26	96.65
	(d)	9.84	69.81	75.95	78.31	81.54	90.21
Repeat-ability	(a)	10.83	20.48	28.47	35.19	32.24	39.60
	(b)	5.77	14.63	25.23	33.64	35.79	42.80
	(c)	3.23	11.12	21.41	20.33	23.82	26.00
	(d)	2.18	6.42	15.52	19.97	17.82	24.80

(a) (b)

Figure 7. *Cont.*

(c) (d)

Figure 7. (**a**–**d**) are different infrared and visible image pairs from CVC datasets.

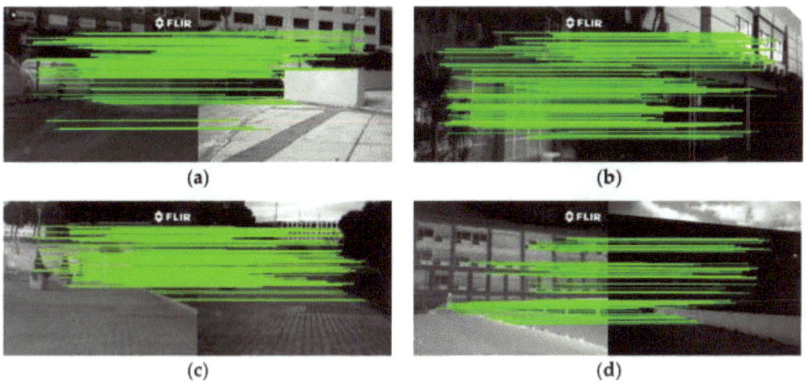

Figure 8. Matching results using the proposed method for Figure 7. (**a**–**d**) are the matching results of the Figure 7a–d respectively.

4.2. Validity Verification Experiments

To test the validity of the proposed method, the second set of images captured from EOP on UAV were used in this section. Those images not only have scale differences caused by focus length changes, but also have a variety of scenes with infrared and visible images with size 1024×768 and 640×512, respectively, and several examples are shown in Figure 9. From that, we can see that (a), (b), (c), and (d) have focus length changes of the visible camera with different scenes, while that of infrared camera keeps to 25 mm. Figure 9e and f have focus length changes of the infrared camera with different scenes, while that of visible camera keeps to 130.2 mm. The six image pairs not only contain large geometric changes, but also have significant differences in contrast.

To ensure the attainment of salient structural features, eight orientations are adopted for different phase congruency images, and $Th = 0.1$ is used to filter the minimum moment image in feature points extraction. Figure 10 shows the matching results of the image pairs in Figure 9 by using the proposed method. In those image pairs, we consider the image that has the larger field of view as the reference image and the other one as the sensed image. It can be seen from those results that the proposed method can achieve good performance whether images have rich texture information (Figure 9a,c,d) or not (Figure 9e,f). In particular, in blurry situations (see Figure 9b) and with large differences in scale (Figure 9e), the proposed method can still get enough correct matching point pairs, while several of the state-of-art methods failed in those cases. For example, MRLSD failed for Figure 9b due to the fact that there are not enough lines to be extracted from the images. MM-SURF and FVS-DR failed for Figure 9e,f, because they cannot get the robust feature descriptors for textures. HOPC failed for Figure 9e as result of the large geometric changes in the image pairs.

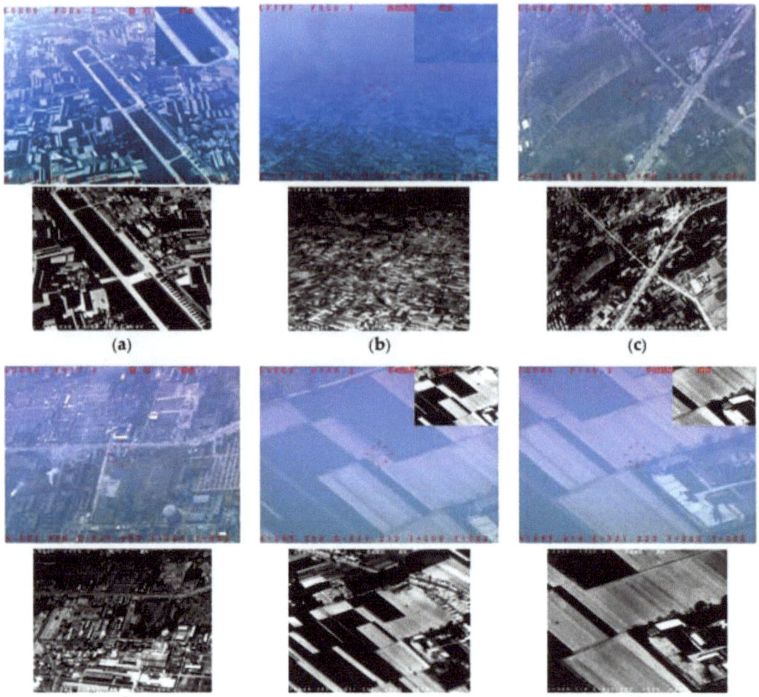

(a) (b) (c)

(d) (e) (f)

Figure 9. (**a**–**f**) are the samples of image pairs captured from electro-optical pod (EOP) on UAV.

(a) (b) (c)

Figure 10. *Cont.*

(d) (e) (f)

Figure 10. Matching results by the proposed method for Figure 9. (**a–f**) are the matching results of the Figure 9a–f respectively.

The matching performance of the proposed method compared with MM-SURF, FVS-DR, LFI, MRLSD, and HOPC are shown in Figure 11. From those results, we can see the proposed method outperforms the other methods both in precision and repeatability. The average precision of our method is higher than 89%, and the average repeatability is higher than 37%, while the average precision and repeatability of the best performances in the other methods are lower than 85% and 33% (failures are not calculated), respectively, which is because of large difference in scale and contrast between the input images. The proposed method can achieve better performance, even in the cases where other methods are invalid for Figure 9b,e,f. In addition to our method, both MRLSD and HOPC achieve better performance (except for the failure case) than the rest of the methods due to the fact that they use phase congruency information and structural features in the feature description. However, linear features do not always exist in the images (Figure 9b) that result in the failure of MRLSD. HOPC cannot deal with large geometric changes, so it failed for Figure 9e. LFI uses the differences of features as the similarity measure directly, which resulted in a matching performance that was worse than our method. Although FVS-DR and MM-SURF have a certain tolerance for geometric changes, they are less able to deal with differences in contrast based on the reversal or symmetry of intensity; therefore, they had a worse matching performance than the proposed method.

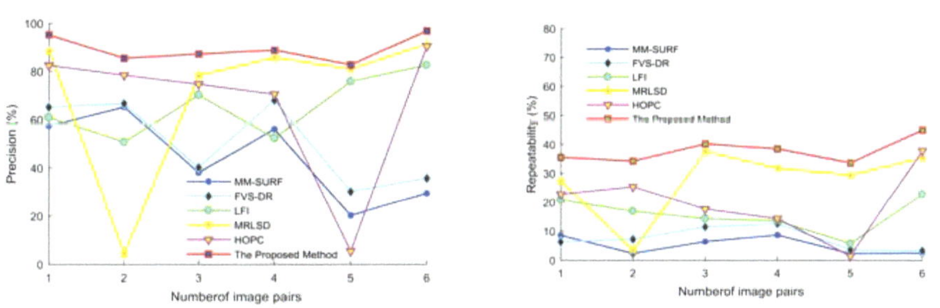

(a) (b)

Figure 11. Comparison of matching performance by the related methods. (**a**) is the matching precision for the six image pairs in Figure 9 by the related methods; (**b**) is repeatability for the six image pairs in Figure 9 by the related methods.

The registration results of using the proposed method for the image pairs in Figure 9 are shown in Figure 12. It can be observed that our method achieves good performance whether the infrared image is used as a reference or not, which indicates that our method is robust against the changes in geometry and contrast. The RMSE of the registration results of using different methods are given in Table 2, where MM-SURF and FVS-DR failed for Figure 10e and f and MRLSD and HOPC failed for Figure 10b,e, respectively, because they could not get enough correct matched point pairs. The proposed method can not only achieve the registration of all the image pairs, but also make the average RMSE less than 2 pixels. Furthermore, the average RMSE in the registration of the second set with 300 images is 1.8 pixels, which is acceptable for practical application.

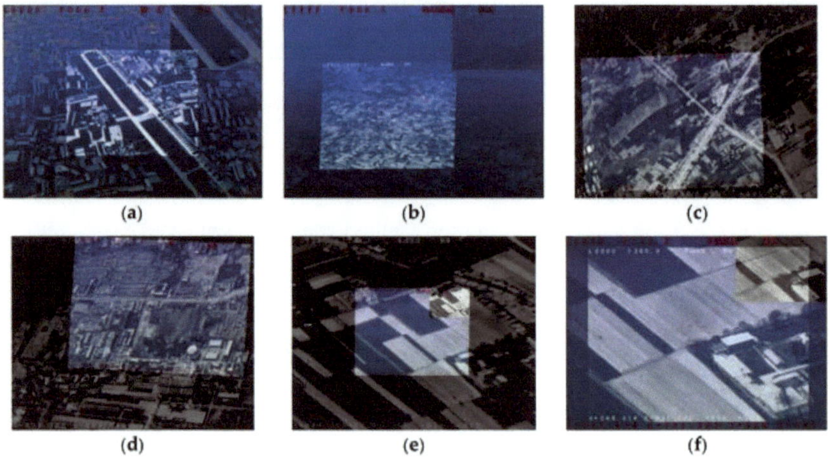

Figure 12. Registration results by the proposed method for Figure 9. **(a–f)** are the registration results of the proposed method for Figure 9a–f respectively.

Table 2. Root-mean-square error (RMSE) of registration results of the related methods in Figure 9.

Image Pairs	MM-SURF	FVS-DR	LFI	MRLSD	HOPC	Our Method
(a)	2.61	2.44	3.54	1.57	2.11	0.82
(b)	3.36	2.88	2.72	—	3.63	1.23
(c)	4.68	3.39	3.66	2.35	4.55	0.76
(d)	3.97	3.73	4.19	2.56	4.62	0.58
(e)	—	—	5.57	3.12	—	1.37
(f)	—	—	4.81	3.38	2.26	1.41

Moreover, the experiments are implemented on computer with Intel Core i7-4810MQ CPU at 2.80 GHz, and the average registration times achieved by the related methods for the six image pairs in Figure 9 are shown in Table 3. From that, we can see that the run time of the proposed method is moderately fast, but the registration performance is significantly improved compared with the other related methods.

Table 3. Average time of registration by the related methods in Figure 9.

Method	MM-SURF	FVS-DR	LFI	MRLSD	HOPC	Our Method
Run time	0.8S	1.85S	2.8S	2.5S	15.8S	2.1S

4.3. Applied Experiments

Finally, we apply the proposed method to UAV trajectory tracking via the registration of the real-time images and the reference image. The real-time images were captured by EOP on UAV, and the reference image was downloaded from Google. To achieve fast registration, the sub-images (300×300) from the real-time images were used to search the matching on the reference image. The reference image (with 1.5 m resolution) is shown in Figure 13, and samples of the sub-image from the real-time images are shown in Figure 14. From that, we can see there are large geometric changes and significant contrast differences existing in those images.

Figure 13. Reference image download from Google.

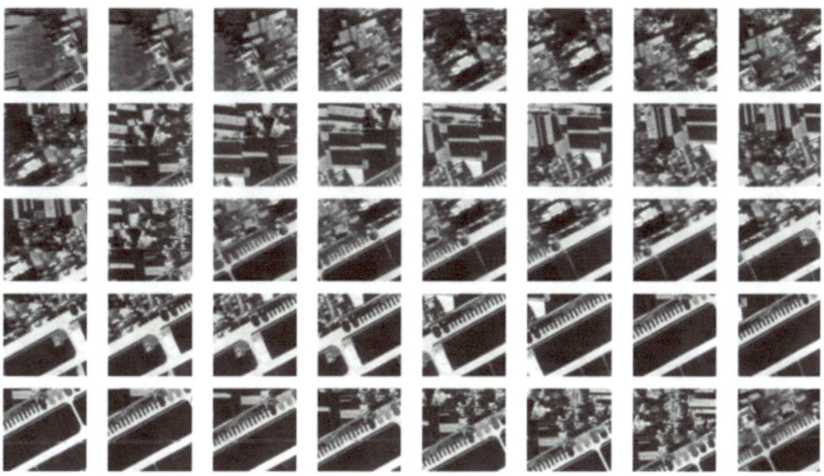

Figure 14. Samples of the sub-images from the real-time images.

Several registration results of samples are given in Figure 15. We can see that the proposed method can deal with large geometric changes, significant differences in contrast, and variance in some structures. The tracking results are shown in Figure 16. It can be seen that the trajectory can be tracked precisely and steadily. In the process of trajectory tracking, the registration time can be shortened to 230 ms when the number of feature points is reduced to 150, which is acceptable in this application. The average RMSE of the registration results is less than 2 pixels, which equals that when the error of tracking is no more than 3 m. Therefore, the proposed method has the potential for practical application.

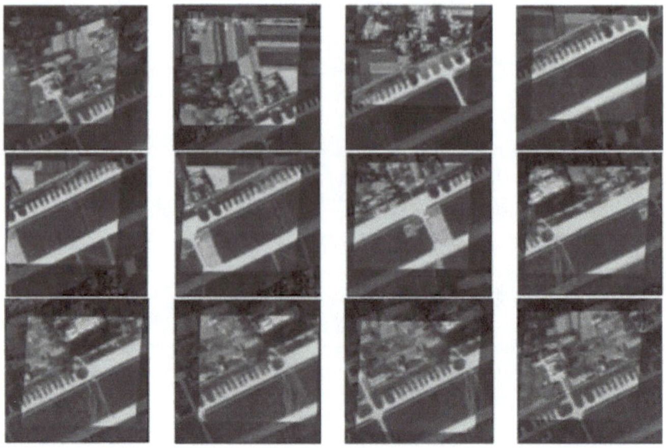

Figure 15. Several registration results of the samples in Figure 14 and the sub-regions of the reference image in Figure 13.

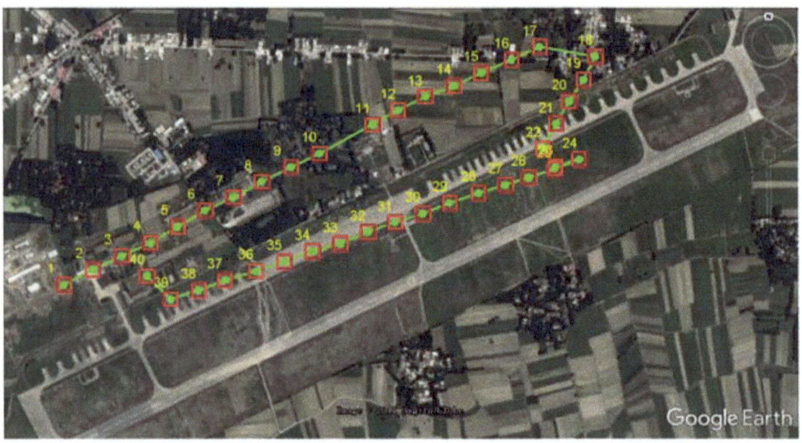

Figure 16. UAV trajectory tracking results of our registration method.

5. Conclusions

In this paper, a robust registration method for infrared and visible image using structural features extracted based on phase congruency is presented. The main contribution of the proposed method is the development of a novel affine and contrast invariant descriptor (MSPC). MSPC firstly uses moment ranking analysis to detect feature points, and then describes structural features by using orientated phase congruency in the regions detected by MSER. Several groups of infrared and visible pairs were used to test the validity and practicality of the proposed method. The experimental results show that our method outperforms several state-of-the-art methods in terms of matching performance and RMSE of registration and also demonstrate its effectiveness in the application of UAV trajectory tracking. For the more than 300 infrared and visible images captured by UAV, the average RMSE of the registration results of the proposed method was less than 2 pixels, which is acceptable for practical application.

Improving the speed of the proposed method and implementing it in the embedded environment is the direction of our future work.

Acknowledgments: This research is supported by the China Postdoctoral Science Foundation (2016M592905XB) and the Ninth Academy of China Aerospace Science and Technology Corporation Innovation Foundation (2016JY06).

Author Contributions: Xiangzeng Liu conceived of and designed the experiments and wrote the paper; Yunfeng Ai performed the experiments; Juli Zhang analyzed the data; and Zhuping Wang supervised the study and reviewed the manuscript.

Conflicts of Interest: The authors declare no conflict of interest.

References

1. Zitova, B.; Flusser, J. Image registration methods: A survey. *Image Vis. Comput.* **2003**, *21*, 977–1000. [CrossRef]
2. Prasad, D.K.; Rajan, D.; Rachmawati, L.; Rajabally, E.; Quek, C. Video processing from Electro-Optical sensors for object detection and tracking in a maritime environment: A Survey. *IEEE Trans. Intell. Transp. Syst.* **2017**, *18*, 1993–2016. [CrossRef]
3. Li, H.; Ding, W.; Cao, X.; Liu, C. Image registration and fusion of visible and infrared integrated camera for medium-altitude unmanned aerial vehicle. *Remote Sens.* **2017**, *9*, 441. [CrossRef]
4. Klimaszewski, J.; Kondej, M.; Kawecki, M.; Putz, B. Registration of infrared and visible images based on edge extraction and phase correlation approaches. In *Image Processing and Communications and Challenges 4. Advances in Intelligent Systems and Computing*; Springer: Berlin/Heidelberg, Germany, 2013; Volume 184, pp. 153–162.
5. Feng, X.; Wu, W.; Li, Z.; Jeon, G.; Pang, Y. Weighted-Hausdorff distance using gradient orientation information for visible and infrared image registration. *Optik* **2015**, *126*, 3823–3829. [CrossRef]
6. Rabatel, G.; Labbe, S. Registration of visible and near infrared unmanned aerial vehicle images based on Fourier-Mellin transform. *Precision Agric.* **2016**, *17*, 564–587. [CrossRef]
7. Sun, M.; Zhang, B.; Liu, J.; Wang, Y.; Yang, Q. The registration of aerial infrared and visible Images. In Proceedings of the International Conference on Educational Information Technology, Chongqing, China, 17–19 September 2010; Volume 1, pp. 438–442.
8. Kuczyński, K.; Stęgierski, R. Problems of infrared and Visible-Light images automatic registration. In *Image Processing and Communications Challenges 5. Advances in Intelligent Systems and Computing*; Springer: Berlin/Heidelberg, Germany, 2014; Volume 233, pp. 125–132.
9. Wang, P.; Qu, Z.; Wang, P.; Gao, Y.; Shen, Z. A coarse-to-fine matching algorithm for FLIR and optical satellite image registration. *IEEE Trans. Geosci. Remote Sens. Lett.* **2012**, *9*, 599–603. [CrossRef]
10. Ye, Y.; Shan, J.; Bruzzone, L.; Li, S. Robust registration of multimodal remote sensing images based on structural similarity. *IEEE Trans. Geosci. Remote Sens.* **2017**, *55*, 2941–2958. [CrossRef]
11. Chen, H.M.; Arora, M.K.; Varshney, P.K. Mutual information-based image registration for remote sensing data. *Int. J. Remote Sens.* **2003**, *24*, 3701–3706. [CrossRef]
12. Yang, F.; Ding, M.; Zhang, X.; Wu, Y.; Hu, J. Two phase non-rigid multi-modal image registration using weber local descriptor-based similarity metrics and normalized mutual information. *Sensors* **2013**, *13*, 7599–7617. [CrossRef] [PubMed]
13. Orchard, J. Efficient least squares multimodal registration with a globally exhaustive alignment search. *IEEE Trans. Image Process.* **2007**, *16*, 2526–2534. [CrossRef] [PubMed]
14. Geng, Y.; Wang, Y. Registration of visible and infrared images based on gradient information. *3D Res.* **2017**, *8*, 1–10. [CrossRef]
15. Zou, Y.; Dong, F.; Lei, B.; Fang, L.; Sun, S. Image thresholding based on template matching with arctangent Hausdorff distance measure. *Opt. Lasers Eng.* **2013**, *51*, 600–609. [CrossRef]
16. Zhu, X.; Hao, Y.G.; Wang, H.Y. Research on infrared and visible images registration algorithm based on graph. In Proceedings of the International Conference on Information Science and Technology, Wuhan, China, 24–26 March 2017; p. 02002.
17. Yi, X.; Wang, B.; Fang, Y.; Liu, S. Registration of infrared and visible images based on the correlation of the edges. In Proceedings of the International Congress on Image and Signal Processing, Hangzhou, China, 16–18 December 2013; pp. 990–994.

18. Han, J.; Pauwels, E.; Zeeuw, P. Visible and infrared image registration employing line-based geometric Analysis. In *Computational Intelligence for Multimedia Understanding*; Springer: Berlin/Heidelberg, Germany, 2011; Volume 7252, pp. 114–125.
19. Li, Y.; Stevenson, R.L. Multimodal image registration with line segments by selective search. *IEEE Trans. Cybern.* **2017**, *47*, 1285–1297. [CrossRef] [PubMed]
20. Wang, Z.; Wu, F.; Hu, Z. MSLD: A robust descriptor for line matching. *Pattern Recognit.* **2009**, *42*, 941–953. [CrossRef]
21. Zhao, C.; Zhao, H.; Lv, J.; Sun, S.; Li, B. Multimodal image matching based on multimodality robust line segment descriptor. *Neurocomputing* **2016**, *177*, 290–303. [CrossRef]
22. Lyu, C.; Jie Jian, J. Remote sensing image registration with line segments and their intersections. *Remote Sens.* **2017**, *9*, 439.
23. Chen, Y.; Dai, J.; Mao, X.; Liu, Y.; Jiang, X. Image registration between visible and infrared images for electrical equipment inspection robots based on quadrilateral features. In Proceedings of the International Conference on Robotics and Automation Engineering, Shanghai, China, 29–31 December 2017; pp. 126–130.
24. Lowe, D. Distinctive image features from scale-invariant keypoints. *Int. J. Comput. Vis.* **2004**, *60*, 91–110. [CrossRef]
25. Bay, H.; Ess, A.; Tuytelaars, T.; Gool, L.V. Speeded-up robust features. *Comput. Vis. Image Understand.* **2008**, *110*, 346–359. [CrossRef]
26. Hossain, M.; Teng, S.W.; Lu, G. Achieving high multi-modal registration performance using simplified Hough-transform with improved symmetric-SIFT. In Proceedings of the International Conference on Digital Image Computing Techniques and Applications, Fremantle, WA, Australia, 3–5 December 2012; pp. 1–7.
27. Zhao, D.; Yang, Y.; Ji, Z.; Hu, X. Rapid multimodality registration based on MM-SURF. *Neurocomputing* **2014**, *131*, 87–97. [CrossRef]
28. Wu, F.; Wang, B.; Yi, X.; Li, M.; Hao, J.; Qin, H.; Zhou, H. Visible and infrared image registration based on visual salient features. *J. Electron. Imaging* **2015**, *24*, 053017. [CrossRef]
29. Wong, A.; Orchard, J. Robust multimodal registration using local phase-coherence representations. *J. Signal Process. Syst.* **2009**, *54*, 89–100. [CrossRef]
30. Xia, R.; Zhao, J.; Liu, Y. A robust feature-based registration method of multimodal image using phase congruency and coherent point drift. In Proceedings of the International Symposium on Multispectral Image Processing and Pattern Recognition, Wuhan, China, 26–27 October 2013; Volume 8919, p. 891903.
31. Liu, X.; Lei, Z.; Yu, Q.; Zhang, X.; Shang, Y.; Hou, W. Multi-modal image matching based on local frequency information. *EURASIP J. Adv. Sig. Process.* **2013**, *3*, 1–11. [CrossRef]
32. Chen, M.; Habib, A.; He, H.; Zhu, Q.; Zhang, W. Robust feature matching method for SAR and optical images by using Gaussian-Gamma-Shaped Bi-Windows-based descriptor and geometric constraint. *Remote Sens.* **2017**, *9*, 882. [CrossRef]
33. Zhang, L.; Dwarikanath, M.; Jeroen, A.W.T.; Jaap, S.; Lucas, J.V.; Frans, M.V. Image registration based on autocorrelation of local structure. *IEEE Trans. Med. Imaging* **2016**, *35*, 63–75.
34. Morrone, M.C.; Ross, J.; Burr, D.C.; Owens, R. Mach bands are phase dependent. *Nature* **1986**, *324*, 250–253. [CrossRef]
35. Kovesi, P. Phase congruency: A low-level image invariant. *Psych. Res.* **2000**, *64*, 136–148. [CrossRef]
36. Kovesi, P. Phase congruency detects corners and edges. In Proceedings of the Conference on Digital Image Computing: Techniques and Applications, 10–12 December 2003; pp. 309–318.
37. Donoser, M.; Bischof, H. Efficient maximally stable extremal region (MSER) tracking. In Proceedings of the IEEE Conference on Computer Vision and Pattern Recognition, New York, NY, USA, 17–22 June 2006; pp. 553–560.

remote sensing

MDPI

Article

Geometry-Based Global Alignment for GSMS Remote Sensing Images

Dan Zeng [1], Rui Fang [1], Shiming Ge [2,*], Shuying Li [3] and Zhijiang Zhang [1]

1 Key Laboratory of Specialty Fiber Optics and Optical Access Networks, Shanghai University, Shanghai 200070, China; dzeng@shu.edu.cn (D.Z.); rui.f.shu@gmail.com (R.F.); zjzhang@shu.edu.cn (Z.Z.)
2 Institute of Information Engineering, Chinese Academy of Sciences, Beijing 100195, China
3 The 16th Institute, China Aerospace Science and Technology Corporation, Shaanxi 710100, China; angle_lisy@163.com
* Correspondence: geshiming@iie.ac.cn; Tel.: +86-10-8254-6269

Academic Editors: Qi Wang, Nicolas H. Younan, Carlos López-Martínez and Prasad S. Thenkabail
Received: 11 March 2017; Accepted: 7 June 2017; Published: 10 June 2017

Abstract: Alignment of latitude and longitude for all pixels is important for geo-stationary meteorological satellite (GSMS) images. To align landmarks and non-landmarks in the GSMS images, we propose a geometry-based global alignment method. Firstly, the Global Self-consistent, Hierarchical, High-resolution Geography (GSHHG) database and GSMS images are expressed as feature maps by geometric coding. According to the geometric and gradient similarity of feature maps, initial feature matching is obtained. Then, neighborhood spatial consistency based local geometric refinement algorithm is utilized to remove outliers. Since the earth is not a standard sphere, polynomial fitting models are used to describe the global relationship between latitude, longitude and coordinates for all pixels in the GSMS images. Finally, with registered landmarks and polynomial fitting models, the latitude and longitude of each pixel in the GSMS images can be calculated. Experimental results show that the proposed method globally align the GSMS images with high accuracy, recall and significantly low computation complexity.

Keywords: image alignment; feature matching; geostationary satellite remote sensing image; GSHHG database

1. Introduction

In many applications, such as weather forecast, environmental monitoring and so on, determining the latitude and longitude of each pixel in the GSMS images is of great importance. However, the GSMS images have the characteristics of round-the-clock, all-weather, long range and high-resolution, which bring new challenges to practical applications.

Remote sensing images matching algorithms are usually divided into two categories: area-based methods and feature-based methods [1,2]. Area-based matching algorithm establishes correspondence between two images by similarity measurements based on correlation functions. There is some classical arithmetic such as cross-correlation [3] and root mean square error (RMSE) [4]. A rough-location method [5] was proposed to locate the remote image with specific physiognomy. By matching the remote sensing image and the digital map, researchers can roughly locate the remote images and the location error is less than 10 km. However, the GSMS images are generally polluted by illumination, scale variation, cloud influence and other factors, and those algorithms do not work well. A feature-based matching algorithm is widely applied to remote sensing images [6–9] because of its robustness. For example, scale-invariant feature transform (SIFT) [10,11] has an excellent performance in most circumstances. However, few feature points can be extracted from the GSMS images with SIFT due to poor textures. In addition, feature-based alignment, which only uses local gradient distribution, will lead to low precision because of too many similar features in the GSMS images.

The challenges of these points-matching methods are removing the outliers. The presence of outliers will have a negative effect on the accuracy of the matching results [12,13]. To remove outliers, many algorithms based on geometric constraint and spatial information are commonly used. Among these algorithms, Random Sample Consensus (RANSAC) [14] is one of the most popular algorithms. It selects a sample randomly from the consensus set in each iteration and finds the largest consensus set to calculate the final model parameters. When the outlier is in the minority, RANSAC performs well and robustly. When the outlier is in the majority, using RANSAC will be time-consuming and unstable. By exploring the spatial relationship of matching points, a matching strategy using spatial consistent matching [8] was proposed to remove outliers. In [15–17], the authors proposed a spatial coding algorithm for image search, which relies on relative position relationship between pairs of matching feature points. It takes into account all matching feature pairs and encodes their coordinates to discover false matches between two images. However, the spatial relationship consistency in this method is too strict for landmark alignment. Since the earth is not a standard sphere, position deviation exists in the GSMS images. Spatial relationship consistency is effective only in a small region, and it also causes lots of correctly matched features to be deleted mistakenly. Furthermore, the number of landmarks is so large that it slows the process of removing outliers. Aguilar et al. [18] proposed a method called Graph Transformation Matching (GTM). It establishes a K-Nearest-Neighbor (KNN) graph to express neighbor geometric structures of the feature points. The mismatching feature points are determined according to the differences between KNN graph established in two images. Shi et al. [19] proposed an image registration algorithm using point structure information. After obtaining robust initial matching point pairs, the final matching results are estimated using GTM based on the local structure information of the point to remove outliers from initial correspondences. On the basis of the GTM algorithm, Weighted Graph Transformation Matching (WGTM) algorithm [20] was proposed. Utilizing the angular distances between edges that connect a feature point to its KNN as the weight, WGTM algorithms can only deal with pseudo isomorphic structures to a certain extent. This arises because angular distance is only invariant with respect to scales and rotations, and shear deformations are not considered in that case. Liu et al. [21] proposed the Restricted Spatial Order Constraints (RSOC) algorithm using a filtering strategy based on two-way geometric order constraints and two decision criteria restrictions. However, when the K-Nearest-Neighbor of the outliers are all the same, RSOC failed to remove such outliers. Zhang et al. [22] proposed a triangle-area representation of the K nearest neighbors (KNN-TAR). It utilizes the descriptor KNN-TAR to find the candidate outliers and removes the real outliers by the local structure and global information. In [23], an algorithm based on integrated spatial structure constraint (ISSC) was proposed for remote sensing image registration. First, a global structure constraint is constructed for each correspondence out of the tentative set to increase the number of inliers and raise the correct rate simultaneously. Then, a local structure constraint based on the triangle area representation is utilized on the neighboring points of each correspondence to remove outliers. Recently, Zhao et al. [24] proposed a vertex trichotomy descriptor. It utilizes the geometrical relations between any of the vertices and lines, which are constructed by mapping each vertex into trichotomy sets. A recovery and filtering vertex trichotomy matching (RFVTM) algorithm was designed to recover some inliers based on identical vertex trichotomy descriptors and restricted transformation errors.

A lot of work has been done toward the images alignment problem. Previous works can be classified in two main categories: direct [25] and feature-based methods [26,27]. Direct approaches minimize pixel-to-pixel dissimilarities. While the feature-based approaches first locate a sparse set of reliable features in the image and then recover the motion parameters considering their correspondences. Miller et al. [28] proposed the congealing method by using an entropy measure to align images with respect to the distribution of the data. Cox et al. [29] proposed a least squares congealing algorithm that minimizes the sum of squared distances between images. Minimization of a log determinant cost function [30] is utilized to align images.

Inspired by these approaches, we propose a geometry-based global alignment method to align GSMS remote sensing images. According to the geometric and gradient similarity of feature maps from the GSHHG and GSMS images, initial feature matching is obtained. Then, feature refinement with a neighborhood spatial consistent matching (NSCM) algorithm is used to remove outliers. Finally, polynomial models are fitted to describe the offsets' tendency according to the matched points set. With the fitted polynomial models, the latitude and longitude of all pixels in the GSMS images can be determined.

2. Materials and Methods

2.1. Local Feature Matching by Geometric Coding

The shorelines of the GSHHG database correspond to the edges of the GSMS images [31], which means that shorelines can be used to simplify alignment of GSHHG and GSMS images.

Since the GSHHG database consists of polygon and line type, the size of the GSHHG database is much smaller than other reference data such as digital elevation model and digital vector map. With sub-satellite point (longitude α_0, latitude γ_0) and satellite height H, the landmarks in GSHHG are mapped to a two-dimensional plane by perspective projection. Therefore, the GSHHG database is quantized to a binary image. As shown in Figure 1a, the white pixels are the landmarks defined in the GSHHG database.

The GSMS image is normalized [32,33] so that the GSHHG and GSMS images have the same size. The edges of the GSMS image extracted by Structured Forests [34] are defined as the edge probability image. As shown in Figure 1b, each element denotes the probability of the pixel being an edge candidate. To distinguish edge candidates from noise, the probability image is binarized to generate the edge binary image as depicted in Figure 1c.

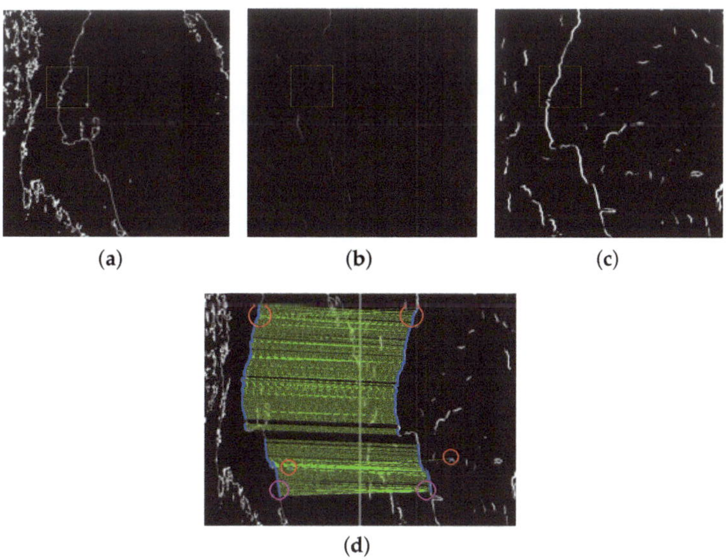

(a) (b) (c) (d)

Figure 1. The landmarks and GSMS images in the southern coastal area of Thailand and their initial matching results. These points in circles are outliers. (**a**) landmarks; (**b**) edge probability image; (**c**) edge binary image; (**d**) initial matching.

For a landmark P_i^G (x_i^G, y_i^G) in the GSHHG image, the neighborhood coding matrix W can be constructed. For a pixel P_i^S (x_i^S, y_i^S) in the edge probability image, the neighborhood coding matrix P can be constructed. Similarly the neighborhood coding matrix P' can be generated with the edge binary image. The matrix W, P and P' all have the same size $(2K+1) \times (2K+1)$.

Then, local features are matched by comparing their geometric similarity and gradient similarity. The geometric similarity between a landmark P_i^G (x_i^G, y_i^G) in the GSHHG image and a pixel P_i^S (x_i^S, y_i^S) in the edge binary image can be calculated as follows:

$$E_{geo}(i, x_i^S, y_i^S) = \sum_{s=-K}^{K} \sum_{t=-K}^{K} W_{s,t}^i \ AND \ P_{s,t}', \tag{1}$$

where the $W_{s,t}^i$ and $P_{s,t}'$ separately denotes the s-th row and t-th column element in matrix W and P'.

Similarly, the gradient similarity between a landmark P_i^G (x_i^G, y_i^G) in the GSHHG image and a pixel P_i^S (x_i^S, y_i^S) in the edge probability image can be calculated by:

$$E_{gra}(i, x_i^S, y_i^S) = \sum_{s=-K}^{K} \sum_{t=-K}^{K} W_{s,t}^i \times P_{s,t}. \tag{2}$$

The number of landmarks located within the template is calculated as follows:

$$C_{geo}(i, x_i^G, y_i^G) = \sum_{s=-K}^{K} \sum_{t=-K}^{K} W_{s,t}^i. \tag{3}$$

Both geometric and gradient similarity are measured to match local features. The procedure of local feature matching between the GSHHG and GSMS image is shown in Algorithm 1. Figure 1d shows the result of initial feature points matching.

Algorithm 1: Local feature matching.

```
Input: W, P, P'; threshold t1, t2 (t1 is set as 0.5, t2 is set as 0.9 based on experience)
Output: the best matching pixel P_i^S for landmark P_i^G
Given landmark P_i^G;
if Max{E_geo(i, x_i^S, y_i^S)} ≥ t1 × C_geo(i, x_i^G, y_i^G) then
    if Max{E_geo(i, x_i^S, y_i^S)} ≥ t2 × SecondMax{E_geo(i, x_i^S, y_i^S)} then
        return the point having Max{E_geo(i, x_i^S, y_i^S)} as P_i^S
    else
        calculate E_gra for the two matching candidates who have bigger E_geo than the other and
          return the one who gets bigger E_gra
    end
else
    could not find the match pixel;
end
```

2.2. Feature Refinement with Neighborhood Spatial Consistent Matching (NSCM)

Since there are lots of similar features in the GSMS image, local feature matching will lead to mismatching. The red circles in Figure 1d show mismatched features. The mauve circles in Figure 1d present many-to-one matched features due to the aperture effect.

The geometric relationship between matched features should not change too much across images. Based on this principle, we propose a neighborhood spatial consistent matching (NSCM) algorithm to remove outliers whose offsets between matched features have sudden mutations.

After Section 2.1, the matched set can be denoted as: $M = (P_i^G,\ P_i^S) = ((x_i^G, y_i^G), (x_i^S, y_i^S))$, $i = 1, 2, 3, \cdots, N$, where the superscripts "G" and "S" refer to the GSHHG and GSMS images, respectively, (P_i^G, P_i^S) denotes a pair of matched features and N is the number of matched features.

Giving one landmark $P_i^G\ (x_i^G, y_i^G)$ in the GSHHG image, the n nearest landmarks can be represented as $NG = \left\{P_{ij}^G\left(x_{ij}^G, y_{ij}^G\right), j = 1, 2, 3, \cdots, n\right\}$ and their corresponding points in the GSMS image are represented as $NS = \left\{P_{ij}^S\left(x_{ij}^S, y_{ij}^S\right), j = 1, 2, 3, \cdots, n\right\}$. Their offsets are represented as $D = \left\{(Dx_{ij}, Dy_{ij}), j = 1, 2, 3, \cdots, n\right\}$ and defined as below:

$$\begin{cases} Dx_{ij} = x_{ij}^G - x_{ij}^S, \\ Dy_{ij} = y_{ij}^G - y_{ij}^S. \end{cases} \tag{4}$$

The neighborhood offsets of the matched feature pair $\left(P_i^G, P_i^S\right)$ can be formulated as:

$$\begin{cases} Dx_i = \sum \mu_j \cdot Dx_{ij}, \\ Dy_i = \sum \mu_j \cdot Dy_{ij}, \end{cases} \tag{5}$$

where $\mu_j = k \cdot exp(-\frac{\left\|P_{ij}^G - P_i^G\right\|^2}{\sigma^2})$ and is constrained to $\sum \mu_j = 1$. In addition, k is a constant normalizing μ_j. When P_{ij}^G is closer to P_i^G, the scalar weight μ_j assigns higher weights to Dx_{ij} and Dy_{ij}.

The offsets between $P_i^G\ (x_i^G, y_i^G)$ and $P_i^S\ (x_i^S, y_i^S)$ in row and column can be calculated by the following formula:

$$\begin{cases} \triangle x_i = x_i^G - x_i^S, \\ \triangle y_i = y_i^G - y_i^S. \end{cases} \tag{6}$$

For the given matched feature pair $\left(P_i^G, P_i^S\right)$, the neighborhood spatial consistent matching indicates that the $\triangle x_i$ and Dx_i should not deviate too much. Similarly, the $\triangle y_i$ and Dy_i also should be close. This constraint can be determined:

$$\begin{cases} |\triangle x_i - Dx_i| < \delta, \\ |\triangle y_i - Dy_i| < \varepsilon, \end{cases} \tag{7}$$

where δ and ε are two thresholds controlling sensitivity on deformations. If their values are large, the incorrect matched features are more likely to be regarded as inliers. They are both set to 0.5 according to experimental results. If $\left(P_i^G, P_i^S\right)$ satisfies the low distortion constraint, it is considered as an inlier.

Figure 2 is the illustration of mismatched features and many-to-one matched features. As shown in Figure 2a, (point 3, point 3′) is a pair of mismatched features. The offsets between them in row and column are −2 and −2. The offsets between other pairs in neighborhood are 1 and 2. Since the offsets of (point 3, point 3′) are over thresholds, they are removed. In Figure 2b, (point 2, point 2′) and (point 3, point 3′) are pairs of many-to-one matched features. The offsets between point 3 and point 3′ in row and column are 2 and 4. The offsets between other pairs in its neighborhood are 1 and 2. (point 3, point 3′) is removed and (point 2, point 2′) is considered an inlier.

The details of the initial matching result and feature refinement in the southern coastal area of Thailand are shown in Figure 3. Figure 3a,c present the details of the top red circles and mauve circles, respectively, in Figure 1d. As shown in Figure 3b,d, these mismatched features are removed.

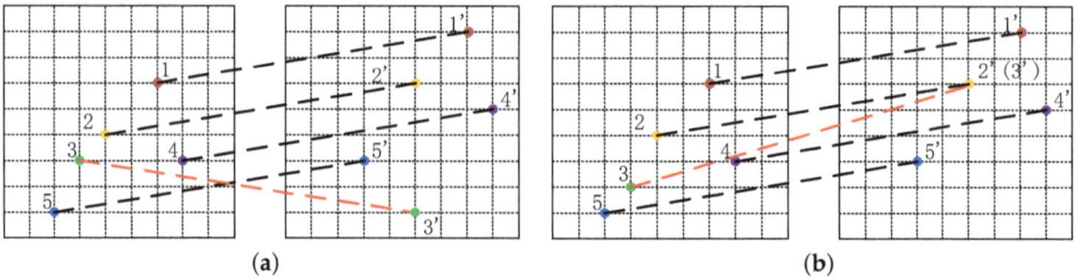

Figure 2. Illustration of mismatched features and many-to-one matched features.(**a**) mismatched features; (**b**) many-to-one matched features.

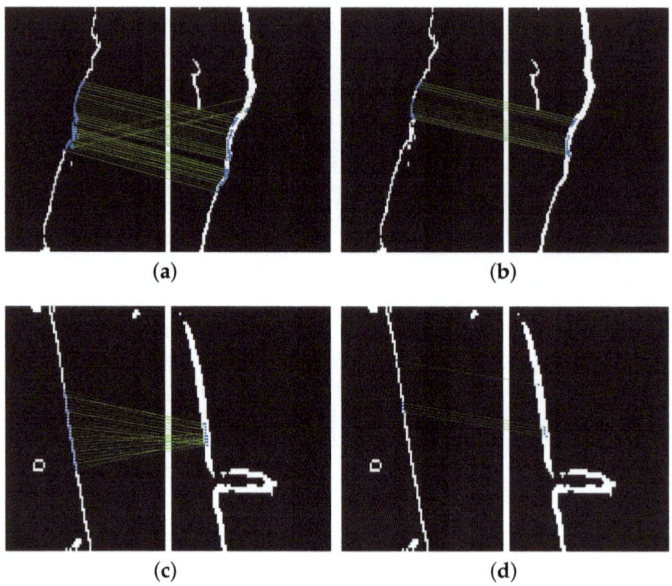

Figure 3. The details of the initial matching and feature refinement in the southern coastal area of Thailand. (**a**) initial matching; (**b**) feature refinement; (**c**) initial matching; and (**d**) feature refinement.

2.3. Pixel Alignment Based on Polynomial Fitting

The earth is not a standard sphere. When using the sphere model to describe the earth, the further the pixel is away from the projection center point, the larger its distance distortion. In this case, the transformation model between sphere and plane is not suitable to describe the projection model of GSMS image.

However, the offsets between the GSHHG and GSMS images in rows and columns are smooth without distortion. For the point $P_i^G\left(x_i^G, y_i^G\right)$ in the GSHHG image and its corresponding point $P_i^S\left(x_i^S, y_i^S\right)$ in the GSMS image, the offsets between them in row and column are presented as $\triangle x_i$ and $\triangle y_i$ according to Equation (6). In order to fit the tendency of offsets in rows and columns, the polynomial functions are applied. Based on the m-th order polynomial function, the fitting functions can be defined as:

$$\begin{cases} f_{\triangle x_i}(x_i^G, y_i^G) = \sum_{k=0}^{m} a_k x_i^{Gk} {y_i}^{G(m-k)} + b_0, \\ f_{\triangle y_i}(x_i^G, y_i^G) = \sum_{k=0}^{m} c_k x_i^{Gk} {y_i}^{G(m-k)} + d_0, \end{cases} \tag{8}$$

where $a_0 \sim a_m$, $c_0 \sim c_m$, b_0 and d_0 are the coefficients treated as the independent variables.

The point P_i^G (x_i^G, y_i^G) in the matched set and its corresponding $\triangle x_i$ are used to estimate the coefficients of polynomial fitting function $f_{\triangle x_i}(x_i^G, y_i^G)$. The correlation coefficient and RMSE are considered to select the optimal coefficients. The fitted function $f_{\triangle x_i}(x_i^G, y_i^G)$ presents the offsets in rows changing with the coordinate (x_i^G, y_i^G). Similarly, the coefficients of polynomial fitting function $f_{\triangle y_i}(x_i^G, y_i^G)$ can also be estimated with the point P_i^G (x_i^G, y_i^G) and its corresponding $\triangle y_i$. In addition, the fitted function $f_{\triangle y_i}(x_i^G, y_i^G)$ describes the offsets in columns changing with the coordinate.

The offsets of pixels between the GSHHG and GSMS images can be obtained by the polynomial fitting functions $f_{\triangle x_i}(x_i^G, y_i^G)$ and $f_{\triangle y_i}(x_i^G, y_i^G)$. For each pixel (x_i^G, y_i^G) in the GSHHG image, the relationship between it and its corresponding point $(x_i^{S'}, y_i^{S'})$ in the GSMS image can be calculated as:

$$\begin{cases} x_i^{S'} = x_i^G - f_{\triangle x_i}(x_i^G, y_i^G), \\ y_i^{S'} = y_i^G - f_{\triangle y_i}(x_i^G, y_i^G). \end{cases} \tag{9}$$

For each pixel in the GSHHG image, the latitude and longitude information is already known. Polynomial fitting functions align all pixels of GSHHG with GSMS images globally. Therefore, the latitude and longitude of all pixels in the GSMS image can be obtained.

3. Results and Discussion

3.1. Dataset and Evaluation Criteria

The remote sensing images used in this experiment are from the FengyunII D meteorological satellite whose sub-satellite point is near (86°E, 0°N). Concerning radial distortion, only landmarks located within ±60° of longitude and ±60° of latitude around sub-satellite point are chosen as reference data. The size of GSMS image is normalized to $10,000 \times 10,000$ pixels. Considering efficiency, both the GSHHG and GSMS images are divided into patches [35–37] whose size is $S1 \times S2$ pixels. Furthermore, feature points are matched in each pair of patches. Some shorelines can not be detected in the GSMS image due to the occlusion of clouds, causing difficulty in matching these shorelines. To reduce this difficulty, 25 patches with relatively more edges in the GSMS image are selected to perform the local feature matching and feature refinement with NSCM.

To evaluate the performance, the ground truth is manually selected from the points with the maximum gradient within their neighborhood. For each landmark in the GSHHG image, we find its corresponding point in the GSMS image as accurately as possible. Since the ground truth is labelled manually, there may be very small errors. If the distance between ground truth and matched point is no bigger than one pixel, this matched point is considered to be correct. Special attention is needed so that our manually labelled ground truth does not contain those landmarks under the clouds and fogs in the GSMS image.

In our experiments, three evaluation criteria including precision, recall and RMSE are mainly used:

$$precision = \frac{N_{inliers}}{N_{inliers} + N_{outliers}},$$
$$recall = \frac{N_{inliers}}{N_{groundtruth}}, \quad (10)$$
$$RMSE = \sqrt{\frac{1}{N_p}\sum_{i=1}^{N_p}[\|P_i - P_i'\|]^2},$$

where $N_{inliers}$ represents the number of inliers in the matched set, $N_{outliers}$ represents the number of outliers in the matched set, $N_{groundtruth}$ represents the number of points of the ground truth, N_p represents the number of matched pairs, P_i represents the matched points and P_i' represents the matched points of the ground truth in the GSMS image.

3.2. Local Feature Matching by Geometric Coding

The size of the template is a key parameter for geometric coding based local feature matching. Figure 4 shows the precision and recall with K varying from 20 to 40. If the size is too small, more points are matched combined with more mismatched points. Therefore, the precision and recall are lower. As K increases, the precision is increasing and finally tends to be stable. If the size is too large, the recall is decreasing since the number of the obtained matched features is decreasing gradually. Considering the tradeoff between precision and recall, K is set to 30 in our experiments.

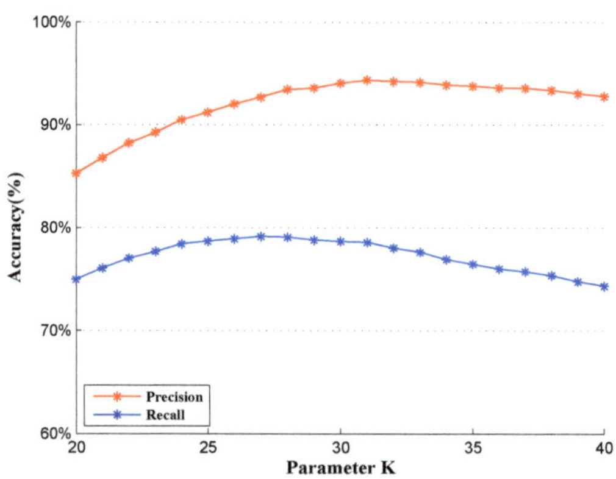

Figure 4. Performance of local feature matching with different Ks.

3.3. Feature Refinement with Neighborhood Spatial Consistent Matching (NSCM)

The NSCM algorithm is applied to remove the outliers caused by similar features and aperture effect. In the NSCM algorithm, the n nearest matched pairs are selected as neighborhood reference pairs. As depicted in Figure 5, with the value of n increasing, more neighborhood spatial consistent information is utilized and more outliers are removed. However, the spatial constraints also become stricter and the recall is decreasing. Considering the tradeoff between precision and recall, the value of n is set as 17 in the feature refinement process.

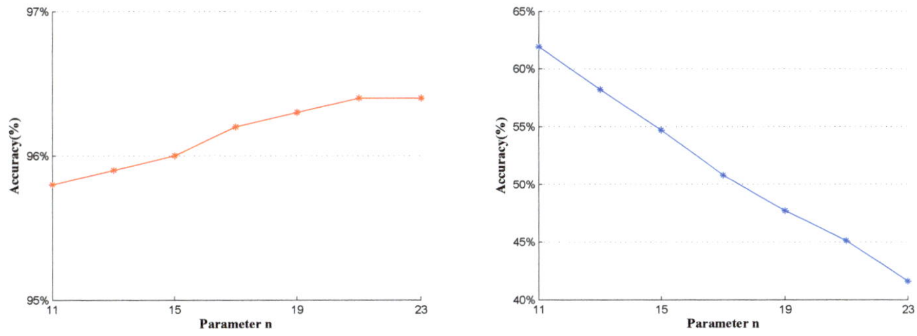

(a) (b)

Figure 5. Mean precision and recall values of feature refinement with different *n*s (the number of candidate matched pairs nearest the seed matched pair). (**a**) mean precision; and (**b**) mean recall.

3.4. Comparison among Feature Matching Algorithms

The proposed NSCM approach is compared with seven refinement algorithms: RANSAC [14], GTM [18], WGTM [20], RSOC [21], KNN-TAR [22], ISSC [23] and RFVTM [24]. Figure 6 presents the performance of these eight algorithms. In addition, the mean of experimental results are shown in Table 1. Table 1 indicates that the average precision of NSCM is the highest and the recall of NSCM algorithm ranks as medium. However, the subsequent processing can improve our recall on the basis of high precision. The RMSE value of NSCM is the smallest as shown in Table 1.

Table 1. Mean precision, recall and RMSE values in NSCM, RANSAC, GTM, WGTM, RSOC, KNN-TAR, ISSC and RFVTM.

Evaluation Criteria	NSCM	RANSAC	GTM	WGTM	RSOC	KNN-TAR	ISSC	RFVTM
precision (%)	96.2	95.2	95.3	96.0	94.2	95.7	95.6	95.2
recall (%)	50.8	61.9	49.5	67.4	61.8	42.8	47.4	63.7
RMSE (pixel)	1.14	1.18	1.15	1.16	1.40	1.34	1.38	1.44
time (s)	0.48	1.08	18.12	16.21	10.92	2.91	2.74	1.89

As shown in Table 1, NSCM significantly outperforms the other algorithms with respect to time efficiency. Assuming that there would be N feature pairs in the matched results. In this paper, n is set to 17, which is much smaller than N. Computation complexity of NSCM is $O(n \times N^2) = O(N^2)$.

3.5. Pixel Alignment Based on Polynomial Fitting

Based on the matched set obtained by feature refinement with NSCM, the offsets between the GSHHG and GSMS images in rows and columns are fitted. The Interpolant, Lowess and Polynomial fitting types are used to get an optimal solution by comparing their precision, recall and RMSE. Table 2 shows the statistical results of the three common fitting functions. The precision of Polynomial fitting is slightly higher compared with Interpolant fitting and Lowess fitting. The recall of Polynomial fitting is far larger than the others, and the RMSE is slightly smaller than the others. In conclusion, the Polynomial fitting outperforms the other methods in all evaluation criteria.

Table 2. Mean precision, recall and RMSE values in Interpolant fitting, Lowess fitting and Polynomial fitting with m set to 3.

Evaluation Criteria	Interpolant Fitting	Lowess Fitting	Polynomial Fitting
precision (%)	92.9	92.9	93.0
recall (%)	68.2	57.5	91.2
RMSE (pixel)	2.33	2.45	2.06

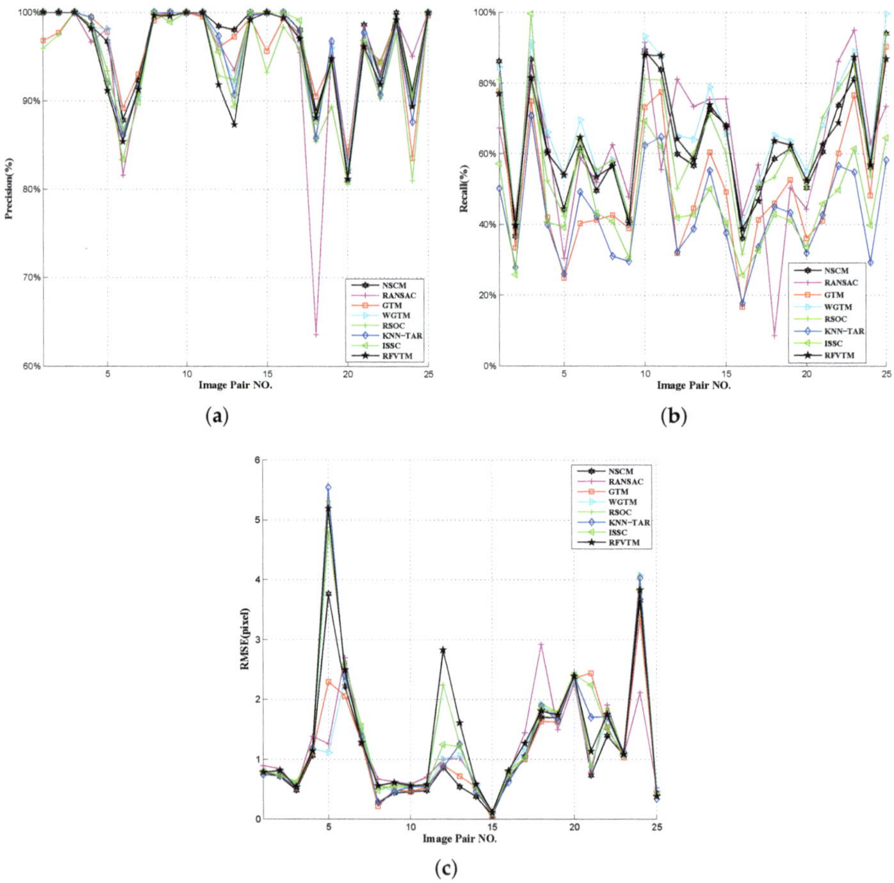

(a) (b) (c)

Figure 6. Performance of eight algorithms on 25 images. NSCM is competitive with RANSAC, GTM, WGTM, RSOC, KNN-TAR, ISSC and RFVTM in precision, recall and RMSE. (**a**) precision; (**b**) recall; (**c**) RMSE.

Figure 7 shows the results of Polynomial fitting functions with different order m from 1 to 5. As shown in Figure 7a,b, when m is smaller, the precision and recall are lower due to under-fitting. However, high-order polynomial leads to over-fitting. When m becomes large, the precision and recall suddenly become very low, but the RMSE becomes very high. Therefore, the third-order Polynomial fitting functions are utilized to fit the offsets' tendency.

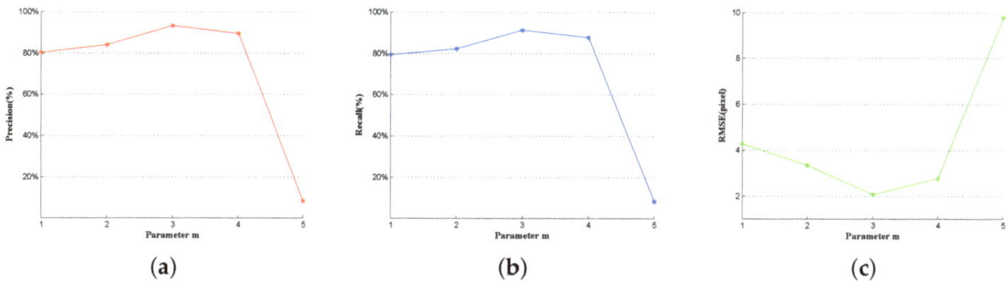

Figure 7. Mean precision, recall and RMSE values of Polynomial fitting with different ms (the order of Polynomial function). (a) mean precision; (b) mean recall; and (c) mean RMSE.

Table 3 gives the three mean values including precision, recall and RMSE before and after pixel alignment. The values of precision are close, but the recall after pixel alignment increases greatly. Figure 8 shows the result of landmark alignment. All pixels in the GSMS remote sensing image are precisely located.

Table 3. Mean precision, recall and RMSE values before and after pixel alignment.

	Before	After
precision (%)	96.2	93.0
recall (%)	50.8	91.2
RMSE (pixel)	1.14	2.06

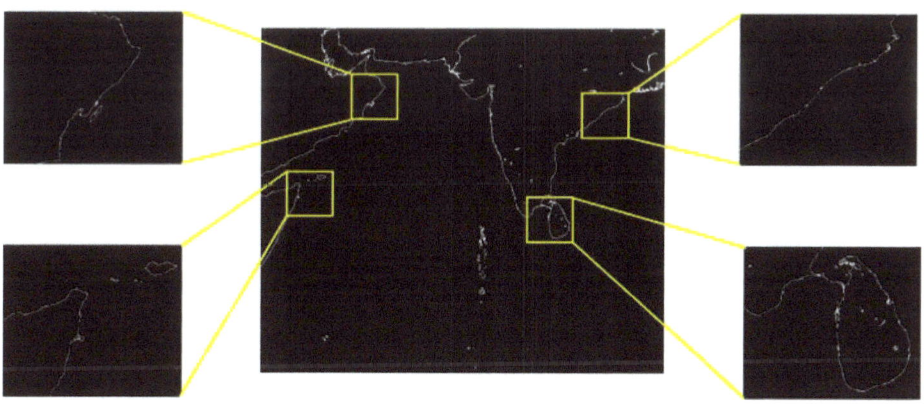

Figure 8. Pixel alignment results.

With pixel alignment, the latitude and longitude of all pixels in the GSMS image can be calculated. For each pixel p_i, the intensity and longitude α_i, latitude γ_i are achieved by NSCM and Polynomial fitting. The coordinate of p_i in the sub-satellite-based earth coordinate system can be represented as:

$$\begin{cases} X_i = Rsin(\gamma_i - \gamma_0)cos(\alpha_i - \alpha_0), \\ Y_i = Rsin(\gamma_i - \gamma_0)sin(\alpha_i - \alpha_0), \\ Z_i = Rcos(\gamma_i - \gamma_0), \end{cases} \tag{11}$$

where R is the radius of the earth; α_0 and γ_0 are the longitude and latitude of the sub-satellite point. With the coordinate (X_i, Y_i, Z_i) and intensity, the GSMS image can be displayed as a 3D earth as shown in Figure 9.

Figure 9. 3D earth.

4. Conclusions

In this paper, we implement global alignment of all pixels in the GSMS images. Before global alignment, we do feature match between the landmarks of GSHHG and the edges of the GSMS images by geometric and gradient similarity measurement. Using spatial consistency of the matched pairs, feature refinement with a neighborhood spatial consistent matching algorithm is proposed to remove outliers. According to the experimental results, compared with other methods, our algorithm can achieve higher accuracy and lower RMSE while its time cost is significantly less than other methods. Based on polynomial fitting, global pixel alignment is applied to obtain the latitude and longitude of all pixels in the GSMS images and improve the recall significantly. The future work will focus on three-dimensional spherical stitching of multi-view remote sensing images.

Acknowledgments: We would like to thank professor Tian Qi (the University of Texas at San Antonio) for his suggestions for this project. We also gratefully acknowledge the support of the National Satellite Meteorological center. This work is supported by the National Natural Science Foundation of China (61572307).

Author Contributions: Dan Zeng supervised and designed the research work, in addition to writing the manuscript; Rui Fang, Shiming Ge and Shuying Li performed the experiments, and participated in experimental designs and data processing; and Zhijiang Zhang helped with writing and revisions.

Conflicts of Interest: The authors declare no conflict of interest.

References

1. Zitova, B.; Flusser, J. Image registration methods: A survey. *Image Vis. Comput.* **2003**, *21*, 977–1000.
2. Govindarajulu, S.; Reddy, K.N.K. Image Registration on satellite Images. *IOSR-JECE* **2012**, *3*, 10–17.
3. Xing, C.; Qiu, P. Intensity-based image registration by nonparametric local smoothing. *IEEE Trans. Pattern Anal. Mach. Intell.* **2011**, *33*, 2081–2092.
4. Jung, J.S.; Song, J.H.; Kwag, Y.K. High precision automatic geocoding method of SAR image using GSHHS. In Proceedings of the 2011 3rd International Asia-Pacific Conference on Synthetic Aperture Radar (APSAR), Seoul, Korea, 26–30 September 2011; pp. 1–4.
5. Jianbin, X.; Wen, H.; Zhe, L.;Yirong, W.; Maosheng, X. The study of rough-location of remote sensing image with coastlines. In Proceedings of the 2003 IEEE International Geoscience and Remote Sensing Symposium (IGARSS'03), Toulouse, France, 21–25 July 2003; Volume 6, pp. 3964–3966.
6. Liu, X.; Tian, Z.; Leng, C.; Duan, X. Remote sensing image registration based on KICA-SIFT descriptors. In Proceedings of the 2010 Seventh International Conference on Fuzzy Systems and Knowledge Discovery (FSKD), Singapore, 10–12 August 2010; Volume 1, pp. 278–282.

7. Wang, G.-H.; Zhang, S.-B.; Wang, H.B.; Li, C.-H.; Tang, X.-M.; Tian, J.J.; Tian, J. An algorithm of parameters adaptive scale-invariant feature for high precision matching of multi-source remote sensing image. In Proceedings of the 2009 Joint Urban Remote Sensing Event, Shanghai, China, 20–22 May 2009; pp. 1–7.
8. Fan, B.; Huo, C.; Pan, C.; Kong, Q. Registration of optical and SAR satellite images by exploring the spatial relationship of the improved SIFT. *IEEE Geosci. Remote Sens. Lett.* **2013**, *10*, 657–661.
9. Wang, X.; Li, Y.; Wei, H.; Liu, F. An ASIFT-based local registration method for satellite imagery. *Remote Sens.* **2015**, *7*, 7044–7061.
10. Lowe, D.G. Distinctive image features from scale-invariant keypoints. *Int. J. Comput. Vis.* **2004**, *60*, 91–110.
11. Wang, Q.; Zhu, G.; Yuan, Y. Statistical quantization for similarity search. *Comput. Vis. Image Underst.* **2014**, *124*, 22–30.
12. Goncalves, H.; Corte-Real, L.; Goncalves, J.A. Automatic image registration through image segmentation and SIFT. *IEEE Trans. Geosci. Remote Sens.* **2011**, *49*, 2589–2600.
13. Ma, J.; Chan, J.C.W.; Canters, F. Fully automatic subpixel image registration of multiangle CHRIS/Proba data. *IEEE Trans. Geosci. Remote Sens.* **2010**, *48*, 2829–2839.
14. Fischler, M.A.; Bolles, R.C. Random sample consensus: A paradigm for model fitting with applications to image analysis and automated cartography. *Commun. ACM* **1981**, *24*, 381–395.
15. Zhou, W.; Lu, Y.; Li, H.; Song, Y.; Tian, Q. Spatial coding for large scale partial-duplicate web image search. In Proceedings of the 18th ACM International Conference on Multimedia, Firenze, Italy, 25–29 October 2010; pp. 511–520.
16. Zhou, W.; Li, H.; Lu, Y.; Tian, Q. Large scale image search with geometric coding. In Proceedings of the 19th ACM International Conference on Multimedia, Scottsdale, AZ, USA, 28 November–1 December 2011; pp. 1349–1352.
17. Zheng, L.; Wang, S. Visual phraselet: Refining spatial constraints for large scale image search. *IEEE Signal Proc. Lett.* **2013**, *20*, 391–394.
18. Aguilar, W.; Frauel, Y.; Escolano, F.; Martinez-Perez, M. E.; Espinosa-Romero, A.; Lozano, M.A. A robust graph transformation matching for non-rigid registration. *Image Vis. Comput.* **2009**, *27*, 897–910.
19. Shi, Q.; Ma, G.; Zhang, F.; Chen, W.; Qin, Q.; Duo, H. Robust image registration using structure features. *IEEE Geosci. Remote Sens. Lett.* **2014**, *11*, 2045–2049.
20. Izadi, M.; Saeedi, P. Robust weighted graph transformation matching for rigid and nonrigid image registration. *IEEE Trans. Image Proc.* **2012**, *21*, 4369–4382.
21. Liu, Z.; An, J.; Jing, Y. A simple and robust feature point matching algorithm based on restricted spatial order constraints for aerial image registration. *IEEE Trans. Geosci. Remote Sens.* **2012**, *50*, 514–527.
22. Zhang, K.; Li, X.Z.; Zhang, J.X. A robust point-matching algorithm for remote sensing image registration. *IEEE Geosci. Remote Sens. Lett.* **2014**, *11*, 469–473.
23. Jiang, J.; Shi, X. A Robust Point-Matching Algorithm Based on Integrated Spatial Structure Constraint for Remote Sensing Image Registration. *IEEE Geosci. Remote Sens. Lett.* **2016**, *13*, 1716–1720.
24. Zhao, M.; An, B.; Wu, Y.; Van Luong, H.; Kaup, A. RFVTM: A Recovery and Filtering Vertex Trichotomy Matching for Remote Sensing Image Registration. *IEEE Trans. Geosci. Remote Sens.* **2017**, *55*, 375–391.
25. Battiato, S.; Bruna, A.R.; Puglisi, G. A robust block-based image/video registration approach for mobile imaging devices. *IEEE Trans. Multimed.* **2010**, *12*, 622–635.
26. Elibol, A. A Two-Step Global Alignment Method for Feature-Based Image Mosaicing. *Math. Comput. Appl.* **2016**, *21*, 30.
27. Adams, A.; Gelfand, N.; Pulli, K. Viewfinder Alignment. In *Computer Graphics Forum*; Blackwell Publishing Ltd.: Oxford, UK, 2008; Volume 27, pp. 597–606.
28. Learned-Miller, E.G. Data driven image models through continuous joint alignment. *IEEE Trans. Pattern Anal. Mach. Intell.* **2006**, *28* 236–250.
29. Cox, M.; Sridharan, S.; Lucey, S.; Cohn, J. Least squares congealing for unsupervised alignment of images. In Proceedings of the 2008 IEEE Conference on Computer Vision and Pattern Recognition (CVPR 2008), Anchorage, AK, USA , 23–28 June 2008; pp. 1–8.
30. Vedaldi, A.; Guidi, G.; Soatto, S. Joint data alignment up to (lossy) transformations. In Proceedings of the 2008 IEEE Conference on Computer Vision and Pattern Recognition (CVPR 2008), Anchorage, AK, USA , 23–28 June 2008; pp. 1–8.

31. Tang, F.; Zou, X.; Yang, H.; Weng, F. Estimation and correction of geolocation errors in FengYun-3C Microwave Radiation Imager Data. *IEEE Trans. Geosci. Remote Sens.* **2016**, *54*, 407–420.
32. Wang, Q.; Zou, C.; Yuan, Y.; Lu, H.; Yan, P. Image registration by normalized mapping. *Neurocomputing* **2013**, *101*, 181–189.
33. Wang, Q.; Yuan, Y.; Yan, P.; Li, X. Saliency detection by multiple-instance learning. *IEEE Trans. Cybern.* **2013**, *43*, 660–672.
34. Dollár, P.; Zitnick, C.L. Structured forests for fast edge detection. In Proceedings of the 2013 IEEE International Conference on Computer Vision (ICCV), Sydney, Australia, 1–8 December 2013; pp. 1841–1848.
35. Gao, J.; Kim, S.J.; Brown, M.S. Constructing image panoramas using dual-homography warping. In Proceedings of the 2011 IEEE Conference on Computer Vision and Pattern Recognition (CVPR), Colorado Springs, CO, USA, 20–25 June 2011; pp. 49–56.
36. Zaragoza, J.; Chin, T.J.; Brown, M.S.; Suter, D. As-projective-as-possible image stitching with moving DLT. In Proceedings of the IEEE Conference on Computer Vision and Pattern Recognition (CVPR), Portland, OR, USA, 23–28 June 2013; pp. 2339–2346.
37. Chang, C.H.; Sato, Y.; Chuang, Y.Y. Shape-preserving half-projective warps for image stitching. In Proceedings of the 2014 IEEE Conference on Computer Vision and Pattern Recognition (CVPR), Toronto, ON, Canada, 23–28 June 2014; pp. 3254–3261.

remote sensing

MDPI

Article

Exploiting Deep Matching and SAR Data for the Geo-Localization Accuracy Improvement of Optical Satellite Images

Nina Merkle [1,*], Wenjie Luo [2], Stefan Auer [1], Rupert Müller [1] and Raquel Urtasun [2]

[1] German Aerospace Center (DLR), Remote Sensing Technology Institute, 82234 Wessling, Germany; stefan.auer@dlr.de (S.A.); rupert.mueller@dlr.de (R.M.)

[2] Department of Computer Science University of Toronto, Toronto, ON M5S 3G, Canada; wenjie@cs.toronto.edu (W.L.); urtasun@cs.toronto.edu (R.U.)

* Correspondence: nina.merkle@dlr.de; Tel.: +49-8153-28-2165

Academic Editors: Qi Wang, Nicolas H. Younan, Carlos López-Martínez and Prasad S. Thenkabail
Received: 20 March 2017; Accepted: 29 May 2017; Published: 10 June 2017

Abstract: Improving the geo-localization of optical satellite images is an important pre-processing step for many remote sensing tasks like monitoring by image time series or scene analysis after sudden events. These tasks require geo-referenced and precisely co-registered multi-sensor data. Images captured by the high resolution synthetic aperture radar (SAR) satellite TerraSAR-X exhibit an absolute geo-location accuracy within a few decimeters. These images represent therefore a reliable source to improve the geo-location accuracy of optical images, which is in the order of tens of meters. In this paper, a deep learning-based approach for the geo-localization accuracy improvement of optical satellite images through SAR reference data is investigated. Image registration between SAR and optical images requires few, but accurate and reliable matching points. These are derived from a Siamese neural network. The network is trained using TerraSAR-X and PRISM image pairs covering greater urban areas spread over Europe, in order to learn the two-dimensional spatial shifts between optical and SAR image patches. Results confirm that accurate and reliable matching points can be generated with higher matching accuracy and precision with respect to state-of-the-art approaches.

Keywords: geo-referencing; multi-sensor image matching; Siamese neural network; satellite images; synthetic aperture radar

---

1. Introduction

1.1. Background and Motivation

Data fusion is important for several applications in the fields of medical imaging, computer vision or remote sensing, allowing the collection of complementary information from different sensors or sources to characterize a specific object or an image. In remote sensing, the combination of multi-sensor data is crucial, e.g., for tasks such as change detection, monitoring or assessment of natural disasters. The fusion of multi-sensor data requires geo-referenced and precisely co-registered images, which are often not available.

Assuming the case of multi-sensor image data where one of the images exhibits a higher absolute geo-localization accuracy, image registration techniques can be employed to improve the localization accuracy of the second image. Images captured by high resolution synthetic aperture radar (SAR) satellites like TerraSAR-X [1] exhibit an absolute geo-localization accuracy in the order of a few decimeters or centimeter for specific targets [2]. Such accuracy is mainly due to the availability of precise orbit information and the SAR imaging principle. Radar satellites have active sensors onboard (emitting electromagnetic signals) and capture images day and night independently from local weather conditions. The principle of synthetic aperture radar relates to collecting backscattered signal energy for ground objects along the sensor flight path and compressing the signal energy in post-processing for a significant increase of the spatial resolution [3]. The visual interpretation of SAR images is a challenging task [4]: the SAR sensor looks sideways (angle typically between 25° to 60° with respect to nadir direction) to be able to solve ambiguities in azimuth related to the targets on ground.

Contrary to radar systems that measure the signal backscattered from the reflecting target to the sensor, optical satellite sensors are passive systems that measure the sunlight reflected from ground objects with a strong dependence on atmospheric and local weather conditions such as cloud and haze. Due to a different image acquisition concept with respect to SAR satellites (active vs. passive sensor), the location accuracy of optical satellites also depends on a precise knowledge of the satellite orientation in space. Inaccurate measurements of the attitude angles in space are the main reason for a lower geo-localization accuracy of optical satellite data. For example the absolute geo-localization accuracy of images from optical satellites like Worldview-2, PRISM or QuickBird ranges from 4 to 30 m. TerraSAR-X images may therefore be employed to improve the localization accuracy of spatially high resolution optical images with less than 5 m ground resolution.

The aim of enhancing the geo-localization accuracy of optical images could be achieved by employing ground control points (GCPs). GCPs can be extracted from high resolution reference images, e.g., from TerraSAR-X, to correctly model the generation process of optical images from the focal plane location of the instrument pixel to the Earth surface location in terms of Earth bound coordinate frames. In Reinartz et al. [5] promising results are archived by using GCPs extracted from high precision orthorectified TerraSAR-X data. Nevertheless, the problem of multi-sensor image to image registration is challenging, and in the specific the precise registration of images from radar and optical sensors is an open problem.

Due to the different acquisition concepts (SAR: synthetic aperture with distance measurements; optical: perspective projection), viewing perspectives (off-nadir; usually near-nadir), wavelengths (radar signal wavelength in cm; optical wavelength in nm) and the speckle effect in SAR images, it is difficult to find complementary features or reliable similarity measures when comparing optical and SAR images. More precisely, the sideways-looking acquisition of SAR sensors causes typical geometric distortion effects (layover, foreshortening) and shadowing for 3D objects such as buildings or trees. These effects have a strong influence on the appearance of all objects above the ground level in SAR images. As a consequence, the boundary of an elevated object in a SAR image does not fit the object boundary in the optical image, even if the imaging perspective is the same for both sensors. Additionally, the different wavelengths measured by the two kinds of sensors lead to different radiometric properties in the optical and SAR images. This is due to the fact that the response of an object depends on the signal properties (wavelength, polarization), the surface properties (roughness, randomness of local reflectors and reflectance properties) and sensor perspective. The same object may therefore appear with high intensity for one sensor and with low intensity in another. The speckle effect further complicates the human and automatic interpretation of SAR imagery and, hence, the matching of optical and SAR images. As an example, Figure 1 shows the difference of an optical and a high resolution SAR image for a selected scene containing man-made structures and vegetation.

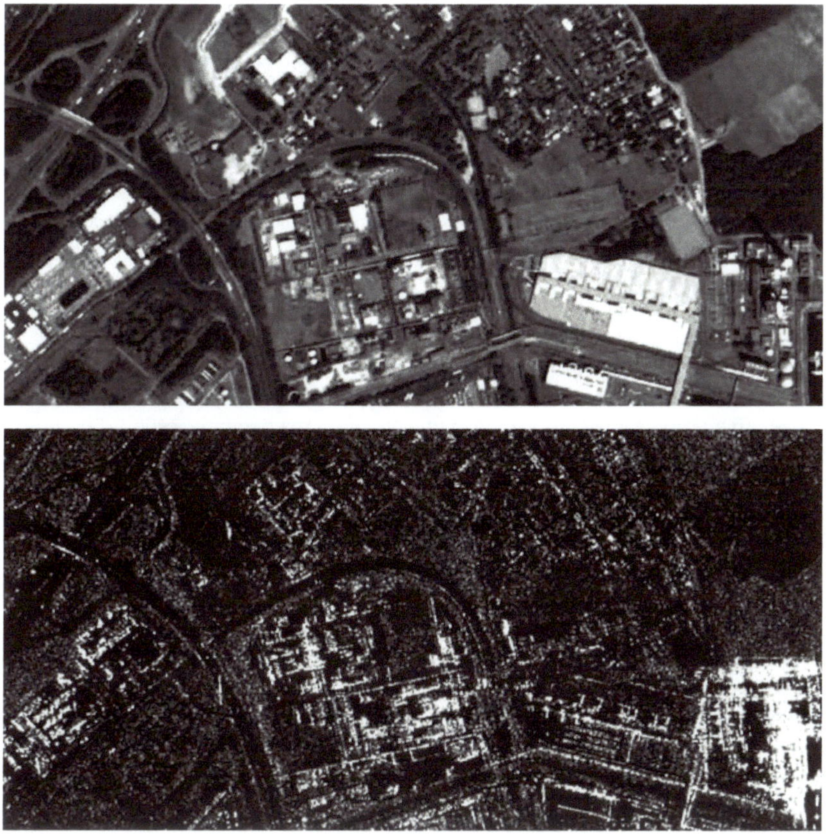

Figure 1. Visual comparison of an optical (**top**) and SAR image (**bottom**) acquired over the same area. Both images have a ground sampling distance of 1.25 m.

1.2. Related Work

To improve the absolute geo-location accuracy of optical satellite images using SAR images as reference, the above-mentioned problems for SAR and optical image registration need to be dealt with. Different research studies investigated the geo-localization accuracy improvement of optical satellite images based on SAR reference data, e.g., [5–7]. The related approaches rely on suitable image registration techniques, which are tailored to the problem of optical and SAR images matching.

The aim of image registration is to estimate the optimal geometric transformation between two images. The most common multi-modal image registration approaches can be divided into two categories. The first category comprises intensity-based approaches, where a transformation between the images can be found by optimizing the corresponding similarity measure. Influenced by the field of medical image processing, similarity measures like normalized cross-correlation [8], mutual information [9,10], cross-cumulative residual entropy [11] and the cluster reward algorithm [12] are frequently used for SAR and optical image registration. A second approach is based on local frequency information and a confidence-aided similarity measure [13]. Li et al. [14] and Ye et al. [15] introduced similarity measures based on the histogram of oriented gradients and the histogram of oriented phase congruency, respectively. However, these approaches are often computationally expensive, suffer from the different radiometric properties of SAR and optical images and are sensitive to speckle in the SAR image.

The second category comprises feature-based approaches, which rely on the detection and matching of robust and accurate features from salient structures. Feature-based approaches are less sensitive to radiometric differences of the images, but have problems in the detection of robust features from SAR images due to the impact of speckle. Early approaches are based on image features like lines [16], contours [17,18] or regions [19]. A combination of different features (points, straight lines, free-form curves or areal regions) is investigated in [20]. The approach shows good performance for the registration of optical and SAR images, but the features from the SAR images have to be selected manually. As the matching between optical and SAR images usually fails using the scale-invariant feature transform (SIFT), Fan [21] introduced a modified version of the algorithm. With the improved SIFT, a fine registration for coarsely-registered images can be achieved, but the approach fails for image pairs with large geometric distortions. To find matching points between area features, a level set segmentation-based approach is introduced in [22]. This approach is limited to images that contain sharp edges from runways, rivers or lakes. Sui et al. [23] and Xu et al. [22] propose iterative matching procedures to overcome the problem of misaligned images caused by imprecise extracted features. In [23], an iterative Voronoi spectral point matching between the line-intersection is proposed, which depends on the presence of salient straight line features in the images.

Other approaches try to overcome the drawbacks of intensity and feature-based approaches by combining them. A global coarse registration using mutual information on selected areas (no dense urban and heterogeneous areas) followed by a fine local registration based on linear features is proposed in [24]. As a drawback, the method highly depends on the coarse registration. If the coarse registration fails, the fine registration will be unreliable.

Besides classical registration approaches, a variety of research studies indicate the high potential of deep learning methods for different applications in remote sensing, such as classification of hyperspectral data [25–27], enhancement of existing road maps [28,29], high-resolution SAR image classification [30] or pansharpening [31]. In the context of image matching, deep matching networks were successfully trained for tasks such as stereo estimation [32,33], optical flow estimation [34,35], aerial image matching [36] or ground to aerial image matching [37]. In [38], a deep learning-based method is proposed to detect and match multiscale keypoints with two separated networks. While the detection network is trained on multiscale patches to identify regions including good keypoints, the description network is trained to match extracted keypoints from different images.

Most of the deep learning image matching methods are based on a Siamese network architecture [39]. The basic idea of these methods is to train a neural network that is composed of two parts: the first part, a Siamese or pseudo-Siamese network, is trained to extract features from image patches, while the second part is trained to measure the similarity between these features. Several types of networks showed a high potential for automatic feature extraction from images, e.g., stacked (denoising) autoencoders [40], restricted Boltzmann machines [41] or convolutional neural networks (CNNs) [42]. From these networks, CNNs have been proven to be efficient for feature extraction and have seen successfully trained for image matching in [32,33,36–38,43–45]. A similarity measure, the L_2 distance [45] or the dot product [32,33], is applied on a fully-connected network [43,44]. The input of the network can be single-resolution image patches [36,43,45], multi-resolution patches [44] or patches that differ in size for the left and right branch of the Siamese network [32,44].

Summarizing, we are tackling the task of absolute geo-location accuracy improvement of optical satellite images by generating few, but very accurate and reliable matching points between SAR and optical images with the help of a neural network. These points serve as input to improve the sensor models for optical image acquisitions. The basis of the approach is a Siamese network, which is trained to learn the spatial shift between optical and SAR image patches. Our network is trained on selected patches where the differences are mostly radiometric, as we try to avoid geometrical ones. The patches for training are semi-manually extracted from TerraSAR-X and PRISM image pairs that capture larger urban areas spread over Europe.

2. Deep Learning for Image Matching

Our research objective is to compute a subset of very accurate and reliable matching points between SAR and optical images. Common optical and SAR image matching approaches are often not applicable to a wide range of images acquired over different cities or at different times of the year. This problem can be handled using a deep learning-based approach. Through training a suitable neural network on a large dataset containing images spread over Europe and acquired at different times of the year, the network will learn to handle radiometric changes of an object over time or at different locations in Europe. To avoid geometrical differences between the SAR and optical patches, we focus our training on patches containing flat surfaces such as streets or runways in rural areas. This is not a strong restriction of our approach as these features frequently appear in nearly every satellite image.

Inspired by the successful use of Siamese networks for the task of image matching, we adopt the same architecture. A Siamese network consists of two parallel networks, which are connected at their output node. If the parameters between the two networks are shared, the Siamese architecture provides the advantage of consistent predictions. As both network branches compute the same function, it is ensured that two similar images will be mapped to a similar location in the feature space. Our Siamese network consists of two CNNs. In contrast to fully-connected or locally-connected networks, a CNN uses filters, which are deployed for the task of feature extraction. Using filters instead of full or local connections reduces the amount of parameters within the network. Less parameters lead to a speed increase in the training procedure and a reduction in the amount of required training data and, hence, reduce the risk of overfitting.

In comparison to common deep learning-based matching approaches, our input images are acquired from different sensors with different radiometric properties. Due to speckle in SAR images, the pre-processing of the images plays an important role during training and for the matching accuracy and precision of the results. Our dataset contains images with a spatial resolution of 2.5 m, and therefore exhibit a lower level of detail in the images compared to the ones used in [32,43–45]. In order to increase the probability of the availability of salient features in the input data, we use large input patches with at least a size of 201 × 201 pixels. The mentioned problems require a careful selection of the network architecture to find the right trade-off between the number of parameters, the number of layers and, more importantly, the receptive field size.

2.1. Dilation

In the context of CNNs, the receptive field refers to the part of the input patches, having an impact on the output of the last convolutional layer. To achieve the whole input patch having an impact on our network output, a receptive field size of 201 × 201 pixels is desired. Standard ways to increase the receptive field size are strided convolutions or pooling (downsampling) layers inside the neural network. Here, the word stride refers to the distance between two consecutive positions of the convolution filters. This would introduce a loss of information as these approaches reduce the resolution of the image features. In contrast, dilated convolutions [46] systematically aggregate information through an exponential growth of the receptive without degradation in resolution. The dilated convolution $*_d$ at a given position p in the image F is defined as:

$$(F *_d k)(p) = \sum_{m=-r}^{r} F(p - d \cdot m)k(m), \tag{1}$$

where k denotes the kernel/filter with size $(2r + 1) \times (2r + 1)$ and d denotes the dilation factor. Instead of looking at local $(2r + 1) \times (2r + 1)$ regions as in the case of standard convolutions, dilated convolutions look at $[d \cdot (2r + 1)] \times [d \cdot (2r + 1)]$ surrounding regions, which lead to an expansion of the receptive field size. Beyond this, dilated convolutions have the same number of network parameters compared to their convolution counterpart.

2.2. Network Architecture

Our matching network is composed of a feature extraction network (a Siamese network) followed by a layer to measure the similarity of the extracted features (the dot product layer). An overview of the network architecture is depicted on the left side of Figure 2. The inputs of the left and right branches of the Siamese network are an optical (left) and a SAR (right) reference image, respectively. The weights of the two branches can be shared (Siamese architecture) or partly shared (pseudo-Siamese architecture).

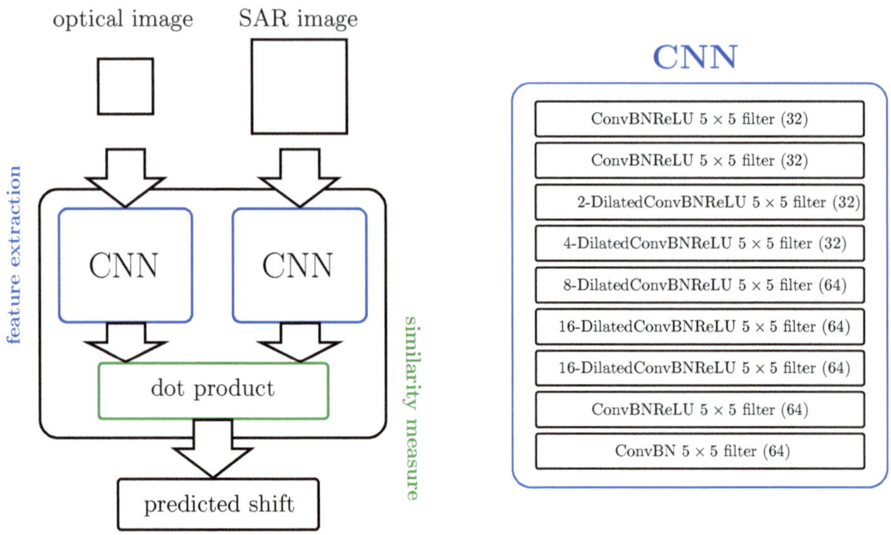

Figure 2. Network architecture (**left**) and a detailed overview of the convolutional layers (**right**). Abbreviations: convolutional neural network (CNN), convolution (Conv), batch normalization (BN) and rectified linear unit (ReLU).

Each layer of the network consists of a spatial convolution (Conv), a spatial batch normalization (BN) [47] and a rectified linear unit (ReLU). The purpose of the convolution layers is to extract spatial features from the input data through trainable filters. The complexity of the features extracted by the layers increases along with the depth. A normalization of the input data is often used as a pre-processing step to increase the learning speed and the performance of the network. By passing the input through the different layers of the network, the distribution of each single layer input changes. Therefore, BN is used in every layer of the network to ensure the consistency in the distribution of the layer inputs, as it provides a form of regularization and reduces the dependency of the network performance on the initialization of the weights. Non-linear activation functions like ReLUs are needed to introduce nonlinearities into the network (otherwise the network can only model linear functions). An Advantage of ReLUs compared to other activation function is a more efficient and faster training of the network.

We removed the ReLU from the last layer to preserve the information encoded in the negative values. In all layers convolutions with a filter size of 5×5 pixels are employed. To overcome the problem of our relatively large input patch size, we adopt dilation convolutions [46] for the layers three to seven with a dilation factor d of 2, 4, 8 and 16 for the last two layers. This setup leads to the desired receptive field size of 201×201 pixels. The number of filters used in layer one to four is 32 and for the others is 64. The overall output is a predicted shift of the optical image within the SAR reference patch and is computed by taking the dot product of the output of the two branches. A detailed overview of one branch of the Siamese network is the depicted on the right side of Figure 2.

2.3. SAR Image Pre-Processing

We use the probabilistic patch-based (PPB) filter proposed in [48] for the pre-processing of the SAR images. This filter is developed to suppress speckle in SAR images by adapting the non-local mean filter by Buades et al. [49] to SAR images. The idea of the non-local mean filter is to estimate the filtered pixel value as the weighted average over all pixels in the image. The weights are measuring the similarity between the pixel values of a patch Δ_s centred around a pixel s and the pixel values of a patch Δ_t centred around a pixel t. The similarity between two patches is estimated through their Euclidean distance. In [48], the noise distribution is modelled using the weighted maximum likelihood estimator, in which the weights express the probability that two patches centred around the pixels s and t have the same noise distribution in a given image. The results of applying this filter and a comparison between SAR and optical patches are shown in Figure 3.

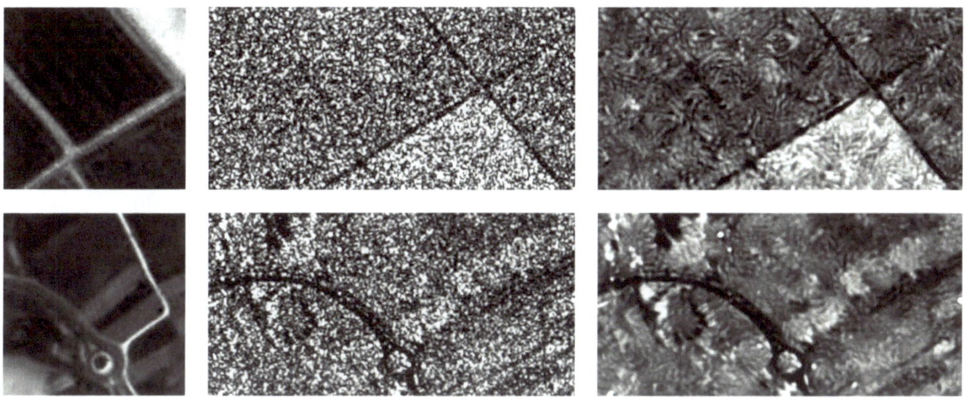

(a) (b) (c)

Figure 3. Visual comparison between optical (**a**), SAR (**b**) and despeckled SAR patches (**c**).

2.4. Matching Point Generation

We generate the matching points by training the network over a large dataset of optical and SAR image patch pairs, which have been manually co-registered. More precisely, the network is trained with smaller left image patches cropped from optical images and larger right images patches cropped from SAR images. Note that given a fixed size $b \times h$ of the left image patch L, the output of the network will depend on the size of the right image patch. The right image patch R has the size $(b+s) \times (h+s)$, where s defines the range over which we perform our search. The output of the network is a two-dimensional scoring map with size $(s+1) \times (s+1)$ over the search space S with size $(b+s) \times (h+s)$.

The scoring map s_i for the i-th input image pair contains a similarity score $s_{i,j}$ for each location $q_{i,j}$ in the search space ($j \in J = \{1, \ldots, |S|\}$, where $|S|$ is the cardinality of S). The search space index J is indexing the two-dimensional search space, where each position $q_{i,j}$ in S corresponds to a specific two-dimensional shift of the left optical patch with respect to the larger SAR patch.

To get the similarity scores for every image pair, we first compute the feature vector f_i for the i-th optical training patch and the feature matrix h_i for the corresponding i-th SAR patch. The feature vector f_i is the output of the left network branches and has a dimension of 64 (as the last convolution layer has 64 filters). The feature matrix h_i is the output of the right network branch with a dimension of $|S| \times 64$ and is composed of the feature vectors $h_{i,j}$ for each location in the search space. We then compute the similarity of the features vectors f_i and $h_{i,j}$ for every position $q_{i,j} \in S$.

To measure the similarity between the two vectors, we use the dot product and obtain the similarity scores $s_{i,j} = f_i \cdot h_{i,j}$ for all $j \in J$. A high value of $s_{i,j}$ indicates a high similarity between the two vectors f_i and $h_{i,j}$ at location $q_{i,j}$ (which is related to a two-dimensional pixel shift). In other words, a high similarity score $s_{i,j}$ indicates a high similarity between the i-th optical patch and the i-th SAR patch at location $q_{i,j}$ in our search space. To get a normalized score over all locations within the search space, we apply the soft-max function at each location $q_{i,j} \in S$:

$$\tilde{s}_{i,j} = \frac{\exp(s_{i,j})}{\sum\limits_{j \in J} \exp(s_{i,j})}. \tag{2}$$

This function is commonly used for multi-class classification problems to compute the probability that a certain training patch belongs to a certain class. In our case, the normalized score $\tilde{s}_{i,j}$ can be interpreted as a probability for the specific shift, which corresponds to location $q_{i,j}$ with index j. Thus, the output of our network (the normalized score map) can be seen as a probability distribution with a probability for every location (shift) of the optical patch within the SAR image patch.

By treating the problem as a multi-class classification problem, where the different classes represent the possible shifts of an optical patch with respect to a larger SAR patch, we train our network by minimizing the cross entropy loss:

$$\min_{w} \sum_{i \in I, j \in J} p_{\text{gt}}(q_{i,j}) \log p_i(q_{i,j}, w) \tag{3}$$

with respect to the weights w, which parametrize our network. Here, $p_i(q_{i,j}, w)$ is the predicted score for sample i at location $q_{i,j}$ in our search space, and p_{gt} is the ground truth target distribution. Instead of a delta function with non-zero probability mass only at the correct location $q_{i,j} = q_i^{\text{gt}}$, we are using a soft ground truth distribution, which is centred around the ground truth location. Therefore, we set p_{gt} to be the discrete approximation of the Gaussian function (with $\sigma = 1$) in an area around q_i^{gt}:

$$p_{\text{gt}}(q_{i,j}) = \begin{cases} \frac{1}{2\pi} \cdot e^{-\frac{\left\|q_{i,j} - q_i^{\text{gt}}\right\|_2^2}{2}} & \text{if} \quad \left\|q_{i,j} - q_i^{\text{gt}}\right\|_2 < 3 \\ 0 & \text{otherwise} \end{cases}, \tag{4}$$

where $\|\cdot\|_2$ denotes the L_2 (Euclidean) distance. We use stochastic gradient descent with Adam [50] to minimize our loss function (3) and, hence, to train our network to learn the matching between optical and SAR patches.

After training, we keep the learned parameters w fixed and decompose the network into two parts: the feature extractor (CNN) and the similarity measure (dot product layer). As the feature extractor is convolutional, we can apply the CNN on images with an arbitrary size. Thus, during the test time, we first give an optical patch as input to the CNN and compute the feature vector f. Then we consider a larger SAR patch which covers the desired search space, and compute the feature matrix h. Afterwards, we use the dot product layer to compute the normalized score map from f and h (in the same way as for the training step). Applying this strategy, we can compute a matching score between optical patches with arbitrary size and SAR images over an arbitrary search space. We obtain the matching points (predicted shifts) by picking for every input image pair the points with the highest value (highest similarity between optical and SAR patch) within the corresponding search space.

2.5. Geo-Localization Accuracy Improvement

The inaccuracy of the absolute geo-localization of the optical satellite data in the geo-referencing process arises mainly from inaccurate measurements of the satellite attitude and thermally-affected mounting angles between the optical sensor and the attitude measurement unit. This insufficient pointing knowledge leads to local geometric distortions of orthorectified images caused by the height variations of the Earth's surface. To achieve higher geometric accuracy of the optical data, ground control information is needed to adjust the parameters of the physical sensor model. We are following the approach described in [51] to estimate the unknown parameters of the sensor model from GCPs by iterative least squares adjustment. In order to get a reliable set of GCP, different levels of point filtering and blunder detection are included in the processing chain. In contrast to [51], where the GCPs are generated from an optical image, we are using the matching points generated by our network.

3. Experimental Evaluation and Discussion

To perform our experiments, we generated a dataset out of 46 orthorectified optical (PRISM) and radar (TerraSAR-X acquired in stripmap mode) satellite image pairs acquired over 13 city areas in Europe. The images include suburban, industrial and rural areas with a total coverage of around $20,000\,km^2$. The spatial resolution of the optical images is 2.5 m, and the pixel spacing of the SAR images is 1.25 m. To have a consistent pixel spacing within the image pairs, we downsampled the SAR images to 2.5 m using bilinear interpolation.

As the ground truth, we are using optical images which were aligned to the corresponding SAR images in the Urban Atlas project [52]. The alignment between the images was achieved by a manual selection of several hundred matching points for every image pair. These matching points are used to improve the sensor model related to the optical images. By using the improved sensor models to orthorectify the optical images, the global alignment error could be reduced from up to 23 m to around 3 m in this project.

To minimize the impact of the different acquisition modes of PRISM and TerraSAR-X, we focus on flat surfaces where only the radiometry between the SAR and optical images is different. Therefore, patches are favored that contain parts of streets or runways in rural areas. The patches are pre-selected using the CORINE land cover [53] from the year 2012 to exclude patches, e.g., containing street segments in city areas. The CORINE layer includes 44 land cover classes and has a pixel size of 100 m. For the pre-selection, the following classes are chosen: airports, non-irrigated arable land, permanently-irrigated land, annual crops associated with permanent crops and complex cultivation patterns, land principally occupied by agriculture, with significant areas of natural vegetation. Note that there are several current global land cover maps available, which enable a similar pre-selection for images outside Europe. The pre-selection was refined manually to ensure that the patches contain streets/runways segments that are visible in the optical and the SAR patches and to avoid patches containing street segments through smaller villages or areas covered by clouds in the optical images.

3.1. Dataset Generation

The training, validation and test datasets are generated by randomly splitting the 46 images into 36 images for training, 4 for validation and 6 for testing. As a form of data augmentation, we use bilinear interpolation to downsample the optical and SAR images, which are used for training, to a pixel spacing of 3.75 m. This leads to a training set with a total number of 92 images for each sensor, where half of the images have a resolution of 2.5 m and the other half of 3.75 m. Data augmentation is commonly used to generate a larger training dataset and, hence, to prevent the network from overfitting.

The training, validation and test patches are cropped from the images of the corresponding sets. The optical patches have a size of 201×201 pixels, and the SAR patches have a size of 221×221 pixels. The final dataset contains 135,000 pairs of training patches, 5000 pairs of validation patches and 14,400 pairs of test patches, and the total number of search locations is 441. Note that the alignment error between the SAR and the optical image is expected to be not larger than 32 m. Therefore, a 21×21 pixel search space with a pixel spacing of 2.5 m in the validation and test case is assumed to be large enough.

3.2. Training Parameters

Our network is trained with 100 rounds, where each round takes 200 iterations over a single batch. The initial learning rate is set to 0.01, and we reduce it by a factor of five at iterations 60 and 80. We train the network in parallel on two Titan X GPUs using a batch size of 100. The weights of the network are initialized with the scheme described in [54], which particularly considers the rectifier nonlinearities. The whole training process takes around 30 h.

3.3. Influence of Speckle Filtering

To find the right setup, we investigated the influence of speckle filtering during training time. Figure 4a illustrates the matching accuracy of the validation set during training with two different network architectures and with and without the speckle filter. Here, the matching accuracy is measured as the percentage of matching points, where the Euclidean (L_2) distance to the ground truth location is less than or equal to 3 pixels. Figure 4b illustrates the average L_2 distance of the matching points to the ground truth location of the validation set in the training. Both images reveal that, independently from the network architectures, speckle filtering helps the network at learning the similarity between optical and SAR patches and, hence, at improving the accuracy of the generated matching points.

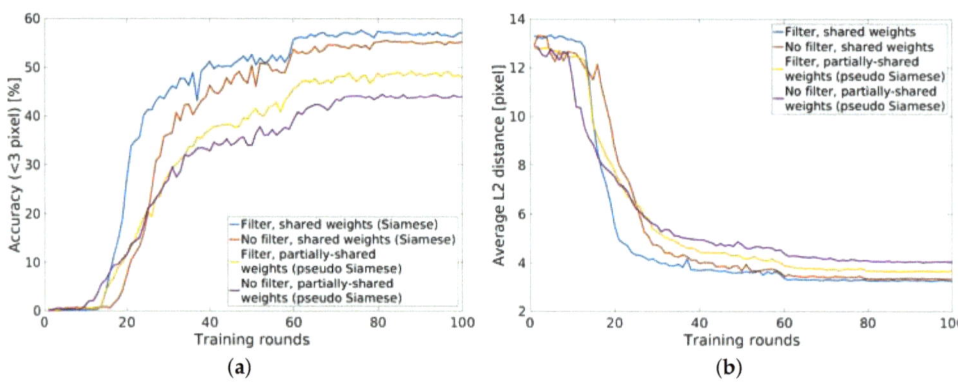

Figure 4. Influence of the speckle filter and comparison of different network architectures during training time (all results are generated from the validation set): (**a**) shows the matching accuracy during training. Here, the matching accuracy is measured as the percentage of matching points, where the L_2 distance to the ground truth location is less than or equal to three pixels; (**b**) shows the average L_2 distance between the matching points and the ground truth location during training.

3.4. Comparison of Network Architectures

The influence of partially-shared (pseudo-Siamese architecture) and shared weights (Siamese architecture) between the two network branches during training was investigated. In the case of the pseudo-Siamese architectures, the weights of the first three layers are different, whereas the remaining layers share their weights. In the case of the Siamese architectures, all weights are shared. Figure 4 shows a comparison of the matching accuracy between the results of Siamese and pseudo-Siamese architecture over the validation set. It can be seen that a full Siamese architecture learns slightly faster and achieves higher matching accuracy in the end. In the following, the results are generated with the best setup: speckle filtering combined with a Siamese architecture.

3.5. Comparison to Baseline Methods

For a better evaluation of our results, we compare our method with three available baseline methods: the similarity measure normalized cross-correlation (NCC) [55], the similarity measure mutual information (MI) [56], and a MI-based method (CAMRI) which is tailored to the problem of optical and SAR matching [10]. To ensure a fair comparison, we applied the pre-processing with the speckle filter [48] to all baseline methods, except for CAMRI [10]. Here, a slightly different speckle filter is implemented internally. Table 1 shows the comparison of our method with the baseline methods. The expression "Ours (score)" denotes our method, where we used a threshold to detect outliers and to generate more precise and reliable matching points (detailed explanation in the next section). "Ours (scores)" achieves higher matching accuracy and precision than NCC, MI and CAMRI [10]. More precisely, the average value over the L_2 distances between the matching points and the ground truth locations is the smallest (measured in pixel units) for our method. Furthermore, the comparison of the matching precisions reveals that our matching points, with a standard deviation σ of 1.14 pixels, are the most reliable ones. The running time of our method during test time is 3.3 m for all 14,000 test patches on a single GPU. The baseline methods are running on CPU, which makes a fair comparison difficult, but CAMRI [10] requires around three days to compute the matching points for the test set.

Table 1. Comparison of the matching accuracy and precision of our method with accuracies of normalized cross-correlation (NCC), mutual information (MI) and CAMRI [10] over the test set. The matching accuracy is measured as the percentage of matching points, having a L_2 distance to the ground truth location smaller than a specific number of pixels and as the average over the L_2 distances between the predicted matching points and the ground truth locations (measured in pixel units). The matching precision is represented by the standard deviation σ (measured in pixel units).

Methods	Matching Accuracy				Matching Precision
	<2 pixels	<3 pixels	<4 pixels	avg L_2 (pixel)	σ (pixel)
NCC	2.94%	7.92%	13.01%	9.92	4.04
MI	18.18%	38.60%	51.99%	4.89	3.64
CAMRI [10]	33.55%	57.06%	79.93%	2.80	2.86
Ours	25.40%	49.60%	64.28%	3.91	3.17
Ours (score)	49.70%	82.80%	94.70%	1.91	1.14

3.6. Outlier Removal

So far, we used the normalized score (after applying the soft-max) and we selected the locations with the highest value (highest probability) within each search area as the predicted matching point after a two-dimensional shift. Another possibility is to use the raw score (before soft-max) as an indicator of the confidence of the prediction. Utilizing this information, we can aggregate the predictions from the network to detect outliers and achieve higher matching performances. Therefore, we investigated the influence of the raw score as a threshold as shown in Figure 5, which enables the detection of correct predicted matching points. A higher threshold on the raw score leads to a better accuracy in terms of correct prediction, as well as a smaller L_2 distance between the predicted matching points and the ground truth locations. Note that the rough shape at the right side of the curves in Figure 5b,c is the result of an outlier. Here, an outlier has a strong influence, since these numbers are computed from less than 20 test patches.

By using only the first 1000 matches with the highest raw score, the average over the L_2 distances between the matching points and the ground truth location can be reduced from 3.91 pixels (using all matches) to 1.91 pixels, and the standard deviation (matching precision) from 3.37 to 1.14 pixels (see Table 1). Note that a higher threshold results in a smaller number of valid matching points, which are more reliable (in terms of the L_2 distance). For a later application, a threshold does not have to be specified. Depending on the number of matching points x needed for an image pair, the best x matching points can be chosen, based on the raw score.

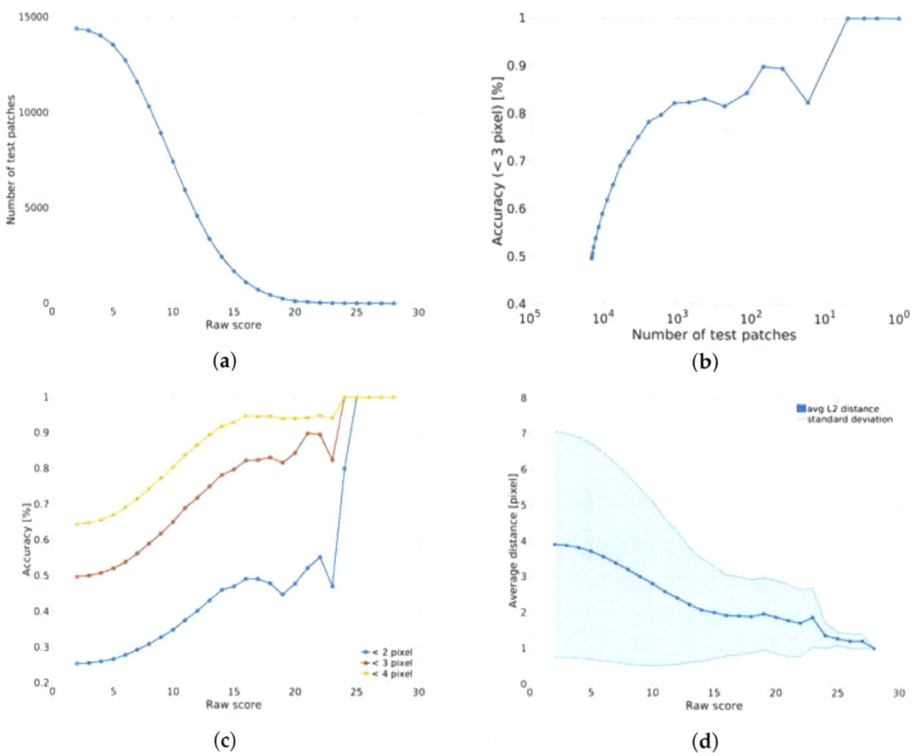

Figure 5. Illustration of influence of the raw score as a threshold: (**a**) the relation between the predicted score and the number of patches; (**b**) relation between the number of patches and the matching accuracy; (**c**) relation between the predicted score and the matching accuracy; and (**d**) relation between the predicted score and the average distance (L_2) between the predicted matching points and the ground truth location. The matching accuracy in Figure 5b is measured as the percentage of matching points, where the L_2 distance to the ground truth location is less than three pixels and in Figure 5c less than 2, 3 and 4 pixels.

3.7. Qualitative Results

In Figure 6, we show a side by side comparison of the score maps of our approach with two baseline methods of sample image patches. Note that CAMRI [10] does not provide a score map as output. Therefore, we perform our search over a 51 × 51 pixels search space, where the used patches have a resolution of 2.5 m. The images in the first column are optical image patches and the images in the last column the despeckled SAR image patches. To generate the images in column 2 to 4 we perform the matching between the corresponding image pairs using NCC, MI and our method. Yellow indicates a higher score, and blue indicates a lower score. The ground truth location is in the center of each patch. Our approach performs consistently better than the corresponding baseline methods. More precisely, the score maps generated with our approach show one high peak at the correct position, except for the last example. Here, two peaks are visible along a line, which corresponds to a street in the SAR patch. In contrast, both baseline methods show a relatively large area with a constantly high score at the wrong positions for most examples.

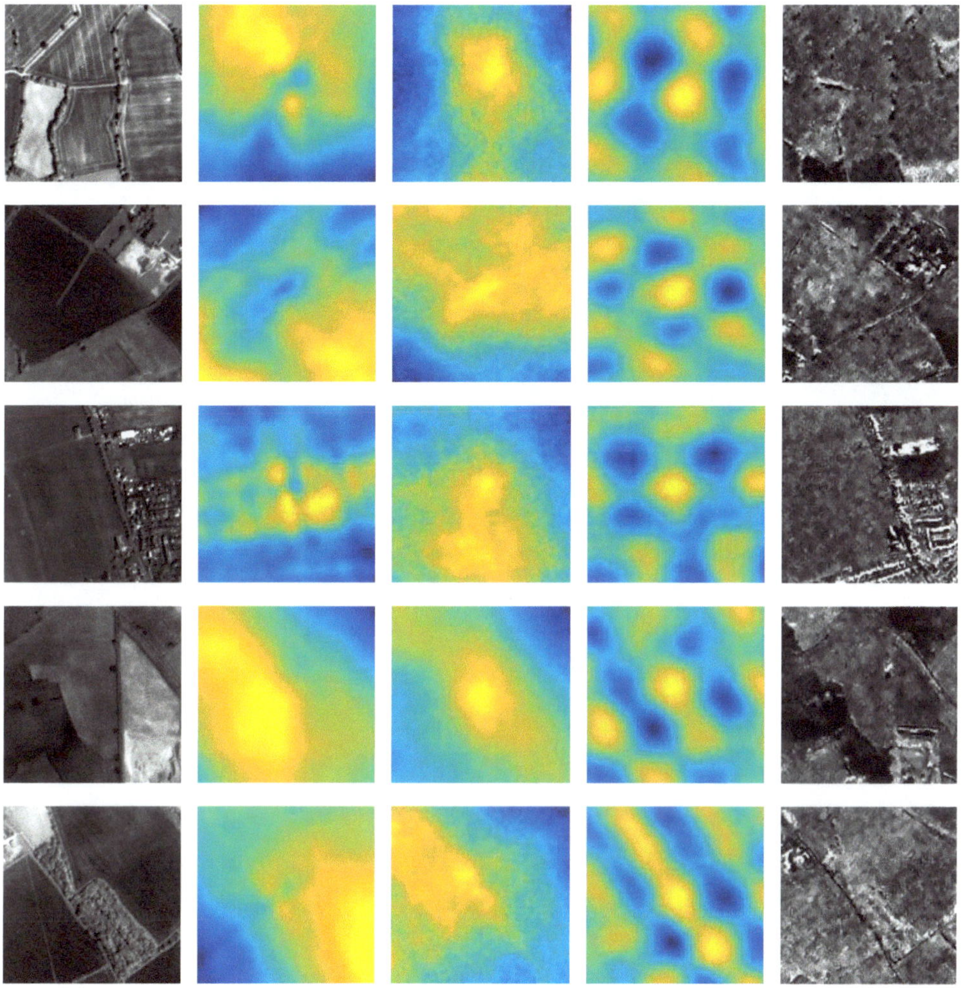

(**a**)Optical image (**b**)NCC (**c**)MI (**d**)Proposed method (**e**)SAR image

Figure 6. Side by side comparison between (**a**) optical patches (201 × 201 pixels), (**b**) the score maps of NCC, (**c**) MI, and (**d**) our method (51 × 51 pixels), and (**e**) the reference despeckled SAR patches (251 × 251 pixels).

In Figure 7, the checkerboard overlay of two optical and SAR image pairs is shown. The residual alignment error between the uncorrected optical and SAR images is clearly visible in the easting direction in Figure 7a. In contrast, the corrected optical and SAR image pair in Figure 7b seems to be aligned. For the correction of the optical image, we used the obtained matching points from the neural network to improve the parameters of the corresponding sensor model and, hence, to improve the geo-location accuracy. In particular, we picked the best 153 matching points (with the highest raw score and with at least a L_2 spatial distance of 50 pixels to each other) as our ground control points (GCPs). We set the empirical distance threshold to 50 pixels to ensure that the points are equally spread over the whole image. Afterwards, the unknown parameters of the sensor model are estimated from these GCPs by iterative least squares adjustment. During this process, a blunder detection removed

11 GCPs. At the end, we used the improved sensor model to generate a new orthorectified optical image with improved absolute geo-localization accuracy. The standard deviation for the remaining 142 GCPs is 1.04 pixels in the easting and 1.28 pixels in the northing direction.

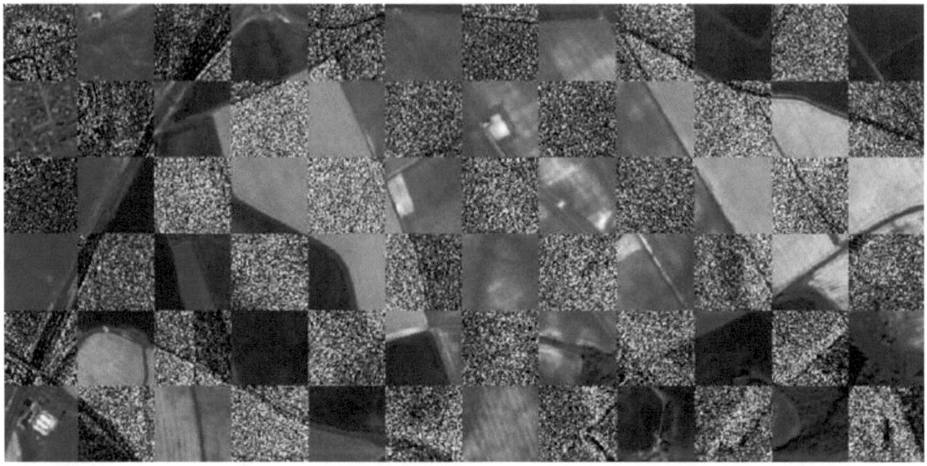

(**a**) Before the geo-localization enhancement of the optical image.

(**b**) After the geo-localization enhancement of the optical image.

Figure 7. Checkerboard overlays of two optical and one SAR image with a pixel spacing of 2.5 m and image tiles size of 100 × 100 m: (**a**) shows the optical image before and (**b**) after the sensor model adjustment (geo-localization enhancement) through the generated matching points.

3.8. Limitations

A drawback of the current network architecture is the restriction to input patches of size 201 × 201 pixels for the left branch of the network. If we were to use the full resolution of the SAR images and upsample the optical images to 1.25 m, our training and test dataset would contain a large amount of image patches, containing just one straight line (street segment). These patches are ambiguous for our two-dimensional search and, hence, not suitable for the training process. As a consequence, we need larger image patches to reduce the amount of ambiguity. Therefore, we downsampled the optical and

SAR images. Due to the memory limits of our available GPUs, it was not possible to increase the input patch size and simultaneously keep a proper batch size. A possible solution could be the investigation of a new network architecture, which enables the use of larger input patches. An alternative solution could be a better selection process of the patches, e.g., only patches containing street crossings.

The processing chain for the generation of our dataset and the relatively small amount of training data represent the main current weaknesses. The selection of the image patches for the dataset was mainly done manually and is limited to one SAR and optical satellite sensor (PRISM and TerraSAR-X). Through the usage of OpenStreetMap and/or a road segmentation network, the generation of the dataset could be done automatically, and our datasets could be quickly extended with new image patches. A larger dataset would help to deal with the problem of overfitting during training, and further improve the network performance.

Additionally, the success of our approach depends on the existence of salient features in the image scene. To generate reliable matching points, these features have to exhibit the same geometric properties in the optical and SAR image, e.g., street-crossings. Therefore, the proposed method is not trained to work on images without such features, e.g., images covering only woodlands, mountainous areas or deserts.

3.9. Strengths

The results prove the potential of our method for the task of geo-localization improvement of optical images through SAR reference data. By interpreting the raw network output as the confidence for predicted matching points (predicted shifts) between optical and SAR patches, we are able to generate matching points with high matching accuracy and precision. Furthermore, the high quality of the matching points does not increase the computation time. After training, we can compute new matching points between arbitrary optical and SAR image pairs within seconds. In contrast, a MI-based approach like CAMRI [10] needs several hours or days to compute the matching points between the same image patches, yielding in less accurate and precise results.

In contrast to other deep learning-based matching approaches, our network is able to match multi-sensor images with different radiometric properties. Our neural network is extendible to images from other optical or radar sensors with little effort, and it is applicable to multi-resolution images. In contrast to other feature-based matching approaches, our method is based on reliable (in terms of equal geometric properties in the optical and SAR image patches) features, e.g., streets and street crossings, which frequently appear in many satellite images. Furthermore, through the variety in our training image pairs, our method is applicable to a wide range of images acquired over different countries or at different times of the year.

4. Conclusions

In this paper, the applicability of a deep learning-based approach for the geo-localization accuracy improvement of optical satellite images through SAR reference data is confirmed for the first time. For this purpose, a neural network has been trained to learn the spatial shift between optical and SAR image patches. The network is composed of a feature extraction part (Siamese network) and a similarity measure part (dot product layer). The network was trained on 134,000 and tested on 14,000 pairs of patches cropped from optical (PRISM) and SAR (TerraSAR-X) satellite image pairs over 13 city areas spread over Europe.

The effectiveness of our approach for the generation of accurate and reliable matching points between optical and SAR images patches has been demonstrated. Our method outperforms state-of-the-art matching approaches, like CAMRI [10]. Particular, matching points can be achieved with an average L_2 distance to the ground truth locations of 1.91 pixels and a precision (standard deviation) of 1.14 pixels. Furthermore, by utilizing the resulting improved sensor model for the geo-referencing and orthorectification processes, we achieve an enhancement of the geo-localization accuracy of the optical images.

In the future, we will further enhance the accuracy and precision of the resulting matching points by using interpolation or polynomial curve fitting techniques to generate sub-pixel two-dimensional shifts. Additionally, we are planning to investigate the influence of alternative network architectures, similarity measures and loss functions on the accuracy and precision of the matching points, as well as the applicability of an automatic processing chain for the dataset generation using OpenStreetMap and a road detection network.

Acknowledgments: We gratefully thank Mathias Schneider for helping with the generation of our dataset, Gerald Baier for helping with the pre-processing of the TerraSAR-X images and Peter Schwind for his valuable suggestions and support during the whole process.

Author Contributions: Nina Merkle, Wenjie Luo, Stefan Auer, Rupert Müller and Raquel Urtasun conceived and designed the experiments. Nina Merkle and Wenjie Luo wrote the source code. Nina Merkle generated the dataset, performed the experiments and wrote the paper. Wenjie Luo, Stefan Auer, Rupert Müller and Raquel Urtasun provided detailed advice during the writing process. Rupert Müller and Raquel Urtasun supervised the whole process and improved the manuscript.

Conflicts of Interest: The authors declare no conflict of interest.

References

1. Werninghaus, R.; Buckreuss, S. The TerraSAR-X Mission and System Design. *IEEE Trans. Geosci. Remote Sens.* **2010**, *48*, 606–614.
2. Eineder, M.; Minet, C.; Steigenberger, P.; Cong, X.; Fritz, T. Imaging Geodesy- Toward Centimeter-Level Ranging Accuracy with TerraSAR-X. *IEEE Trans. Geosci. Remote Sens.* **2011**, *49*, 661–671.
3. Cumming, I.; Wong, F. *Digital Processing of Synthetic Aperture Radar Data: Algorithms and Implementation*; Number Bd. 1 in Artech House Remote Sensing Library; Artech House: Boston, MA, USA; London, UK, 2005.
4. Auer, S.; Gernhardt, S. Linear Signatures in Urban SAR Images—Partly Misinterpreted? *IEEE Geosci. Remote Sens. Lett.* **2014**, *11*, 1762–1766.
5. Reinartz, P.; Müller, R.; Schwind, P.; Suri, S.; Bamler, R. Orthorectification of VHR Optical Satellite Data Exploiting the Geometric Accuracy of TerraSAR-X Data. *ISPRS J. Photogramm. Remote Sens.* **2011**, *66*, 124–132.
6. Merkle, N.; Müller, R.; Reinartz, P. Registration of Optical and SAR Satellite Images based on Geometric Feature Templates. In Proceedings of the International Archives of the Photogrammetry, Remote Sensing and Spatial Information Sciences, International Conference on Sensors & Models in Remote Sensing & Photogrammetry, Kish Island, Iran, 25–27 February 2015; Volume XL-1/W5, pp. 23–25.
7. Perko, R.; Raggam, H.; Gutjahr, K.; Schardt, M. Using Worldwide Available TerraSAR-X Data to Calibrate the Geo-location Accuracy of Optical Sensors. In Proceedings of the 2011 IEEE International Geoscience and Remote Sensing Symposium, IGARSS 2011, Vancouver, BC, Canada, 24–29 July 2011; pp. 2551–2554.
8. Shi, W.; Su, F.; Wang, R.; Fan, J. A Visual Circle Based Image Registration Algorithm for Optical and SAR Imagery. In Proceedings of the 2012 IEEE International Geoscience and Remote Sensing Symposium, Munich, Germany, 22–27 July 2012; pp. 2109–2112.
9. Siddique, M.A.; Sarfraz, M.S.; Bornemann, D.; Hellwich, O. Automatic Registration of SAR and Optical Images Based on Mutual Information Assisted Monte Carlo. In Proceedings of the 2012 IEEE International Geoscience and Remote Sensing Symposium, Munich, Germany, 22–27 July 2012; pp. 1813–1816.
10. Suri, S.; Reinartz, P. Mutual-Information-Based Registration of TerraSAR-X and Ikonos Imagery in Urban Areas. *IEEE Trans. Geosci. Remote Sens.* **2010**, *48*, 939–949.
11. Hasan, M.; Pickering, M.R.; Jia, X. Robust Automatic Registration of Multimodal Satellite Images Using CCRE with Partial Volume Interpolation. *IEEE Trans. Geosci. Remote Sens.* **2012**, *50*, 4050–4061.
12. Inglada, J.; Giros, A. On the Possibility of Automatic Multisensor Image Registration. *IEEE Trans. Geosci. Remote Sens.* **2004**, *42*, 2104–2120.
13. Liu, X.; Lei, Z.; Yu, Q.; Zhang, X.; Shang, Y.; Hou, W. Multi-Modal Image Matching Based on Local Frequency Information. *EURASIP J. Adv. Signal Process.* **2013**, *2013*, 1–11.
14. Li, Q.; Qu, G.; Li, Z. Matching Between SAR Images and Optical Images Based on HOG Descriptor. In Proceedings of the IET International Radar Conference, Xi'an, China, 14–16 April 2013; pp. 1–4.
15. Ye, Y.; Shen, L. HOPC: A Novel Similarity Metric Based on Geometric Structural Properties for Multi-modal Remote Sensing Image Matching. *ISPRS Ann. Photogramm. Remote Sens. Spat. Inf. Sci.* **2016**, *III-1*, 9–16.

16. Hong, T.D.; Schowengerdt, R.A. A Robust Technique for Precise Registration of Radar and Optical Satellite Images. *Photogramm. Eng. Remote Sens.* **2005**, *71*, 585–593.
17. Li, H.; Manjunath, B.S.; Mitra, S.K. A Contour-Based Approach to Multisensor Image Registration. *IEEE Trans. Image Process.* **1995**, *4*, 320–334.
18. Pan, C.; Zhang, Z.; Yan, H.; Wu, G.; Ma, S. Multisource Data Registration Based on NURBS Description of Contours. *Int. J. Remote Sens.* **2008**, *29*, 569–591.
19. Dare, P.; Dowmanb, I. An Improved Model for Automatic Feature-Based Registration of SAR and SPOT Images. *ISPRS J. Photogramm. Remote Sens.* **2001**, *56*, 13–28.
20. Long, T.; Jiaoa, W.; Hea, G.; Zhanga, Z.; Chenga, B.; Wanga, W. A Generic Framework for Image Rectification Using Multiple Types of Feature. *ISPRS J. Photogramm. Remote Sens.* **2015**, *102*, 161–171.
21. Fan, B.; Huo, C.; Pan, C.; Kong, Q. Registration of Optical and SAR Satellite Images by Exploring the Spatial Relationship of the Improved SIFT. *IEEE Geosci. Remote Sens. Lett.* **2013**, *10*, 657–661.
22. Xu, C.; Sui, H.; Li, H.; Liu, J. An Automatic Optical and SAR Image Registration Method with Iterative Level Set Segmentation and SIFT. *Int. J. Remote Sens.* **2015**, *36*, 3997–4017.
23. Sui, H.; Xu, C.; Liu, J.; Hua, F. Automatic Optical-to-SAR Image Registration by Iterative Line Extraction and Voronoi Integrated Spectral Point Matching. *IEEE Trans. Geosci. Remote Sens.* **2015**, *53*, 6058–6072.
24. Han, Y.; Byun, Y. Automatic and Accurate Registration of VHR Optical and SAR Images Using a Quadtree Structure. *Int. J. Remote Sens.* **2015**, *36*, 2277–2295.
25. Chen, Y.; Lin, Z.; Zhao, X.; Wang, G.; Gu, Y. Deep Learning-Based Classification of Hyperspectral Data. *Sel. Top. Appl. Earth Obs. Remote Sens. IEEE J.* **2014**, *7*, 2094–2107.
26. Liang, H.; Li, Q. Hyperspectral Imagery Classification Using Sparse Representations of Convolutional Neural Network Features. *Remote Sens.* **2016**, *8*, 99.
27. Wang, Q.; Lin, J.; Yuan, Y. Salient Band Selection for Hyperspectral Image Classification via Manifold Ranking. *IEEE Trans. Neural Netw. Learn. Syst.* **2016**, *27*, 1279–1289.
28. Mnih, V.; Hinton, G.E. Learning to Detect Roads in High-Resolution Aerial Images. In Proceedings of the 11th European Conference on Computer Vision, Heraklion, Crete, Greece, 5–11 September 2010; Springer: Berlin/Heidelberg, Germany, 2010; pp. 210–223.
29. Matthyus, G.; Wang, S.; Fidler, S.; Urtasun, R. HD Maps: Fine-grained Road Segmentation by Parsing Ground and Aerial Images. In Proceedings of the IEEE International Conference on Computer Vision (ICCV), Las Vegas, NV, USA, 27–30 June 2016; pp. 3611–3619.
30. Geng, J.; Fan, J.; Wang, H.; Ma, X.; Li, B.; Chen, F. High-Resolution SAR Image Classification via Deep Convolutional Autoencoders. *Geosci. Remote Sens. Lett. IEEE* **2015**, *12*, 2351–2355.
31. Masi, G.; Cozzolino, D.; Verdoliva, L.; Scarpa, G. Pansharpening by Convolutional Neural Networks. *Remote Sens.* **2016**, *8*, 594.
32. Luo, W.; Schwing, A.G.; Urtasun, R. Efficient Deep Learning for Stereo Matching. In Proceedings of the IEEE Conference on Computer Vision and Pattern Recognition (CVPR), Las Vegas, NV, USA, 27–30 June 2016.
33. Zbontar, J.; LeCun, Y. Stereo Matching by Training a Convolutional Neural Network to Compare Image Patches. *J. Mach. Learn. Res.* **2016**, *17*, 1–32.
34. Bai, M.; Luo, W.; Kundu, K.; Urtasun, R. Exploiting Semantic Information and Deep Matching for Optical Flow. In Proceedings of the European Conference on Computer Vision (ECCV), Amsterdam, The Netherlands, 8–16 October 2016.
35. Weinzaepfel, P.; Revaud, J.; Harchaoui, Z.; Schmid, C. DeepFlow: Large Displacement OpticalFlow with Deep Matching. In Proceedings of the IEEE Intenational Conference on Computer Vision (ICCV), Sydney, Australia, 1–8 December 2013 .
36. Altwaijry, H.; Trulls, E.; Hays, J.; Fua, P.; Belongie, S. Learning to Match Aerial Images with Deep Attentive Architectures. In Proceedings of the IEEE Conference on Computer Vision and Pattern Recognition (CVPR), Las Vegas, NV, USA, 27–30 June 2016.
37. Lin, T.Y.; Cui, Y.; Belongie, S.; Hays, J. Learning Deep Representations for Ground-to-Aerial Geolocalization. In Proceedings of the IEEE Conference on Computer Vision and Pattern Recognition (CVPR), Boston, MA, USA, 7–12 June 2015; pp. 5007–5015.
38. Altwaijry, H.; Veit, A.; Belongie, S. Learning to Detect and Match Keypoints with Deep Architectures. In Proceedings of the British Machine Vision Conference (BMVC), York, UK, 19–22 September 2016.

39. Bromley, J.; Guyon, I.; LeCun, Y.; Säckinger, E.; Shah, R. Signature Verification using a "Siamese" Time Delay Neural Network. In Proceedings of the Conference on Neural Information Processing Systems (NIPS), Denver, CO, USA, 28 November–1 December1994.
40. Vincent, P.; Larochelle, H.; Lajoie, I.; Bengio, Y.; Manzagol, P.A. Stacked Denoising Autoencoders: Learning Useful Representations in a Deep Network with a Local Denoising Criterion. *J. Mach. Learn. Res.* **2010**, *11*, 3371–3408.
41. Fischer, A.; Igel, C. An Introduction to Restricted Boltzmann Machines. In *Progress in Pattern Recognition, Image Analysis, Computer Vision, and Applications, Proceedings of the 17th Iberoamerican Congress, CIARP 2012, Buenos Aires, Argentina, 3–6 September 2012*; Springer: Berlin/Heidelberg, Germany, 2012; pp. 14–36.
42. Krizhevsky, A.; Sutskever, I.; Hinton, G.E. ImageNet Classification with Deep Convolutional Neural Networks. In Proceedings of the Advances in Neural Information Processing Systems, Lake Tahoe, NV, USA, 3–8 December 2012.
43. Han, X.; Leung, T.; Jia, Y.; Sukthankar, R.; Berg, A.C. MatchNet: Unifying Feature and Metric Learning for Patch-Based Matching. In Proceedings of the IEEE Conference on Computer Vision and Pattern Recognition (CVPR), Boston, MA, USA, 7–12 June 2015 .
44. Zagoruyko, S.; Komodakis, N. Learning to Compare Image Patches via Convolutional Neural Networks. In Proceedings of the IEEE Conference on Computer Vision and Pattern Recognition (CVPR), Boston, MA, USA, 7–12 June 2015.
45. Simo-Serra, E.; Trulls, E.; Ferraz, L.; Kokkinos, I.; Fua, P.; Moreno-Noguer, F. Discriminative Learning of Deep Convolutional Feature Point Descriptors. In Proceedings of the International Conference on Computer Vision (ICCV), Santiago, Chile, 11–18 December 2015.
46. Yu, F.; Koltun, V. Multi-Scale Context Aggregation by Dilated Convolutions. In Proceedings of the International Conference on Learning Representations (ICCV), San Juan, Puerto Rico, 2–4 May 2016.
47. Ioffe, S.; Szegedy, C. Batch Normalization: Accelerating Deep Network Training by Reducing Internal Covariate Shift. In Proceedings of the 32nd International Conference on Machine Learning (ICML-15), Lille, France, 6–11 July 2015.
48. Deledalle, C.; Denis, L.; Tupin, F. Iterative Weighted Maximum Likelihood Denoising with Probabilistic Patch-Based Weights. *IEEE Trans. Image Process.* **2009**, *18*, 2661–2672.
49. Buades, A.; Coll, B. A Non-Local Algorithm for Image Denoising. In Proceedings of the IEEE Conference on Computer Vision and Pattern Recognition (CVPR), San Diego, CA, USA, 20–26 June 2005.
50. Kingma, D.; Ba, J. Adam: A Method for Stochastic Optimization. *arXiv* **2014**, arXiv:1412.6980.
51. Müller, R.; Krauß, T.; Schneider, M.; Reinartz, P. Automated Georeferencing of Optical Satellite Data with Integrated Sensor Model Improvement. *Photogramm. Eng. Remote Sens.* **2012**, *78*, 61–74.
52. Schneider, M.; Müller, R.; Krauss, T.; Reinartz, P.; Hörsch, B.; Schmuck, S. Urban Atlas—DLR Processing Chain for Orthorectification of PRISM and AVNIR-2 Images and TerraSAR-X as possible GCP Source. In Proceedings of the International Proceedings: 3rd ALOS PI Symposium, Kona, HI, USA, 9–13 November 2009.
53. Bossard, M.; Feranec, J.; Otahel, J. *CORINE Land Cover Technical Guide—Addendum 2000;* European Environmental Agency: Copenhagen, Denmark, 2000.
54. He, K.; Zhang, X.; Ren, S.; Sun, J. Delving Deep Into Rectifiers: Surpassing Human-Level Performance on ImageNet Classification. In Proceedings of the IEEE International Conference on Computer Vision, Santiago, Chile, 11–18 December 2015.
55. Burger, W.; Burge, M.J. *Principles of Digital Image Processing: Core Algorithms*, 1st ed.; Springer Publishing Company: London, UK, 2009.
56. Walters-Williams, J.; Li, Y. Estimation of Mutual Information: A Survey. In *Rough Sets and Knowledge Technology, Proceedings of the 4th International Conference, RSKT 2009, Gold Coast, Australia, 14–16 July 2009*; Springer: Berlin/Heidelberg, Germany, 2009; pp. 389–396.

remote sensing

MDPI

Article

Automatic Color Correction for Multisource Remote Sensing Images with Wasserstein CNN

Jiayi Guo [1,2,3], Zongxu Pan [1,2,3], Bin Lei [1,2,3,*] and Chibiao Ding [1,2,3]

1 School of Electronic, Electrical and Communication Engineering, University of Chinese Academy of Sciences, Huairou District, Beijing 101408, China; guojiayi14@mails.ucas.ac.cn (J.G.); zxpan@mail.ie.ac.cn (Z.P.); cbding@mail.ie.ac.cn (C.D.)
2 Institute of Electronics, Chinese Academy of Sciences, Beijing 100190, China
3 Key Laboratory of Geo-spatial Information Processing and Application System Technology, Beijing 100190, China
* Correspondence: leibin@mail.ie.ac.cn; Tel.: +86-135-8167-8087

Academic Editors: Qi Wang, Nicolas H. Younan, Carlos López-Martínez, Sangram Ganguly and Prasad S. Thenkabail
Received: 19 March 2017; Accepted: 12 May 2017; Published: 15 May 2017

Abstract: In this paper a non-parametric model based on Wasserstein CNN is proposed for color correction. It is suitable for large-scale remote sensing image preprocessing from multiple sources under various viewing conditions, including illumination variances, atmosphere disturbances, and sensor and aspect angles. Color correction aims to alter the color palette of an input image to a standard reference which does not suffer from the mentioned disturbances. Most of current methods highly depend on the similarity between the inputs and the references, with respect to both the contents and the conditions, such as illumination and atmosphere condition. Segmentation is usually necessary to alleviate the color leakage effect on the edges. Different from the previous studies, the proposed method matches the color distribution of the input dataset with the references in a probabilistic optimal transportation framework. Multi-scale features are extracted from the intermediate layers of the lightweight CNN model and are utilized to infer the undisturbed distribution. The Wasserstein distance is utilized to calculate the cost function to measure the discrepancy between two color distributions. The advantage of the method is that no registration or segmentation processes are needed, benefiting from the local texture processing potential of the CNN models. Experimental results demonstrate that the proposed method is effective when the input and reference images are of different sources, resolutions, and under different illumination and atmosphere conditions.

Keywords: remote sensing image correction; color matching; optimal transport; CNN

1. Introduction

Large-scale remote sensing content providers aggregate remote sensing imagery from different platforms, providing a vast geographical coverage with a range of spatial and temporal resolutions. One of the challenges is that the color correction task becomes more complicated due to the wide difference in viewing angles, platform characteristics, and light and atmosphere conditions (see Figure 1). For further processing purposes, it is often desired to perform color correction to the images. Histogram matching [1,2] is a cheap way to address this when a reference image with no color errors is available that shares the same coverage of land and reflectance distribution.

To gain a deeper insight, first we would like to place histogram matching in a broader context as the simplest form of color matching [3]. These methods try to match the color distribution of the input images to a reference, also known as color transferring. They can either work by matching low order statistics [3–5] or by transferring the exact distribution [6–8]. Matching the low order statistics is sensitive to the color space selected [9]. The performances of both methods are highly related to the similarity between the contents of the input and the reference. Picking an appropriate reference requires manual intervention and may become the bottle neck for processing. A drawback of such methods is that the colors on the edges of the targets would be mixed up [10–12]. Methods exploiting the spatial information were proposed to migrate the problem, but segmentation, spatial matching, and alignment are required [13,14]. Matching the exact distribution is not sensitive to the color space selection, but has to work in an iterative fashion [8]. Both the segmentation and the iteration increase the computation burden and are not suitable for online viewing and querying. For video and stereo cases, extra information from the correlation between frames can be exploited to achieve better color harmony [15,16]. The holography method is introduced into color transfer to eliminate the artifacts [17]. Manifold learning is an interesting framework to find the similarity between the pixels, so that the output color can be more natural and it can suppress the color leakage as well [18]. Another perspective to comprehend the problem is image-to-image translation. Convolutional neural networks have proven to be successful for such applications [19], for example, the auto colorization of grayscale images [20,21]. Recently, deep learning shows its potential and power in hyper-spectral image understanding applications [22].

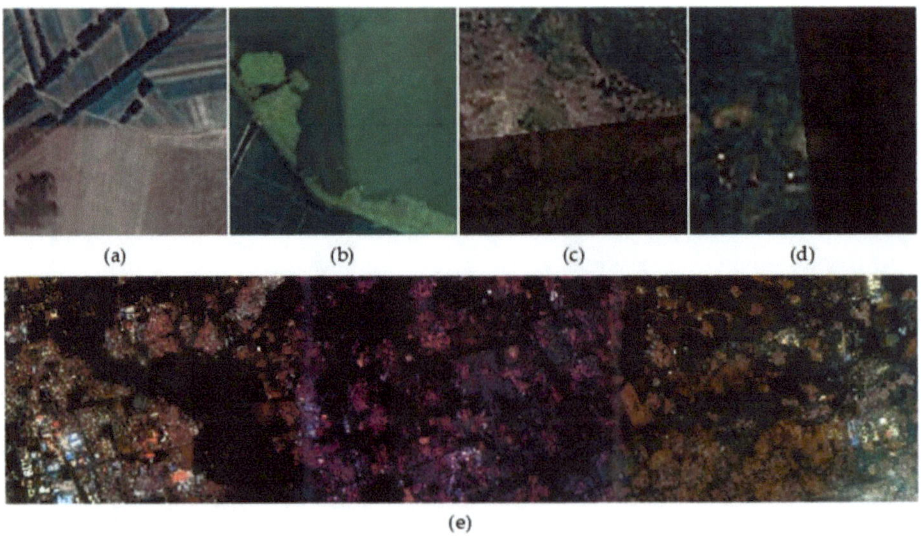

Figure 1. Color discrepancy in remote sensing images. (**a**,**b**) Digital Globe images on different dates from Google Earth; (**c**,**d**) Digital Globe (bottom, right) and NASA (National Aeronautics and Space Administration) Copernicus (**top**, **left**) images on the same date from Google Earth; (**e**) GF1 (Gaofen-1) images from different sensors, same area and date.

Unfortunately, for large-scale applications, it is too strict a requirement that the whole reflectance distribution should be the same between the reference image and the ones to be processed. As a result, such reference histograms are usually not available and have greatly restricted the applications of these sample-based color matching methods. In [23] the authors choose a color correction plan that minimizes the color discrepancy between it and both the input image and the reference image. This is a good solution in stitching applications. However, the purpose of this paper is to eliminate the errors raised by atmosphere, light, etc., so that the result can be further employed in ground reflectance retrieval or atmosphere parameters retrieval. We hope that the output is as close as possible to the reference images, rather than modifying the ground truth values as in [23]. Since it is usually infeasible to find such a reference, a natural question is, can we develop a universal function which can automatically determine the references directly according to the input images? Once this function is obtained, we can combine it with simple histogram matching or other color transfer methods into a very powerful algorithm. In this paper, a Wasserstein CNN model is built to infer the reference histograms for remote sensing image color correction applications. The model is completely data driven, and no registration or segmentation is needed in both the training phase and the inferring phase. Besides, as will be explained in Section 2, the input and the reference can be of different scales and sources. In Section 2, the details of the proposed method are elaborated in an optimal transporting framework [24,25]. In Section 3, the experiments are conducted to validate the feasibility of the proposed method, in which images from the GF1 and GF2 satellites are used as the input and the reference datasets accordingly. Section 4 comprises the discussions and comparisons with other color matching (correcting) methods. And finally, Section 5 gives the conclusion and points out our future works.

2. Materials and Methods

2.1. Analysis

Given an input image I and a reference image I' with N_c channels, an automatic color matching algorithm aims to alter the color palette of I to that of I', the reference. Some of the algorithms require that the reference image is known, which are called sample-based methods. Of course an ideal algorithm should work without knowing I'. The matching can be operated either in the N_c-dimensional color space at once, or in each dimension separately [8,26]. The influence of the light and the atmosphere conditions and other factors can be included into a function $h(I', x, y)$ that acts on the grayscale value of the pixel located at (x, y). Under such circumstances, the problem is converted to learning an inverse transfer function $f(I, x, y)$ that maps the grayscale values of the input image I back to that of the reference image I', where (x, y) denotes the location of the target pixel inside I.

When the input image is divided into patches that each possess a relatively small geographical coverage, the spatial variance of the color discrepancy inside each patch is usually small enough to be neglected. Thus $h(I', x, y)$ should be the same with $h(I', x', y')$ as long as (x, y) and (x', y') share the same grayscale values. Let $u_{x,y}$ and $v_{x,y}$ be the grayscale values of the pixels located at (x, y) in I and I' accordingly, and $h(I', x, y)$ can be rewritten as $h(I', v_{x,y})$, because the color discrepancy function is not related to the location of the pixel but only to its value. The three assumptions of the transformation from the input images to the reference images are made as follows, and some properties which f should satisfy can be derived from them.

Assumption 1: $v_{x,y} = v_{x',y'} \Rightarrow u_{x,y} = u_{x',y'}$

Assumption 1 suggests that when two pixels in I' have the same grayscale value, so do the corresponding pixels in I. This assumption is straight forward since in general cases the cameras are well calibrated and the inhomogeneity of light and atmosphere is usually small within a small geographical coverage. It is true that when severe sensor errors occur this assumption may not hold, however that is not the focus of this paper.

Assumption 2: $u_{x,y} = u_{x',y'} \Rightarrow v_{x,y} = v_{x',y'}$

Assumption 2 indicates that when two pixels in I have the same grayscale, so are their corresponding pixels in I'. The assumption is based on the fact that the pixel value the sensor recorded is not related to its context or location, but only to its raw physical intensity.

Assumption 3: $u_{x,y} > u_{x',y'} \Leftrightarrow v_{x,y} > v_{x',y'}$

Assumption 3 implies that the transformation is order preserving, or a brighter pixel in I should also be brighter in I', and vice versa.

According to the above assumptions, we expect the transfer function f to possess the following properties.

Property 1: $u_{x,y} = u_{x',y'} \Rightarrow f(I, u_{x,y}) = f(I, u_{x',y'})$

Property 2: $u_{x,y} > u_{x',y'} \Leftrightarrow f(I, u_{x,y}) > f(I, u_{x',y'})$, *or f is order-preserving*

Property 3: $I_1 \neq I_2 \Rightarrow f(I_1, \bullet) \neq f(I_2, \bullet)$

Consider that even when two pixels inside I_1 and I_2 share the same grayscale values, the corrected values can still be different according to their ground truth values in the references. Property 3 is to say that f should be content related. In other words, for different input images, the transfer function values should be different to maintain the content consistency. To better explain the point, consider that two input images having different contents, the grassland and the lake so to speak, happen to be of similar color distributions. The pixel in the lake should be darker and the other pixel in the grassland should be brighter in the corresponding reference images. If f is only related to the grayscale values while discarding the input images (the contexts of the pixels), this cannot be done because similar pixels in different input images have to be mapped to similar output levels.

An issue to take into account is whether the raw image or its histogram of the input and reference images should be made use of for the matching. Table 1 lists all possible cases, each of which will be discussed.

Table 1. Different color matching schemes according to the input form and the reference form.

Input	Reference	Scheme
Histogram	Histogram	A
Image	Image	B
Image	Histogram	C
Histogram	Image	D

Scheme A is the case when both the input and reference are histograms, and this is essentially histogram matching. Many previous studies employ this scheme for simplicity, for example, histogram matching and low order statistics matching in various color spaces. Since histograms do not contain the content information, the corresponding histogram matching is not content related. Concretely speaking, two pixels that belong to two regions with different contents but with the same grayscale fall into the same bin of the histogram, and have to be assigned to the same grayscale value in the output image, which does not meet Property 3. In order for one distribution with different contexts to be correctly matched to different corresponding distributions, we cannot enclose different transformations in one unified mapping (see Figure 2). This should not be appropriate for large scale datasets that demand a high degree of automation.

Scheme B corresponds to the case where both the input and output are images, which is usually referred to as image to image translation. The image certainly contains much more information than its histogram, thus providing a possibility that the mapping is content related. Although Property 3 can be satisfied, this scheme emphasizes the content of the image, and the consequence is that the pixels with same grayscales may be mapped to different grayscales as their contexts could be different, and in this case Property 1 is violated (see Figure 3).

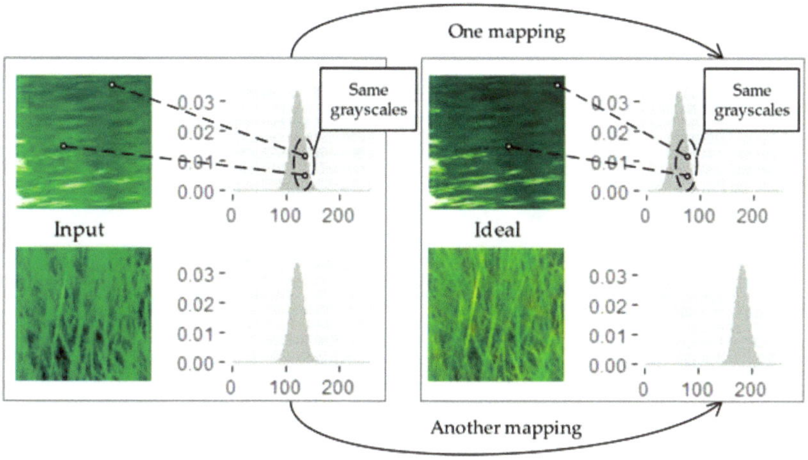

Figure 2. Matching algorithms of "scheme A" take both input and reference in the form of histograms. As this scheme is not content related, two similar distributions with different contexts could be not be mapped to their corresponding reference with one unified mapping.

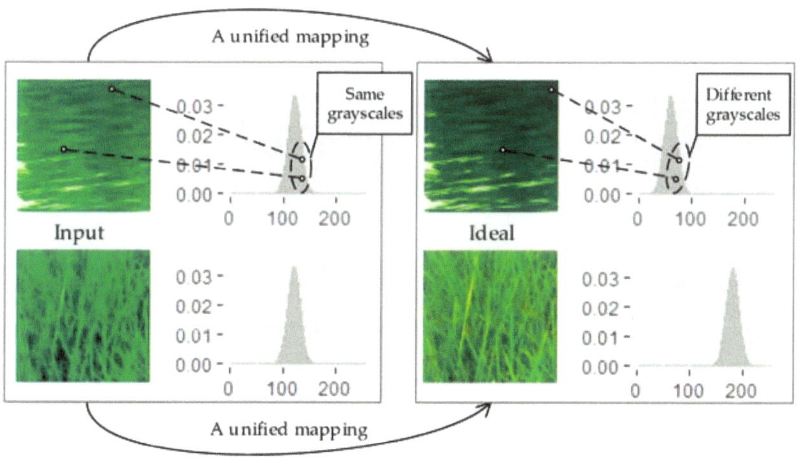

Figure 3. Matching algorithms of "scheme B" take both input and reference in the form of images. Similar distributions could be mapped to different corresponding references, as the scheme is content based. However, the same grayscales could be mapped to different grayscales when they are in different contexts, violating Property 1.

Scheme C is the case where the input is an image and the output is a histogram. As mentioned above, scheme A does not satisfy Property 3 because the context of the image is not used, while scheme B violates Property 1. Mapping one image to another, with constraints that the pixels with the same grayscales also have the same grayscale values in the output, is essentially a grayscale to grayscale transforming process. Under such circumstances, the output of scheme B is always equivalent to that of scheme C. Since scheme C automatically possesses Properties 1 and 3, the task has been now converted to devise the algorithm so that it possesses Property 2 as well (see Figure 4). The task is addressed under an optimal transporting framework, which will be elaborated in Section 2.2.

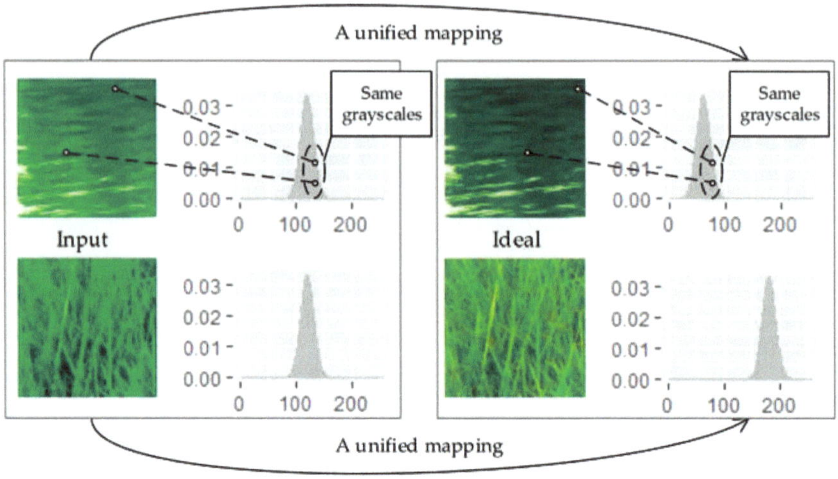

Figure 4. Matching algorithms of "scheme C" take images as inputs and histograms as references in the form of images. Similar distributions could be mapped to different corresponding references, as the scheme is content related.

The scheme of type D corresponds to the case where the input is the histogram and the output is the image. Since it is nearly impossible to determine a transformation mapping of a histogram to an image, we do not take this case into consideration.

2.2. Optimal Transporting Perspective of View

Denote u and v as the input and the reference color distributions, then $T : \mathbb{R}^{N_c} \rightarrow \mathbb{R}^{N_c}$ is a mapping that transforms u to v. The total cost of $T(u, v)$ can be defined as $C(u, v)$ [25–27]:

$$C(u, v) = \inf_{\pi \in \Pi(u,v)} \int c(x, y) \, d\pi(x, y) \tag{1}$$

where $c(x, y)$ is the cost of transporting one unit of mass from x to y, and $\pi(u, v)$ is the joint probability measure of $\mathbb{R}_+^{N_c} \times \mathbb{R}_+^{N_c}$, having u and v as its marginal distributions. Again, N_c indicates the number of color channels and $\Pi(u, v)$ is the collection of every feasible $\pi(u, v)$.

When $c(x, y)$ is defined as a distance $d(x, y)$, the p-order Wasserstein distance can be defined as [25,27]:

$$W_p(u, v) = \left(\inf_{\pi \in \Pi(u,v)} \int d(x, y)^p d\pi(x, y) \right)^{1/p} \tag{2}$$

Finding the transformation $T(u, v)$ that minimizes the total cost $C(u, v)$ is known as the Monge's optimal transportation problem, or the MK problem. The solution to the problem is the gradient of some convex function [25,27,28]:

$$T = \nabla \phi, \text{ where } \phi : \mathbb{R}^{N_c} \rightarrow \mathbb{R} \text{ is convex} \tag{3}$$

Specifically in one dimensional cases, this statement is equivalent to monotonicity, as consequence meets Property 2.

For high dimensional problems, the solution of the MK problem is intractable. In this paper, the distributions of the N_c channels are matched separately. The Wasserstein distance between the inferred values and the ideal values can be calculated in the following way: first sort the pixels on a 1-D axis, and then calculate the distance between each pair of inferred pixels and the ideal pixels accordingly. This is equivalent to using a stacked histogram (see Figure 5). The Wasserstein distance when p equals 2 can be formulated as:

$$W_2 = \left(\int \left(h_{pred}(f) - h_{ref}(f) \right)^2 df \right)^{1/2}, \text{ where } f \text{ is the cumulative frequency} \quad (4)$$

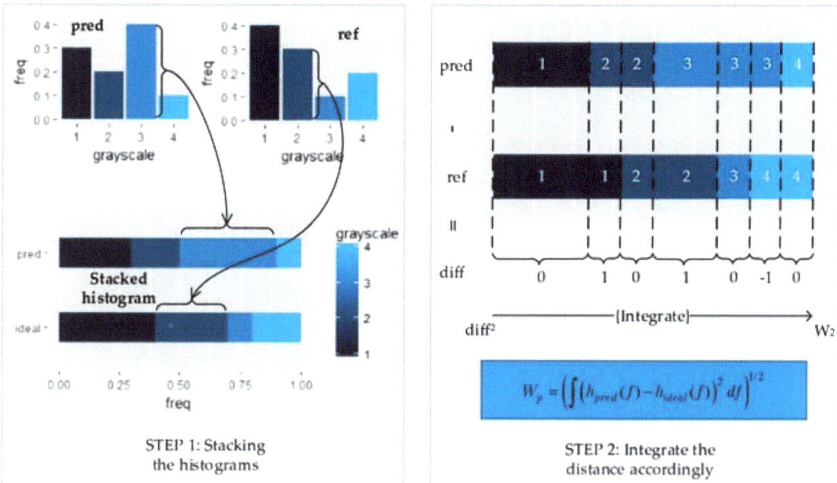

Figure 5. Calculation method of the Wasserstein distance between the inferred histograms and the ground-truth reference. STEP 1: stack the histograms on the frequency axis; STEP 2: subtract the stacked histograms, and integrate with respect to the cumulative frequency.

2.3. The Model Structure

The transformation can be fitted by a CNN model, where the Wasserstein distance plays the role of the loss function. To reduce the memory and computation burden, we used a modified version of Squeeze-net v1.1 [29] (see Figures 6 and 7). In this section we will first introduce the basic modules and then go on to state the major modifications.

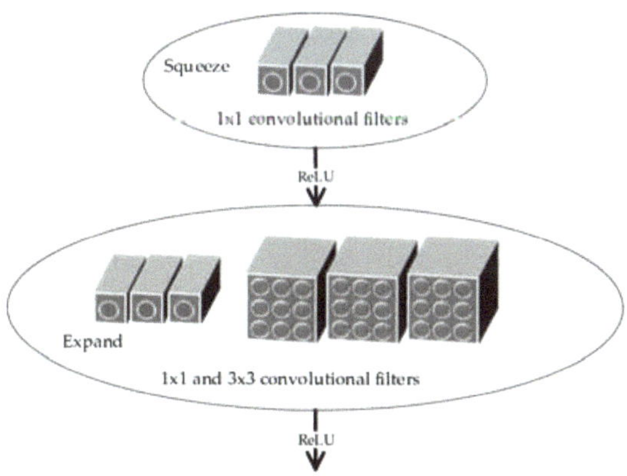

Figure 6. Structure of the "fire module" in the Squeeze-net.

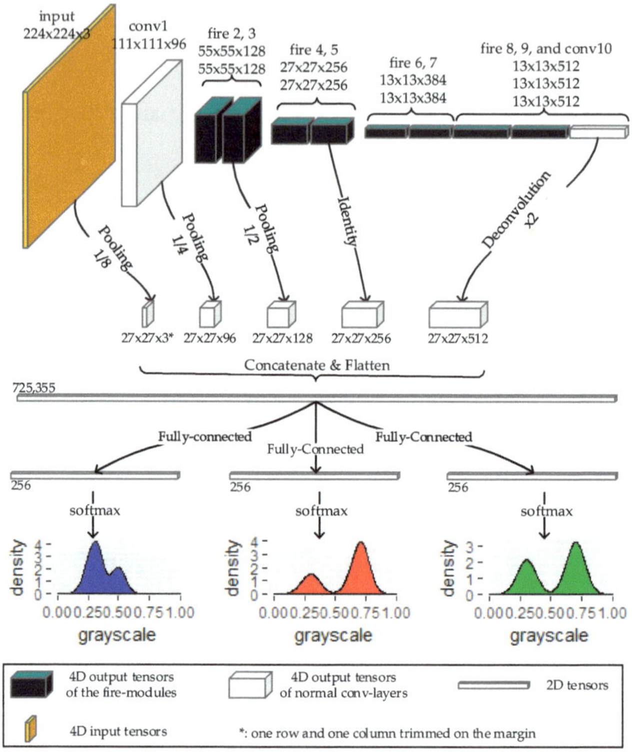

Figure 7. Model structure of the proposed model.

2.3.1. Basic Modules

The Squeeze-net is a light-weight convolutional neural network. The basic modules of the squeeze-net are called the "fire" modules [29], and each consists of two convolution layers, the "squeeze" layer and the "expand" layer. The kernels in the "squeeze" layers are all of 1×1 sizes to maximally lessen the parameters inside the model and reduce the computational burden. Two types of kernels, 1×1 and 3×3 filters, comprise the "expand" layer. The "fire" modules prove to be computationally efficient, and also make the network less likely to be over fitted, as it "squeezes" the amount of parameters to a much smaller scale. In our experiment, the final global average pooling layer and the softmax layer of the squeeze-net was removed, and the rest of the parts were used to extract the features from the raw input images.

2.3.2. The Multi-Scale Concatenation and the Histogram Predictors

As stated in Section 2.3.1, we used a modified version of Squeeze-net to extract features from the input images. The layers at different levels in the CNN model extract features at different scales, and each level has its own characteristics. In general, the former layers in the CNN model are more associated with the raw pixels, while the latter ones are more meaningful in semantic senses [30,31]. Besides, the scales of the former feature maps are also different from the latter ones.

To utilize the information from different scales and semantic levels, we used a concatenating structure. In order for the feature maps to be concatenated, average pooling and deconvolution operations were applied to resize them to a unified shape (27×27). All the padding modes in the pooling layers were "valid", so that the residual parts which could not fill up the pooling kernel were discarded. The strides and kernel sizes within each pooling layer were the same. All the resized shapes were 27 × 27, except for the input, whose output was 28 × 28. Its last row and column were trimmed in order to be consistent with the other tensors to be concatenated. The concatenated feature maps were then flattened into a 2-dimensional tensor of 725,355 length, and then was fed into three fully-connected layers separately, one for each channel (blue, green, and red). The fully-connected layer was then attached by a softmax head each to infer the corrected color distribution.

2.4. Data Augmentation

Data augmentation was performed on the original inputs to avoid over fitting as well as to enclose more patterns of color discrepancy into the model. The augmentation operations include:

1. Random cropping: A patch of 227 × 227 is cropped at a random position from each 256 × 256 sample. It is worth noting that this implies that no registration is needed in the training process.
2. Random flipping: Each sample in the input batch is randomly horizontally and vertically flipped by a chance of 50%.
3. Random color augmentation: The brightness, saturation, and gamma values of the input color are randomly shifted. Small perturbations are added to each color channel. Figure 8 shows an example of such transformation of the color distribution.

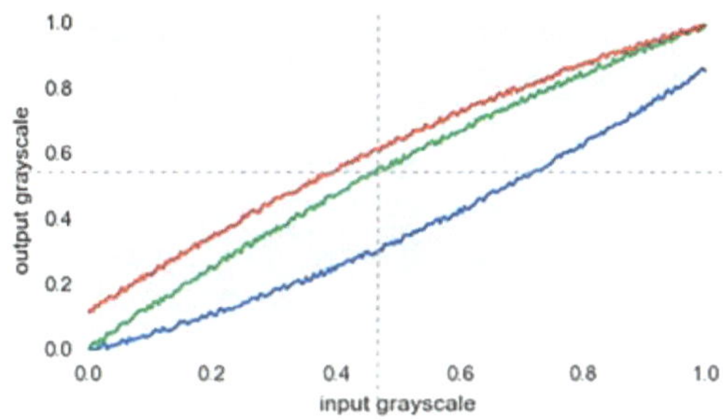

Figure 8. Color transforming curves in the random augmentation process.

2.5. Algorithm Flow Chart

The entire model can be trained in an end-to-end fashion with the gradient descent algorithm, as displayed in Algorithm 1 (Algorithm flow of the training process).

Algorithm 1. Training Process of the Automatic Color Matching WCNN, Our Proposed Algorithm.

Notations: θ, the parameters in the WCNN model; g_θ, the gradients w.r.t. θ; $h(\bullet)$, the predicted color distribution; r, the reference color distribution; $\mathbb{L}_w(\bullet, \bullet)$, the Wasserstein loss.
Required constants: α, the learning rate; m, the batch size.
Required initial values: θ_0, the initial parameters.

while θ has not converged **do**
  Sample $\left\{x^{(i)}\right\}_{i=1}^{m} \sim \mathbb{P}_{in}$ a batch from the input data
  Sample $\left\{y^{(i)}\right\}_{i=1}^{m} \sim \mathbb{P}_{ref}$ a batch from the reference data
  Apply random augmentation to $\left\{x^{(i)}\right\}_{i=1}^{m}$
  $g_\theta \leftarrow \nabla_\theta [\frac{1}{m} \sum_{i=1}^{m} \mathbb{L}_w(h(x_i), y_i)]$
  $\theta \leftarrow \theta - \alpha \cdot SGD(\theta, g_\theta)$
end while

3. Results

We had our algorithm evaluated with satellite images from GF1 and GF2 that cover the same areas. The GF2 images were chosen as the reference. The parameters of the data are listed in Table 2.

Table 2. Parameters of the GF1 and GF2 data in the experiment.

Resolution	GF1	GF2
	8 m	4 m
Band1	0.45–0.52 μm	0.45–0.52 μm
Band2	0.52–0.59 μm	0.52–0.59 μm
Band3	0.63–0.69 μm	0.63–0.69 μm

The direct outputs of WCNN are the inferred distributions (or histograms, see Figure 9) based on the contents of the input images. The corrected images are obtained by histogram matching (see Figure 10). The reference images are only used in the training process and are unnecessary in practical applications, as the purpose of the WCNN model is to generate the reference histogram when there are no available ones. It is worth noting that the patches were only roughly sliced according to the longitude and the latitude information within the GeoTIFF files, so registration was not necessary, and neither was pre-segmentation.

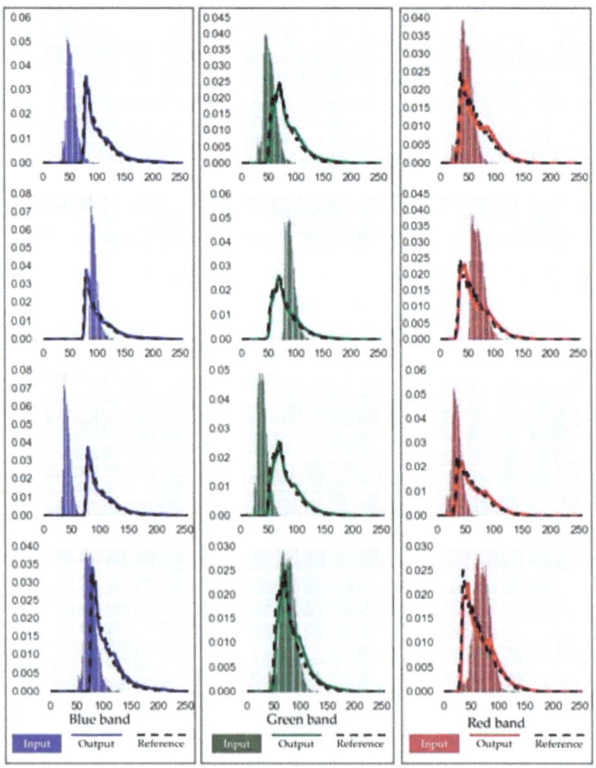

Figure 9. Results of matching the color palette of GF1 to GF2. Bars: histograms of input patches; solid lines with color: predicted histograms of our model; dashed lines in black: histograms of reference images; from top to bottom: histograms of images of the same area, but under different illumination and atmospheric conditions.

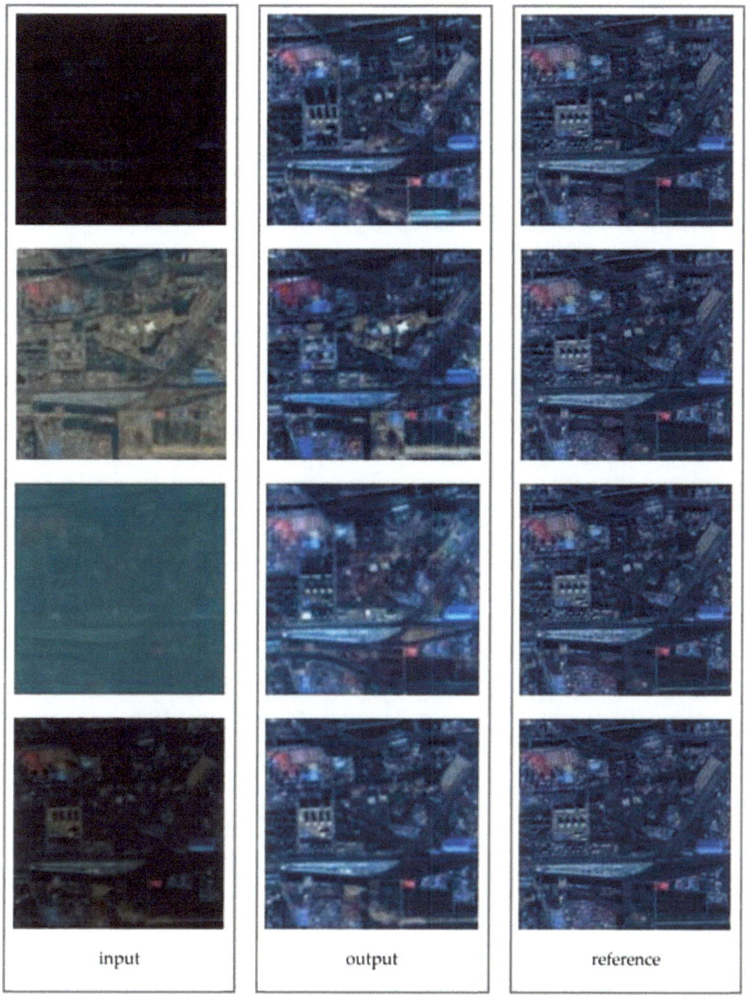

Figure 10. Color matching results of GF1 and GF2. From top to bottom: satellite images of the same area, but under different illumination and atmospheric conditions; left: input images; middle: output images with the predicted color palette; right: reference images, only needed in the training process to calculate the loss function. The model is able to infer the corrected color palette based on the content of the input images in the absence of a reference, when the model is fully trained.

4. Discussion

4.1. Comparison between KL Divergence and Wasserstein Distance

As has been mentioned in Section 2.2, the Wasserstein distance is a natural choice to represent the difference between two color distributions. The Kullback–Leibler divergence (also known as KL divergence) is another commonly used measure (but not a metric) in such circumstances. The definition of KL divergence [27] is:

$$D_{KL}(u \parallel v) = \int u(x) \log\left(\frac{u(x)}{v(x)}\right) dx \tag{5}$$

and the definition of 2-Wasserstein distance is:

$$W_p(u,v) = \left(\inf_{\pi \in \Pi(u,v)} \int \|x-y\|^2 \mathrm{d}\pi(x,y) \right)^{1/2} \tag{6}$$

Consider two simple distributions, $u_1 \sim U(-0.5, 0.5)$ and $u_2 \sim U(-0.5+a, 0.5+a)$, as shown in Figure 11. The Kullback–Leibler divergence should be:

$$D_{KL}(u_1 \parallel u_2) = \begin{cases} a & if\ |a| \leq 1 \\ +\infty & if\ |a| > 1 \end{cases} \tag{7}$$

And the Wasserstein distance is:

$$W_2(u_1 \parallel u_2) = a, \quad where\ a \in [-\infty, +\infty] \tag{8}$$

Because both the Wasserstein metric and the KL divergence are fully differentiable, there is no difference in the back-propagation pipeline between the two losses. From the above discussion, however, we could see that the Wasserstein distance is more numerically stable compared to the KL divergence.

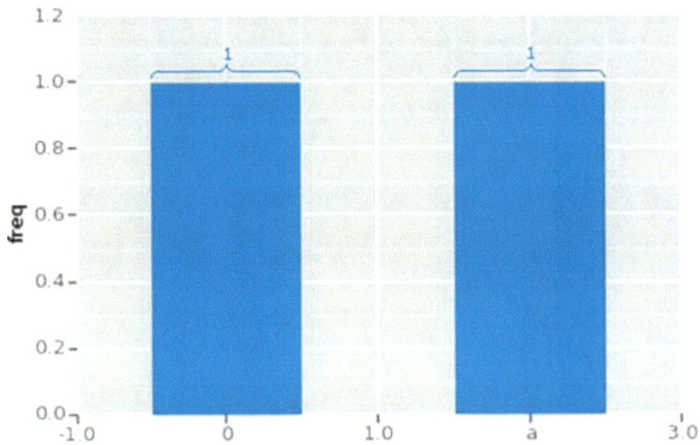

Figure 11. Two one-dimensional uniform distributions.

4.2. Connection and Comparison with Other Color Matching Methods

Histogram matching can be regarded as the simplest case of color matching. It is widely used in seamless mosaic workflows. The method requires that a reference image is selected for each input, which certainly puts restriction on the applications with large scale datasets. Wasserstein CNN is able to directly predict the corrected color distribution, and the histogram matching is the final step in the workflow of our proposed method (but not the only choice, other sample-based color matching methods would also do).

Matching low order statistics faces similar problems. Its performance is closely related to the similarity between the input images and the reference images. To handle low similarity cases, the images may have to be segmented and the color needs to be transferred part to part. Besides, for images with complex contents, color leakage on the edges could be a problem, and the image quality will degrade. Considering these restrictions, such methods may not be appropriate for automatic color matching in remote sensing applications. Matching the exact distribution is more precise than just matching the low order statistics, but is also more complex and computationally expensive. To match two non-Gaussian distributions, iterative approaches have to be exploited, as there are no closed-form solutions [8]. The Wasserstein CNN method is non-iterative, and is more suitable for large scale processing.

Poisson image editing (PIE) is another well-known color matching method. Rather than directly matching the color distributions, the PIE method tries to preserve the gradients of the input image and matches the pixel values on the border to those in the reference image. The problem is equivalent to solving a Poisson equation. However, in our case, this idea might not be very appropriate, because the gradients between the input image and the reference image can be very different, especially when the atmosphere visibility is low (see the PIE result in Figure 12).

Comparisons between the color matching methods are displayed in Figure 12. The ground truth was not included in the training set, as it was supposed as an unknown in the color matching problem. Because the PIE, statistics transferring, and the histogram matching methods are all sample-based, an image must be selected from the training set to act as the reference. However, all that the WCNN model needs is the input image, thus it can operate without selecting such a reference. As the reference is not likely to be exactly the same as the ground truth, we can see the color discrepancy between the output and the ground truth in the results of PIE, statistics transferring, and histogram matching in Figure 12. Also, several features and descriptors were computed for all input images, output images, and the ground truth images in the test set, including the Oriented FAST and Rotated BRIEF (ORB) descriptor, the Scale-Invariant Feature Transform (SIFT) descriptor, and the Binary Robust Invariant Scalable Keypoints (BRISK) descriptor. To be a representation of similarity, the distances between the features of the output and the ground truth are computed, and are displayed in the boxplots in Figure 13.

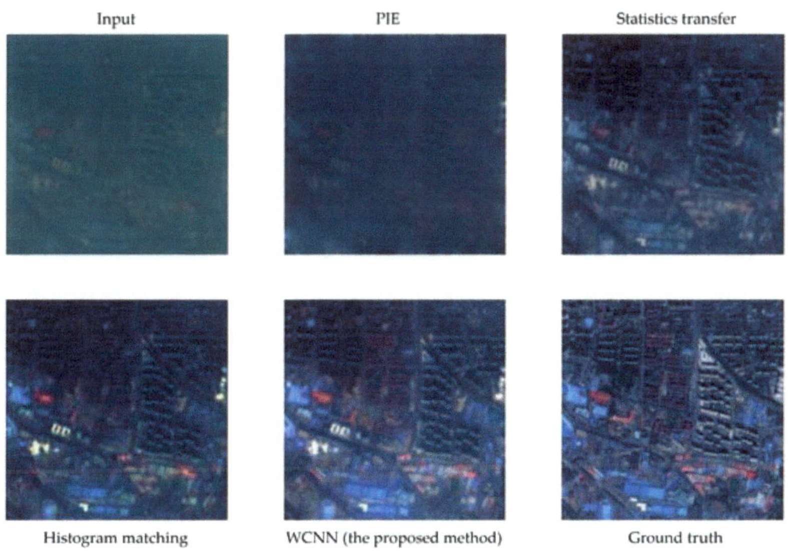

Figure 12. Comparisons between color matching methods.

From Figure 13 we can see that generally the processed images are closer to the ground truth, in regards to the distances of the feature descriptors, except for the PIE method. One of the reasons why PIE fails to generate high quality results is that the low atmosphere visibility may deteriorate the gradients, resulting in a significant difference between the gradients of the input image and the ground truth. The WCNN model results achieve the maximum similarity to the ground truth, and the model is also the most stable one.

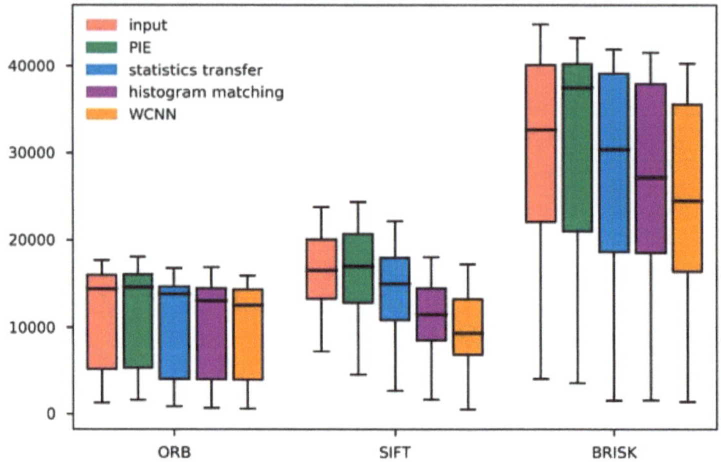

Figure 13. Boxplots of L1-norm distances between the processed images and the ground truth with respect to left: ORB; middle: SIFT, and right: BRISK feature descriptors. The distances represent the dissimilarity between the processed results and the ground truth (the smaller the better). There are five horizontal line segments in each patch, indicating five percentiles of the distances within the processed images by the corresponding method; from top to bottom: the maximum (worst) distance, the worst-25% distance, the median distance, the best-25% distance, and the minimum (best) distance.

4.3. Processing Time and Memory Comsumption

The processing time of 512 patches with a size of 227 × 227 × 3 on a single NVIDIA® GeForce® GTX 1080 graphics processing unit is 0.408 s, or 0.8×10^{-3} s for a single patch, which means that the method could handle images as large as 2000 × 2000 in real time. A total of 6990 MB memory is consumed for 512 patches, or 13.7 MB for each.

5. Conclusions

This paper presents a nonparametric color correcting scheme in a probabilistic optimal transport framework, based on the Wasserstein CNN model. The multi-scale features are first to be extracted from the intermediate layers, and then are used to infer the corrected color distribution which minimizes the errors with respect to the ground truth. The experimental results demonstrate that the method is able to handle images of different sources, resolutions, and illumination and atmosphere conditions. With high efficiency in computing speed and memory consumption, the proposed method shows its prospects for utilization in real time processing of large-scale remote sensing datasets.

We are currently extending the global color matching algorithm to take the local inhomogeneity of illumination into consideration, in order to enhance the precision. Local histogram matching of each band could serve for reflectance retrieval and atmospheric parameter retrieval purposes, and the preliminary results are encouraging.

Acknowledgments: This work was supported by the National Natural Science Foundation of China (Grant No. 61331017.)

Author Contributions: Jiayi Guo and Bin Lei conceived and designed the experiments; Jiayi Guo performed the experiments; Jiayi Guo analyzed the data; Bin Lei and Chibiao Ding contributed materials and computing resources; Jiayi Guo and Zongxu Pan wrote the paper.

Conflicts of Interest: The authors declare no conflict of interest.

References

1. Haichao, L.; Shengyong, H.; Qi, Z. Fast seamless mosaic algorithm for multiple remote sensing images. *Infrared Laser Eng.* **2011**, *40*, 1381–1386.
2. Rau, J.; Chen, N.-Y.; Chen, L.-C. True orthophoto generation of built-up areas using multi-view images. *Photogramm. Eng. Remote Sens.* **2002**, *68*, 581–588.
3. Reinhard, E.; Adhikhmin, M.; Gooch, B.; Shirley, P. Color transfer between images. *IEEE Comput. Graph. Appl.* **2001**, *21*, 34–41. [CrossRef]
4. Abadpour, A.; Kasaei, S. A fast and efficient fuzzy color transfer method. In Proceedings of the IEEE Fourth International Symposium on Signal Processing and Information Technology, Rome, Italy, 18–21 Dcember 2004; pp. 491–494.
5. Kotera, H. A scene-referred color transfer for pleasant imaging on display. In Proceedings of the IEEE International Conference on Image Processing, Genova, Italy, 14 November 2005.
6. Morovic, J.; Sun, P.-L. Accurate 3d image colour histogram transformation. *Pattern Recognit. Lett.* **2003**, *24*, 1725–1735. [CrossRef]
7. Neumann, L.; Neumann, A. Color style transfer techniques using hue, lightness and saturation histogram matching. In Proceedings of the Computational Aesthetics in Graphics, Visualization and Imaging, Girona, Spain, 18–20 May 2005; pp. 111–122.
8. Pitie, F.; Kokaram, A.C.; Dahyot, R. N-dimensional probability density function transfer and its application to color transfer. In Proceedings of the Tenth IEEE International Conference on Computer Vision, Beijing, China, 17–21 October 2005; pp. 1434–1439.
9. Reinhard, E.; Pouli, T. Colour spaces for colour transfer. In Proceedings of the International Workshop on Computational Color Imaging, Milan, Italy, 20–21 April 2011; Springer: Berlin/Heidelberg, Germany; pp. 1–15.
10. An, X.; Pellacini, F. User-controllable color transfer. In *Computer Graphics Forum*; Wiley Online Library: Hoboken, NJ, USA, 2010; pp. 263–271.
11. Pouli, T.; Reinhard, E. Progressive color transfer for images of arbitrary dynamic range. *Comput. Graph.* **2011**, *35*, 67–80. [CrossRef]
12. Tai, Y.-W.; Jia, J.; Tang, C.-K. Local color transfer via probabilistic segmentation by expectation-maximization. In Proceedings of the IEEE Computer Society Conference on Computer Vision and Pattern Recognition, San Diego, CA, USA, 20–26 June 2005; pp. 747–754.
13. HaCohen, Y.; Shechtman, E.; Goldman, D.B.; Lischinski, D. Non-rigid dense correspondence with applications for image enhancement. *ACM Trans. Graph.* **2011**, *30*, 70. [CrossRef]
14. Kagarlitsky, S.; Moses, Y.; Hel-Or, Y. Piecewise consistent color mappings of images acquired under various conditions. In Proceedings of the IEEE 12th International Conference on Computer Vision, Kyoto, Japan, 29 September–2 October 2009; pp. 2311–2318.
15. Bonneel, N.; Sunkavalli, K.; Paris, S.; Pfister, H. Example-based video color grading. *ACM Trans. Graph.* **2013**, *32*, 39:1–39:12. [CrossRef]
16. Wang, Q.; Yan, P.; Yuan, Y.; Li, X. Robust color correction in stereo vision. In Proceedings of the 2011 18th IEEE International Conference on Image Processing, Brussels, Belgium, 11–14 September 2011; pp. 965–968.
17. Gong, H.; Finlayson, G.D.; Fisher, R.B. Recoding color transfer as a color homography. *arXiv* **2016**, arXiv:1608.01505.
18. Liao, D.; Qian, Y.; Li, Z.-N. Semisupervised manifold learning for color transfer between multiview images. In Proceedings of the 2016 23rd International Conference on Pattern Recognition, Cancun, Mexico, 4–8 December 2016; pp. 259–264.
19. Isola, P.; Zhu, J.-Y.; Zhou, T.; Efros, A.A. Image-to-image translation with conditional adversarial networks. *arXiv* **2016**, arXiv:1611.07004.
20. Larsson, G.; Maire, M.; Shakhnarovich, G. Learning representations for automatic colorization. In Proceedings of the European Conference on Computer Vision, Amsterdam, The Netherlands, 8–16 October 2016; pp. 577–593.
21. Zhang, R.; Isola, P.; Efros, A.A. Colorful image colorization. In Proceedings of the European Conference on Computer Vision, Amsterdam, The Netherlands, 8–16 October 2016; pp. 649–666.
22. Wang, Q.; Lin, J.; Yuan, Y. Salient band selection for hyperspectral image classification via manifold ranking. *IEEE Trans. Neural Netw. Learn. Syst.* **2016**, *27*, 1279–1289. [CrossRef] [PubMed]

23. Vallet, B.; Lelégard, L. Partial iterates for symmetrizing non-parametric color correction. *ISPRS J. Photogramm. Remote Sens.* **2013**, *82*, 93–101. [CrossRef]
24. Danila, B.; Yu, Y.; Marsh, J.A.; Bassler, K.E. Optimal transport on complex networks. *Phys. Rev. E Stat. Nomlin. Soft. Matter Phys.* **2006**, *74*, 046106. [CrossRef] [PubMed]
25. Villani, C. *Optimal Transport: Old and New*; Springer: Berlin, Germany, 2008.
26. Pitié, F.; Kokaram, A. The linear monge-kantorovitch linear colour mapping for example-based colour transfer. In Proceedings of the European Conference on Visual Media Production, London, UK, 27–28 November 2007.
27. Frogner, C.; Zhang, C.; Mobahi, H.; Araya, M.; Poggio, T.A. Learning with a wasserstein loss. In Proceedings of the Advances in Neural Information Processing Systems, Montreal, QC, Canada, 2015; pp. 2053–2061.
28. Cuturi, M.; Avis, D. Ground metric learning. *J. Mach. Learn. Res.* **2014**, *15*, 533–564.
29. Iandola, F.N.; Han, S.; Moskewicz, M.W.; Ashraf, K.; Dally, W.J.; Keutzer, K. Squeezenet: Alexnet-level accuracy with 50x fewer parameters and <0.5 mb model size. *arXiv* **2016**, arXiv:1602.07360.
30. Long, J.; Shelhamer, E.; Darrell, T. Fully convolutional networks for semantic segmentation. In Proceedings of the IEEE Conference on Computer Vision and Pattern Recognition, Boston, MA, USA, 2015; pp. 3431–3440.
31. Chen, L.-C.; Papandreou, G.; Kokkinos, I.; Murphy, K.; Yuille, A.L. Semantic image segmentation with deep convolutional nets and fully connected crfs. *arXiv* **2014**, arXiv:1412.7062.

remote sensing

Article

Image Registration and Fusion of Visible and Infrared Integrated Camera for Medium-Altitude Unmanned Aerial Vehicle Remote Sensing

Hongguang Li [1,2], Wenrui Ding [1,3,4,*], Xianbin Cao [3] and Chunlei Liu [3]

1 Unmanned Systems Research Institute, Beihang University, Beijing 100191, China; lihongguang@buaa.edu.cn
2 School of Mechanical Engineering and Automation, Beihang University, Beijing 100191, China
3 School of Electronic and Information Engineering, Beihang University, Beijing 100191, China; xbcao@buaa.edu.cn (X.C.); lchl_buaa@163.com (C.L.)
4 Collaborative Innovation Centre of Geospatial Technology, Wuhan 430000, China
* Correspondence: ding_buaa@163.com or dwr_buaa@163.com; Tel.: +86-10-8233-9906

Academic Editors: Qi Wang, Nicolas H. Younan, Carlos López-Martínez, Gonzalo Pajares Martinsanz and Prasad S. Thenkabail
Received: 27 March 2017; Accepted: 28 April 2017; Published: 5 May 2017

Abstract: This study proposes a novel method for image registration and fusion via commonly used visible light and infrared integrated cameras mounted on medium-altitude unmanned aerial vehicles (UAVs).The innovation of image registration lies in three aspects. First, it reveals how complex perspective transformation can be converted to simple scale transformation and translation transformation between two sensor images under long-distance and parallel imaging conditions. Second, with the introduction of metadata, a scale calculation algorithm is designed according to spatial geometry, and a coarse translation estimation algorithm is presented based on coordinate transformation. Third, the problem of non-strictly aligned edges in precise translation estimation is solved via edge–distance field transformation. A searching algorithm based on particle swarm optimization is introduced to improve efficiency. Additionally, a new image fusion algorithm is designed based on a pulse coupled neural network and nonsubsampled contourlet transform to meet the special requirements of preserving color information, adding infrared brightness information, improving spatial resolution, and highlighting target areas for unmanned aerial vehicle (UAV) applications. A medium-altitude UAV is employed to collect datasets. The result is promising, especially in applications that involve other medium-altitude or high-altitude UAVs with similar system structures.

Keywords: image registration; image fusion; UAV; metadata; visible light and infrared integrated camera

1. Introduction

1.1. Background

1.1.1. Medium-Altitude UAV and Multi-Sensor-Based Remote Sensing

Medium-altitude unmanned aerial vehicles (UAVs) are an important information acquisition platform in the integrated Earth observation network [1]. UAVs offer the advantages of flexibility and rapid response. Compared with manned aerial vehicles, medium-altitude UAVs can work in high-risk areas to accomplish detection missions. They are also capable of flying long distances and feature a wide detection range and an operation time that lasts longer than that of low-altitude

UAVs. Medium-altitude UAVs play an irreplaceable role in normal observation, disaster monitoring, and battlefield detection applications.

Visible light cameras and infrared cameras are the most commonly used imaging devices in medium-altitude UAVs. Visible light imaging offers the advantages of intuitive impression, rich information, and high resolution, but it is susceptible to low-visibility atmospheric conditions. By contrast, infrared imaging is not significantly affected by atmospheric conditions, and it can identify hidden or disguised heat source targets. Given the complementarity of these two types of cameras, most UAVs are equipped with visible light and infrared integrated cameras.

1.1.2. Utility of Visible and Infrared Image Fusion

With the development of imaging sensors, image fusion has become a hot research topic in image processing, pattern recognition, and computer vision. Image fusion combines different sets of information from two or more images of a given scene acquired at different situations with one or multiple sensors [2]. In the past decade, visible and infrared image fusion was widely used in both military and civil applications. In the military, visible and infrared image fusion plays an increasingly important role in UAV autonomous navigation [3], target detection [4], environment perception [5], and military information monitoring [6]. In the civilian realm, many applications, including national environmental protection [7], agricultural remote sensing [8], wildlife multispecies remote sensing [9], safety surveillance [10], and saliency detection [11,12], significantly benefited from information enhancement after visible and infrared image fusion.

1.1.3. Problems of Visible and Infrared Image Registration and Fusion for UAV Applications

Registration and fusion are two of the most crucial technologies in the applications of image fusion mentioned.

Image registration [13] is the process of matching two or more images obtained at different times by different sensors (imaging equipment) or under different conditions (weather, illumination, position, and perspective); this technology has been widely used in computer vision, pattern recognition, medical image analysis, and remote sensing image analysis. Compared with homologous image registration, the registration of visible and infrared images involves certain difficulty and particularity. First, the remote sensing images of the same area obtained by different sensors show different resolutions, pixel values, spectral phases, and scene characteristics because of different imaging mechanisms. Second, the particularity of medium-altitude UAV imaging brings some adverse effects to image registration. Visible images may be degraded under long-distance imaging conditions because of atmospheric effects, which could reduce the number of extracted image features. Large motion between image frames could increase the time consumption of image search.

The purpose of image fusion is to process multi-source redundant data in space and time according to certain algorithms, obtain more accurate and more abundant information than any single dataset, and generate combination images with new space, spectrum, and time characteristics. Image fusion is not only a simple combination of data, but it also emphasizes the optimization of information to highlight useful and interesting information and eliminate or suppress irrelevant information. Despite the availability of many image fusion algorithms, improving the resulting image resolution and enhancing the saliency of interesting areas in images remain problematic.

1.2. Related Work

1.2.1. Image Registration

Popular registration methods usually depend on image information. These methods can be divided into the following two categories according to various similarity measures: intensity-based methods and feature-based methods. Intensity-based methods include gray information-based methods and transform domain-based methods.

Gray information-based methods measure similarity using the gray statistical information of an image itself. These algorithms are convenient to implement, but the application scope is narrow, and the computation is significantly large. The correlation method can match input images with similar scale and gray information based on gray information [14,15]. A novel and robust statistic as a similarity measure for robust image registration was proposed in [14]. The statistic is called the increment sign correlation because it is based on the average evaluation of the incremental tendency of brightness in adjacent pixels. Tsin and Kanade [15] extended the correlation technique to point set registration using a method called kernel correlation. Another classical registration algorithm is based on mutual information. Mutual information is obtained by calculating the entropy of two variables and their joint entropy, which can be used in image registration. On the basis of traditional mutual information registration, Zhuang et al. [16] proposed a novel hybrid algorithm that combines the particle swarm optimization (PSO) algorithm and Powell search method to obtain improved performance in terms of time and precision. In [17], a novel infrared and visual image registration method based on phase grouping and mutual information of gradient orientation was presented. The visible and infrared registration method proposed in [18] combines a bilateral filter and cross-cumulative residual entropy.

Image registration methods based on the transform domain mostly use Fourier transform. They are limited by the invariance of the Fourier transform, which is only suitable for the images of corresponding definitions (such as rotation, translation, etc.) in Fourier transform. Pohitand Sharma [19] developed an algorithm based on Fourier slice theorem to measure the simultaneous rotation and translation of an object in a 2D plane. Niu H. et al. [20] proposed a novel method based on the combination of fractional Fourier transform (FRFT) and a conventional phase correlation technique. Compared with conventional fast Fourier transform-based methods, the proposed method employs called FRFT contains both spatial and frequency information. Li, Zhang, and Hu [21] proposed a registration scheme for multispectral systems using phase correlation and scale invariant feature matching. This scheme uses phase correlation method to calculate the parameters of a coarse-offset relationship between different band images and then detects the scale invariant feature transform (SIFT) points for image matching. In addition to the Fourier transform, a uniform space was used in a new registration method for non-rigid images proposed in [22]. The key point is normalized mapping, which transforms any image into an intermediate space. Under a uniform space, the anatomical feature points of different images are matched via rotation and scaling.

Feature-based methods are the most common category in image registration. These methods depend on image points [23–26], line segments [27,28], regions [29], and other features [30], and they show a wide range of applications. SIFT [23,24] is one of the most widely used features with satisfactory performance. Based on SIFT, several studies [25] conducted improved, extended, and in-depth research on visible and infrared image registration. An image registration method based on speeded up robust features was proposed in view of the slow speed of SIFT [26]. In [27], a new general registration method for images of varying nature was presented. Edge images are processed to extract straight linear segments, which are then grouped to form triangles. To solve the feature matching problem, wherein the interest points extracted from both images are not always identical, Han et al. [28] emphasized the geometric structure alignment of features (lines) instead of focusing on descriptor-based individual feature matching. In [29], Liu et al. proposed an edge-enhanced, maximally stable extremal region method in multi-spectral image registration. An image registration method based on visually salient (VS) features was introduced [30]. A VS feature detector based on a modified visual attention model was presented to extract VS points. This detector combines the information of infrared images and its negative image to overcome the contrast reverse problem between visible and infrared images, thereby facilitating the search for corresponding points on visible/infrared images.

Other new methods emerged in addition to these three methods, and they include diffusion map-based method [31], alignment metric-based method [32], hybrid image feature-based method [33], nonsubsampled contourlet transform (NSCT) and gradient mirroring-based method [34], and the random projection and sparse representation-based method [35]. Some of these studies achieved good results in visible and infrared image registration and they provide new ideas to solve the problem of multimodal image registration.

These studies achieved great successes in the area of image registration. However, most of them are based only on image information and attempt to establish correspondence between visible and infrared images, thereby establishing matching transformation between the two images. In fact, they explore two vital issues of homonymy feature detection and feature matching. Given the different spectra and imaging mechanisms, homonymy feature detection is a difficult problem for multimodal images. From the aerial perspective, the transformation between two sets of image features is required to meet perspective invariance, which increases the difficulty of image feature matching.

For UAV applications, image registration methods still depend on image information despite the rapid development of visible and infrared sensors. Rich metadata from imaging sensors and other equipment of UAV systems are insufficiently exploited.

1.2.2. Image Fusion

Image fusion can be conducted at three different levels, namely, the pixel layer, feature layer, and decision level [36]. This study mainly explores pixel layer-based fusion methods.

Image fusion methods based on pixel levels are traditionally divided into spatial domain methods and transform domain methods. Spatial domain-based methods operate directly on the gray values of images; they mainly include the gray weighted method, principal component analysis (PCA) method [37], color mapping method [38], contrast or gray adjustment method, Markov random field method [39], Bayesian optimization method [40], double modal neural network method [41], and pulse coupled neural network (PCNN) method [42]. In the transform domain fusion, the images should be transformed into the transform domain space before the fusion of the coefficients is conducted. This type of methods mainly include the Laplace pyramid transform-based method [43], wavelet transform-based method [44], ridgelet transform-based method [45], contourlet transform-based method [46], NSCT-based method [47], compressed sensing-based method [48], and sparse representation-based method [49].

In recent years, several scholars introduced effective methods for multi-modality image fusion. Zhang et al. [50] proposed a systematic review of sparse representation-based multi-sensor image fusion literature, which highlighted the pros and cons of each category of approaches. Han et al. [51] presented a saliency-aware fusion algorithm for integrating infrared and visible light images (or videos) to enhance the visualization of the latter. The algorithm involves saliency detection followed by biased fusion. The goal of saliency detection is to generate a saliency map for the infrared image to highlight the co-occurrence of high brightness values and motion. Markov random fields are used to combine these two sources of information. Liu et al. [52] introduced a novel method to fuse infrared and visible light images based on region segmentation. Region segmentation is used to determine important regions and background information in input images.

For UAV applications, visible light sensors can capture relatively abundant spectral information with clear texture and high spatial resolution, but in poor light conditions, image quality declines significantly. By contrast, infrared sensors can penetrate smoke and fog and perform effective detection under poor light conditions; however, the obtained image shows low contrast, fuzzy scene, and poor details. Based on the requirements of UAV applications, the fusion of visible and infrared images need to combine the two types of image feature data. This method can obtain a high spatial resolution of scene information and interesting target areas can be highlighted.

1.3. Present Work

This study aims to develop a method of visible and infrared image registration and fusion for medium-altitude UAV applications. The research scope is applicable to widely used visible light and infrared integrated cameras, which include two aspects of registration and fusion.

In image registration, our method attempts to solve the problem from the UAV system level instead of using image information alone. Three main problems are studied. The first problem is the transformation between two images under long distance aerial imaging with visible light and infrared integrated cameras. In addition to image information, the second problem is the use of the rich metadata of UAV systems to estimate the transformation between visible and infrared images. The third problem is the detection and matching of homonymy features in multimodal images to obtain precise image registration with the aid of metadata.

Based on image registration, image fusion for UAV applications should not only obtain high spatial resolution and extensive scene information and highlight interesting target areas. Thus, a new pixel layer-based image fusion method using PCNN and NSCT is examined in this study.

2. Methodology

2.1. UAV System with a Visible Light and Infrared Integrated Camera

In this study, we employ a medium-altitude UAV, which is used in earthquake emergency and rescue to collect images of disaster areas effectively and accurately with the aid of imaging devices (Figure 1). The specific parameters are described in Table 1.

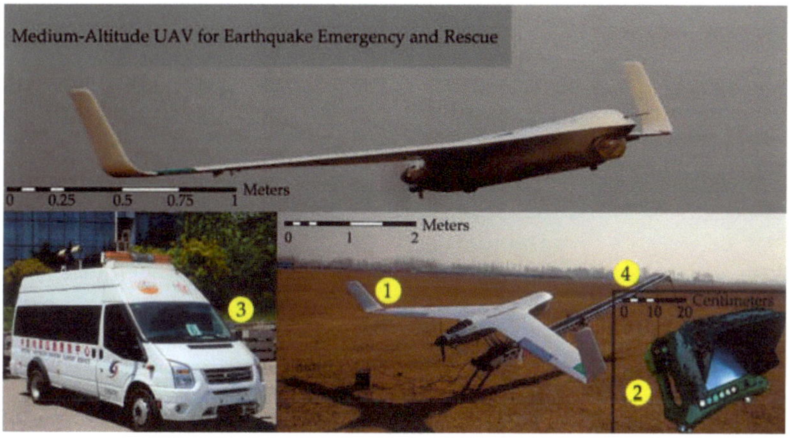

Figure 1. UAV system for earthquake emergency and rescue including: (1) unmanned aerial vehicle (UAV); (2) ground control system; (3) information processing center; and (4) launcher.

Table 1. Main parameters of employed medium-altitude UAV.

Item	Description
Wing Span	4.0 m
Length	1.85 m
Height	0.7 m
Service Ceiling	5000 m
Maximum Payload	5 kg
Maximum Takeoff Weight	35 kg
Flight Speed	80–140 km h^{-1}
Control Radius	60 km
Endurance	5 h
ImagingDevice	VisibleLight and Infrared
Control Mode	Remote, Program or Autonomous
Takeoff Mode	Catapulted Launching
Recovery	Parachute
Engine	Piston Engine
Navigation Mode	BD2/GPS and INS

A visible light and infrared integrated camera platform is mounted on the front belly of the UAV, as shown in Figure 2. The two optical axes of the visible and infrared imaging sensors are parallel. The visible image resolution is 1392 × 1040, and the infrared image resolution is 640 × 512. The UAV features three degrees of freedom (DOF), and the imaging device features two DOF relative to the UAV body. Equipped with GPS (Global Position System), INS (Inertial Navigation System), and an altimeter, the UAV can measure position and orientation.

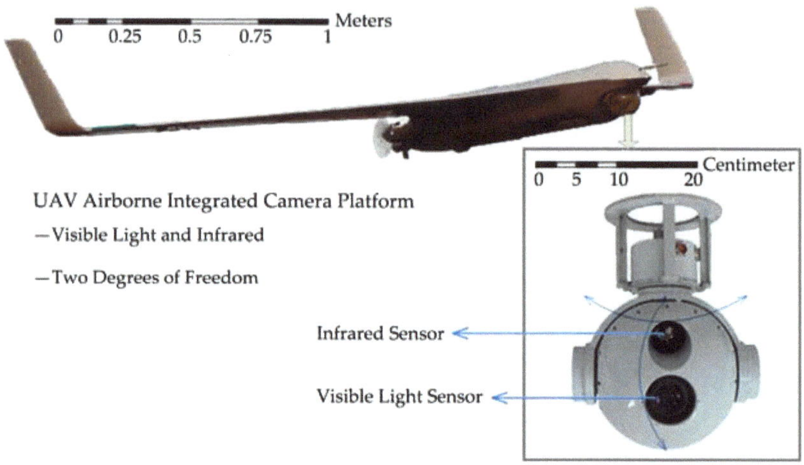

Figure 2. UAV airborne visible light and infrared integrated camera platform with two degrees of freedom.

These types of visible light and infrared integrated cameras have been widely used for medium-altitude UAVs. Therefore, our research shows extensive application potential and practical value.

2.2. Scheme of Visibleand Infrared Image Registration and Fusion

2.2.1. Long-Distance Integrated Parallel Vision

According to the visible light and infrared integrated camera of a medium-altitude UAV, this study attempts to reveal the principle of integrated parallel vision. Most medium-altitude UAV systems are mounted with visible light and infrared integrated cameras, which integrate two types of sensors, as shown in Figure 2. In the integrated structure, the optical axes of the visible sensor and infrared sensor are parallel to each other, and the imaging model can be approximated as a pinhole model [53] under the condition of long-distance imaging over thousands of meters.

Figure 3 shows that the image planes of the two sensors are parallel to each other and the two optical axes are also parallel. With camera rotation, the two sensors always point in the same direction and they have a common field of view (FOV), which is reflected as an overlapping area in the two images. In aerial images, this transformation between two image planes should be described using a perspective transformation. However, under long-distance imaging conditions, only scale transformation and translation transformation exist between the visible and infrared images obtained from the integrated camera at the same moment.

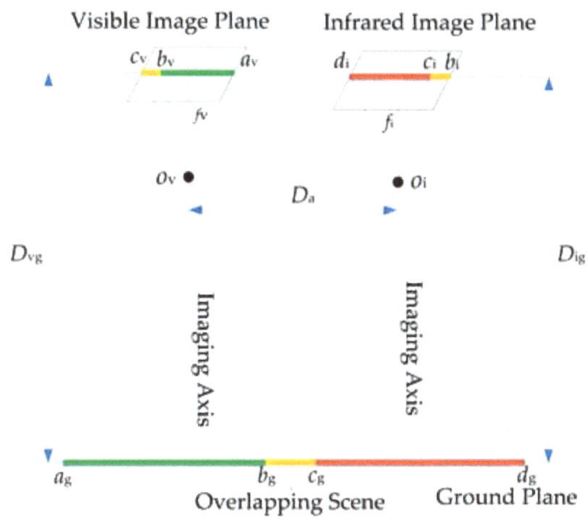

Figure 3. Visible light and infrared integrated camera, in which the two imaging axes are parallel to each other.

The assumption is that the visible and infrared image planes are parallel to the ground, similar to the imaging relationship principle. Line $a_g b_g c_g d_g$ represents the FOV of the two sensors, and line $b_g c_g$ is the common FOV. f_v and f_i are the focal lengths of the two sensors. O_v and O_i are the two foci. D_a is the distance between two imaging axes. D_{vg} and D_{ig} denote the distances from the image plane to the ground. Based on the pinhole imaging principle, Equations (1) and (2) are obtained according to triangle similarity.

$$\frac{b_g c_g}{c_v b_v} = \frac{D_{vg} - f_v}{f_v} \tag{1}$$

$$\frac{b_g c_g}{c_i b_i} = \frac{D_{ig} - f_i}{f_i} \tag{2}$$

D_{vg} and D_{ig} are approximately equal under long-distance imaging conditions. D_g could be introduced to represent the distance from the image plane to the ground in Equation (3).

$$\frac{c_i b_i}{c_v b_v} = \frac{D_g - f_v}{f_v} \times \frac{f_i}{D_g - f_i} \tag{3}$$

Then, Equation (4) can be inferred as

$$\begin{cases} c_i b_i = k c_v b_v \\ k = \frac{D_g - f_v}{D_g - f_i} \times \frac{f_i}{f_v} \end{cases} \tag{4}$$

where k is a constant. This equation proves that the overlapping regions of $c_i b_i$ and $c_v b_v$ have the same direction and scale size. Hence, only translation transformation and scale transformation exist between the two image planes.

According to the above analysis, a complex perspective transformation of image registration could be converted to scale and translation transformation under long-distance integrated parallel vision. This principle is applicable to all of the visible light and infrared integrated cameras of medium-altitude UAVs. This equation breaks the conventional problem of perspective transformation through a direct solution via image feature detection and matching, which is difficult in most cases and sometimes impossible due to the different imaging mechanisms of multimodal images.

2.2.2. Visibleand Infrared Image Registration

According to the long-distance integrated parallel vision in Section 2.2.1, only scale transformationand translation transformation exist between the visible image and infrared image. The transformation from the infrared image to the visible image can be expressed as Equation (5)

$$\begin{cases} I_{\mathrm{v}} = \mathbf{M} I_{\mathrm{i}} \\ \mathbf{M} = \mathbf{M}_{\mathrm{T}} \mathbf{M}_{\mathrm{S}} \end{cases} \tag{5}$$

where I_{v} denotes a visible image and I_{i} denotes an infrared image. $\mathbf{M}$ is the transformation matrix from the infrared image to the visible image; it is composed of two parts, namely, the scale matrix $\mathbf{M}_{\mathrm{S}}$ and translation matrix $\mathbf{M}_{\mathrm{T}}$, which are defined in Equations (6) and (7).

$$\mathbf{M}_{\mathrm{S}} = \begin{bmatrix} s_{\mathrm{x}} & 0 & 0 \\ 0 & s_{\mathrm{y}} & 0 \\ 0 & 0 & 1 \end{bmatrix} \tag{6}$$

$$\mathbf{M}_{\mathrm{T}} = \begin{bmatrix} 1 & 0 & t_{\mathrm{x}} \\ 0 & 1 & t_{y} \\ 0 & 0 & 1 \end{bmatrix} \tag{7}$$

where s_{x}, s_{y}, t_{x}, and t_{y} are transformation parameters. The translation matrix $\mathbf{M}_{\mathrm{T}}$ is solved in two steps of Equation (8) to improve efficiency and accuracy.

$$\mathbf{M}_{\mathrm{T}} = \mathbf{M}_{\mathrm{Tp}} \mathbf{M}_{\mathrm{Tc}} \tag{8}$$

where $\mathbf{M}_{\mathrm{Tc}}$ is the coarse registration matrix from the visible image to the infrared image based on metadata and $\mathbf{M}_{\mathrm{Tp}}$ is the precise registration matrix based on the image matching method.

Accordingly, the problem of visible and infrared image registration can be decomposed into scale calculation, coarse translation estimation, and precise translation estimation. The overall solution process is shown in Figure 4.

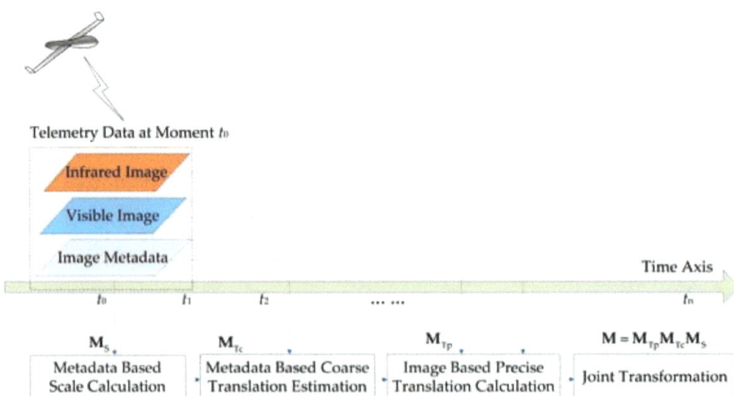

Figure 4. Process of visible and infrared image registration, including scale calculation, coarse translation estimation, and precise translation estimation.

Scale calculation is based on spatial geometry using pixel size and the focal length of two sensors. Translation calculation is divided into metadata-based coarse translation estimation and image-based precise translation estimation. In coarse translation estimation, the transformation from the image plane to the ground plane is established according to the theory of photogrammetry and coordinate transformation. We then attempt to detect the same name points of two images in the ground coordinate system through geographical information and obtain the translation from the infrared image center to the visible image center. Precise translation estimation is based on image features. Edge features are selected for good structure expression in multimodal images to ensure the accuracy and computation efficiency in registration.

2.2.3. Visible and Infrared Image Fusion

To meet the four requirements of UAV image fusion, namely, preserving color information, adding infrared brightness information, improving spatial resolution, and highlighting target areas, this study presents a new image fusion method based on NSCT and PCNN. The main features of the method include the following:

1. The IHS transform is used to extract H and S to preserve the color information, and the NSCT multi-scale decomposition is designed to resolve the declining resolution of fusion images caused by the direct substitution of the I channel.
2. The lowpass sub-band of the infrared image obtained via NSCT decomposition is processed by gray stretch to enhance the contrast between the target and the background and highlight the interesting areas.
3. In view of the PCNN neuron with synchronous pulse and global coupling characteristics, which can realize automatic information transmission and fusion, an algorithm of visible and infrared bandpass sub-band fusion-based PCNN model is proposed.

The process of visible and infrared image fusion based on PCNN and NSCT is shown in Figure 5.

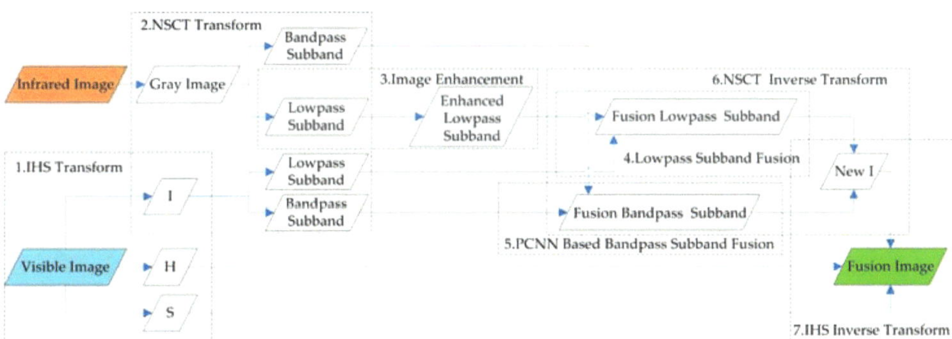

Figure 5. Process of visible and infrared image fusion based on PCNN and NSCT.

The fusion algorithm is implemented in seven steps: (1) IHS transform of visible image; (2) NSCT transform of infrared image and I channel of visible image; (3) enhancement of lowpass subband of infrared image; (4) lowpass subband fusion; (5) bandpass subband fusion; (6) NSCT inverse transform using fusion lowpass subband and fusion bandpass subband; and (7) IHS inverse transform using H channel, S channel, and new I channel.

2.3. Metadata-Based Scale Calculation

2.3.1. Metadata

Metadata represents a type of telemetry data produced simultaneously with images in a UAV system. The most useful parameters are listed in Table 2. The parameter of terrain height is acquired from the geographic information system installed in a ground or airborne computer. Camera installation translations are measured with special equipment before flight. Other parameters come from airborne position and orientation sensors, such as GPS, INS, and altimeter.

Table 2. Useful metadata.

Name	Notation	Source	Description	Accuracy
Longitude	L	GPS	Unit: °	2.5 m
Latitude	B	GPS	Unit: °	2.5 m
Altitude	H_a	Altimeter	Unit: m	0.1 m
Terrain Height	H_g	GIS	Unit: m	1.0 m
Vehicle Heading	h_V	INS	Unit: °	1°
Vehicle Roll	r_V	INS	Unit: °	0.2°
Vehicle Pitch	p_V	INS	Unit: °	0.2°
Camera Installation Translation	t_C^x, t_C^y, t_C^z	Measuring Equipment	Unit: m	0.01 m
Camera Pan	p_C	Camera	Unit: °	0.2°
Camera Tilt	t_C	Camera	Unit: °	0.2°
Resolution	$u \times v$	Camera	u: Image Row v: Image Column	—
Focal Length	f	Camera	Unit: m	—
Pixel Size	s	Camera	Unit: m	—

2.3.2. Spatial Geometry-Based Scale Calculation

For image matching, one image should be scaled to the other. According to spatial geometry, the scale transformation $\mathbf{M}_S$ is only related to the pixel size and focal length, which can be expressed as Equation (9)

$$\mathbf{M}_S = \begin{bmatrix} \frac{s_i}{s_v} \times \frac{f_v}{f_i} & 0 & 0 \\ 0 & \frac{s_i}{s_v} \times \frac{f_v}{f_i} & 0 \\ 0 & 0 & 1 \end{bmatrix} \tag{9}$$

where s_i and s_v denote the pixel sizes of the infrared sensor and visible light sensor, respectively; and f_i and f_v represent the two focal lengths. Using $\mathbf{M}_S$, the infrared image $I^i(x^i, y^i)$ could be transformed to the scale-transformed image $I^{iS}(x^{iS}, y^{iS})$, which is on the same plane of the visible image $I^v(x^v, y^v)$, by employing Equation (10)

$$I^{iS} = \mathbf{M}_S I^i \tag{10}$$

2.4. *Metadata-Based Coarse Translation Estimation*

Based on the theory of coordinate transformation [54,55], this section proposes a method for estimating the transformation between the visible image and the infrared image using image metadata. This estimation is coarse, but it could eliminate the global motion between the frames, reduce the matching range of image registration, and greatly improve the efficiency.

2.4.1. Five Coordinate Systems

Coordinate transformation is the key aspect in the whole process of coarse translation estimation. The following five coordinate systems are used as basis, as shown in Figure 6.

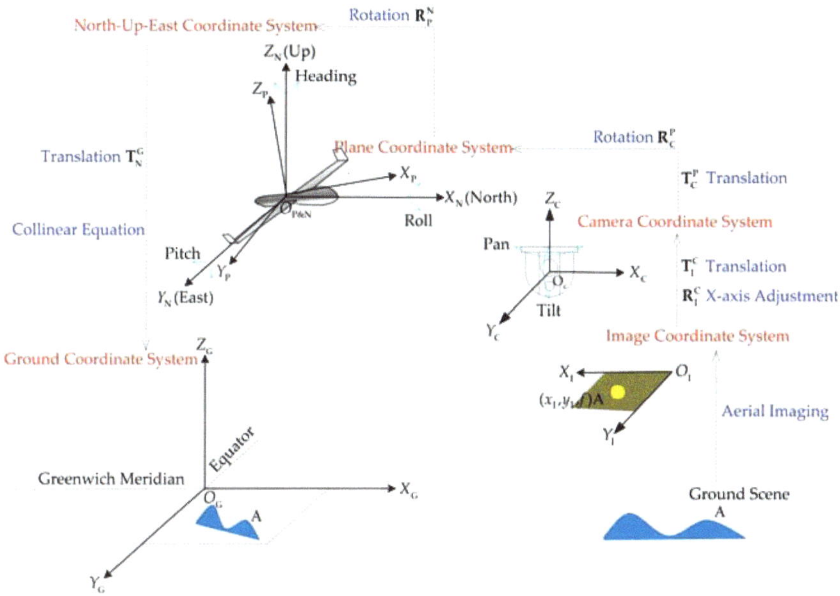

Figure 6. Five coordinate systems of coarse translation estimation.

- Image Coordinate System (ICS) $O_I - X_I Y_I Z_I$

ICS is defined as a rectangular coordinate system, which is related to pixels. The top left corner of the image is considered the coordinate system origin. The values of x_I, y_I are related to the physical size of the row u and column v of the image. The relationship is established by pixel size s. According to different calculation modes, the value of z_I could be set as the focal length of camera f or $-f$.

- Camera Coordinate System (CCS) $O_C - X_C Y_C Z_C$

CCS is the image coordinate system represented by physical units with respect to the center of the image as the origin of the coordinate system, in which axis X_C and axis Y_C are parallel to axis X_I and the axis Y_I. Axis Z_C is upward along the optical axis direction. In this system, the unit is generally in meters.

- Plane Coordinate System (PCS) $O_P - X_P Y_P Z_P$

The origin of the PCS is the center of the GPS device. In PCS, the direction of the axis X_P is positive when it points to the head of the plane, axis Y_P is perpendicular to axis X_p on the body plane, and Z_P is positive when it points upward.

- North–East–Up Coordinate System (NCS) $O_N - X_N Y_N Z_N$

The origin of the NCS is coincident with the origin of the PCS. The direction of axis X_N is positive when it points north, the direction of axis Y_N is positive when it points to the east, and axis Z_N points up.

- Ground Coordinate System (GCS) $O_G - X_G Y_G Z_G$

The Gauss–Kruger surface projection is used in the GCS. The coordinate system (x_G, y_G) is the plane rectangular coordinate system in which national mapping involves the use of Gauss–Kruger 3° or 6° to project and z_G is the absolute altitude. The system consists of a rectangular space and a left-handed coordinate system.

2.4.2. Metadata-Based Coordinate Transformation

Based on the five coordinate systems, the transformation from image I_I in the ICS to image I_G in the GCS should be implemented according to the coordinate system transformation. The process is as follows: ICS $\rightarrow$ CCS $\rightarrow$ PCS $\rightarrow$ NCS $\rightarrow$ GCS. The transformations between the above coordinate systems present translations and rotations, which can be expressed as Equations (11) and (12), respectively.

$$\mathbf{T} = \begin{bmatrix} 1 & 0 & T_x \\ 0 & 1 & T_y \\ 0 & 0 & 1 \end{bmatrix} \tag{11}$$

$$\mathbf{R} = \begin{bmatrix} \cos(\gamma) & -\sin(\gamma) & 0 \\ \sin(\gamma) & \cos(\gamma) & 0 \\ 0 & 0 & 1 \end{bmatrix} \begin{bmatrix} \cos(\beta) & 0 & \sin(\beta) \\ 0 & 1 & 0 \\ -\sin(\beta) & 0 & \cos(\beta) \end{bmatrix} \begin{bmatrix} 1 & 0 & 0 \\ 0 & \cos(\alpha) & -\sin(\alpha) \\ 0 & \sin(\alpha) & \cos(\alpha) \end{bmatrix} \tag{12}$$

where T_x and T_y are translation parameters; and α, β, and γ are the three rotation parameters of the X, Y, and Z axes.

The coordinate transformations in our UAV system are listed in Table 3. They can be calculated with Equations (11) and (12) using relevant metadata.

Table 3. Coordinate transformations and relevant metadata.

Transformation	Notation	Description	Relevant Metadata
ICS to CCS	$\mathbf{R}_I^C$	Direction rotation of coordinate axis	None
	$\mathbf{T}_I^C$	Translation of coordinate system center	u, v, s
CCS to PCS	$\mathbf{T}_C^P$	Translation of installation error	t_C^x, t_C^y, t_C^z
	$\mathbf{R}_C^P$	Rotation of two angles	p_C,t_C
PCS to NCS	$\mathbf{R}_P^N$	Rotation of three angles	h_V, r_V, p_V
NCS to GCS	$\mathbf{T}_N^G$	Translation of coordinate system center	L, B, H_a, H_g

Assuming that any ground point in the ICS, NCS, and GCS could be denoted as (x_I, y_I, z_I), (x_N, y_N, z_N), and (x_G, y_G, z_G), respectively, and the imaging center O in the ICS, NCS, and GCS are denoted as (x_I^O, y_I^O, z_I^O), (x_N^O, y_N^O, z_N^O), and (x_G^O, y_G^O, z_G^O), respectively, the values can be computed via coordinate transformation. Given that the NCS is parallel to the GCS, we can obtain the following formula using the collinear equation according to the central projection model shown in Equation (13).

$$\begin{bmatrix} x_N - x_N^O \\ y_N - y_N^O \\ z_N - z_N^O \end{bmatrix} = \frac{1}{\lambda} \begin{bmatrix} x_G - x_G^O \\ y_G - y_G^O \\ z_G - z_G^O \end{bmatrix} \tag{13}$$

Then, we can obtain any point transformation from the ICS to the GCS via Equations (14) and (15).

$$\begin{bmatrix} x_G \\ y_G \\ z_G \end{bmatrix} = \lambda \mathbf{M}_I^N \left(\begin{bmatrix} x_I \\ y_I \\ z_I \end{bmatrix} - \begin{bmatrix} x_I^O \\ y_I^O \\ z_I^O \end{bmatrix} \right) + \mathbf{M}_I^G \begin{bmatrix} x_I^O \\ y_I^O \\ z_I^O \end{bmatrix} \tag{14}$$

$$\mathrm{f_T}(\mathbf{X}_I) = \left\{ \mathbf{X}_G \middle| \mathbf{X}_G = \lambda \mathbf{M}_I^N \left(\mathbf{X}_I - \mathbf{X}_I^O \right) + \mathbf{M}_I^G \mathbf{X}_I^O \right\} \tag{15}$$

where $\mathbf{M}_I^N = \mathbf{R}_P^N \mathbf{R}_C^P \mathbf{T}_C^P \mathbf{T}_I^C \mathbf{R}_I^C$, $\mathbf{M}_I^G = \mathbf{T}_N^G \mathbf{R}_P^N \mathbf{R}_C^P \mathbf{T}_C^P \mathbf{T}_I^C \mathbf{R}_I^C$, and $Z_I = -f$. λ is a coefficient and could be eliminated during computation. $\mathrm{f_T}$ represents the transformation from image I_I in the ICS to image I_G in the GCS.

2.4.3. Coordinate Transformation-Based Coarse Translation Estimation

Given the same mode of center projection, the coordinate transformation is applicable to both the visible image and infrared image. According to the inverse process of Equation (16), we can conveniently obtain the corresponding pixel positions in the visible image and infrared image of any point in the GCS. The overlapping image of the two sensors in the GCS could be denoted as $I_G^{iv}(x_G^{iv}, y_G^{iv})$, and the corresponding visible image and infrared image in the ICS are denoted as $I_I^v(x_I^v, y_I^v)$ and $I_I^i(x_I^i, y_I^i)$, respectively. The following equation could then be established as Equation (16):

$$\begin{cases} I_I^v(x_I^v, y_I^v) = \mathrm{f_{Tv}}^{-1}(I_G^{iv}(x_G^{iv}, y_G^{iv})) \\ I_I^i(x_I^i, y_I^i) = \mathrm{f_{Ti}}^{-1}(I_G^{iv}(x_G^{iv}, y_G^{iv})) \end{cases} \tag{16}$$

where $\mathrm{f_{Tv}}^{-1}$ and $\mathrm{f_{Ti}}^{-1}$ represent the transform from the GCS to the ICS of the two sensors; they show different expressions because of the different parameters of the two sensors. Accordingly, the coarse translation estimation $\mathbf{M}_{\mathrm{Tc}}$ from the scale-transformed infrared image to the visible image can be calculated using Equation (17).

$$\mathbf{M}_{\mathrm{Tc}} = \begin{bmatrix} 1 & 0 & x_I^v - x_I^i \\ 0 & 1 & y_I^v - y_I^i \\ 0 & 0 & 1 \end{bmatrix} \tag{17}$$

Based on the scale calculation in Section 2.3.2, $\mathbf{M}_{\mathrm{Tc}}$ can be considered as the translation from the center of the infrared scale-transformed image $I_I^{iS}(x_I^i, y_I^i)$ to the center of the original visible image $I_I^v(x_I^v, y_I^v)$.

2.5. Image-Based Precise Translation Estimation

2.5.1. Edge Detection of Visible and Infrared Images

According to current studies, line and edge are robust features for the good representation of scene structure information, and they are widely applied to scene registration and modeling. As described in a study on video analysis [56], line features play an important role in fast 3D camera modeling. In the present study, edge features are used in visible and infrared image registration. The Canny operator [57] is one of the most popular edge detection algorithms. As the scene and illumination of visible and infrared images change frequently, the high and low thresholds of the Canny operator often change thereby leading to poor self-adaptation. In many cases, the conventional Canny operator cannot obtain a satisfying detection result. In the present work, a self-adaptive threshold Canny operator is used to detect enough real edges and avoid disconnected or false edges in detection [58].

2.5.2. Edge Distance Field Transformation of Visible Image

As a result of different imaging mechanisms, the edge features of visible and infrared images show different characteristics. In the visible image, the edges appear relatively smooth, complete, and less noisy. In the infrared image, the edges appear to be incomplete, rough, and noisy, as shown in Figure 7. This characteristic indicates that the edges of the visible and infrared images are roughly the same. However, some details are slightly biased, and they could be defined as the non-strictly aligned characteristics of edges.

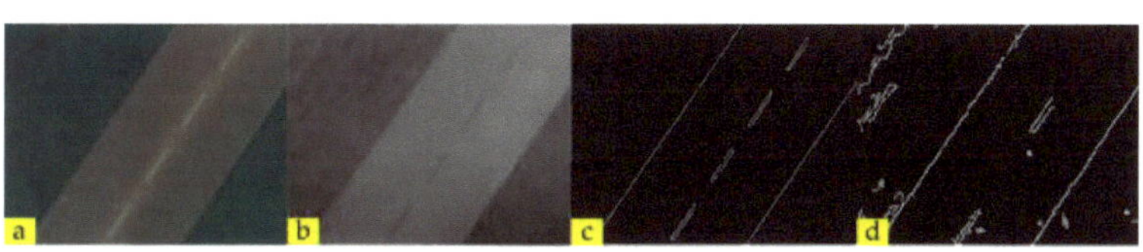

Figure 7. Non-strictly aligned characteristics of edges: (**a**) original visible image; (**b**) original infrared image; (**c**) visible edge image; and (**d**) infrared edge image.

To adapt to the non-strictly aligned characteristics of edges, this study proposes a new registration method based on a Gaussian distance field. This method can extend the edge range with a certain weight and convert the conventional edge-to-edge registration to the edge-to-field registration, which is effective for non-strict matching.

Using the edge detection algorithm of Section 2.4.1, we can extract the edge feature image I^{ve} from the original visible image I^{v}, with the edge pixel value being 255 and the non-edge pixel value being 0. In the edge feature image, the distance transformation of a point is defined as the distance from the nearest edge point to the point itself, as shown in Equation (18).

$$D(p) = \min_{e}(d(p, p^{e})) \tag{18}$$

where $d(p, p^{e})$ represents the distance between two points of p in the distance field map of the visible image and p^{e} in the visible edge image I^{ve}. Given that the points away from the edge exert little effect on edge registration, distance transformation should only be performed in an edge-centered band region. Specifically, the band threshold is set to R, and the distance transformation values of all pixels larger than R are set to $R+1$ via Equation (19).

$$D(p) = \begin{cases} R+1 & D(p) > R \\ D(p) & D(p) \leq R \end{cases} \tag{19}$$

In image matching, $D(p)$ can be used to measure the similarity of the point in the infrared image and the point in the visible image. A small value equates to great matching probability, which could be expressed with a Gaussian model shown in Equation (20):

$$\mathrm{f}(D(p)) = \frac{1}{\sqrt{2\pi}\sigma}\mathrm{e}^{-\frac{D^{2}(p)}{2\sigma^{2}}} \tag{20}$$

where $\mathrm{f}(D(p))$ represents the matching probability. Standard deviation is set to $\sigma = R/3$. In this paper, $R = 10$, which could be different in specific situations. Based on Equation (20), the distance field map I^{vef} of the visible image is established, as shown in Figure 8.

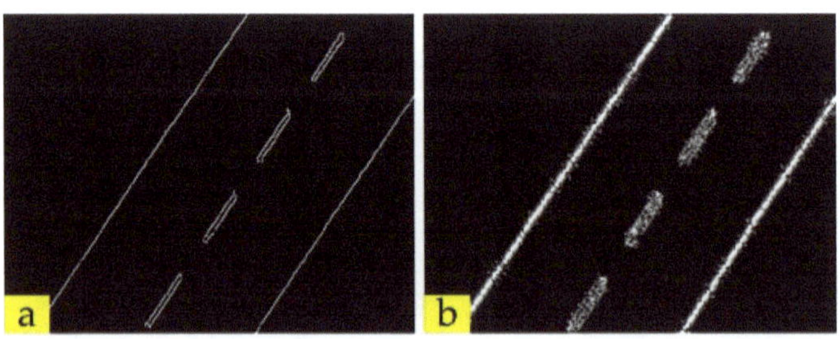

Figure 8. Edge distance field transformation based on Gaussian: (**a**) visible edge image; and (**b**) distance field map of visible edge.

2.5.3. Non-Strict Registration Based on the Edge Distance Field

Similarity for Registration

Assuming that the template image to be registered I^{iet} is extracted from the infrared edge image I^{ie}, then the similarity between I^{iet} and the corresponding template image I^{veft} from the visible distance field map I^{vef} can be expressed using Equation (21):

$$S = \sum \int_{D(p)}^{R} \mathrm{f}(D(p))d(D(p)) \tag{21}$$

where $p(x, y)$ is any point in I^{iet}, and $\mathrm{f}(D(p))$ is the function of the distance field transformation [59].

Infrared Template Image Extraction

Given that the edge distribution of the infrared image is unknown, the infrared template image I^{iet} should be automatically extracted for matching. The position of I^{iet} can be calculated using Equation (22):

$$\begin{cases} x^{iet} = \sum x/N \\ y^{iet} = \sum y/N \end{cases} \tag{22}$$

where N is the number of edge pixels in the infrared edge map I^{ie} and (x, y) is any edge point.

As shown in Figure 9, the width and height of I^{iet} are defined as w and h, respectively. On the x-axis, the edge pixels of the interval $[x^{iet} - 0.5w, x^{iet} + 0.5w]$ occupy a certain proportion of the total pixels of I^{ie}. The edge pixels of the interval $[y^{iet} - 0.5h, y^{iet} + 0.5h]$ account for the same proportion on the y-axis.

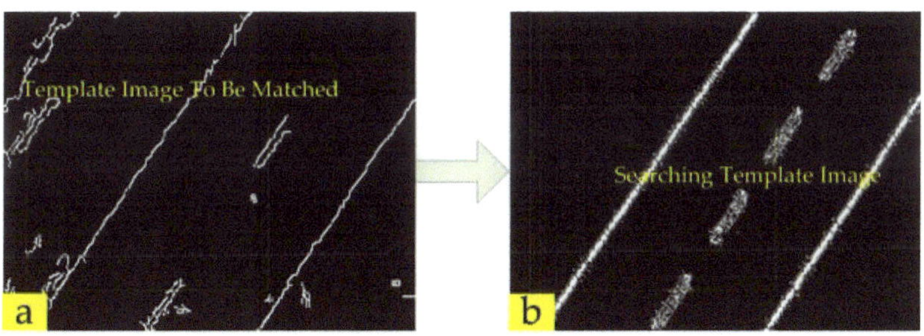

Figure 9. Infrared template image extraction and template image searching in the distance field map of visible edge: (**a**) infrared edge image; and (**b**) distance field map of visible edge.

Searching Algorithm Based on Particle Swarm Optimization

As shown in Figure 9, a searching algorithm is used to find the best matching position in the distance field map of visible edge I^{vef} according to the similarity of the template image I^{iet} and the template image I^{veft} extracted from I^{vef}. The time-consuming performance of the algorithm relative to conventional window searching should be improved, and the occasional accuracy deviation of the metadata attributed to the large motion of the UAV body or camera should be addressed. A novel searching algorithm with a time-varying inertia weight is proposed based on particle swarm optimization (PSO) [60,61].

PSO is a relatively new population-based evolutionary computation technique. This approach uses M particles to construct a group of particles and search for the optimal solution via iteration in the D dimensional space. Each particle comprises several parameters, including current position, velocity, and the best position found by the particles. For a D dimensional search space, these parameters are represented with D dimensional vectors. The position and velocity of the k particle are presented in Equation (23):

$$\begin{cases} x_k = (x_{k1}, x_{k2}, ..., x_{kD}) \\ v_k = (v_{k1}, v_{k2}, ..., v_{kD}) \end{cases} \quad (23)$$

At the n iteration step, the position and velocity of particle i are updated according to Equation (24).

$$\begin{cases} x_i(n) = x_i(n-1) + v_i(n) \\ v_k(n) = \omega v_k(n-1) + c_1 r_1 (p_i - x_i(n-1)) + c_2 r_2 (p_g - x_i(n-1)) \end{cases} \quad (24)$$

where ω is the inertia weight; r_1 and r_2 are two distinct random values between 0 and 1; c_1 and c_2 are the acceleration constants known as cognitive and social scaling parameters, respectively; p_i is the best previous position of the particle itself; and p_g denotes the best previous position of all particles of the swarm. A large value of ω facilitates global exploration with increased diversity, whereas a small value promotes local exploitation [62].

In terms of image registration, $x_k(x_{k1}, x_{k2})$ is the center of image I^{veft}, and p_g is the searching result serving as the best matching position of image I^{iet} and image I^{veft}. As a result of the complex motion of medium-altitude UAVs and cameras, the translational motion between the visible image and the infrared image presents a certain vibration, which requires the search algorithm to automatically adjust the inertia weight ω. A time-varying ω is then proposed in Equation (25):

$$\omega(t) = \omega_0 + r\omega_1 + (|x_{g1}(t-1) - x_{g1}(t-2)| + |x_{g2}(t-1) - x_{g2}(t-2)|)/(4u_v + 4v_v) \quad (25)$$

where t represents the time of image capture. The first item ω_0 is the constant inertia weight, which denotes the confirmed global and local searching ability. The second item $r\omega_1$ is the stochastic inertia weight. This item could allow the algorithm to jump out of local optimization to maintain diversity and global exploration; r is a distinct random value between 0 and 1. The third item is the motion adaptive inertia weight to balance global searching and local searching according to the translation motion between the visible image and the infrared image. $p_g^{t-1}(x_{g1}(t-1), x_{g2}(t-1))$ and $p_g^{t-2}(x_{g1}(t-2), x_{g2}(t-2))$ are the two best previous positions of all particles of the swarm at moments $t-1$ and $t-2$, respectively. u_v and v_v are the row and column of the visible image, respectively. In this study, $\omega_0 = 0.5$, and $\omega_1 = 0.2$.

As the result of the searching algorithm, $p_g^t(x_{g1}(t), x_{g2}(t))$ is the best position at which the similarity of image I^{iet} and image I^{veft} is the highest. The precise translation from scale and the coarse translation-transformed infrared image to the visible image can then be expressed as Equation (26).

$$\mathbf{M}_{Tp} = \begin{bmatrix} 1 & 0 & x_{g1}(t) - x^{iet} \\ 0 & 1 & x_{g2}(t) - y^{iet} \\ 0 & 0 & 1 \end{bmatrix} \quad (26)$$

2.6. PCNN- and NSCT-Based Visibleand Infrared Image Fusion

2.6.1. Simplified PCNN Model

PCNN is a type of feedback network used to explain the characteristics of the neurons in the visual cortex of a cat. As a result of synchronous pulse and global coupling, PCNN neurons can realize automatic information transmission and achieve good results in the field of image fusion. PCNN is connected by a number of neurons, and each neuron corresponds to a pixel of the image. Owing to the complexity of the original PCNN model, a simplified PCNN model [63] is adopted in this study. The mathematical equation is described in Equation (27).

$$\begin{cases} F_{ij}(n) = I_{ij}(n) \\ L_{ij}(n) = \exp(-a_L)L_{ij}(n-1) + \sum\limits_{p,q} W_{ij,pq}Y_{pq} \\ U_{ij}(n) = F_{ij}(n) \times (1 + \beta L_{ij}(n)) \\ Yij = \begin{cases} 1, U_{ij}(n) > \theta_{ij}(n) \\ 0, U_{ij}(n) \leq \theta_{ij}(n) \end{cases} \\ \theta_{ij}(n) = \exp(-a_\theta)\theta_{ij}(n) + V_\theta Y_{ij}(n) \end{cases} \tag{27}$$

where n denotes the iteration times. $F_{ij}(n)$, $L_{ij}(n)$, and $Y_{ij}(n)$ represent the feedback input, link input, and output of the (i,j) neuron in the nth iteration, respectively. I_{ij}, U_{ij}, and θ_{ij} are the external input signal, internal activity term, and output of variable threshold function, respectively. β, W, V_θ, a_L, and a_θ are the link strength, link weight coefficient matrix, threshold magnification factor, link input, and time decay constant, respectively.

2.6.2. NSCT-Based Image Decomposition

Nonsubsampled contourlet transformation (NSCT) is developed based on contourlet transformation. NSCT consists of two parts, namely, nonsubsampled pyramid filter banks (NSPFBs) and nonsubsampled directional filter banks (NSDFBs). NSPFBs enable NSCT to acquire multiscale characteristics. Through decomposition, the image can produce a lowpass subband and a bandpass subband, and then each decomposition level is iterated on the lowpass subband. A nonsubsampled directional filter bank (NSDFB) is a set of two channel nonsampled filter banks based on the sector directional filter bank designed by Bamberger and Smit [64]. NSDFB can be used to carry out the level direction decomposition of the bandpass subband gained by the NSPFB and obtain the direction subband images with the same size as the original image. Three levels of NSCT transform are shown in Figure 10. The number of subbands in each direction increases by up to two times.

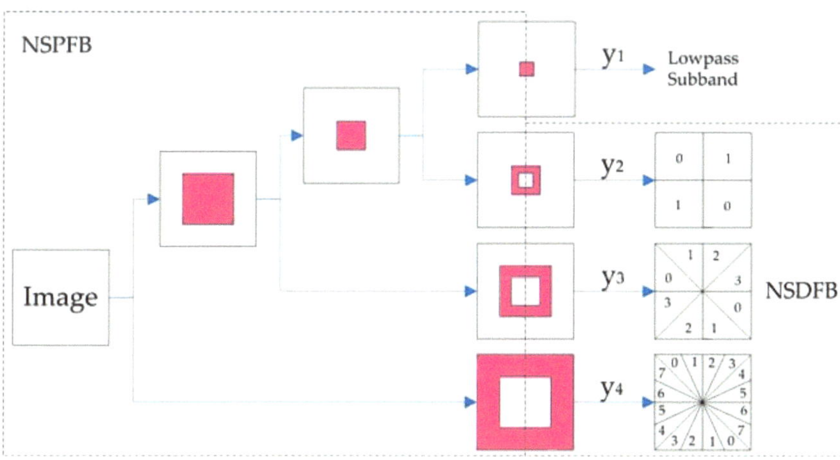

Figure 10. NSPFB and NSDFB of NSCT transform. The left-hand portion is the image decomposition based on NSPFB. The right-hand portion shows the decomposition of each subband in different directions based on NSDFB.

2.6.3. Fusion Algorithm

Based on PCNN and NSCT, the scheme of the visible and infrared image fusion algorithm is introduced in Section 2.2.3. The specific steps of the method are as follows.

1. IHS transform of visible image.

The IHS transform is used to preserve the color information of visible images, which could convert an image from the RGB color space to the IHS color space with the aid of Equations (28)–(30):

$$\begin{pmatrix} I \\ v_1 \\ v_2 \end{pmatrix} = \begin{pmatrix} 1/\sqrt{3} & 1/\sqrt{3} & 1/\sqrt{3} \\ 1/\sqrt{6} & 1/\sqrt{6} & -2/\sqrt{6} \\ 1/\sqrt{2} & 1/\sqrt{2} & 0 \end{pmatrix} \begin{pmatrix} R \\ G \\ B \end{pmatrix} \tag{28}$$

$$H = \tan^{-1}(v_2/v_1) \tag{29}$$

$$S = \sqrt{{v_1}^2 + {v_2}^2} \tag{30}$$

where I denotes intensity, H denotes hue, and S denotes saturation. H and S are preserved for finial IHS inverse transform, and I is used to fuse with the infrared image.

2. NSCT transform of infrared image and I channel of visible image.

As the infrared sensor and visible light sensor can zoom individually, the spatial resolution of the infrared image may be lower than that of the visible light image. Thus, the method of directly replacing the I channel of the visible image with the infrared image may cause the spatial resolution of the fusion image to decline. The NSCT multi-scale decomposition is used to solve this problem. The gray image (8 bit) of the infrared image and the I channel (8 bit) of the visible image are decomposed by three levels through the NSCT transform. One image can be decomposed into one lowpass sub-band and some bandpass subbands. The lowpass represents the outline of the original image, and the bandpass sub-bands represent the edges and textures of the image.

3. Enhancement of lowpass subband of infrared image

Based on NSCT transform, the lowpass subband of the infrared image is processed via histogram equalization to enhance the contrast between the target and the background and to highlight the interesting areas.

4. Lowpass subband fusion

During the lowpass sub-band fusion of the visible light and infrared image, the coefficients are selected according to the principle of the maximum absolute value.

5. Bandpass sub-band fusion

The bandpass sub-band fusion of the visible light and infrared image is based on PCNN. The method chooses the regional energy that can reflect the local phase characteristics of the image as the link strength β of the neuron. Assuming that (i, j) is the center of the region size of $M \times N$, the regional energy E_{ij}^k is expressed as Equation (31):

$$E_{ij}^k = \sum_{m \in M, n \in N} \left[D_{ij}^k(i+m, j+n) \right]^2 \tag{31}$$

where D_{ij}^k represents the bandpass subband coefficient of the kth level at (i, j) of the image.

6. NSCT inverse transform using fusion lowpass subband and fusion bandpass sub-band

New fusion lowpass sub-band and bandpass subbands are generated based on Equations (4) and (5). Then, a new I channel can be obtained according to the NSCT inverse transform.

7. IHS inverse transform using H channel, S channel, and new I channel

Using the new I channel and the preserved H channel and S channel, the fusion image of the RGB color space can be calculated with Equations (32)–(34):

$$\begin{pmatrix} R \\ G \\ B \end{pmatrix} = \begin{pmatrix} 1/\sqrt{3} & 1/\sqrt{6} & 1/\sqrt{2} \\ 1/\sqrt{3} & 1/\sqrt{6} & -/\sqrt{2} \\ 1/\sqrt{3} & -2/\sqrt{6} & 0 \end{pmatrix} \begin{pmatrix} I \\ v_1 \\ v_2 \end{pmatrix} \tag{32}$$

$$v_1 = S \cdot \cos(H) \tag{33}$$

$$v_2 = S \cdot \sin(H) \tag{34}$$

3. Result and Discussion

3.1. Study Area and Dataset

The study area is located inland in Eastern China, as shown in Figure 11. The main types of landforms include cities, villages, and open fields. After performing a number of flights, a database that includes one hundred hours of visible light and infrared videos and metadata was established.

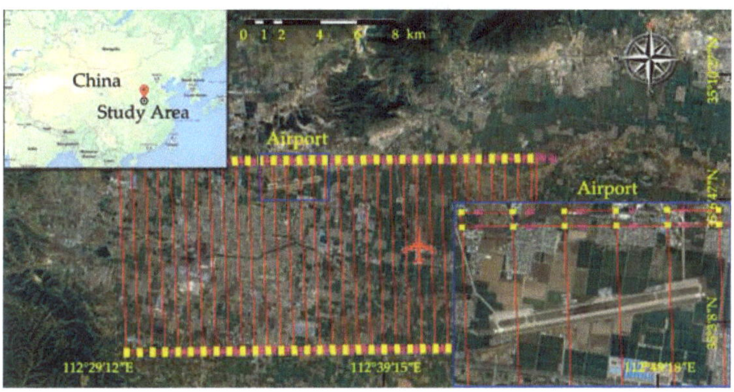

Figure 11. Study area and flight course covering about 300 km^2 in Eastern China.

3.2. Spatial Geometry-Based Scale Calculation

According to Section 2.3.2, the scale transformation from the infrared image to the visible image is determined by pixel size and focal length of the two sensors. In the visible light and infrared integrated camera, the focal length of the visible light sensor can be varied continuously in a certain range, whereas the focal length of the infrared sensor has only two fixed values of 540 mm and 135 mm. In this section, three experiments with different focal lengthsare designed to test the performance of the spatial geometry-based scale calculation. The source data are shown in Table 4, and the results are shown in Table 5 and Figures 12–14.

Table 4. Source data for scale calculation.

Item	Resolution			Focal Length (mm)			Pixel Size (μm)		
Group ID	1	2	3	1	2	3	1	2	3
Visible image		1392 × 1040		172	65.4	50.4		4.65	
Infrared image		720 × 576		540	135	135		25	

Table 5. Infrared image after scale transformation.

Group ID	1	2	3
Result image resolution	1042 × 834	1666 × 1333	1284 × 1027

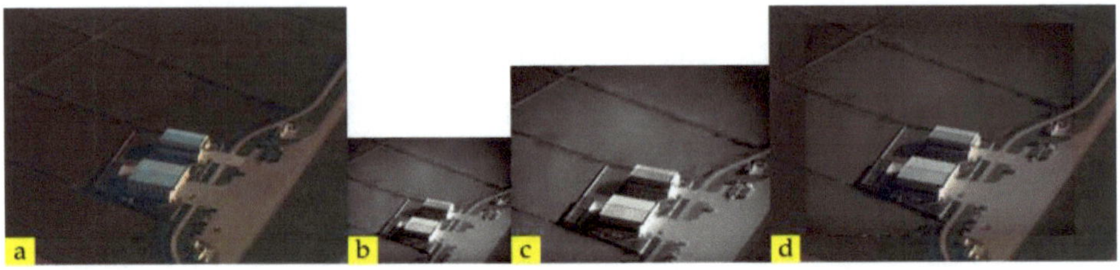

Figure 12. First experiment of scale calculation: (**a**) original image; (**b**) original infrared image; (**c**) scale-transformed result of image (**b**); and (**d**) fusion image of images (**a**) and (**c**).

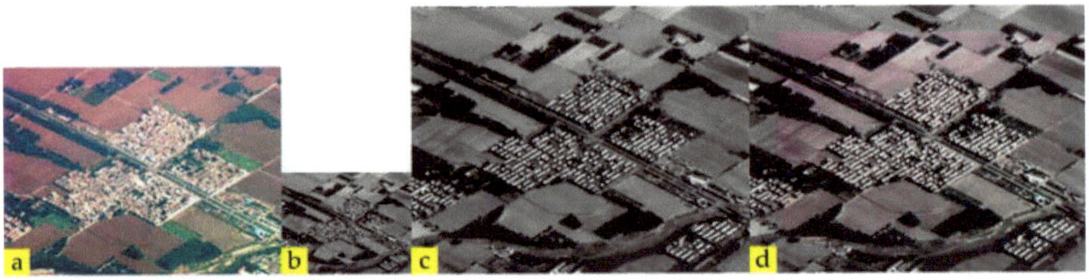

Figure 13. Second experiment of scale calculation. (**a**) Original image; (**b**) original infrared image; (**c**) scale-transformed result of image (**b**); and (**d**) fusion image of images (**a**) and (**c**).

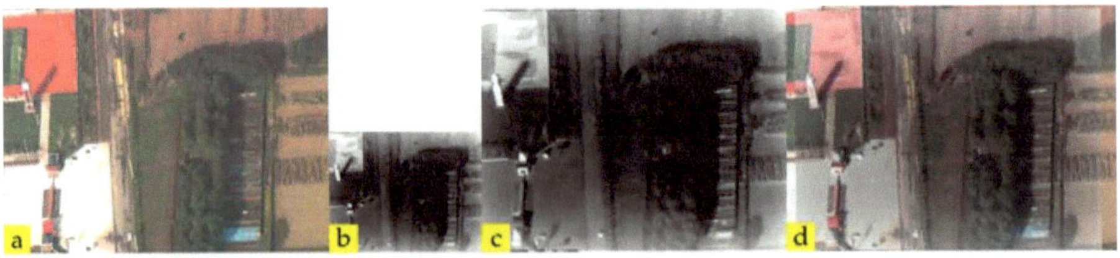

Figure 14. Third experiment of scale calculation: (**a**) original image; (**b**) original infrared image; (**c**) scale-transformed result of image (**b**); and (**d**) fusion image of images (**a**) and (**c**).

In Figures 12–14, Figures 12c, 13c and 14c are the scale-transformed result of Figures 12b, 13b and 14b, respectively, which could be obtained with Equation (10) in Section 2.3.2. Based on the artificial registration of Figures 12a, 13a, 14a and Figures 12c, 13c, 14c, the fusion images of Figures 12d, 13d, 14d are obtained with Equation (35), with C^v and C^i, which represent R, G, and B channels of the visible image and infrared image and with C representing the responding channel of the fusion image.

$$C = (C^v + C^i)/2 \tag{35}$$

According to the fusion results, the two images maintain consistency in shape and size, as indicated by the clarity and lack of aliasing in the overlapping pixels. This result proves the validity of the spatial geometry-based scale calculation.

3.3. Coordinate Transformation-Based Coarse Translation Estimation

After scale calculation, the infrared image is converted to the same plane of the visible image. According to Section 2.4, coarse translation estimation can calculate the translation $\mathbf{M}_{\text{Tc}}$ from the infrared scale-transformed image I^{iS} to the original visible image I^{v}. Then, the infrared image after coarse translation transformation can be obtained with Equation (36).

$$I^{\text{iST}_\text{c}} = \mathbf{M}_{\text{Tc}} I^{\text{iS}} \tag{36}$$

Figure 15 shows the fusion image of the coarse translation-transformed infrared image I^{iST_c} and the original visible image I^{v} obtained with Equation (36).

Figure 15. Fusion image of coarse translation-transformed infrared image and original visible image: (**a**) first experiment image; (**b**) second experiment image; and (**c**) third experiment image.

As shown in Figure 15 and Table 6, the coarse translation shows a positive effect on the registration of the infrared image and visible image, but the result fails to reach high levels of accuracy. Moreover, the error has some fluctuations.

Table 6. Results of coarse translation estimation.

Image Sequence	Translation		
Group ID	1	2	3
Actual Translation	(−31,−29)	(−6,−15)	(−21,11)
Translation Estimation	(−36,−37)	(−20,−10)	(−8,2)
Error	9.43	14.87	15.81

3.4. Image Edge-Based Translation Estimation

Precise translation estimation is performed based on image edge features to achieve an accurate registration. In such estimation, the coarse translation-transformed infrared image I^{iST_c} is converted to the precise translation-transformed image $I^{\text{iST}_\text{c}\text{T}_\text{p}}$ with Equation (37).

$$I^{\text{iST}_\text{c}\text{T}_\text{p}} = \mathbf{M}_{\text{Tp}} I^{\text{iST}_\text{c}} \tag{37}$$

where $\mathbf{M}_{\text{Tp}}$ can be obtained following the description in Section 2.5.

Figure 16 shows the fusion image of the precise translation-transformed infrared image $I^{\text{iST}_\text{c}\text{T}_\text{p}}$ and the original visible image I^{v} obtained with Equation (37).

Figure 16. Fusion image of the precise translation-transformed infrared image and the original visible image: (**a**) first experiment image; (**b**) second experiment image; and (**c**) third experiment image.

Comparing Figures 15 and 16 indicates that the fusion image based on precise translation is better than the fusion image based on coarse translation because of its clear edges in the overlapping region and absence of aliasing. As indicated in Table 7, image registration accuracy is significantly improved.

Table 7. Results of precise translation estimation.

Image Sequence	Translation		
Group ID	1	2	3
Actual Translation	(−31,−29)	(−6,−15)	(−21,11)
Translation Estimation	(−30,−27)	(−8,−13)	(−20,9)
Error	2.24	2.83	2.24

3.5. PCNN- and NSCT-Based Image Fusion

3.5.1. Fusion of Visible Image and Low Spatial Infrared Image

When the spatial resolution of the infrared image (Figure 17b) is low, the method of directly replacing the I channel of the visible image (Figure 17a) with the infrared image causes the spatial resolution of the fusion image to decline (Figure 17c). The proposed NSCT- and PCNN-based method can generate a fusion image with satisfactory spatial resolution (Figure 17d). As shown in Figure 17, the spatial resolution of Figure 17 dis higher than that of Figure 17c.

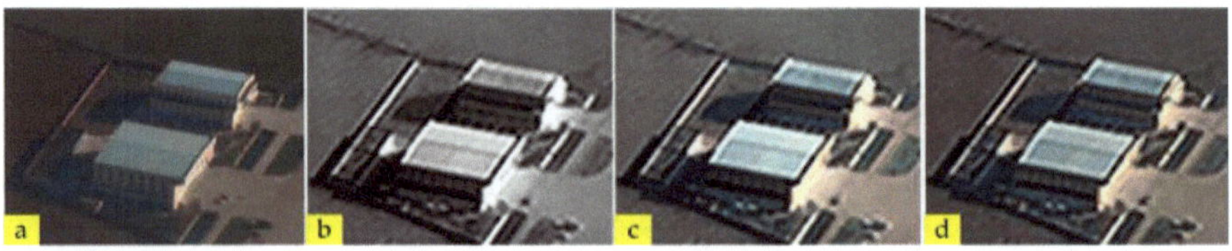

Figure 17. Fusion of visible image and low spatial infrared image: (**a**) Visible image; (**b**) infrared image; (**c**) fusion image based on IHS; and (**d**) fusion image based on the proposed method.

3.5.2. Fusion of Interesting Areas

Another important purpose of image fusion is to highlight target information. Figure 18 shows the saliency analysis between the original image and the fusion image in two scenes. Figure 18a,b,d,e shows the original images. Figure 18c,f shows the fusion results of the proposed method. The yellow frame area represents the low salient areas in the visible image. The fusion results show that these areas become increasingly salient.

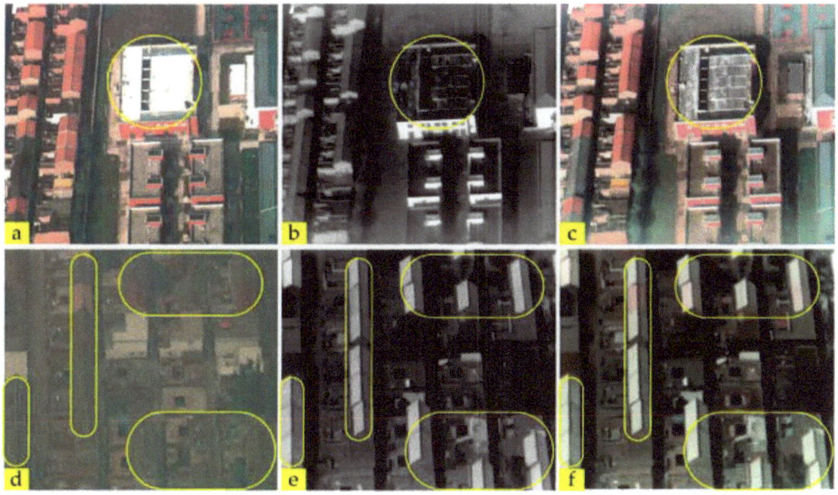

Figure 18. Fusion of interesting areas in two scenes: (**a**,**b**,**d**,**e**) original image; and (**c**,**f**) fusion image based on the proposed method.

3.6. Performance Analysis

3.6.1. Performance Analysis of Image Registration

In the performance test experiments, we choose 257 groups of images and corresponding metadata with three typical types of motions: translation, rotation, and scale. Based on the result of the scale transformation, we tested the performance of the five methods: the proposed method of integrated parallel vision-based registration (IPVBR), alignment metric-based registration (AMBR) [32], mutual information-based registration (MIBR) [16], peak signal-to-noise ratio-based registration (PSNRBR), and structural similarity-based registration (SSIMBR). PSNRBR and SSIMBR are two registration methods that use PSNR and SSIM as the similarity standard [65].

Under each motion condition, the values of root mean square error (RMSE) are calculated using Equation (38):

$$\begin{cases} RMSE = \sqrt{\frac{E_1^2+E_2^2+......+E_n^2}{n}} \\ E_i = \sqrt{(x_a - x_c)^2 + (y_a - y_c)^2} (i = 1,2,3,......) \end{cases} \quad (38)$$

where the measurement error E_i denotes the pixel distance from the corresponding calculated matching point (x_c, y_c) to the actual matching point (x_a, y_a) in the visible image. The error analysis results of the three experiments are shown in Figures 19–21.

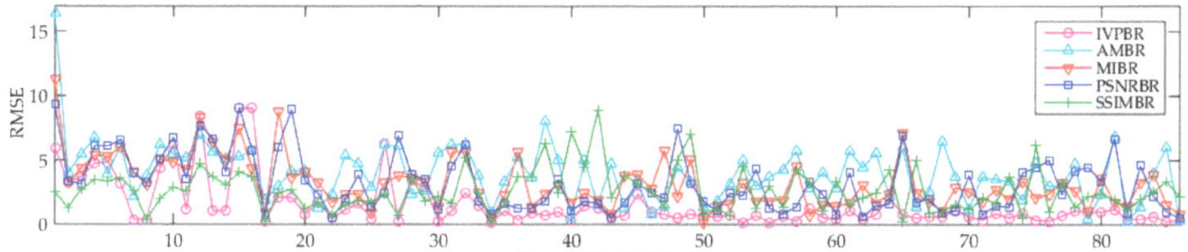

Figure 19. Performance analysis of the first experiment under translation conditions.

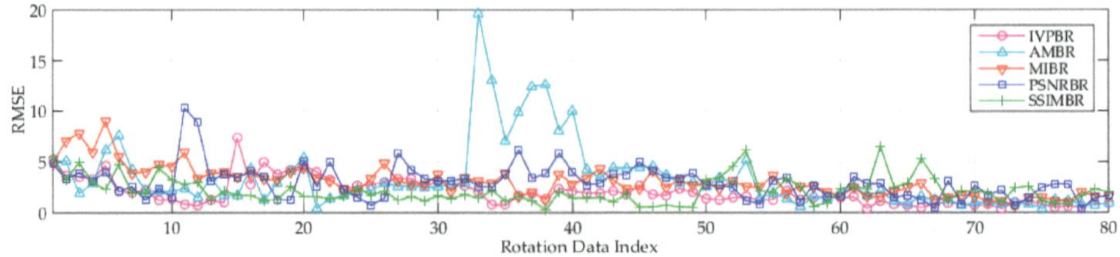

Figure 20. Performance analysis of the second experiment under rotation conditions.

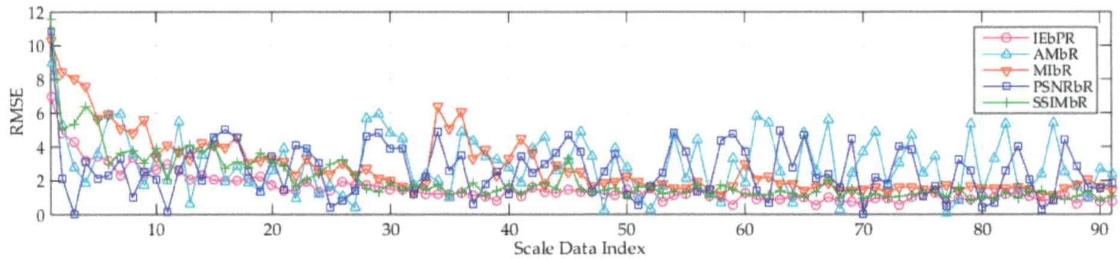

Figure 21. Performance analysis of the third experiment under scale conditions.

The average RMSE values of the five methods in the three experiments are shown in Table 8.

Table 8. Average RMSE of the five methods.

Index	Test Data (Frame Number)	AMBR (RMSE)	MIBR (RMSE)	PSNRBR (RMSE)	SSIMBR (RMSE)	Propose IVPBR (RMSE)
1	Translation (86)	3.98	3.19	3.24	2.63	1.55
2	Rotation (80)	3.37	3.04	2.97	2.16	2.01
3	Scale (91)	3.00	2.61	2.94	1.90	1.54
Average RMSE		3.45	2.95	3.05	2.23	1.70

As shown in Figures 19–21, the RMSE curve of IPVBR remains stable and low. The four other curves present different performances. The curve of SSIMBR presents good performance in Experiments 2 and 3, but it shows high vibration in Experiment 1. The curve of PSNRBR always maintains a certain vibration in Experiments1 and 3. The curve of AMBR indicates some high errors in Experiment 2 and presents high vibrations in Experiments 1 and 3. The curve of MIBR shows no good or bad performance. As shown in Table 8, the proposed IPVBR achieves the minimum average RMSE in the three experiments. SSIMBP also has a low average RMSE, along with IPVBR.

Three points can be concluded from these three experiments.

1. Compared with the four other methods, the proposed IPVBR presents a stable and low MSER. This result shows the high stability and precision of the proposed method.
2. SSIMBP is better than PSNRBP, which indicates that structure information is more reliable than pixel information for multimodal image registration.
3. The two representative conventional methods of AMBR and MIBR fail to achieve good results under the three motion conditions for medium-altitude UAV applications.

Three experiments are conducted based on the fact that all five algorithms can obtain nearly correct results. In some cases, the compared image-based algorithms fail to solve the perspective transform, and the proposed edge feature extraction and matching method is effective in translation calculation. At this point, the result reflects the obvious advantages of the proposed method.

3.6.2. Performance Analysis of Image Fusion

To analyze the performance, this study introduces three other methods: IHS transform-based fusion (IHSBF), PCA-based fusion (PCABF) [66], and SIDWT-based fusion (SIDWTBF) [67]. These methods are compared with the proposed method in the experiment.

Using 10 sets of visible and infrared images of different scenes as the experiment data, we select the average gradient (Equation (39)) and Shannon value (Equation (40)) as the evaluation indexes of the four methods. The average gradient can sensitively reflect the ability of the image to express the smallest details and can be used to evaluate the clarity of the image. A high average gradient equates to a clear image. A high Shannon value equates to a large amount of information in the image:

$$G = \frac{1}{(M-1)(N-1)} \sum_{m=1}^{M} \sum_{n=1}^{N} \sqrt{\frac{(f(x+1,y)-f(x,y))^2 + (f(x,y+1)-f(x,y))^2}{2}} \tag{39}$$

where $f(x,y)$ denotes the pixel value at (x,y) and $M \times N$ denotes the image resolution.

$$H = -\sum_{0}^{255} P_i \log_2 P_i \tag{40}$$

where i represents a sample in the image and P_i represents the probability of the sample.

The average gradient and Shannon results are shown in Figure 22, and the average values of the four image fusion methods are listed in Table 9.

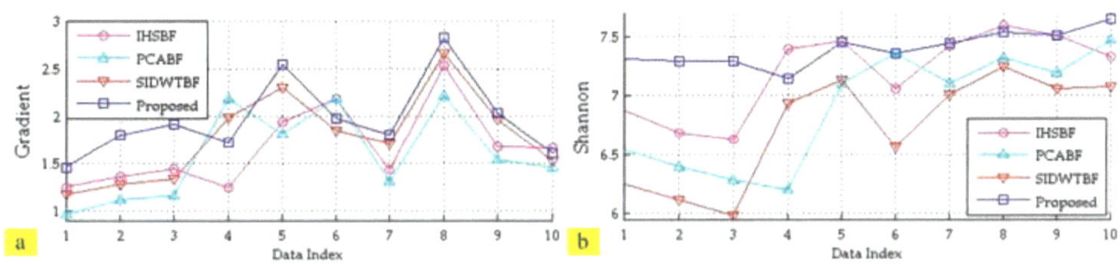

Figure 22. Average gradient and Shannon values of the four image fusion methods: (**a**) average gradient; and (**b**) Shannon value.

Table 9. Average gradient and Shannon value of the four methods.

Index	Evaluation Index	IHSBF	PCABF	SIDWTBF	Proposed
1	Average Gradient	1.67	1.59	1.78	**1.97**
2	Shannon	7.20	6.90	6.74	**7.40**

As shown in Figure 22a,b, the two group curves of our method are high and stable. Table 9 shows that the average values of our method are higher than those of the other three methods. The results also show that the fusion image obtained by our method has higher contrast, better details, and more information than the images obtained with the other methods.

4. Conclusions

Visible and infrared image registration is a difficult problem in medium-altitude UAVs because of different imaging mechanisms, poor image quality, and large amounts of motion in videos. For the special requirements of UAV applications, an appropriate image fusion method becomes a key technology.

This study proposed a novel image registration method that uses both metadata and image based on the imaging characteristic analysis of the most common visible light and infrared integrated camera. The main contributions of this work are reflected in three aspects. First, we reveal the principle of long-distance integrated parallel vision, which provides the theoretical foundation of the conversion from a perspective transformation to scale and translation transformations. Second, two new algorithms for scale calculation and coarse translation estimation are presented using the image metadata of the UAV system according to spatial geometry and coordinate transformation. Third, an edge distance field-based registration is proposed in precise translation estimation to solve the non-strict edge alignment of the visible image and infrared image. A searching algorithm based on PSO is also put forward to improve efficiency. In image fusion, this study designs a new method based on PCNN and NSCT. This method can meet the four requirements of preserving color information, adding infrared brightness information, improving spatial resolution, and highlighting target areas for UAV applications.

A medium-altitude UAV is employed to collect experimental data, including three typical groups of translation, rotation, and scale. Results show that the proposed method achieves encouraging performance in image registration and fusion. These results can be applied to other medium-altitude or high-altitude UAVs with a similar system structure. However, future work should focus on analysis and experiments, such as the improved transformation of edge distance field and real time optimization of image fusion.

Acknowledgments: This work was supported by the National Key Research and Development Project (No. 2016YFB1200100) and the National Natural Science Foundation of China (No. 61601014).

Author Contributions: Hongguang Li wrote the program and the manuscript. Wenrui Ding and Xianbin Cao revised the paper. Chunlei Liu performed the experiments and analyzed the data.

Conflicts of Interest: The authors declare no conflict of interest.

References

1. Deren, L.I. On space-air-ground integrated earth observation network. *J. Geo-Inf. Sci.* **2012**, *14*, 419–425.
2. Zhao, J.; Zhou, Q.; Chen, Y.; Feng, H.; Xu, Z.; Li, Q. Fusion of visible and infrared images using saliency analysis and detail preserving based image decomposition. *Infrared Phys. Technol.* **2013**, *56*, 93–99. [CrossRef]
3. Zhou, D.; Zhong, Z.; Zhang, D.; Shen, L.; Yan, C. Autonomous landing of a helicopter UAV with a ground-based multisensory fusion system. In Proceedings of the International Conference on Machine Vision, Koto-ku, Japan, 18–22 May 2015.
4. Ulusoy, I.; Yuruk, H. New method for the fusion of complementary information from infrared and visual images for object detection. *IET Image Process.* **2011**, *5*, 36–48. [CrossRef]
5. Niu, Y.F.; Xu, S.T.; Hu, W.D. Fusion of infrared and visible image based on target regions for environment perception. *Appl. Mech. Mater.* **2011**, *128–129*, 589–593. [CrossRef]
6. Pulpea, B.G. Aspects regarding the development of pyrotechnic obscurant systems for visible and infrared protection of military vehicles. In Proceedings of the International Conference Knowledge-Based Organization, Land Forces Academy, Sibiu, Romania, 11–13 June 2015; pp. 731–736.
7. Teng, H.; Viscarra Rossel, R.A.; Shi, Z.; Behrens, T.; Chappell, A.; Bui, E. Assimilating satellite imagery and visible-near infrared spectroscopy to model and map soil loss by water erosion in australia. *Environ. Model. Softw.* **2016**, *77*, 156–167. [CrossRef]
8. Peña, J.M.; Torres-Sánchez, J.; Serrano-Pérez, A.; de Castro, A.I.; López-Granados, F. Quantifying efficacy and limits of unmanned aerial vehicle (UAV) technology for weed seedling detection as affected by sensor resolution. *Sensors* **2015**, *15*, 5609–5626. [CrossRef] [PubMed]
9. Chrétien, L.P.; Théau, J.; Ménard, P. Visible and thermal infrared remote sensing for the detection of white-tailed deer using an unmanned aerial system. *Wildl. Soc. Bull.* **2016**, *40*, 181–191. [CrossRef]
10. Zhao, B.; Li, Z.; Liu, M.; Cao, W.; Liu, H. Infrared and visible imagery fusion based on region saliency detection for 24-h-surveillance systems. In Proceedings of the IEEE International Conference on Robotics and Biomimetics, Shenzhen, China, 12–14 December 2013; pp. 1083–1088.

11. Wang, Q.; Yan, P.; Yuan, Y.; Li, X. Multi-spectral saliency detection. *Pattern Recognit. Lett.* **2013**, *34*, 34–41. [CrossRef]
12. Wang, Q.; Zhu, G.; Yuan, Y. Multi-spectral dataset and its application in saliency detection. *Comput. Vis. Image Underst.* **2013**, *117*, 1748–1754. [CrossRef]
13. Berenstein, R.; Hočevar, M.; Godeša, T.; Edan, Y.; Benshahar, O. Distance-dependent multimodal image registration for agriculture tasks. *Sensors* **2014**, *15*, 20845–20862. [CrossRef] [PubMed]
14. Kaneko, S.I.; Murase, I.; Igarashi, S. Robust image registration by increment sign correlation. *Pattern Recognit.* **2010**, *35*, 2223–2234. [CrossRef]
15. Tsin, Y.; Kanade, T. A correlation-based approach to robust point set registration. In Proceedings of the Computer Vision—ECCV 2004, European Conference on Computer Vision, Prague, Czech Republic, 11–14 May 2004; pp. 558–569.
16. Zhuang, Y.; Gao, K.; Miu, X.; Han, L.; Gong, X. Infrared and visual image registration based on mutual information with a combined particle swarm optimization—Powell search algorithm. *Optik—Int. J. Light Electron Opt.* **2016**, *127*, 188–191. [CrossRef]
17. Zhang, Z.; Yang, G.; Chen, D.; Li, J.; Yang, W. Registration of infrared and visual images based on phase grouping and mutual information of gradient orientation. In Proceedings of the SPIE Photonics Europe, Brussels, Belgium, 4–7 April 2016.
18. Li, C.; Chen, Q. Ir and visible images registration method based on cross cumulative residual entropy. *Proc. SPIE—Int. Soc. Opt. Eng.* **2013**, *8704*, 145–223.
19. Pohit, M.; Sharma, J. Image registration under translation and rotation in two-dimensional planes using fourier slice theorem. *Appl. Opt.* **2015**, *54*, 4514–4519. [CrossRef] [PubMed]
20. Niu, H.; Chen, E.; Qi, L.; Guo, X. Image registration based on fractional fourier transform. *Optik—Int. J. Light Electron Opt.* **2015**, *126*, 3889–3893. [CrossRef]
21. Li, H.; Zhang, A.; Hu, S. A registration scheme for multispectral systems using phase correlation and scale invariant feature matching. *J. Sens.* **2016**, *2016*, 1–9. [CrossRef]
22. Wang, Q.; Zou, C.; Yuan, Y.; Lu, H.; Yan, P. Image registration by normalized mapping. *Neurocomputing* **2013**, *101*, 181–189. [CrossRef]
23. Lowe, D.G. Object recognition from local scale-invariant features. In Proceedings of the IEEE International Conference on Computer Vision, Kerkyra, Greece, 20–27 September 1999; p. 1150.
24. Lowe, D.G.; Lowe, D.G. Distinctive image features from scale-invariant keypoints. *Int. J. Comput. Vis.* **2004**, *60*, 91–110. [CrossRef]
25. Huang, Q.; Yang, J.; Wang, C.; Chen, J.; Meng, Y. Improved registration method for infrared and visible remote sensing image using nsct and sift. In Proceedings of the Geoscience and Remote Sensing Symposium, Munich, Germany, 22–27 July 2012; pp. 2360–2363.
26. Li, D. An Infrared and Visible Image Registration Based on Surf. 2012, 19–25. Available online: https://datahub.io/dataset/an-infrared-and-visible-image-registration-based-on-surf (accessed on 5 May 2017).
27. Coiras, E.; Santamaria, J.; Miravet, C. Segment-based registration technique for visual-infrared images. *Opt. Eng.* **2000**, *39*, 202–207. [CrossRef]
28. Han, J.; Pauwels, E.; Zeeuw, P.D. *Visible and Infrared Image Registration Employing Line-Based Geometric Analysis*; Springer: Berlin/Heidelberg, Germany, 2012; pp. 114–125.
29. Liu, L.; Tuo, H.Y.; Xu, T.; Jing, Z.L. Multi-spectral image registration and evaluation based on edge-enhanced mser. *Imaging Sci. J.* **2014**, *62*, 228–235. [CrossRef]
30. Qin, H. Visible and infrared image registration based on visual salient features. *J. Electron. Imaging* **2015**, *24*, 053017.
31. Piella, G. Diffusion maps for multimodal registration. *Sensors* **2014**, *14*, 10562–10577. [CrossRef] [PubMed]
32. Xia, Y.J.; Yin, J.Q.; Chen, R. An automatic registration method for multi-modal images based on alignment metric. *Appl. Mech. Mater.* **2012**, *182–183*, 1308–1312. [CrossRef]
33. Han, J.; Pauwels, E.J.; Zeeuw, P.D. Visible and infrared image registration in man-made environments employing hybrid visual features. *Pattern Recognit. Lett.* **2013**, *34*, 42–51. [CrossRef]
34. Huang, Q.; Yang, J.; Chen, J.; Gao, Q.; Song, Z. Visible and infrared image registration algorithm based on nsct and gradient mirroring. *Proc. SPIE—Multispectr. Hyperspectr. Ultraspectr. Remote Sens. Technol. Tech. Appl.* **2014**. [CrossRef]

35. Wang, R.; Du, L. Infrared and visible image fusion based on random projection and sparse representation. *Int. J. Remote Sens.* **2014**, *35*, 1640–1652. [CrossRef]
36. Pohl, C.; Genderen, J.L.V. Review article multisensor image fusion in remote sensing: Concepts, methods and applications. *Int. J. Remote Sens.* **1998**, *19*, 823–854. [CrossRef]
37. Nawaz, Q.; Bin, X.; Weisheng, L.; Jiao, D.; Hamid, I. Multi-modal medical image fusion using RGB-principal component analysis. *J. Med. Imaging Health Inf.* **2016**, *6*, 1349–1356. [CrossRef]
38. Toet, A.; Walraven, J. New false color mapping for image fusion. *Opt. Eng.* **1996**, *35*, 650–658. [CrossRef]
39. Kadar, I. Quick markov random field image fusion. *Proc. SPIE—Int. Soc. Opt. Eng.* **1998**, *3374*, 302–308.
40. Sharma, R.K.; Leen, T.K.; Pavel, M. Bayesian sensor image fusion using local linear generative models. *Opt. Eng.* **2002**, *40*, 1364–1376.
41. Zhang, Z.L.; Sun, S.H.; Zheng, F.C. Image fusion based on median filters and sofm neural networks: A three-step scheme. *Signal Process.* **2001**, *81*, 1325–1330. [CrossRef]
42. Zhang, Y.X.; Chen, L.; Zhao, Z. A novel pulse coupled neural network based method for multi-focus image fusion. *Int. J. Signal Process. Image Process. Pattern Recognit.* **2014**, *12*, 357–366. [CrossRef]
43. Guan, W.; Li, L.; Jin, W.; Qiu, S.; Zou, Y. Research on hdr image fusion algorithm based on laplace pyramid weight transform with extreme low-light CMOS. In Proceedings of the Applied Optics and Photonics China, Beijing, China, 5–7 May 2015; p. 967524.
44. Li, H.; Manjunath, B.S.; Mitra, S.K. Multisensor image fusion using the wavelet transform. *Graph. Models Image Process.* **1995**, *57*, 235–245. [CrossRef]
45. Chen, T.; Zhang, J.; Zhang, Y. In Remote sensing image fusion based on ridgelet transform. In Proceedings of the 2005 IEEE International Geoscience and Remote Sensing Symposium, Seoul, South Korea, 25–29 July 2005; pp. 1150–1153.
46. Lutz, A.; Giansiracusa, M.; Messer, N.; Ezekiel, S.; Blasch, E.; Alford, M. Optimal multi-focus contourlet-based image fusion algorithm selection. In Proceedings of the SPIE Defense + Security, Baltimore, MD, USA, 17–21 April 2016; p. 98410E.
47. Zhang, Q.; Guo, B.L. Multifocus image fusion using the nonsubsampled contourlet transform. *Signal Process.* **2009**, *89*, 1334–1346. [CrossRef]
48. Han, J.; Loffeld, O.; Hartmann, K.; Wang, R. Multi image fusion based on compressive sensing. In Proceedings of the International Conference on Audio Language and Image Processing, Shanghai, China, 23–25 November 2010; pp. 1463–1469.
49. Wei, Q.; Bioucas-Dias, J.; Dobigeon, N.; Tourneret, J.Y. Hyperspectral and multispectral image fusion based on a sparse representation. *IEEE Trans. Geosci. Remote Sens.* **2014**, *53*, 3658–3668. [CrossRef]
50. Zhang, Q.; Liu, Y.; Blum, R.S.; Han, J.; Tao, D. Sparse Representation Based Multi-Sensor Image Fusion: A Review. Available online: https://arxiv.org/abs/1702.03515 (accessed on 4 May 2017).
51. Han, J.; Pauwels, E.J.; De Zeeuw, P. Fast saliency-aware multi-modality image fusion. *Neurocomputing* **2013**, *111*, 70–80. [CrossRef]
52. Liu, K.; Guo, L.; Li, H.; Chen, J. Fusion of infrared and visible light images based on region segmentation. *Chin. J. Aeronaut.* **2009**, *22*, 75–80.
53. Sturm, P. *Pinhole Camera Model*; Springer: Washington, DC, USA, 2014; pp. 300–321.
54. Hartley, R.; Zisserman, A. Multiple view geometry in computer vision. *Kybernetes* **2001**, *30*, 1865–1872.
55. Li, H.; Li, X.; Ding, W.; Huang, Y. Metadata-assisted global motion estimation for medium-altitude unmanned aerial vehicle video applications. *Remote Sens.* **2015**, *7*, 12606–12634. [CrossRef]
56. Han, J.; Farin, D.; De With, P. Broadcast court-net sports video analysis using fast 3-d camera modeling. *IEEE Trans. Circuits Syst. Video Technol.* **2008**, *18*, 1628–1638.
57. Canny, J. A computational approach to edge detection. *IEEE Trans. Pattern Anal. Mach. Intell.* **1986**, *8*, 679–698. [CrossRef] [PubMed]
58. Jie, G.; Ning, L. An improved adaptive threshold canny edge detection algorithm. In Proceedings of the International Conference on Computer Science and Electronics Engineering, Colchester, UK, 28–30 September 2012; pp. 164–168.
59. Li, Z.; Zhu, X. Matching Multi—Sensor Images Based on Edge Similarity. *J. Spacecr. TTC Technol.* **2011**, *30*, 37–41.
60. Kennedy, J.; Eberhart, R. Particle swarm optimization. In Proceedings of the IEEE International Conference on Neural Networks, Perth, Australia, 27 November–1 December 1995; Volume 1944, pp. 1942–1948.

61. Song, Y.; Eberhart, R. A modified particle swarm optimizer. In Proceedings of the 1998 IEEE International Conference on Evolutionary Computation Proceedings, Anchorage, AK, USA, 4–9 May 1998.
62. Arasomwan, M.A.; Adewumi, A.O. On the performance of linear decreasing inertia weight particle swarm optimization for global optimization. *Sci. World J.* **2013**, *78*, 1648–1653. [CrossRef] [PubMed]
63. Kuntimad, G.; Ranganath, H.S. Perfect image segmentation using pulse coupled neural networks. *IEEE Trans. Neural Netw.* **1999**, *10*, 591. [CrossRef] [PubMed]
64. Bamberger, R.H.; Smith, M.J.T. A filter bank for the directional decomposition of images: Theory and design. *IEEE Trans. Signal Process.* **1992**, *40*, 882–893. [CrossRef]
65. Hore, A.; Ziou, D. Image quality metrics: PSNR vs. SSIM. In Proceedings of the 2010 20th International Conference on Pattern Recognition (ICPR), Istanbul, Turkey, 23–26 August 2010; pp. 2366–2369.
66. He, C.; Liu, Q.; Li, H.; Wang, H. Multimodal medical image fusion based on IHS and PCA. *Procedia Eng.* **2010**, *7*, 280–285. [CrossRef]
67. Xin, W.; Wei, Y.L.; Fu, L. A new multi-source image sequence fusion algorithm based on sidwt. In Proceedings of the Seventh International Conference on Image and Graphics, Qingdao, China, 26–28 July 2013; pp. 568–571.

remote sensing

MDPI

Article

Learning a Dilated Residual Network for SAR Image Despeckling

Qiang Zhang [1], Qiangqiang Yuan [1,*], Jie Li [2], Zhen Yang [3] and Xiaoshuang Ma [4]

1. School of Geodesy and Geomatics, Wuhan University, Wuhan 430079, China; zhang_qiang@whu.edu.cn
2. International School of Software, Wuhan University, Wuhan 430079, China; aaronleecool@whu.edu.cn
3. School of Resource and Environmental Science, Wuhan University, Wuhan 430079, China; SendimageYZ@whu.edu.cn
4. School of Resources and Environmental Engineering, Anhui University, Hefei 230000, China; mxs.88@whu.edu.cn

* Correspondence: qqyuan@sgg.whu.edu.cn; Tel.: +86-159-7217-1792

Received: 13 November 2017; Accepted: 24 January 2018; Published: 29 January 2018

Abstract: In this paper, to break the limit of the traditional linear models for synthetic aperture radar (SAR) image despeckling, we propose a novel deep learning approach by learning a non-linear end-to-end mapping between the noisy and clean SAR images with a dilated residual network (SAR-DRN). SAR-DRN is based on dilated convolutions, which can both enlarge the receptive field and maintain the filter size and layer depth with a lightweight structure. In addition, skip connections and a residual learning strategy are added to the despeckling model to maintain the image details and reduce the vanishing gradient problem. Compared with the traditional despeckling methods, the proposed method shows a superior performance over the state-of-the-art methods in both quantitative and visual assessments, especially for strong speckle noise.

Keywords: SAR image; despeckling; dilated convolution; skip connection; residual learning

1. Introduction

A synthetic aperture radar (SAR) is a coherent imaging sensor, which can access a wide range of high-quality massive surface data. Moreover, with the ability to operate at night and in adverse weather conditions such as thin clouds and haze, SAR has gradually become a significant source of remote sensing data in the fields of geographic mapping, resource surveying, and military reconnaissance. However, SAR images are inherently affected by multiplicative noise, i.e., speckle noise, which is caused by the coherent nature of the scattering phenomena [1]. The presence of speckle severely affects the quality of SAR images, and greatly reduces the utilization efficiency in SAR image interpretation, retrieval, and other applications [2–4]. Consequently, SAR image speckle reduction is an essential preprocessing step and has become a hot research topic.

For the purpose of removing the speckle noise of SAR images, scholars firstly proposed spatial linear filters such as the Lee filter [5], Kuan filter [6], and Frost filter [7]. These methods usually assume that the image filtering result values have a linear relationship with the original image, through searching for a relevant combination of the central pixel intensity in a moving window with a mean intensity of the filter window. Thus, the spatial linear filters achieve a trade-off between balancing in homogeneous areas and a constant all-pass identity filter in edge included areas. The results have confirmed that spatial-domain filters are adept at suppressing speckle noise for some critical features. However, due to the nature of local processing, the spatial linear filter methods often fail to integrally preserve edges and details, which exhibit the following deficiencies: (1) unable to preserve the average value, especially when the equivalent number of look (ENL) of the original SAR image is small; (2) the powerfully reflective specific targets like points and small surficial features are easily blurred or erased; and (3) speckle noise in dark scenes is not removed [8].

Except for the spatial-domain filters above, wavelet theory has also been applied to speckle reduction. Starck et al. [9] primarily employed ridgelet transform as a component step, and implemented curvelet sub-bands using a filter bank of the discrete wavelet transform (DWT) filters for image denoising. For the case of speckle noise, Solbo et al. [10] utilized the DWT of the log-transformed speckled image in homomorphic filtering, which is empirically convergent in a self-adaptive strategy and calculated in the Fourier space. In summary, the major weaknesses of this type of approach are the backscatter mean preservation in homogeneous areas, details preservation, and producing an artificial effect that is incorporated into the results, such as ring effects [11].

Aimed at overcoming these deficiencies, the nonlocal means (NLM) algorithm [12–14] has provided a breakthrough in detail preservation in SAR image despeckling. The basic idea of the NLM-based methods [12] is that natural images have self-similarity and there are similar patches repeating over and over throughout the whole image. For SAR images, Deledalle et al. [13] modified the choice of weights, which can be iteratively determined based on both the similarity between noisy patches and the similarity of patches extracted from the previous estimate. Besides, Parrilli et al. [14] used the local linear minimum mean square error (LLMMSE) criterion and undecimated wavelet transform considering the peculiarities of SAR images, allowing for a sparse Wiener filtering representation and an effective separation between original signal and speckle noise through predefined thresholding, which has become one of the most effective SAR despeckling methods. However, the low computational efficiency of the similar patch searching restricts its application.

In addition, the variational-based methods [15–18] have gradually been utilized for SAR image despeckling because of their stability and flexibility, which break through the traditional idea of filters by solving the problem of energy optimization. Then, the despeckling task is cast as the inverse problem of recovering the original noise-free image based upon reasonable assumptions or prior knowledge of the noise observation model with log-transform, such as the total variation (TV) model [15], sparse representation [16], and so on. Although these variational methods have achieved a good reduction of speckle noise, the result is usually dependent on the choice of model parameters and prior information, and is often time-consuming. In addition, the variational-based methods cannot accurately describe the distribution of speckle noise, which also constraints the performance of speckle noise reduction.

In general, although many SAR despeckling methods have been proposed, they sometimes fail to preserve sharp features in domains of a complicated texture, or even create some block artifacts in the speckled image. In this paper, considering that image speckle noise can be expressed more accurately through non-linear models than linear models, and to overcome the above-mentioned limitations of the linear models, we propose a novel deep neural network-based approach for SAR image despeckling, learning a non-linear end-to-end mapping between the speckled and clean SAR images by a dilated residual network (SAR-DRN). Our despeckling model employs dilated convolutions, which can both enlarge the receptive field and maintain the filter size and layer depth with a lightweight structure. Furthermore, skip connections are added to the despeckling model to maintain the image details and avoid the vanishing gradient problem. Compared with the traditional despeckling methods in both simulated and real SAR experiments, the proposed approach shows a state-of-the-art performance in both quantitative and visual assessments, especially for strong speckle noise.

The rest of this paper is organized as follows. The SAR image speckling noise degradation model and the related deep convolution neural network method are introduced in Section 2. The network architecture of the proposed SAR-DRN and details of its structure are described in Section 3. Then, the results of the despeckling assessment in both simulated and real SAR image experiments are presented in Section 4. Finally, the conclusions and future research are summarized in Section 5.

2. Related Work

2.1. SAR Image Speckling Noise Degradation Model

For SAR images, the main reason for the degradation of the image quality is multiplicative speckle noise. Differing from additive white Gaussian noise (AWGN) in nature or hyperspectral images [19,20], speckle noise is described by the multiplicative noise model:

$$y = x \cdot n \tag{1}$$

where y is the speckled noise image, x is the clean image, and n represents the speckle noise. It is well-known that, for SAR amplitude images, the speckle follows a Gamma distribution [21]:

$$\rho_n(n) = \frac{L^L n^{L-1} \exp(-nL)}{\Gamma(L)} \tag{2}$$

where $L \geq 1$, $n \geq 0$, Γ is the Gamma function, and L is the equivalent number of looks (*ENL*), as defined in Equation (3), which is usually regarded as the quantitative evaluation index for real SAR image despeckling experiments in the homogeneous areas.

$$ENL = \frac{\overline{x}}{\text{var}} \tag{3}$$

where $\overline{x}$ and var, respectively, represent the image mean and variance.

Therefore, for this non-linear multiplicative noise, choosing a non-linear expression for speckle reduction is an important strategy. In the following, we briefly introduce the use of convolutional neural networks (CNNs) for SAR image despeckling, considering both the low-level features as the bottom level and the output feature representation from the top level of the network.

2.2. CNNs for SAR Image Despeckling

With recent advances made by deep learning for computer vision and image processing applications, it has gradually become an efficient tool which has been successfully applied to many computer vision tasks such as image classification, segmentation, object recognition, scene classification, and so on [22–24]. CNNs can extract the internal and underlying features of images and avoid complex priori constraints, organized in the j-th feature map $O_j^{(l)} (j = 1, 2, \ldots M^{(l)})$ of l-th layer, within which each unit is connected to local patches of the previous layer $O_j^{(l-1)} (j = 1, 2, \ldots M^{(l-1)})$ through a set of weight parameters $W_j^{(l)}$ and bias parameters $b_j^{(l)}$. The output feature map is:

$$L_j^{(l)}(m,n) = F(O_j^{(l)}(m,n)) \tag{4}$$

And

$$O_j^{(l)}(m,n) = \sum_{i=1}^{M^{(l)}} \sum_{u,v=0}^{S-1} W_{ji}^{(l)}(u,v) \cdot L_i^{(l-1)}(m-u, n-v) + b_j^{(l)} \tag{5}$$

where $F(\cdot)$ is the nonlinear activation function, and $O_j^{(l)}(m,n)$ represents the convolutional weighted sum of the previous layer's results, to the j-th output feature map at pixel (m,n). Besides, the special parameters in the convolution layer contain the number of output feature maps j, and filter kernel size $S \times S$. Particularly, the network parameters W and b need to be regenerated through the back-propagation (BP) algorithm and the chain rule of derivation [25].

To ensure that the output of the CNNs is a non-linear combination of the input, due to the relationship between the input data and the output label usually being a highly nonlinear mapping, a non-linear function is introduced as an excitation function, such as the rectified linear unit (ReLU), which is defined as:

$$F(O_j^{(l)}) = \max(0, O_j^{(l)}) \tag{6}$$

After finishing each process of forward propagation, the BP algorithm starts to perform for update trainable parameters of networks, to better learn the relationships between label data and reconstructing data. From the top layer of the network to the bottom, BP updates the trainable parameters of the l-th layer through the outputs of the $l+1$-th layer. The partial derivative of loss function with respect to convolution kernels $W_{ji}^{(l)}$ and bias $b_j^{(l)}$ of the l-th convolution layer is respectively calculated as follows:

$$\frac{\partial L}{\partial W_{ji}^{(l)}} = \sum_{m,n} \delta_j^{(l)}(m,n) \cdot L_j^{(l)}(m-u, y-v) \tag{7}$$

$$\frac{\partial L}{\partial b_j^{(l)}} = \sum_{m,n} \delta_j^{(l)}(m,n) \tag{8}$$

where the error map $\delta_j^{(l)}$ is defined as

$$\delta_j^{(l)} = \sum_j \sum_{u,v=0}^{S-1} W_{ji}^{(l+1)}(u,v) \cdot \delta_j^{(l+1)}(m+u, n+v) \tag{9}$$

The iterative training rule for updating the network parameters $W_{ji}^{(l)}$ and $b_j^{(l)}$ is through the gradient descent strategy as follows:

$$W_{ji}^{(l)} = W_{ji}^{(l)} - \alpha \cdot \frac{\partial L}{\partial W_{ji}^{(l)}} \tag{10}$$

$$b_j^{(l)} = b_j^{(l)} - \alpha \cdot \frac{\partial L}{\partial b_j^{(l)}} \tag{11}$$

where α is a preset hyperparameter for the whole network, which is also named the learning rate in a deep learning framework and controls the sampling interval of the trainable parameter.

For natural Gaussian noise reduction, a new method named the feed-forward denoising convolutional neural network (DnCNN) [26] has recently shown excellent performances, in contrast with the traditional methods which employ a deep convolutional neural network. DnCNN employs a 20 convolutional layers structure, a learning strategy of residual learning to remove the latent original image in the hidden layers, and an output data regularization method of batch normalization [27], which can deal with several universal image restoration tasks such as blind or non-blind image Gaussian denoising, and single image super-resolution and JPEG image deblocking.

Recently, borrowing the thought of the DnCNN model, Chierchia et al. [28] also employed a set of convolutional layers named SAR-CNN, along with batch normalization (BN) and ReLU activation function, and a component-wise division residual layer to estimate the speckled image. As an alternative way of dealing with the multiplicative noise of SAR images, SAR-CNN uses the homomorphic approach with coupled logarithm and exponent transforms in combination with a similarity measure for speckle noise distribution. In addition, Wang et al. [29] also used a similar structure like DnCNN, with eight-layers of the Conv-BN-ReLU block, and replaced residual mean square error (MSE) with a combination of Euclidean loss and total variation loss, which is incorporated into the total loss function to facilitate more smooth results.

3. Proposed Method

In this paper, rather than using log-transform [28] or modifying training loss function like [29], we propose a novel network for SAR image despeckling with a dilated residual network (SAR-DRN), which is trained in an end-to-end fashion using a combination of dilated convolutions and skip connections with a residual learning structure. Instead of relying on a pre-determined image, a *priori* knowledge, or a noise description model, the main superiority of using the deep neural network strategy for SAR image despeckling is that the model can directly acquire and update the network parameters from the training data and the corresponding labels, which need not manually adjust critical parameters and can automatically learn the complex internal non-linear relations with trainable network parameters from the massive training simulative data.

The proposed holistic neural network model (SAR-DRN) for SAR image despeckling contains seven dilated convolution layers and two skip connections, as illustrated in Figure 1. In addition, the proposed model uses a residual learning strategy to predict the speckled image, which adequately utilizes the non-linear expression ability of deep learning. The details of the algorithm are described in the following.

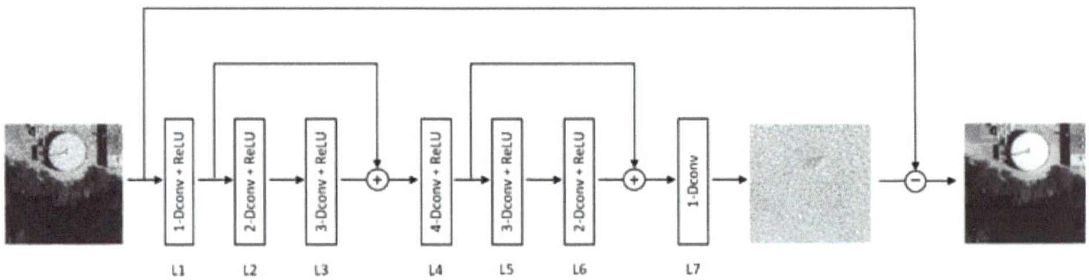

Figure 1. The architecture of the proposed SAR-DRN.

3.1. Dilated Convolutions

In image restoration problems such as single-image super-resolution (SISR) [30], denoising [31], and deblurring [32], contextual information can effectively facilitate the recovery of degraded regions. In deep convolutional networks, the contextual information is mainly augmented through enlarging the receptive field. Generically, there are two ways to achieve this purpose: (1) increasing the network depth; and (2) enlarging the filter size. Nevertheless, as the network depth increases, the accuracy becomes "saturated" and then degrades rapidly. Enlarging the filter size can also lead to more convolution parameters, which greatly increases the calculative burden and training times.

To solve this problem effectively, dilated convolutions were first proposed in [33], which can both enlarge the receptive field and maintain the filter size. Let C be an input discrete two-dimensional matrix such as an image, and let k be a discrete convolution filter of size $(2r+1) \times (2r+1)$. Then, the original discrete convolution operator $*$ can be given as

$$(C * k)(p) = \sum_{i+j=p} C(i) \cdot k(j) \tag{12}$$

After defined this convolution operator $*$, let d be a dilation factor and let $*_d$ be equivalent to

$$(C *_d k)(p) = \sum_{i+d \cdot j=p} C(i) \cdot k(j) \tag{13}$$

where $*_d$ is served as the dilated convolution or a d-dilated convolution. Particularly, the common discrete convolution $*$ can be regarded as the l-dilated convolution. Setting the size of the convolutional kernel with 3×3 as an example, let k_l be the discrete 3×3 convolution filters. Consider applying the filters with exponentially increasing dilation as

$$R_{l+1} = R_l *_{\phi} k_l \tag{14}$$

where $l = 0, 1, \ldots, n-2$, $\phi = 2^l$, and R_l represents the size of the receptive field. The common convolution receptive field has a linear correlation with the layer depth, in that the receptive field size: $R_l^c = (2l+1) \times (2l+1)$. By contrast, the dilated convolution receptive field has an exponential correlation with the layer depth, where the receptive field size: $R_l^d = (2^{l+1}-1) \times (2^{l+1}-1)$. For instance, when $l = 4$, $R_l^c = 9 \times 9$, while $R_l^d = 31 \times 31$ with the same layer depth. Figure 2 illustrates the dilated convolution receptive field size, which: (a) corresponds to the one-dilated convolution, which is equivalent to the common convolution operation at this point; (b) corresponds to the two-dilated convolution; and (c) corresponds to the four-dilated convolution.

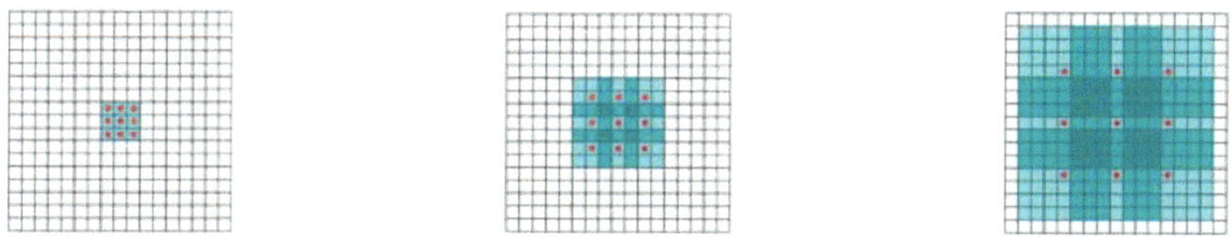

(**a**) 1-dilated convolution. (**b**) 2-dilated convolution. (**c**) 4-dilated convolution.

Figure 2. Receptive field size of different dilated convolution. (d = 1, 2, and 4, where the dark color regions represent the receptive field).

In the proposed SAR-DRN model, considering that trade-off between feature extraction ability and reducing training time, the dilation factors of the 3×3 dilated convolutions from layer 1 to layer 7 are respectively set to 1, 2, 3, 4, 3, 2, and 1, empirically. Compared with other deep neural networks, we propose a lightweight model with only seven dilated convolution layers, as shown in Figure 3.

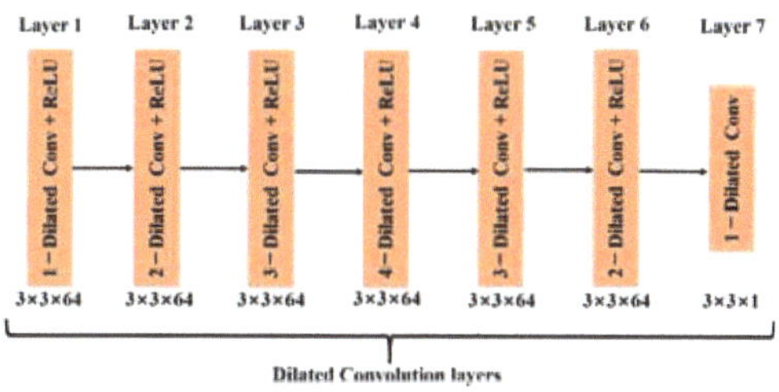

Figure 3. Dilated convolution in the proposed model.

3.2. Skip Connections

Although the increase of network layer depth can help to obtain more data feature expressions, it often results in the vanishing gradient problem, which makes the training of the model much harder. To solve this problem, a new structure called skip connection [34] has been created for the DCNNs, to obtain better training results. The skip connection can pass the previous layer's feature information to its posterior layer, maintaining the image details and avoiding or reducing the vanishing gradient problem. For the l-th layer, let $L^{(l)}$ be the input data, and let $f(L^{(l)}, \{W, b\})$ be its feed-forward propagation with trainable parameters. The output of the $(l+k)$-th layer with k-interval skip connection is recursively defined as follows:

$$L^{(l+k)} = f(L^{(l)}, \{W, b\}_{l+1 \to l+k}) + L^{(l)} \tag{15}$$

For clarity, in the proposed SAR-DRN model, two skip connections are employed to connect layer 1 to layer 3 (as shown in Figure 4a) and layer 4 to layer 7 (as shown in Figure 4b), whose effects are compared with no skip connections in the discussion section.

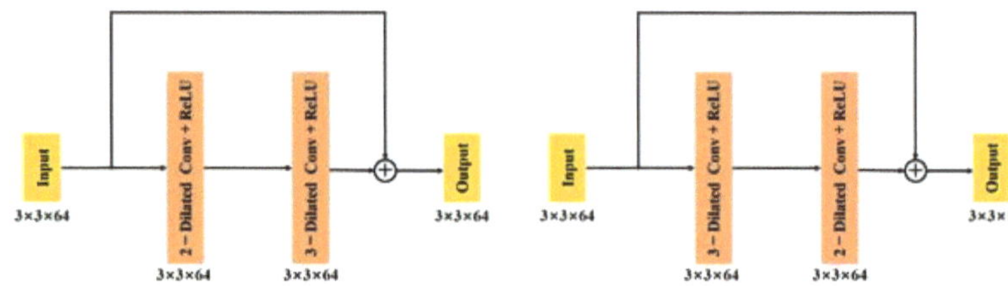

(a) (b)

Figure 4. Diagram of skip connection structure in the proposed model. (**a**) Connecting dilated convolution layer 1 to dilated convolution layer 3. (**b**) Dilated convolution layer 4 to dilated convolution layer 7.

3.3. Residual Learning

Compared with traditional data mapping, He et al. [35] found that residual mapping can acquire a more effective learning effect and rapidly reduce the training loss after passing through a multi-layer network, which has achieved a state-of-the-art performance in object detection [36], image super-resolution [37], and so on. Essentially, Szegedy et al. [38] demonstrated that residual networks take full advantage of identity shortcut connections, which can efficiently transfer various levels of feature information between not directly connected layers without attenuation. In the proposed SAR-DRN model, the residual image φ is defined as follows:

$$\varphi = y_i - x_i \tag{16}$$

As the layer depth increases, the degradation phenomenon manifests that common deep networks might have difficulties in approximating identical mappings by stacked non-linear layers like the Conv-BN-ReLU block. By contrast, it is reasonable to consider that most pixel values in residual image φ are very close to zero, and the spatial distribution of the residual feature maps should be very sparse, which can transfer the gradient descent process to a much smoother hyper-surface of loss to filtering parameters. Thus, searching for an allocation which is on the verge of the optimal for the network's parameters becomes much quicker and easier, allowing us to add more trainable layers to the network and improve its performance. The learning procedure with a residual unit is easier to approximate to the original multiplicative speckle noise through the deeper and intrinsic non-linear feature extraction and expression, which can better weaken the range difference between optical images and SAR images.

Specifically for the proposed SAR-DRN, we choose a collection of N training image pairs $\{x_i, y_i\}_N$ from the training data sets as described in 4.1 below, where y_i is the speckled image, and θ is the network parameters. Our model uses the mean squared error (MSE) as the loss function:

$$loss(\Theta) = \frac{1}{2N} \sum_{i=1}^{N} \| \phi(y_i, \theta) - \varphi \|_2^2 \tag{17}$$

In summary, with the dilated convolution, skip connections and residual learning structure, the flowchart of learning a deep network for the SAR image despeckling process is described in Figure 5. To learn the complicated non-linear relation between the speckled image y and original image x, the proposed SAR-DRN model is employed with converged loss between the residual image φ and the output $\phi(y,\theta)$, then preparing for real speckle SAR image processing as illuminated in Figure 5.

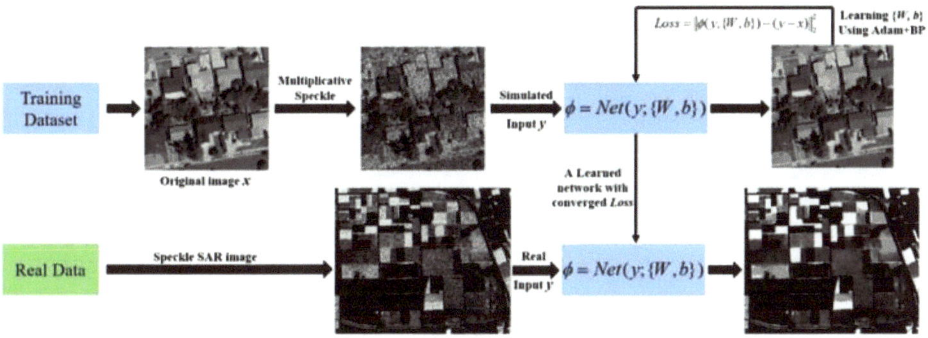

Figure 5. The framework of SAR image despeckling based on deep learning.

4. Experimental Results and Analysis

4.1. Implementation Details

4.1.1. Training and Test Datasets

Considering that it is quite hard to obtain clean reference training SAR images without speckle at all, we used the *UC Merced* land-use dataset [39] as our training dataset with different numbers of looks for simulating SAR image despeckling, which contains 21 scene classes with 100 images per class. Because the optical images and SAR images are statistically different, the amplitude information of optical images is processed before training for single-polarization SAR data despeckling, to better accord with the data distribution property of SAR images. To train the proposed SAR-DRN, we chose 400 images of size 256 × 256 from this dataset and set each patch size as 40 × 40 and stride equal to 10. Then, 193,664 patches are cropped for training SAR-DRN with a batch size of 128 for parallel computing. Additionally, the number of looks L was set to noise levels of 1, 2, 4, and 8 for adding multiplicative speckle noise, respectively.

To test the performance of the proposed model, three examples of the Airplanes, Buildings, and Rivers classes were respectively set up as simulated images. For the real SAR image despeckling experiments, we used the classic *Flevoland* SAR image (cropped to 500 × 600), *Deathvalley* SAR image (cropped to 600 × 600), and *San Francisco* SAR image (cropped to 400 × 400), which are commonly used in real SAR data image despeckling.

4.1.2. Parameter Setting and Network Training

Table 1 lists the network parameters of each layer for SAR-DRN. The proposed model was trained using the Adam [40] algorithm as the gradient descent optimization method, with momentum $\beta_1 = 0.9$, momentum $\beta_2 = 0.999$, and $\varepsilon = 10^{-8}$, where the learning rate α was initialized to 0.01 for the whole network. The optimization procedure is given below.

$$m_t = \beta_1 \cdot m_{t-1} + (1 - \beta_1) \cdot \frac{\partial L}{\partial \theta_t} \tag{18}$$

$$n_t = \beta_2 \cdot n_{t-1} + (1 - \beta_2) \cdot (\frac{\partial L}{\partial \theta_t})^2 \quad (19)$$

$$\Delta\theta_t = -\alpha \cdot \frac{m_t}{\sqrt{n_t} + \varepsilon} \quad (20)$$

where θ is the trainable parameter in the network of the t-th iteration. The training process of SAR-DRN took 50 epochs (about 1500 iterations), and after every 10 epochs, the learning rate was reduced through being multiplied by a descending factor $gamma = 0.5$. We used the *Caffe* [41] framework to train the proposed SAR-DRN in the Windows 7 environment, 16 GB-RAM, with an Nvidia Titan-X (Pascal) GPU. The total training time costs about 4 h 30 min, which is less than SAR-CNN [28] with about 9 h 45 min under the same computational environment.

Table 1. The network configuration of the SAR-DRN model.

Layer Number	Network Configurations
Layer 1	Dilated Conv + ReLU: 64 × 3 × 3, dilate = 1, stride = 1, pad = 1
Layer 2	Dilated Conv + ReLU: 64 × 3 × 3, dilate = 2, stride = 1, pad = 2
Layer 3	Dilated Conv + ReLU: 64 × 3 × 3, dilate = 3, stride = 1, pad = 3
Layer 4	Dilated Conv + ReLU: 64 × 3 × 3, dilate = 4, stride = 1, pad = 4
Layer 5	Dilated Conv + ReLU: 64 × 3 × 3, dilate = 3, stride = 1, pad = 3
Layer 6	Dilated Conv + ReLU: 64 × 3 × 3, dilate = 2, stride = 1, pad = 2
Layer 7	Dilated Conv: 64 × 3 × 3, dilate = 1, stride = 1, pad = 1

4.1.3. Compared Algorithms and Quantitative Evaluations

To verify the proposed method, we compared the SAR-DRN method with four mainstream despeckling methods: The probabilistic patch-based (PPB) filter [13] based on patch matching, SAR-BM3D [14] based on 3-D patch matching and wavelet, SAR-POTDF [16] based on sparse representation, and SAR-CNN [28] based on the deep neural network. In the simulated-image experiments, the peak signal-to-noise ratio (PSNR) and structural similarity (SSIM) were employed as the quantitative evaluation indexes. In the real-image experiments, the *ENL* was considered as the smoothness of a homogeneous region after SAR image despeckling (the *ENL* is commonly regarded as the quantitative evaluation index for real SAR image despeckling experiments), whose value is larger, demonstrating that the homogeneous region is smoother, as defined in Equation (3).

4.2. Simulated-Data Experiments

To verify the effectiveness of the proposed SAR-DRN model in SAR image despeckling, four different speckle noise levels of looks L = 1, 2, 4, and 8 were set up for the three simulated images for PPB, SAR-BM3D, SAR-POTDF, SAR-CNN, and ours. The PSNR and SSIM evaluation indexes and their standard deviations of the 10 simulated experiments with the three images are listed in Tables 2–4, respectively, where the best performance is marked in bold.

Table 2. Mean and Stand Deviation Results of PSNR (dB) and SSIM for Airplane with L = 1, 2, 4, and 8.

Looks	Index	PPB	SAR-BM3D	SAR-POTDF	SAR-CNN	SAR-DRN
$L = 1$	PSNR	20.11 ± 0.065	21.83 ± 0.051	21.75 ± 0.061	22.06 ± 0.053	**22.97 ± 0.052**
	SSIM	0.512 ± 0.001	0.623 ± 0.003	0.604 ± 0.003	0.623 ± 0.002	**0.656 ± 0.001**
$L = 2$	PSNR	21.72 ± 0.055	23.59 ± 0.062	23.79 ± 0.041	24.13 ± 0.048	**24.54 ± 0.043**
	SSIM	0.601 ± 0.001	0.693 ± 0.004	0.686 ± 0.003	0.710 ± 0.002	**0.726 ± 0.002**
$L = 4$	PSNR	23.48 ± 0.073	25.51 ± 0.079	25.84 ± 0.047	25.97 ± 0.051	**26.52 ± 0.046**
	SSIM	0.678 ± 0.003	0.755 ± 0.002	0.752 ± 0.002	0.748 ± 0.003	**0.763 ± 0.002**
$L = 8$	PSNR	24.98 ± 0.084	27.17 ± 0.064	27.56 ± 0.060	27.89 ± 0.062	**28.01 ± 0.058**
	SSIM	0.743 ± 0.003	0.800 ± 0.003	0.794 ± 0.004	0.801 ± 0.002	**0.819 ± 0.003**

Table 3. Mean and Stand Deviation Results of PSNR (dB) and SSIM for Building with L = 1, 2, 4, and 8.

Looks	Index	PPB	SAR-BM3D	SAR-POTDF	SAR-CNN	SAR-DRN
$L = 1$	PSNR	25.05 ± 0.036	26.14 ± 0.059	25.10 ± 0.035	26.25 ± 0.052	**26.80 ± 0.044**
	SSIM	0.715 ± 0.002	0.786 ± 0.005	0.731 ± 0.001	0.775 ± 0.002	**0.796 ± 0.003**
$L = 2$	PSNR	26.36 ± 0.064	27.95 ± 0.046	27.44 ± 0.041	27.98 ± 0.058	**28.39 ± 0.045**
	SSIM	0.778 ± 0.003	0.831 ± 0.004	0.811 ± 0.003	0.826 ± 0.003	**0.838 ± 0.002**
$L = 4$	PSNR	28.05 ± 0.053	29.84 ± 0.033	29.56 ± 0.066	29.96 ± 0.057	**30.14 ± 0.048**
	SSIM	0.833 ± 0.002	**0.879 ± 0.002**	0.866 ± 0.002	0.869 ± 0.003	0.870 ± 0.002
$L = 8$	PSNR	29.50 ± 0.069	31.36 ± 0.070	31.55 ± 0.051	31.63 ± 0.054	**31.78 ± 0.058**
	SSIM	0.871 ± 0.00	**0.902 ± 0.001**	0.900 ± 0.002	0.901 ± 0.002	0.901 ± 0.001

Table 4. Mean and Stand Deviation Results of PSNR (dB) and SSIM for Highway with L = 1, 2, 4, and 8.

Looks	Index	PPB	SAR-BM3D	SAR-POTDF	SAR-CNN	SAR-DRN
$L = 1$	PSNR	20.13 ± 0.059	21.12 ± 0.031	20.63 ± 0.047	21.07 ± 0.036	**21.71 ± 0.024**
	SSIM	0.472 ± 0.002	0.558 ± 0.002	0.530 ± 0.002	0.552 ± 0.003	**0.613 ± 0.003**
$L = 2$	PSNR	21.40 ± 0.073	22.62 ± 0.028	22.51 ± 0.063	22.88 ± 0.062	**22.96 ± 0.057**
	SSIM	0.572 ± 0.002	**0.646 ± 0.002**	0.637 ± 0.003	0.641 ± 0.002	0.644 ± 0.003
$L = 4$	PSNR	22.61 ± 0.037	24.29 ± 0.049	24.39 ± 0.071	24.46 ± 0.061	**24.64 ± 0.063**
	SSIM	0.674 ± 0.002	0.765 ± 0.003	0.768 ± 0.004	0.762 ± 0.003	**0.772 ± 0.002**
$L = 8$	PSNR	24.90 ± 0.045	26.41 ± 0.075	26.37 ± 0.044	26.48 ± 0.058	**26.53 ± 0.046**
	SSIM	0.764 ± 0.005	0.834 ± 0.002	0.837 ± 0.002	0.834 ± 0.003	**0.836 ± 0.002**

As shown in Tables 2–4, the proposed SAR-DRN model obtains all the best PSNR results and nine of the twelve best SSIM results in the four noise levels. When L = 1, the proposed method outperforms SAR-BM3D by about 0.9 dB/0.6 dB/0.6 dB for Airplane, Building, and Highway images, respectively. When L = 2 and 4, SAR-DRN outperforms PPB, SAR-POTDF, SAR-BM3D, and SAR-CNN by at least 0.5 dB/0.7 dB/0.3 dB and 0.4 dB/0.3 dB/0.2 dB for Airplane/Building/Highway, respectively. Compared with the traditional despeckling methods above, the proposed method shows a superior performance over the state-of-the-art methods in both quantitative and visual assessments, especially for strong speckle noise.

Figures 6–8 correspondingly show the filtered images for the Airplane/Building/Highway images contaminated by two-look speckle, four-look speckle, and four-look speckle, respectively. It can be clearly seen that PPB has a good speckle-reduction ability, but PPB simultaneously creates many texture distortions, especially around the edges of the airplane, building, and highway. SAR-BM3D and SAR-POTDF perform better than PPB for the Airplane, Building, and Highway images, especially for strong speckle noise such as L = 1, 2, or 4, which reveals an excellent speckle-reduction ability and local detail preservation ability. Furthermore, they generate fewer texture distortions, as shown in Figures 6–8. However, SAR-BM3D and SAR-POTDF also simultaneously result in over-smoothing, to some degree, as they mainly concentrate on some complex geometric features. SAR-CNN also shows a good speckle-reduction ability and local detail preservation ability, but introduces some radiation distortions in homogeneous regions. Compared with the other algorithms above, SAR-DRN achieves the best performance in speckle reduction, concurrently avoiding introducing radiation and geometric distortion. In addition, from the red boxes of the Airplane and Building images in Figures 6–8, respectively, it can be clearly seen that SAR-DRN also shows the best local detail preservation ability, while the other methods either miss partial texture details or produce blurry results, to some extent.

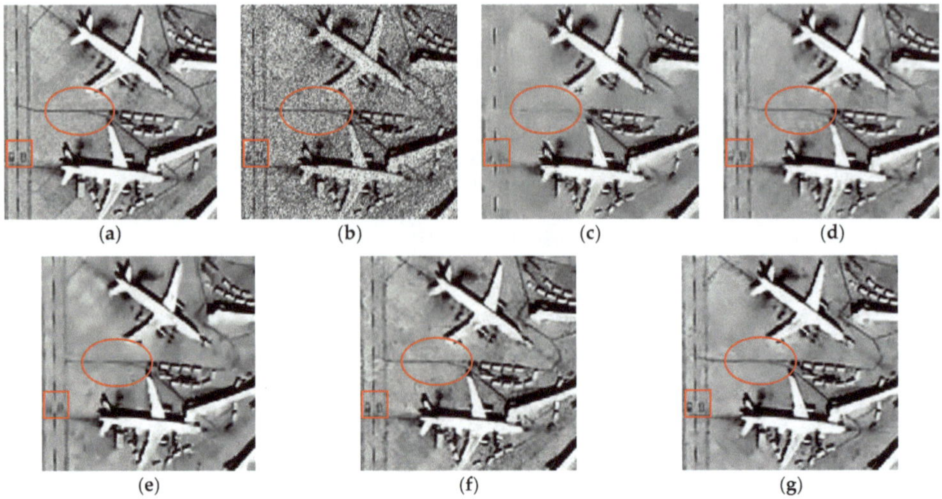

Figure 6. Filtered images for the Airplane image contaminated by two-look speckle. (a) Original image. (b) Speckled image. (c) PPB [13]. (d) SAR-BM3D [14]. (e) SAR-POTDF [16]. (f) SAR-CNN [28]. (g) SAR-DRN.

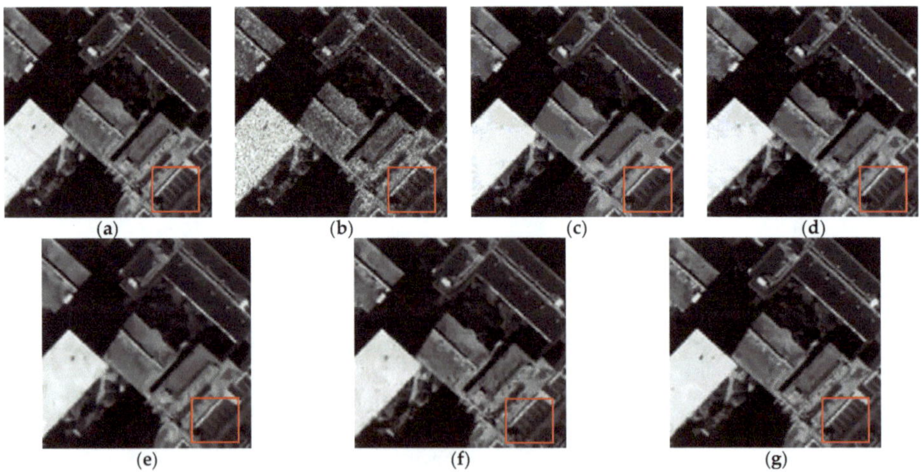

Figure 7. Filtered images for the Building image contaminated by four-look speckle. (a) Original image. (b) Speckled image. (c) PPB [13]. (d) SAR-BM3D [14]. (e) SAR-POTDF [16]. (f) SAR-CNN [28]. (g) SAR-DRN.

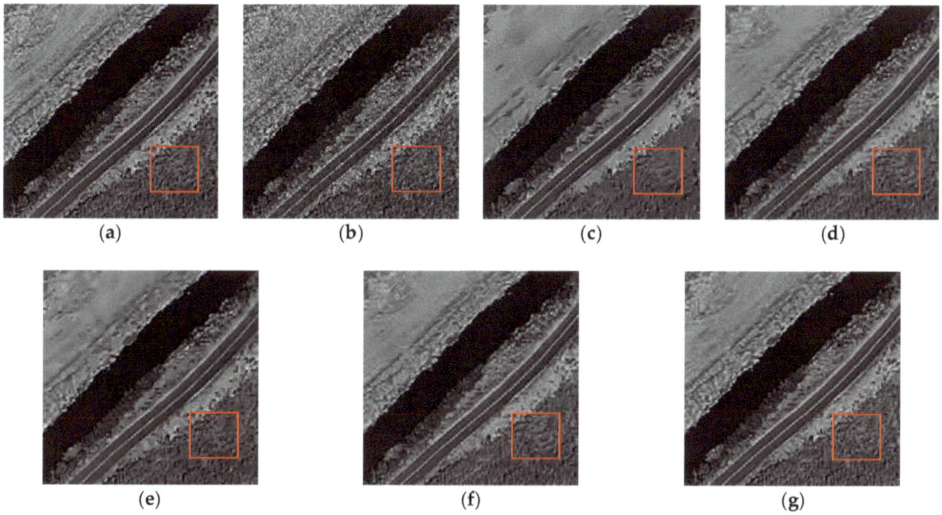

Figure 8. Filtered images for the Highway image contaminated by four-look speckle. (a) Original image. (b) Speckled image. (c) PPB [13]. (d) SAR-BM3D [14]. (e) SAR-POTDF [16]. (f) SAR-CNN [28]. (g) SAR-DRN.

4.3. Real-Data Experiments

As shown in Figures 9–11, we also compared the proposed method with the four state-of-the-art methods described above for three real SAR images. These three SAR images are all acquired by the Airborne Synthetic Aperture Radar (AIRSAR), which are all four-look data. In Figure 9, it can be clearly seen that the result of SAR-BM3D still contains a great deal of residual speckle noise, while the results of PPB, SAR-POTDF, SAR-CNN, and the proposed SAR-DRN method reveal a good speckle-reduction ability. PPB performs very well in speckle reduction, but it generates a few texture distortions in the edges of prominent objects. In homogeneous regions, SAR-POTDF does not perform as well in speckle reduction as the proposed SAR-DRN. As for SAR-CNN, its edge-preserving ability is weaker than that of SAR-DRN. Visually, SAR-DRN achieves the best performance in speckle reduction and local detail preservation, performing better than the other mainstream methods; in Figure 10, all the five methods can reduce the speckle noise well, but PPB obviously results in an over-smoothing phenomenon. Besides, in Figure 11, the result of SAR-CNN still contains some residual speckle noise. Simultaneously, PPB, SAR-BM3D, and SAR-POTDF also result in an over-smoothing phenomenon, to some degree, as shown in the marked regions with complex geometric features. It can be clearly seen that the proposed method has both a well speckled noise reduction ability and preserving detail ability for the edge and texture information.

In addition, we also evaluated the filtered results, through *ENL* in Table 5 and EPD-ROA [15] in Table 6 to measure the speckle-reduction and edge-preserving ability [42], respectively. Because it is difficult to find homogeneous regions in Figure 11, the *ENL* values were respectively estimated from four chosen homogeneous regions of Figures 9 and 10 (the red boxes in Figures 9a and 10a). Clearly, SAR-DRN has a much better speckle-reduction ability than the other methods, which is consistent with the visual observation.

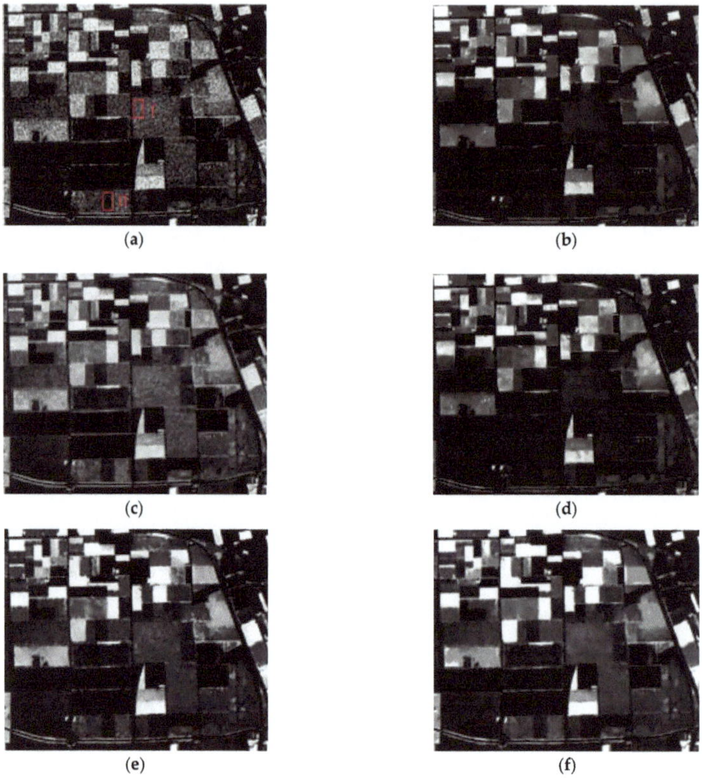

Figure 9. Filtered images for the *Flevoland* SAR image contaminated by four-look speckle. (**a**) Original image. (**b**) PPB [13]. (**c**) SAR-BM3D [14]. (**d**) SAR-POTDF [16]. (**e**) SAR-CNN [28]. (**f**) SAR-DRN.

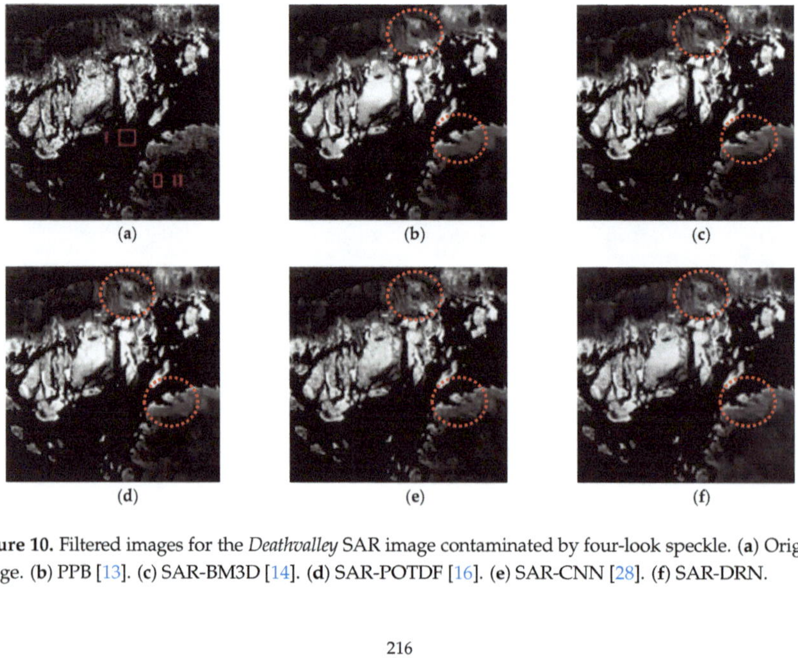

Figure 10. Filtered images for the *Deathvalley* SAR image contaminated by four-look speckle. (**a**) Original image. (**b**) PPB [13]. (**c**) SAR-BM3D [14]. (**d**) SAR-POTDF [16]. (**e**) SAR-CNN [28]. (**f**) SAR-DRN.

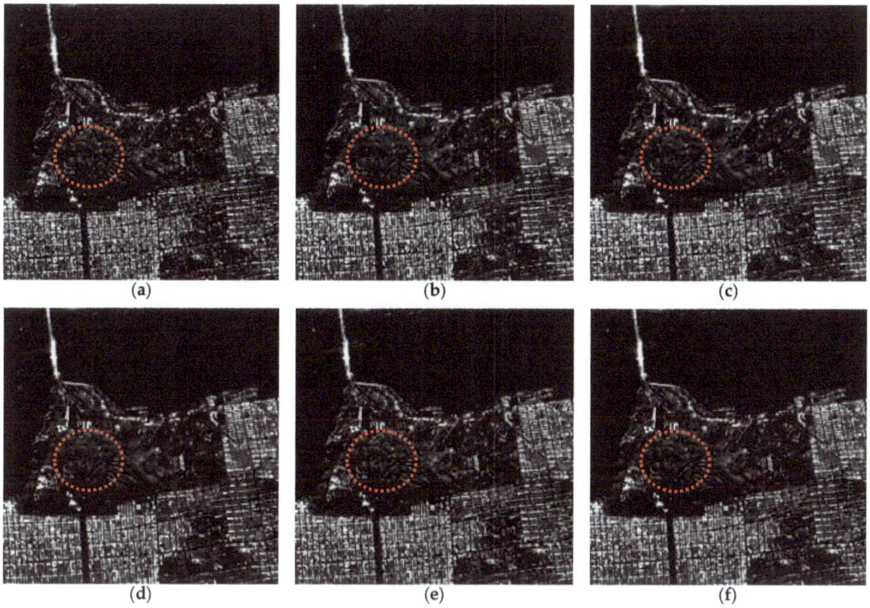

Figure 11. Filtered images for the *San Francisco* SAR image contaminated by four-look speckle. (**a**) Original image. (**b**) PPB [13]. (**c**) SAR-BM3D [14]. (**d**) SAR-POTDF [16]. (**e**) SAR-CNN [28]. (**f**) SAR-DRN.

Table 5. *ENL* results for the *Flevoland* and *Deathvalley* images.

Data		Original	PPB	SAR-BM3D	SAR-POTDF	SAR-CNN	SAR-DRN
Figure 9	Region I	4.36	122.24	67.43	120.32	86.29	**137.63**
	Region II	4.11	**56.89**	24.96	38. 90	23.38	45.64
Figure 10	Region I	5.76	14.37	12.65	12.72	13.26	**14.58**
	Region II	4.52	43.97	**55.76**	44.87	37.45	48.32

Table 6. EPD-ROA indexes for the real despeckling results.

Data	PPB	SAR-BM3D	SAR-POTDF	SAR-CNN	SAR-DRN
Figure 9	0.619	0.733	0.714	0.748	**0.754**
Figure 10	0.587	0.714	0.702	0.698	**0.723**
Figure 11	0.632	**0.685**	0.654	0.621	0.673

4.4. Discussion

4.4.1. Dilated Convolutions and Skip Connections

As mentioned in Section III, dilated convolutions are employed in the proposed method, which can both enlarge the receptive field and maintain the filter size and layer depth with a lightweight structure. In addition, skip connections are also added to the despeckling model to maintain the image details and reduce the vanishing gradient problem. To verify the effectiveness of the dilated convolutions and skip connections, we implemented four sets of experiments in the same environment as that shown in Figure 12: (1) with dilated convolutions and skip connections (the red line); (2) with dilated convolutions but without skip connections (the green line); (3) without dilated convolutions but with skip connections (the blue line); and (4) without dilated convolutions and skip connections (the black line).

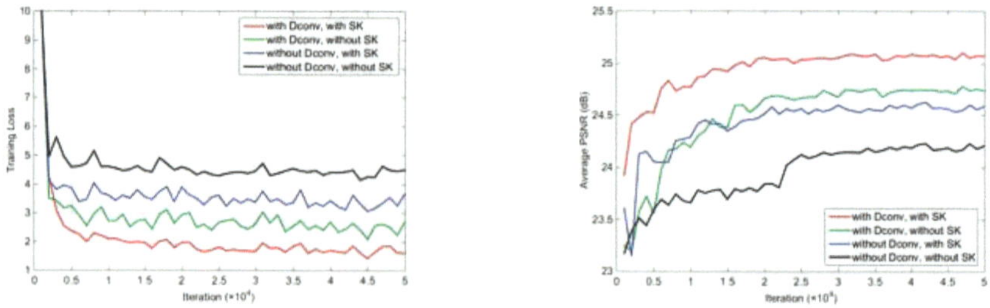

(**a**) Training loss (**b**) Average PSNR

Figure 12. The simulated SAR image despeckling results of the four specific models in (**a**) training loss and (**b**) average PSNR, with respect to iterations. The four specific models were different combinations of dilated convolutions (Dconv) and skip connections (SK), and were trained with one-look images in the same environment. The results were evaluated for the *Set14* [43] dataset.

As Figure 12 implies, the dilated convolutions can effectively reduce the training loss and enhance the despeckling performance (the less training Loss and the best PSNR), which also testifies that augmenting the contextual information through enlarging the receptive field is effective for recovering the degraded image, as demonstrated in Section III for dilated convolution. Meanwhile, the skip connections also accelerate the convergence speed of the network and enhance the model stability, as is shown by the comparison with or without skip connection in Figure 12. Besides, the combination of dilated convolution and skip connections can promote each other's effect, up from about 1.1 dB in PSNR compared with the combination of without dilated convolution and without skip connections.

4.4.2. With or without Batch Normalization (BN) in the Network

Unlike the methods proposed in [28,29], which utilize batch normalization to normalize the output features, SAR-DRN does not add this preprocessing layer, considering that the skip connections can also maintain the outputs of the data distribution in the different dilated convolution layers. The quantitative comparison of the two structures for SAR image despeckling is provided in Section IV. Furthermore, getting rid of the BN layers can simultaneously reduce the amount of computation, saving about 3 h of training time in the same environment. Figure 13 shows that this modification improves the despeckling performance and reduces the complexity of the model. Regarding this phenomenon, we suggest that a probable reason is that the input and output have a highly similar spatial distribution for this regression problem, while the BN layers normalize the hidden layers' output, which destroys the representation of the original space [44].

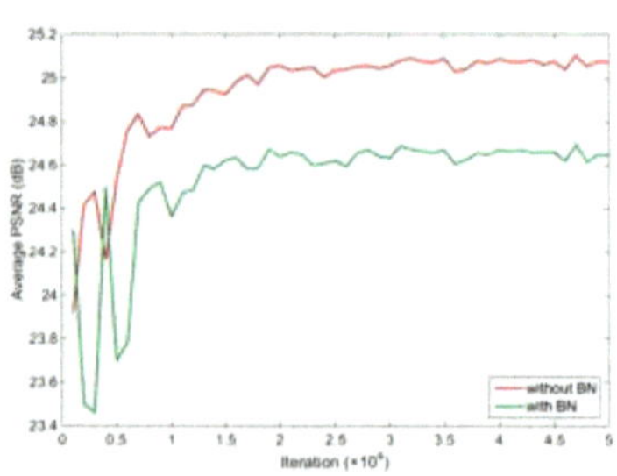

Figure 13. The simulated SAR image despeckling results of the two specific models with/without batch normalization (BN). The two specific models were trained with one-look images in the same environment, and the results were evaluated for the *Set14* [43] dataset.

4.4.3. Runtime Comparisons

For evaluating the efficiency of despeckling algorithms, we make statistics of runtime under the same environment with MALAB R2014b, as listed in Table 7. Distinctly, SAR-DRN exhibits the lowest run-time complexity than other algorithms, because of the lightweight model with only seven layers than other deep learning methods like SAR-CNN [28] with 17 layers.

Table 7. Runtime comparisons for five despeckling methods with an image of size 256 × 256 (s).

Method	PPB	SAR-BM3D	SAR-POTDF	SAR-CNN	Ours
Runtime	10.13	16.48	12.83	1.13	**0.38**

5. Conclusions

In this paper, we have proposed a novel deep learning approach for the SAR image despeckling task, learning an end-to-end mapping between the noisy and clean SAR images. Differently from common convolutions operation, the presented approach is based on dilated convolutions, which can both enlarge the receptive field and maintain the filter size with a lightweight structure. Furthermore, skip connections are added to the despeckling model to maintain the image details and avoid the vanishing gradient problem. Compared with the traditional despeckling methods, the proposed SAR-DRN approach shows a state-of-the-art performance in both simulated and real SAR image despeckling experiments, especially for strong speckle noise.

In our future work, we will investigate more powerful learning models to deal with the complex real scenes in SAR images. Considering that the training of our current method performed for each number of looks, we will explore an integrated model to solve this problem. Furthermore, the proposed approach will be extended to polarimetric SAR image despeckling, whose noise model is much more complicated than that of single-polarization SAR. Besides, for better reducing speckle noise in more complex real SAR image data, some *prior* constraint like multi-channel patch matching, band selection, location *prior*, and locality adaptive discriminant analysis [45–48], can also be considered to improve the precision of despeckling results. In addition, we will try to collect enough SAR images and then train the model with multi-temporal data [49] for SAR image despeckling, which will be sequentially explored in future studies.

Acknowledgments: This work was supported by the National Key Research and Development Program of China under Grant 2016YFB0501403, the National Natural Science Foundation of China under Grants 61671334, the Fundamental Research Funds for the Central Universities under Grant 2042017kf0180, and the Natural Science Foundation of Hubei Province under Grant ZRMS2016000241.

Author Contributions: Qiang Zhang proposed the method and performed the experiments; Qiang Zhang, Qiangqiang Yuan., Jie Li., and Zhen Yang conceived and designed the experiments; Qiang Zhang, Qiangqiang Yuan., Jie Li. Zhen Yang, and Xiaoshuang Ma wrote the manuscript. All the authors read and approved the final manuscript.

Conflicts of Interest: The authors declare no conflict of interest.

References

1. Goodman, J. Some fundamental properties of speckle. *J. Opt. Soc. Am.* **1976**, *66*, 1145–1150. [CrossRef]
2. Li, H.; Hong, W.; Wu, Y.; Fan, P. Bayesian wavelet shrinkage with heterogeneity-adaptive threshold for SAR image despeckling based on generalized gamma distribution. *IEEE Trans. Geosci. Remote Sens.* **2013**, *51*, 2388–2402. [CrossRef]
3. Xu, B.; Cui, Y.; Li, Z.; Yang, J. An iterative SAR image filtering method using nonlocal sparse model. *IEEE Geosci. Remote Sens. Lett.* **2015**, *12*, 1635–1639.
4. Wu, J.; Liu, F.; Hao, H.; Li, L.; Jiao, L.; Zhang, X. A nonlocal means for speckle reduction of SAR image with multiscale-fusion-based steerable kernel function. *IEEE Geosci. Remote Sens. Lett.* **2016**, *13*, 1646–1650. [CrossRef]

5. Lee, J. Digital image enhancement and noise filtering by use of local statistics. *IEEE Trans. Pattern Anal. Mach. Intell.* **1980**, *2*, 165–168. [CrossRef] [PubMed]
6. Kuan, D.; Sawchuk, A.; Strand, T.; Chavel, P. Adaptive noise smoothing filter for images with signal-dependent noise. *IEEE Trans. Pattern Anal. Mach. Intell.* **1985**, *2*, 165–177. [CrossRef]
7. Frost, V.; Stiles, J.; Shanmugan, K.; Holtzman, J. A model for radar images and its application to adaptive digital filtering of multiplicative noise. *IEEE Trans. Pattern Anal. Mach. Intell.* **1982**, *2*, 157–166. [CrossRef]
8. Yahya, N.; Kamel, N.S.; Malik, A.S. Subspace-based technique for speckle noise reduction in SAR images. *IEEE Trans. Geosci. Remote Sens.* **2014**, *52*, 6257–6271. [CrossRef]
9. Starck, J.; Candès, E.; Donoho, D. The curvelet transform for image denoising. *IEEE Trans. Image Process.* **2002**, *11*, 670–684. [CrossRef] [PubMed]
10. Solbo, S.; Eltoft, T. Homomorphic wavelet-based statistical despeckling of SAR images. *IEEE Trans. Geosci. Remote Sens.* **2004**, *42*, 711–721. [CrossRef]
11. López, C.M.; Fàbregas, X.M. Reduction of SAR interferometric phase noise in the wavelet domain. *IEEE Trans. Geosci. Remote Sens.* **2002**, *40*, 2553–2566. [CrossRef]
12. Buades, A.; Coll, B.; Morel, J.M. A non-local algorithm for image denoising. In Proceedings of the IEEE Computer Society Conference on Computer Vision and Pattern Recognition, San Diego, CA, USA, 20–25 June 2005; pp. 60–65.
13. Deledalle, C.A.; Denis, L.; Tupin, F. Iterative weighted maximum likelihood denoising with probabilistic patch-based weights. *IEEE Trans. Image Process.* **2009**, *18*, 2661–2672. [CrossRef] [PubMed]
14. Parrilli, S.; Poderico, M.; Angelino, C.V.; Verdoliva, L. A nonlocal SAR image denoising algorithm based on LLMMSE wavelet shrinkage. *IEEE Trans. Geosci. Remote Sens.* **2012**, *50*, 606–616. [CrossRef]
15. Ma, X.; Shen, H.; Zhao, X.; Zhang, L. SAR image despeckling by the use of variational methods with adaptive nonlocal functionals. *IEEE Trans. Geosci. Remote Sens.* **2016**, *54*, 3421–3435. [CrossRef]
16. Xu, B.; Cui, Y.; Li, Z.; Zuo, B.; Yang, J.; Song, J. Patch ordering-based SAR image despeckling via transform-domain filtering. *IEEE J. Sel. Top. Appl. Earth Obs. Remote Sens.* **2015**, *8*, 1682–1695. [CrossRef]
17. Feng, W.; Lei, H.; Gao, Y. Speckle reduction via higher order total variation approach. *IEEE Trans. Image Process.* **2014**, *23*, 1831–1843. [CrossRef] [PubMed]
18. Zhao, Y.; Liu, J.; Zhang, B.; Hong, W.; Wu, Y. Adaptive total variation regularization based SAR image despeckling and despeckling evaluation index. *IEEE Trans. Geosci. Remote Sens.* **2015**, *53*, 2765–2774. [CrossRef]
19. Yuan, Q.; Zhang, L.; Shen, H. Hyperspectral image denoising employing a spectral-spatial adaptive total variation model. *IEEE Trans. Geosci. Remote Sens.* **2012**, *10*, 3660–3677. [CrossRef]
20. Li, J.; Yuan, Q.; Shen, H.; Zhang, L. Noise removal from hyperspectral image with joint spectral-spatial distributed sparse representation. *IEEE Trans. Geosci. Remote Sens.* **2016**, *54*, 5425–5439. [CrossRef]
21. Ranjani, J.J.; Thiruvengadam, S.J. Dual-tree complex wavelet transform based SAR despeckling using interscale dependence. *IEEE Trans. Geosci. Remote Sens.* **2010**, *48*, 2723–2731. [CrossRef]
22. LeCun, Y.A.; Bengio, Y.; Hinton, G. Deep learning. *Nature* **2015**, *521*, 436–444. [CrossRef] [PubMed]
23. Zhang, L.; Zhang, L.; Du, B. Deep Learning for Remote Sensing Data: A Technical Tutorial on the State of the Art. *IEEE Geosci. Remote Sens. Mag.* **2016**, *4*, 22–40. [CrossRef]
24. Xia, G.; Hu, J.; Hu, F.; Shi, B.; Bai, X.; Zhong, Y. AID: A Benchmark Data Set for Performance Evaluation of Aerial Scene Classification. *IEEE Trans. Geosci. Remote Sens.* **2017**, *55*, 3965–3981. [CrossRef]
25. LeCun, Y.A.; Boser, B.; Denker, J.S.; Howard, R.E.; Habbard, W.; Jackel, L.D. Handwritten digit recognition with a back-propagation network. In *Advances in Neural Information Processing Systems*; Morgan Kaufmann Publishers Inc.: San Francisco, CA, USA, 1990; pp. 396–404.
26. Zhang, K.; Zuo, W.; Chen, Y.; Meng, D.; Zhang, L. Beyond a gaussian denoiser: Residual learning of deep CNN for image denoising. *IEEE Trans. Image Process.* **2017**, *26*, 3142–3155. [CrossRef] [PubMed]
27. Ioffe, S.; Szegedy, C. Batch normalization: Accelerating deep network training by reducing internal covariate shift. In Proceedings of the 32nd International Conference on Machine Learning, Lille, France, 6–11 July 2015; pp. 448–456.
28. Chierchia, G.; Cozzolino, D.; Poggi, G.; Verdoliva, L. SAR image despeckling through convolutional neural networks. In Proceedings of the 2017 IEEE International Geoscience and Remote Sensing Symposium, Fort Worth, TX, USA, 23–28 July 2017.

29. Wang, P.; Zhang, H.; Patel, V.M. SAR image despeckling using a convolutional neural network. *IEEE Signal Process. Lett.* **2017**, *24*, 1763–1767. [CrossRef]
30. Dong, C.; Loy, C.; He, K.; Tang, X. Image super-resolution using deep convolutional networks. *IEEE Trans. Pattern Anal. Mach. Intell.* **2016**, *38*, 295–307. [CrossRef] [PubMed]
31. Zhang, L.; Zuo, W. Image restoration: From Sparse and Low-Rank Priors to Deep Priors [Lecture Notes]. *IEEE Signal Process. Mag.* **2017**, *34*, 172–179. [CrossRef]
32. Chakrabarti, A. A neural approach to blind motion deblurring. In Proceedings of the European Conference on Computer Vision, Amsterdam, The Netherlands, 8–16 October 2016; pp. 221–235.
33. Yu, F.; Koltun, V. Multi-scale context aggregation by dilated convolutions. In Proceedings of the 2016 International Conference on Learning Representations, San Juan, Puerto Rico, 2–4 May 2016.
34. Mao, X.; Shen, C.; Yang, Y.-B. Image restoration using very deep convolutional encoder-decoder networks with symmetric skip connections. In Proceedings of the Advances in Neural Information Processing Systems, Barcelona, Spain, 5–10 December 2016; pp. 2802–2810.
35. He, K.; Zhang, X.; Ren, S.; Sun, J. Deep residual learning for image recognition. In Proceedings of the IEEE Conference Computer Vision and Pattern Recognition, Seattle, WA, USA, 27–30 June 2016; pp. 770–778.
36. Zhang, X.; Zou, J.; He, K.; Sun, J. Accelerating very deep convolutional networks for classification and detection. *IEEE Trans. Pattern Anal. Mach. Intell.* **2016**, *38*, 1943–1955. [CrossRef] [PubMed]
37. Kim, J.; Kwon, L.J.; Mu, L.K. Accurate image super-resolution using very deep convolutional networks. In Proceedings of the IEEE Conference Computer Vision and Pattern Recognition, Seattle, WA, USA, 27–30 June 2016; pp. 1646–1654.
38. Szegedy, C.; Ioffe, S.; Vanhoucke, V. Inception-v4, Inception-ResNet and the Impact of Residual Connections on Learning. In Proceedings of the 31st AAAI Conference on Artificial Intelligence, San Francisco, CA, USA, 4–9 February 2017; pp. 4278–4284.
39. Yang, Y.; Newsam, S. Bag-of-visual-words and spatial extensions for land-use classification. In Proceedings of the 18th SIGSPATIAL International Conference on Advances in Geographic Information Systems, San Jose, CA, USA, 2–5 November 2010; pp. 270–279.
40. Kingma, D.; Ba, J. Adam: A method for stochastic optimization. In Proceedings of the 3rd International Conference for Learning Representations, San Diego, CA, USA, 7–9 May 2015.
41. Jia, Y.; Shelhamer, E.; Donahue, J.; Karayev, S.; Long, J.; Girshick, R. Caffe: Convolutional architecture for fast feature embedding. In Proceedings of the 22nd ACM International Conference on Multimedia, Orlando, FL, USA, 3–7 November 2014; pp. 675–678.
42. Luis, G.; Maria, E.B.; Julio, C.; Marta, E. A new image quality index for objectively evaluating despeckling filtering in SAR images. *IEEE J. Sel. Top. Appl. Earth Obs. Remote Sens.* **2016**, *9*, 1297–1307.
43. Zeyde, R.; Elad, M.; Protter, M. On single image scale-up using sparse-representations. In Proceedings of the International Conference on Curves and Surfaces, Avignon, France, 24–30 June 2010; pp. 711–730.
44. Lim, B.; Son, S.; Kim, H.; Nah, S.; Lee, K.M. Enhanced deep residual networks for single image super-resolution. In Proceedings of the IEEE Conference Computer Vision and Pattern Recognition Workshops, Honolulu, HI, USA, 21–26 July 2017.
45. Li, J.; Yuan, Q.; Shen, H.; Zhang, L. Hyperspectral image recovery employing a multidimensional nonlocal total variation model. *Signal Process.* **2015**, *111*, 230–248. [CrossRef]
46. Wang, Q.; Lin, J.; Yuan, Y. Salient band selection for hyperspectral image classification via manifold ranking. *IEEE Trans. Neural Netw. Learn. Syst.* **2016**, *27*, 1279–1289. [CrossRef] [PubMed]
47. Wang, Q.; Meng, Z.; Li, X. Locality adaptive discriminant analysis for spectral-spatial classification of hyperspectral images. *IEEE Geosci. Remote Sens. Lett.* **2017**, *14*, 2077–2081. [CrossRef]
48. Wang, Q.; Gao, J.; Yuan, Y. Embedding structured contour and location prior in siamesed fully convolutional networks for road detection. *IEEE Trans. Intell. Transp. Syst.* **2017**, *99*, 230–241. [CrossRef]
49. Ma, X.; Wu, P.; Wu, Y.; Shen, H. A review on recent developments in fully polarimetric SAR image despeckling. *IEEE J. Sel. Top. Appl. Earth Obs. Remote Sens.* **2017**, *99*, 1–16. [CrossRef]

remote sensing

MDPI

Article

Unassisted Quantitative Evaluation of Despeckling Filters

Luis Gomez [1,*,†], **Raydonal Ospina** [2,†] and **Alejandro C. Frery** [3,†]

1 Department of Electronic Engineering and Automation, University of Las Palmas de G.C., Las Palmas 35001, Spain
2 Departamento de Estatística. Universidade Federal de Pernambuco, Recife, PE 50670-901, Brazil; rayospina@gmail.com
3 LaCCAN—Laboratório de Computação Científica e Análise Numérica, Universidade Federal de Alagoas, Maceió, AL 57072-900, Brazil; acfrery@laccan.ufal.br
* Correspondence: luis.gomez@ulpgc.es; Tel.: +34-928-451-254
† These authors contributed equally to this work.

Academic Editors: Qi Wang, Nicolas H. Younan, Carlos López-Martínez, Xiaofeng Li and Prasad S. Thenkabail
Received: 27 February 2017; Accepted: 13 April 2017; Published: 20 April 2017

Abstract: SAR (Synthetic Aperture Radar) imaging plays a central role in Remote Sensing due to, among other important features, its ability to provide high-resolution, day-and-night and almost weather-independent images. SAR images are affected from a granular contamination, speckle, that can be described by a multiplicative model. Many despeckling techniques have been proposed in the literature, as well as measures of the quality of the results they provide. Assuming the multiplicative model, the observed image Z is the product of two independent fields: the backscatter X and the speckle Y. The result of any speckle filter is $\widehat{X}$, an estimator of the backscatter X, based solely on the observed data Z. An ideal estimator would be the one for which the ratio of the observed image to the filtered one $I = Z/\widehat{X}$ is only speckle: a collection of independent identically distributed samples from Gamma variates. We, then, assess the quality of a filter by the closeness of I to the hypothesis that it is adherent to the statistical properties of pure speckle. We analyze filters through the ratio image they produce with regards to first- and second-order statistics: the former check marginal properties, while the latter verifies lack of structure. A new quantitative image-quality index is then defined, and applied to state-of-the-art despeckling filters. This new measure provides consistent results with commonly used quality measures (equivalent number of looks, PSNR, MSSIM, β edge correlation, and preservation of the mean), and ranks the filters results also in agreement with their visual analysis. We conclude our study showing that the proposed measure can be successfully used to optimize the (often many) parameters that define a speckle filter.

Keywords: quality assessment; ratio images; Synthetic Aperture Radar (SAR); speckle; speckle filters

1. Introduction

Speckle reduction has occupied both the scientific literature and the production software industry since the deployment of SAR platforms. Good speckle filters are expected to improve the perceived image quality while preserving the scene reflectivity. The former requires, at the same time, preservation of details in heterogeneous areas and constancy in homogeneous targets.

Early works assessed the performance of despeckling techniques by visual inspection of the filtered images; cf. references [1,2]. Since then, speckle filtering has reached such a level of sophistication [3] that forthcoming improvements are likely to be incremental, and assessing them quantitatively is, at the same time, desirable and hard. Also, as filters are often defined with many parameters, e.g., window size, thresholds, etc., finding an optimal setting is also an issue.

The Equivalent Number of Looks (ENL) is among the simplest and most spread measures of quality of despeckling filters. It can be estimated, in textureless areas and intensity format, as the ratio of the squared sample mean to the sample variance, i.e., the reciprocal of the squared coefficient of variation (see [4] for other methods for the estimation of ENL). Being proportional to the signal-to-noise ratio, the higher ENL is, the better the quality of the image is in terms of speckle reduction. However, it is well known that large ENL values are easily obtained just by overfiltering an image, which severely degrades details and gives the filtered image an undesirable blurred appearance. In particular, ENL $= \infty$ is obtained in completely flat areas where the sample variance is null. Testing a filter merely by its performance over textureless areas, where a simple generic filter as the Boxcar, would perform well, is bound to produce misleading results.

Other measures of quality commonly used for speckle filter assessment enhance certain characteristics, but suffer from shortcomings. The proposal and assessment of a new filter is frequently supported by a plethora of measures. As such, it is hard to used them to optimize the parameters that often specify a filter.

An alternative approach for assessing the performance of despeckling methods is the analysis of ratio images, as proposed in [4]. This is becoming a standard procedure in the SAR community [5–8]. It consists of checking by visual inspection whether patterns appear in the ratio image $I = Z/\widehat{X}$, where Z is the original image and $\widehat{X}$ is its filtered version. Under the multiplicative model, the ratio image from the ideal filter should be pure speckle with no visible patterns. The presence of geometric structures, changes of statistical properties, or any detail correlated to the original image Z in I indicates poor filter performance, i.e., not only speckle but also other possible relevant information has been removed from the original image. The visual interpretation of ratio images, being subjective, is qualitative and irreproducible.

Figure 1 illustrates this idea. This image is part of a single-look HH SAR data set obtained over Oberpfaffenhofen, Germany, with textureless areas, bright scatterers, and urban areas with geometric content as buildings and roads. Figure 1 (top) is the original speckled image, and below left is its filtered version obtained with the SRAD (speckle anisotropic diffusion) filter [9]. The filtered image is acceptable in terms of edge and details preservation: textureless areas look smooth, as expected after a successful despeckling. The middle row right is the resulting ratio image, with a ROI (region of interest) in the urban area. The third row of Figure 1 (left) presents a zoom of the highlighted area. It shows remaining structures in the ratio image, an evidence that the SRAD filter is not ideal for this case.

The quantitative assessment of such residual geometrical content is a challenging task because, besides being subtle, it has similar properties to the rest of the ratio image: brightness, marginal distribution etc. That is, areas with and without geometrical structure (even narrow edges) are extremely noisy and, therefore, simple algorithms as, for instance, those based on edge detection, fail at detecting them; cf. the result of applying the Canny edge detector in the third row of Figure 1 (right). Also, the better the filter is, the harder will be identifying and quantifying remaining structures in the ratio image.

This work proposes a new measure of quality that does not require any ground reference. Using only the original image, an estimate of its number of looks, and the filtered image, we measure the deviation from the ideal filter as a combination of deviations from the ideal marginal properties with a measure of remaining structure in the ratio image. We test this unassisted measure of quality in both simulated data and on images obtained by an actual SAR sensor, and we show it is able to rank with a single value the results produced by four state-of-the-art filters in a way that captures other measures of quality. We also show it can be used to fine-tune filter parameters.

The remainder of this article is organized as follows. Section 2 recalls the basic assumptions underlying this proposal: the multiplicative model. With this in view, we discuss the properties to be measured in a ratio image. Section 3 presents our proposal of an unassisted quantitative measure for assessing the quality of despeckling filters. In Section 4 we present the results observed on both simulated and actual SAR images, and show an example of filter parameter tuning. Section 5 concludes the article.

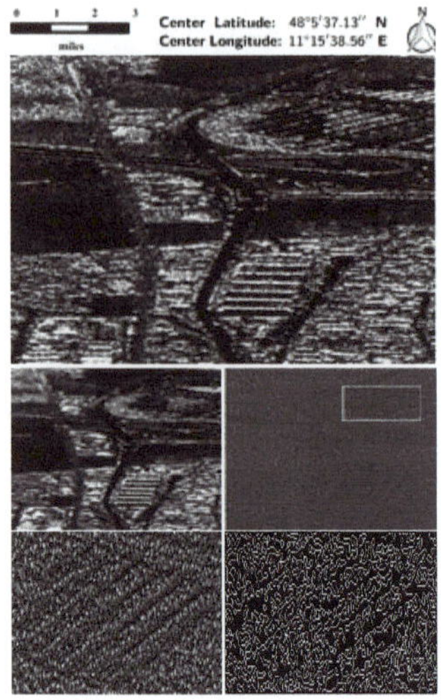

Figure 1. (**Top**): original SAR image; (**Middle**): SRAD ($T = 50$) filtered image and ratio image; (**Bottom**): zoom of a selected area within the ratio image and extracted edges by Canny's edge detector.

2. SAR Image Formation and Ratio Images

Although we recognize the nature of SAR data depends of many system parameters, our work starts by assuming the multiplicative model for the observations. Observations can be, thus, described by the product of two independent variables, X and Y that model, respectively, the (desired but unobserved) backscatter and the speckle noise. So, $Z = XY$ models the observed data, and one aims at obtaining $\widehat{X}$, a good estimator of X. Appendix B Extension to the Gaussian Additive Noise Model.

Without loss of generality, we will assume the available data is in intensity format, i.e., power. Amplitude data should be squared before applying our method.

The usual assumption is that Y is a collection of independent identically distributed Gamma random variables with unitary mean, and shape parameter equal to the number of looks. The backscatter is constant in textureless areas, and otherwise can be described by another random variable.

Our main aim is assessing the quality of despeckling filters by measuring how the ratio images they produce deviate from the idealized result.

The perfect filtered image is $\widetilde{X} = X$ and, thus, produces a ratio image $Z/\widetilde{X} = Y$ which consists of pure speckle. Based on this observation, our measure of quality captures departures from the following hypothesis: "the perfect speckle filter leads to a ratio image formed by a collection of independent identically distributed Gamma random variables with unitary mean and shape parameter equal to the (equivalent) number of looks the original image has".

In the following, we illustrate our idea with images and one-dimensional slices. We elaborate three situations to make our point on the usefulness of ratio images for detecting the performance of a speckle filter.

Firstly, we will see the effect of oversmoothing textured areas.

Figure 2a shows a step function in pink (the backscatter), and the observed return from this backscatter in single look fully developed speckle, i.e., exponential deviates with mean equal to 11 (left half) and 1 (right half).

Figure 2b shows a similar situation, but when the backscatter is no longer constant. In this case, the backscatter is textured with mean 11 and 1, as in the previous example, but varying according to exponential deviates. The textured step backscatter is shown in pink. When speckle enters the scene, modeled here again as unitary mean exponential random variables, the observed data obeys a $\mathcal{K}$ distribution; shown in lavender.

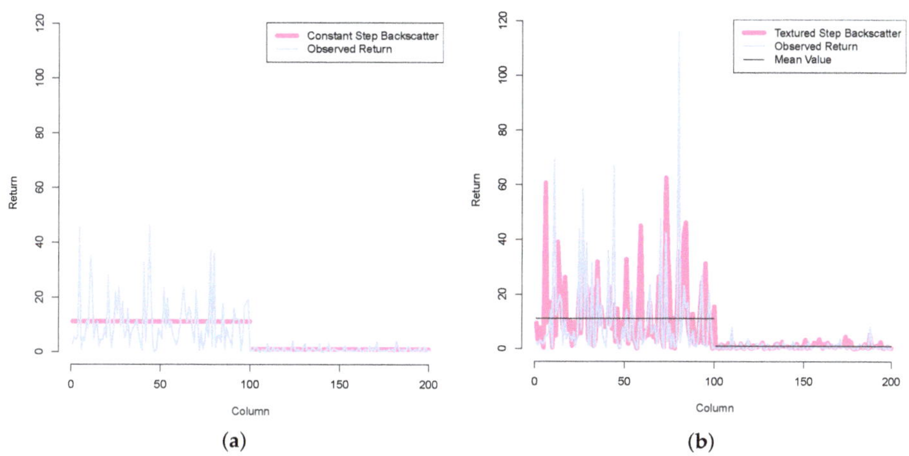

Figure 2. A step: constant and textured versions, and their return. (**a**) Constant step and speckled return; (**b**) Textured step and speckled return.

What should the ideal filter return? It is our understanding that $\widetilde{X}$ should be the underlying backscatter, i.e., either the step function in the case where there is no texture, or the textured observations without speckle (both depicted in pink in Figure 2).

A filter that returns the step function in the textured case (thin black line in Figure 2b) is oversmoothing. Figure 3 shows, in semilogarithmic scale, the estimated speckle as produced by the ideal filter (pink) and by oversmoothing (lavender); these are the resulting ratio images from the ideal and a poor filter, respectively. This last estimate is the result of dividing the observed return from Figure 2b by the step function.

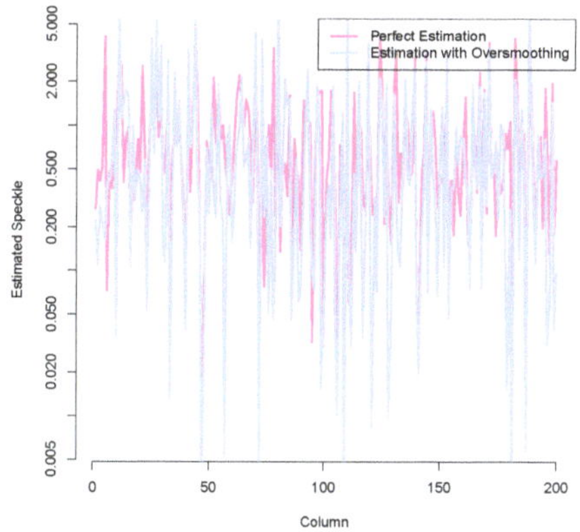

Figure 3. Estimated speckle by the ideal filter and by overmoothing.

The effect of oversmoothing is noticeable: the speckle produced by the ideal filter has less variability than the one resulting from returning the step function as estimator. While the sample variance of pure speckle is $s^2 = 0.80$, that of the speckle with remaining structure is $s^2 = 2.12$. Although numerically detectable by first-order statistics, this effect is seldom visible.

Secondly, we will see how neglecting structures impacts on ratio images.

Figure 4 shows the situation of fully developed speckle, in this case with three looks. It affects an structure seen as slowly-varying backscatter, the sine curve depicted in pink. The observed return, obtained as the point-by-point product of the speckle with the backscatter is shown in lavender; Figure 4a.

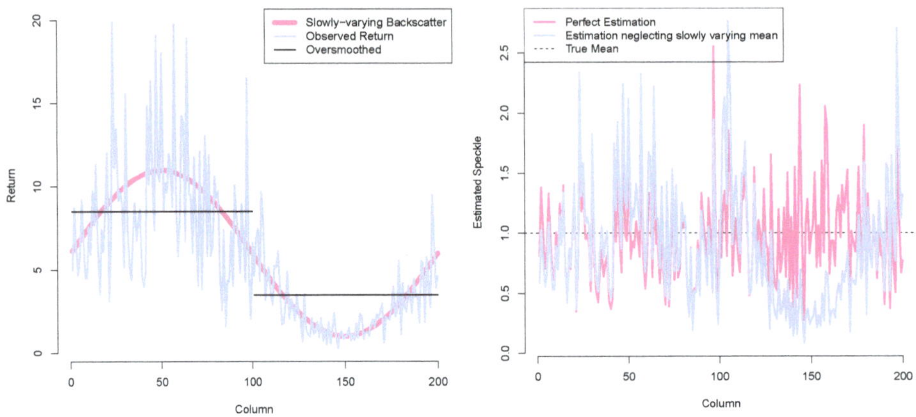

(**a**) (**b**)

Figure 4. Slowly varying backscatter, fully developed speckle, and estimated speckle. (**a**) Slowly-varying mean value and its return; (**b**) Estimated speckle.

On the one hand if, as we postulate, the ideal filter retrieves the true backscatter, the ratio image or estimated speckle will coincide with the true speckle (in pink in Figure 4b). On the other hand, if the filter oversmooths the backscatter and returns a step function (in black in Figure 4a), the resulting estimated speckle will retain part of the missing estructure; cf. Figure 4b in lavender.

Figures 3 and 4b also show that detecting departures from the ideal situation is a hard task. Figure 5 shows how the ratio image obtained from neglecting the slowly varying structure looks like. We postulate and show evidence that this remaining structure can be effectively detected and quantified with second-order statistics.

Finally, we will see how a poor filter will render a ratio image with detectable structure when dealing with edges.

Figure 6a shows a line of the strips image typical of articles that analyze the performance of speckle filters with simulated data; cf. [10,11]. The strips take two values: 1 and 20 (pink), the speckle is a collection of i.i.d. Gamma variates with three looks and unitary mean, and the observed data (in lavender) is the product of the strips and speckle.

Figure 6b shows, again, the strips and the estimated backscatter as returned by a simple filter: the local mean using eleven observations. The oversmoothing is noticeable. It not only degrades the sharpness of the edges, but also reduces observed value. This last effect is more noticeable over narrow strips (to the left of the figure).

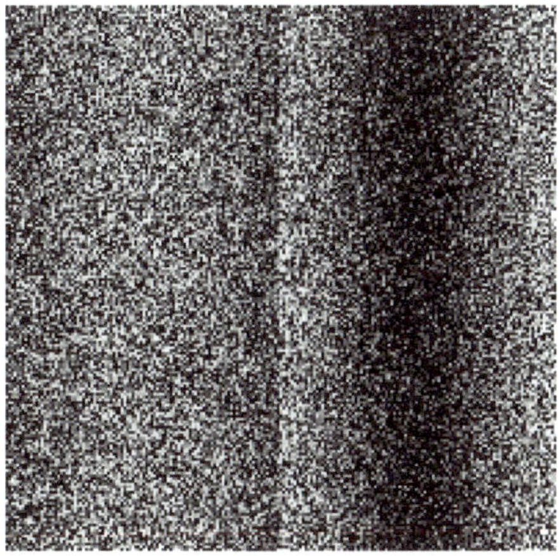

Figure 5. Ratio image resulting from neglecting a slowly varying structure under fully developed speckle.

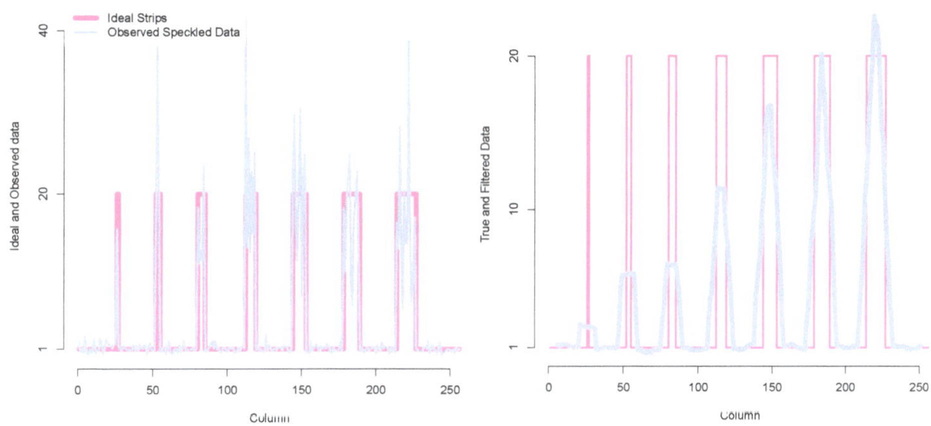

(a) (b)

Figure 6. The effect of oversmoothing on an image of strips of varying width. (**a**) Strips and speckle; (**b**) Filtered strips with oversmoothing.

The estimated speckle, as expected, will be affected by the poor result returned by the local mean filter, as shown in Figure 7. The true speckle is shown in pink, while the one estimated using the oversmoothed backscatter tends to have peaks where the smaller strips are (cf. the lavender signal). This will affect the ratio images rendering data whose behavior departs from the ideal situation, which is a collection of i.i.d. deviates from the a Gamma distribution with unitary mean and shape parameter equal to the equivalent number of looks of the original image.

Figure 8 shows these effects in the strips image. Again, we postulate that identifying and quantifying the departure from the ideal filter, i.e., the remaining structure visible in Figure 8c, is feasible with both first- and second-order statistics.

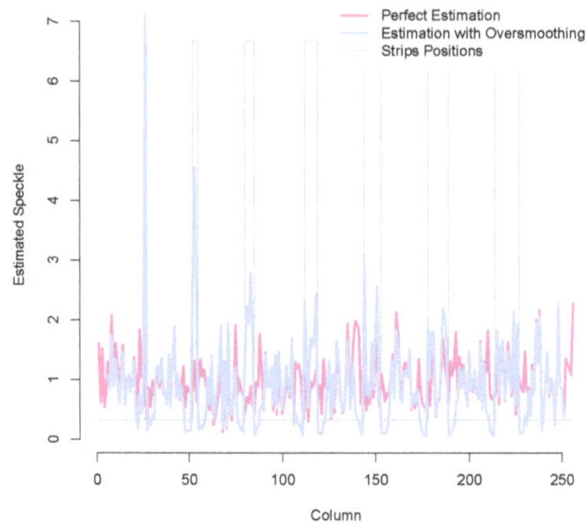

Figure 7. Estimated speckle: ideal and oversmoothing filters.

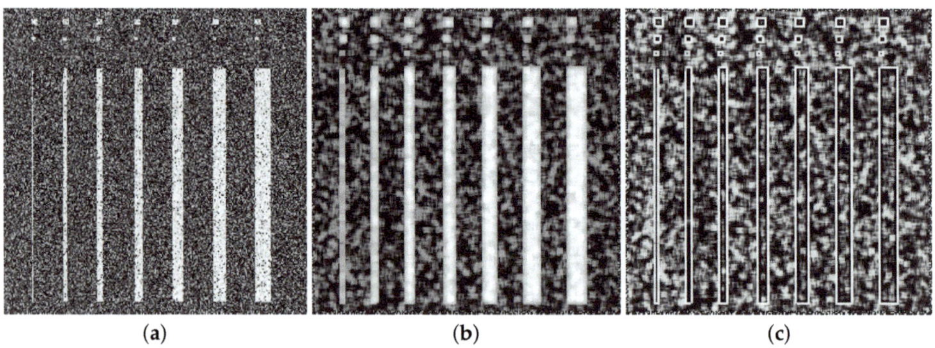

Figure 8. Speckled strips, result of applying a 5 × 5 BoxCar filter, ratio image. (**a**) Speckled strips; (**b**) Filtered strips; (**c**) Ratio image.

3. Unassisted Measure of Quality Based on First- and Second-Order Descriptors

We propose an evaluation based on two components. A statistical measure of the quality of the remaining speckle is the first-order component of the quality measure. This component is comprised of two terms: one for mean preservation, and another for preservation of the equivalent number of looks The second-order component measures the remaining geometrical content within the ratio image. The three elements that comprise our measure of quality are relative, in order to make them comparable.

As pointed out before, the usual approach to evaluate ratio images consists of, after the visual inspection, to estimate the ENL within an homogeneous area. Then, the best filter is the one for which the ratio image has the mean value closest to unity and the equivalent number of looks closest to the ENL of the original (noisy) image (see for instance [5]).

To avoid user intervention, which is one of the requirements of our proposal, we automatically select suitable textureless areas. First, we estimate the local mean and standard deviation on sliding windows of side w over the original image. With these values, we compute the local ENL ($\widehat{\text{ENL}}_{\text{noisy}}$) as the reciprocal of the squared coefficient of variation. Then, we also compute the local mean and standard deviation on the ratio image with the same window, obtaining $\widehat{\mu}_{\text{ratio}}$ and $\widehat{\text{ENL}}_{\text{ratio}}$.

We select as textureless areas those where both $\widehat{\mathrm{ENL}}_{\mathrm{ratio}}$ is close enough to $\widehat{\mathrm{ENL}}_{\mathrm{noisy}}$ and $\widehat{\mu}_{\mathrm{ratio}}$ is close enough to 1. We stipulate a tolerance for the absolute relative error, and with this we select n areas. This procedure is illustrated in Figure 9.

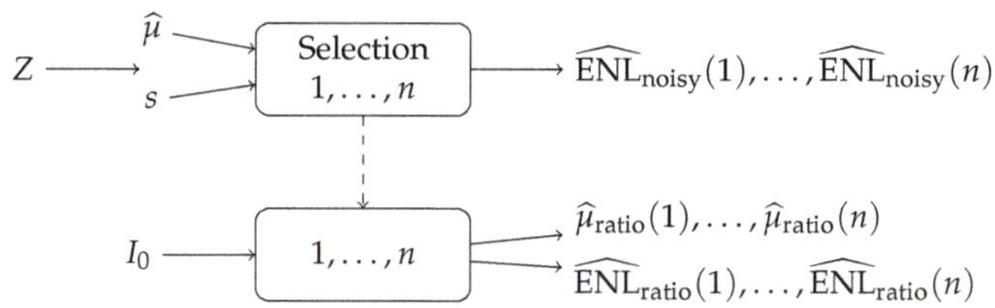

Figure 9. Selection of mean and ENL values for the first-order measure.

Then, for the n selected homogeneous areas, we calculate the first-order residual as

$$r_{\widehat{\mathrm{ENL}},\widehat{\mu}} = \frac{1}{2}\sum_{i=1}^{n}\left(r_{\widehat{\mathrm{ENL}}}(i) + r_{\widehat{\mu}}(i)\right), \tag{1}$$

where, for each homogeneous area i,

$$r_{\widehat{\mathrm{ENL}}}(i) = \frac{|\widehat{\mathrm{ENL}}_{\mathrm{noisy}}(i) - \widehat{\mathrm{ENL}}_{\mathrm{ratio}}(i)|}{\widehat{\mathrm{ENL}}_{\mathrm{noisy}}(i)}$$

is the absolute value of the relative residual due to deviations from the ideal ENL, and

$$r_{\widehat{\mu}}(i) = |1 - \widehat{\mu}_{\mathrm{ratio}}(i)|$$

is the absolute value of the relative residual due to deviations from the ideal mean (which is 1). An ideal despeckling operation would yield $r_{\widehat{\mathrm{ENL}},\widehat{\mu}} = 0$.

We measure the remaining geometrical content with the inverse difference moment (also called *homogeneity*) from Haralik's co-ocurrence matrices [12,13]. Low values are associated with low textural variations and vice versa. Let $P(i,j)$ be a co-occurence matrix at an arbitrary position, and $p(i,j) = P(i,j)/K$ its normalized version, with K a constant. The homogeneity, our second-order measure, is

$$h = \sum_i\sum_j \frac{1}{1+(i-j)^2}\cdot p(i,j). \tag{2}$$

This is computed for every coordinate, yielding measures of the remaining structure, but we need a reference to compare it with.

The null hypothesis implies that the probability distribution of the ratio image I is invariant under random permutations, i.e., if $I_1, I_2, \ldots, I_M$ are independent identically distributed random variables, also are $g(I_1, I_2, \ldots, I_M)$, any random permutation. Applying this idea, we measure the geometric content in a ratio image evaluating h on the ratio image and then on a shuffled versions of it. If there is no structure in I, h will not change after shuffling, but if I has structure, then shuffling will tend to destroy it.

Let h_o and h_g be the mean of all values of homogeneity obtained from the original ratio image I_o and from the result of randomly permuting all its values I_g, respectively. We use $\delta h = 100|h_o - \overline{h_g}|/h_o$, the absolute value of the relative variation of h_o in percentage as a measure of the departure from the null hypothesis: the larger this variation is, the greater the amount of structure relies on the ratio image. Here $\overline{h_g}$ is the average over $p \geq 1$ samples of I_g.

Since the spatial structure is subtle in ratio images produced by state-of-the-art filters, δh requires being scaled to be comparable with $r_{\widehat{\mathrm{ENL}}}$. After careful experimentation with both simulated data and images from operational sensors, we found that 100 produces sensible and consistent results. This value was then fixed as part of our proposal, requiring no further tuning. Note that δh provides an objective measure for ranking despeckled results regarding solely the remaining geometrical content within the related ratio images.

The proposed estimator combines the measures of the remaining structure and of deviations from the statistical properties of the ratio image:

$$\mathcal{M} = r_{\widehat{\mathrm{ENL}},\widehat{\mu}} + \delta h. \quad (3)$$

The perfect despeckling filter will produce $\mathcal{M} = 0$, and the larger $\mathcal{M}$ is, the further the filter is from the ideal.

In the following, we will show that the proposed measure of quality is expressive and able to translate into a single value a number of measures of quality, both objective and subjective.

4. Experimental Setup

In this section we present the results of using the new metric for evaluating the quality of widely-used despeckling filters. We employ both simulated data and images from operational SAR systems, and we conclude with an application of our metric for filter optimization.

We used the following filters: E-Lee (Enhanced Lee [14]), SRAD (Speckle Reducing Anisotropic Diffusion [9]), PPB (Probabilistic Patch Based [15]), and FANS (Fast Adaptive Nonlocal SAR [16]). All of them provide good results and may be considered state-of-the art despeckling filters. E-Lee filter is an improved version of the classical adaptive Lee filter [2]. SRAD belongs to the category of PDE-based (Partial Differential Equations) filters, while the other two belong to the category of nonlocal means filters. In particular, FANS employs a set of wavelet transforms in its collaborative filtering stage.

The filters were tuned to the recommended designs as provided by their authors, with slight modifications (mask size and related threshold values) for PPB and FANS that yielded improved mean and ENL preservation. This was done for a fair comparison with SRAD and E-Lee filters which perform particularly well on preserving those features.

The E-Lee filter uses a 9×9 search window, and all the other parameters are as in [14]. The diffusion time for SRAD is $T = 300$, and the other parameters are as recommended in [9]. The PPB filter uses 7×7 patches and 21×21 search windows, and 25 iterations. The FANS filter uses 8×8 blocks, and 39×39 pixels search area; the remaining parameters are set as specified in [16]. The E-Lee and the SRAD filters are our own implementation. The source codes of PPB and FANS are available at [17,18], respectively.

For all the experiments, the co-occurrence matrices were computed after quantizing the observations to eight values, $p = 100$ independent samples were obtained for each image, and the tolerance and window side for Equation (1) were set to 0.03 and $w = 25$, respectively. The window side does not have a strong impact on the proposed measure; smaller windows will detect larger textureless patches with less observations, while larger windows will produce the opposite effect.

We will show that usual measures of quality are unable to provide enough evidence for the choice of a filter and, oftentimes, these quantities are conflicting in both simulated data and images from a SAR sensor. We will also see that our proposed measure is able to provide a sensible score of filter performance, and to guide in the choice of optimal parameters.

4.1. Simulation Results

Figure 10a shows the phantom with which we simulated images. This phantom has both large flat areas, linear edges between them and small pointwise-like details of 2×4 and 4×4 pixels (Appendix A). Figure 10b shows the result of injecting single-look speckle to this phantom.

The mean, variance and ENL are also computed within the four squares and the background. Good despeckling must preserve the mean value while significantly reducing the variance in these textureless areas increasing, thus, ENL.

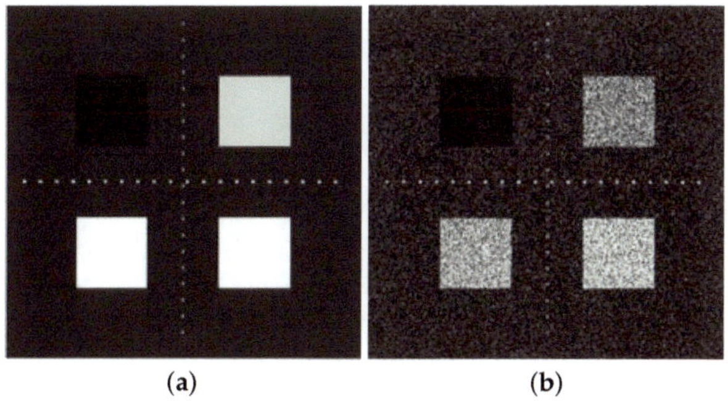

Figure 10. Blocks and points phantom, and 500 × 500 pixels simulated single-look intensity image. (**a**) Blocks and points phantom; (**b**) Speckled version, single look.

The data shown in Figure 10 allows measuring the ability of speckle filters at reducing noise (it presents large textureless areas), and at preserving small details [14,19]. The background intensity is 10, while that of the four squares is: 2 (top left), 40 (top right), 60 (bottom left), and 80 (bottom right). There are two sets of bright scatterers (intensity 240): twenty of size 4×4 along the horizontal direction, and twenty of size 4×2 along the vertical. The simulated data are obtained by multiplying these values by iid exponential deviates with unitary mean.

Figure 11 shows the results of applying the four filters on the simulated image, and their ratio images (first and second column respectively).

The four filters perform well since they preserve edges and bright scatterers, and also make textureless areas smoother. The ratio images reveal that the SRAD, and the E-Lee filters seem to be the least effective in terms of remaining structure as the squares edges are still visible (more for the SRAD filter). This remaining geometric content seems minimum for the PPB and FANS filter, although a careful observation reveals structures in all ratio images. See details in Figure 12.

It is expected that this subjective assessment be confirmed by the quantitative results provided by our proposal.

An objective assessment can be performed with respect to the ground reference. To that aim, we computed the Mean Structural Similarity Index MSSIM [20], the Peak Signal-to-Noise Ratio PSNR, and the measure of correlation between edges β [21].

MSSIM measures the similarity between the simulated and the despeckled images with local statistics (mean, variance and covariance between the unfiltered and despeckled pixel values) [20,22]. This measure is bounded in $(-1, 1)$, and a good similarity produces values close to 1. The β estimator is useful for assessing edge preservation. It evaluates the correlation between edges in the ground reference and the denoised images; edges are detected by either the Laplacian or the Canny filter. This parameter ranges between 0 and 1, and the bigger it is, the better the filter is; ideal edge preservation yields $\beta = 1$. PSNR is a global measure of quality, as it measures the ratio of the maximum value and the square root of the total error. High PSNR indicates a well-filtered image.

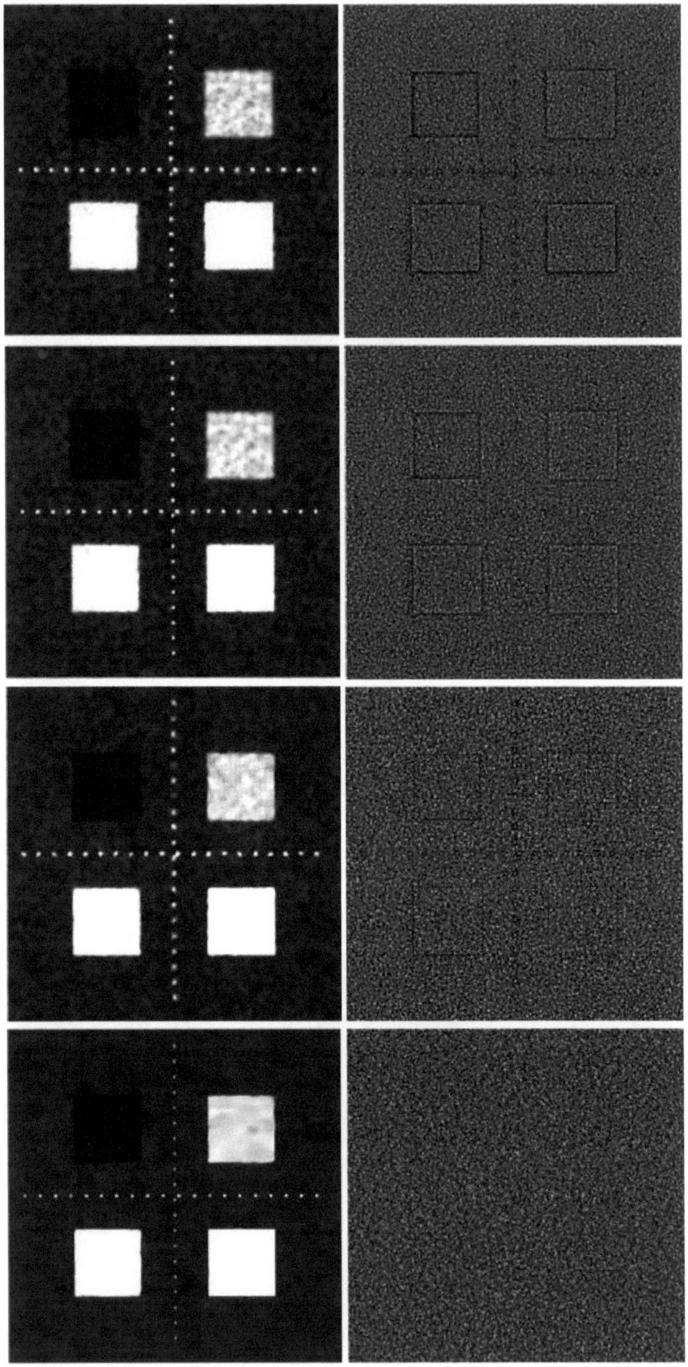

Figure 11. Results for the simulated single-look intensity data. Top to bottom, (**left**) results of applying the SRAD, the E-Lee, the PPB and the FANS filters. Top to bottom (**right**), their ratio images.

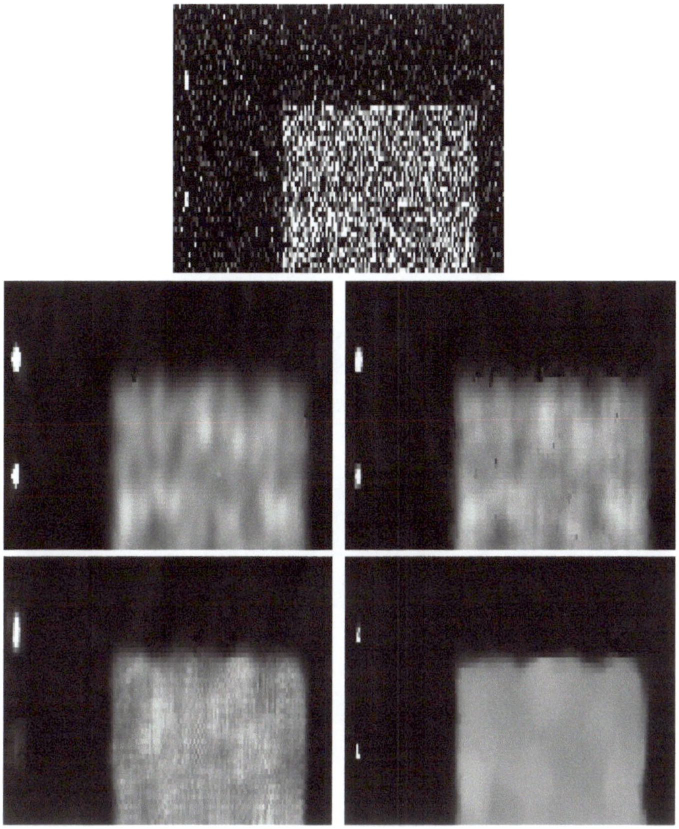

Figure 12. Zoom of the results for synthetic data: (**top**) Noisy image, (**first row, left**) SRAD filter, (first row, right) E-Lee filter, (**second row, left**) PPB filter and, (**second row, right**) FANS filter.

Table 1 presents the measures of quality as estimated in the simulated image ROIs (four squares and background), and also in the complete image. From this table, SRAD, E-Lee and PPB performances are comparable and quite acceptable. However, FANS obtains most of the best scores (mainly for variance reduction and ENL) while preserving reasonably well mean values. MSSIM and β are also better (for instance, $\beta = 0.40$ for FANS and $\beta = 0.22$ for the E-Lee filter). The zoom in Figure 12 corroborates this numerical assessment.

Table 1 also shows the values for ENL and the estimated μ within the background of the ratio image. All are close to the ideal (ENL ≈ 1, $\mu \approx 1$), although the best results are for FANS (ENL $= 1.0028$ and for E-Lee ($\mu = 1.0019$)).

Table 2 shows that the proposed measure provides significantly different values for each filter. According to $\mathcal{M}$, FANS is the best filter, followed by SRAD, E-Lee and PPB. The results are consistent with both the quantitative and qualitative visual assessment of the filtered images and their ratio. Note that FANS is the one with least geometric content within the ratio image ($\delta h = 6.26$), and also with lowest $r_{\widehat{\text{ENL}},\widehat{\mu}}$ residual. The opposite behavior is observed in PPB, although less residual content is visible in the ratio image (compared to SRAD and E-Lee filters) it obtains the highest (worst) $\mathcal{M}$ score (7.0371). Note that this result agrees with the commonly accepted criteria of evaluation of a despeckling filter: mean and ENL must be preserved. Due to that high score in the $r_{\widehat{\text{ENL}},\widehat{\mu}}$ residual, PPB is strongly penalized.

Table 1. Quantitative evaluation of filters on the simulated SAR image (best values in boldface).

Simulated SAR Data		True	Simulated	SRAD	E-Lee	PPB	FANS
Background	μ	10	9.93	**9.94**	9.91	10.12	10.13
	s	10	9.99	0.96	0.92	1.02	**0.45**
	ENL	1	0.98	105.86	115.13	90.87	**489.38**
Top left square	μ	2	1.96	1.97	1.96	**1.99**	2.01
	s	2	1.93	0.19	0.19	0.20	**0.08**
	ENL	1	1.03	101.80	106.17	94.64	**640.55**
Top right square	μ	40	40.07	**39.98**	39.85	40.69	40.59
	s	40	39.83	4.41	4.24	3.89	**2.04**
	ENL	1	1.01	82.09	88.10	109.20	**394.84**
Bottom left square	μ	60	59.92	60.12	**59.93**	60.17	61.54
	s	60	60.00	6.76	5.78	5.67	**2.88**
	ENL	1	0.99	78.88	107.49	112.50	**455.83**
Bottom right square	μ	80	**79.32**	**79.35**	78.99	81.53	81.61
	s	80	78.89	8.76	7.63	8.20	**3.68**
	ENL	1	1.01	81.88	106.99	96.68	**490.45**
Whole image	PSNR	—	73.87	**80.30**	78.72	79.07	77.85
	MSSIM	—	0.38	0.95	0.95	0.95	**0.98**
	β	—	0.14	0.22	0.27	0.30	**0.40**
Ratio image	$\widehat{\text{ENL}}_{\text{ratio}}$	1	—	1.0744	1.0346	1.0858	**1.0028**
	$\widehat{\mu}_{\text{ratio}}$	1	—	0.9914	**1.0019**	0.9775	0.9974

Table 2. Quantitative evaluation of ratio images for the simulated data (best value in boldface), computed on $n = 83$ automatically detected homogeneous areas.

Filter	h_o	$\overline{h_g}$	δh	$r_{\widehat{\text{ENL}},\widehat{\mu}}$	$\mathcal{M}$
SRAD	0.3026	0.3023	9.41	4.6634	7.0371
E-Lee	0.3465	0.3460	14.30	2.8781	8.5910
PPB	0.5551	0.5543	14.56	5.7751	10.1704
FANS	0.3827	0.3829	**6.26**	**2.0944**	**4.1816**

4.2. Results for Actual SAR Images

We show the benefits of our proposal on two SAR images obtained by the AIRSAR sensor in HH polarization, three looks in intensity format; cf. Figure 13.

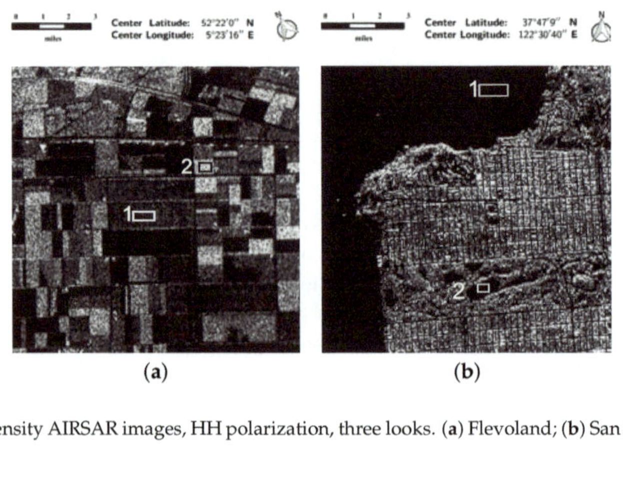

(a) (b)

Figure 13. Intensity AIRSAR images, HH polarization, three looks. (**a**) Flevoland; (**b**) San Francisco bay.

Figure 13a shows a subregion of 500 × 500 pixels from the image of Flevoland, The Netherlands. It corresponds to a flat area made up of reclaimed land used for agriculture and forestry. The image contains numerous crop types grown in large rectangular fields which are very appropriate to evaluate mean and variance values. There are also bright scatterers which allow evaluating the filters ability at preserving them. Figure 14 shows the filtered images in the first column, and their ratio images in the second.

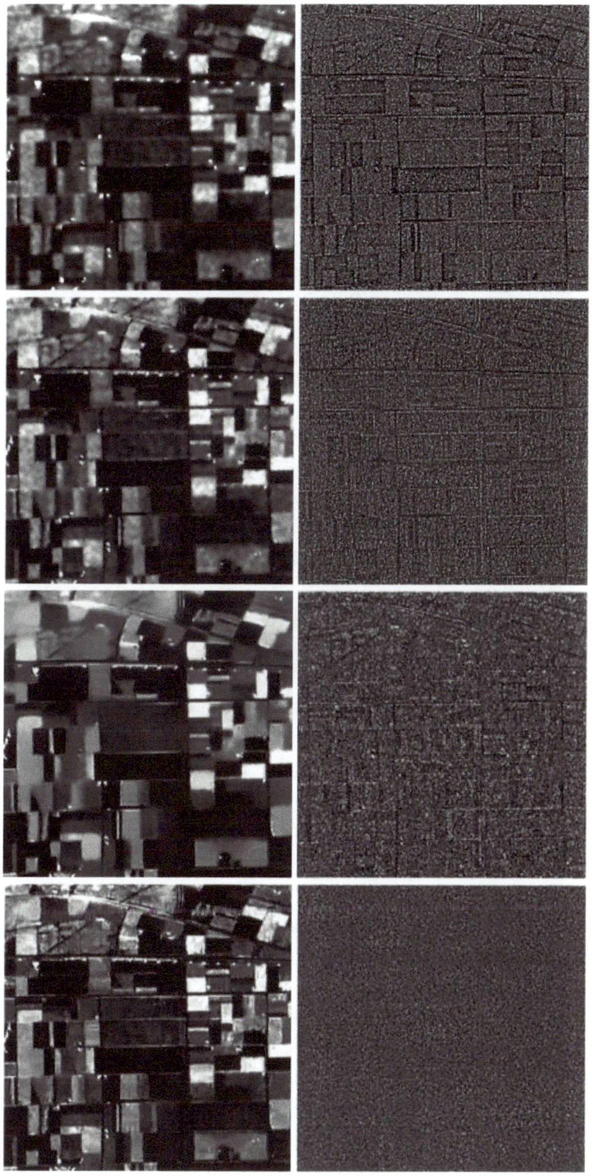

Figure 14. Results for the Flevoland image. Top to bottom, (**left**) results of applying SRAD, E-Lee, PPB and FANS filters. Top to bottom (**right**), their ratio images.

As expected, the filters perform well in terms of variance reduction and edge and bright scatterers preservation. FANS (bottom) provides the best visual result, outperforming the other filters: homogeneous areas are notably more homogeneous. SRAD blurs a little the image. PPB gets a fine visual result but it seems also overfiltered although patch homogeneity outperforms to the other filters. Edge preservation is better for FANS too as it can be appreciated in the images shown in Figure 15.

FANS is also the best with respect to structural content in the ratio image, and SRAD is the one leaving most structure within it. However, as for the simulated image, minute geometrical content still remains after applying FANS.

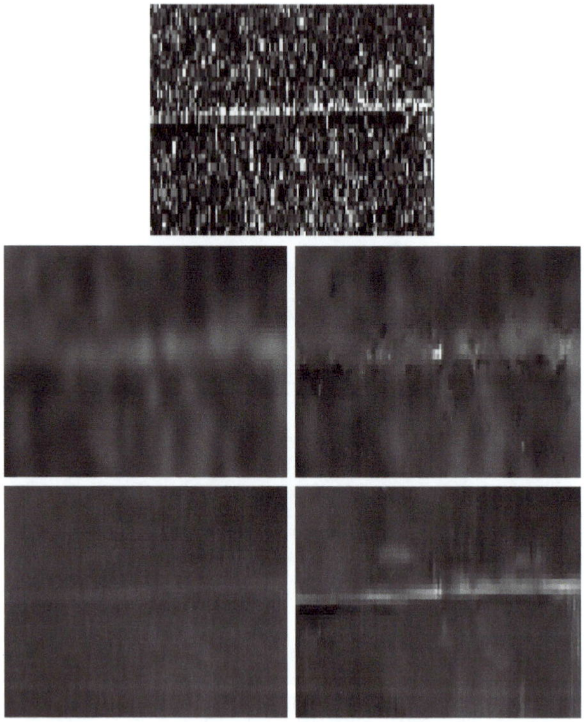

Figure 15. Zoom of the results for Flevoland image: (**top**) Noisy image, (**first row, left**) SRAD filter, (**first row, right**) E-Lee filter, (**second row, left**) PPB filter and, (**second row, right**) FANS filter.

Table 3 presents the mean, standard deviation and ENL values estimated in the boxed regions identified in Figure 13 (left). FANS is the best with respect to the mean preservation in both regions, although all filters obtain competitive values. The best variance reduction and ENL values are obtained with PPB, notably in ROI-2.

The analysis of the ratio images (see Table 4) is not conclusive: no filter gets the best values for all estimators. PPB produced a poor ENL result in both ROI-1 and ROI-2 (2.8048 and 3.8159, resp., instead of 3). However, all results are acceptable with small differences and, based on the solely analysis of these estimations within the ratio images one can hardly decide if a filter performs better than the others.

Table 3. Quantitative assessment of Flevoland filtered data in selected ROIs (best values in boldface).

Filter	ROI-1			ROI-2		
	$\widehat{\mu}$	s	**ENL**	$\widehat{\mu}$	s	**ENL**
Original	0.0047	0.0030	2.5000	0.0208	0.0110	3.5441
SRAD	**0.0047**	7.3561×10^{-4}	41.2367	0.0204	0.0012	283.7539
E-Lee	**0.0047**	7.4516×10^{-4}	39.1870	0.0206	0.0012	276.1669
PPB	0.0048	$\mathbf{3.3690} \times 10^{-4}$	**200.6540**	0.0212	$\mathbf{5.8933} \times 10^{-4}$	$\mathbf{1.2918} \times 10^{3}$
FANS	**0.0047**	5.0309×10^{-4}	86.1449	**0.0209**	6.2295×10^{-4}	1.1290×10^{3}

Table 4. Quantitative assessment of ratio images for Flevoland filtered data in selected ROIs (best values in boldface).

Filter	ROI-1		ROI-2	
	$\widehat{\mu}$	**ENL**	$\widehat{\mu}$	**ENL**
SRAD	0.9862	**2.9836**	1.0152	3.6287
E-Lee	**0.9981**	2.9824	**1.0097**	**3.5755**
PPB	0.9720	2.8048	0.9729	3.8159
FANS	0.9942	2.8822	0.9874	3.7082

In agreement with the visual inspection, FANS has the best $\mathcal{M}$ score (see Table 5). For these data, E-Lee obtains the worst score (86.2818) showing also a high $r_{\widehat{\text{ENL}},\widehat{\mu}}$ residual (11.2636). It is interesting to point out that, although the best preservation of $r_{\widehat{\text{ENL}},\widehat{\mu}}$ within the ratio image is provided by SRAD ($r_{\widehat{\text{ENL}},\widehat{\mu}} = 8.2782$), its final $\mathcal{M}$ score is heavily penalized by $\delta h = 66.81$ which accounts for the remaining structural content, as expected. Notice that $\delta h = 1.09$ for FANS.

Table 5. Quantitative evaluation of ratio images for Flevoland data (best value in boldface), computed on $n = 8$ automatically detected homogeneous areas.

Filter	h_o	$\overline{h_g}$	δh	$r_{\widehat{\text{ENL}},\widehat{\mu}}$	$\mathcal{M}$
SRAD	0.2043	0.2029	66.81	**8.2782**	37.5450
E-Lee	0.2247	0.2212	161.30	11.2636	86.2818
PPB	0.6210	0.6140	114.30	10.2211	5.6174
FANS	0.8944	0.8943	**1.09**	8.8547	**4.9771**

In the following, we present the results for the other AIRSAR image.

Figure 13b shows a subregion of 500 × 500 pixels from the three-look intensity AIRSAR, HH polarization, over the San Francisco Bay. This image contains mostly urban areas and sea, parks and hills covered by vegetation. There are few textureless areas except for the ocean.

Figure 16 presents the results obtained with SRAD, E-Lee, PPB and FANS (top to bottom, left). The corresponding ratio images are also shown (second column).

SRAD clearly overfiltered and, consequently much structure is found within its ratio image. Notice that we have applied the recommended filter parameters [9] that provided acceptable results for the simulated case and for the previous actual case (Flevoland) but, as showed, another more suitable set is required for this image. E-Lee preserves well the bright scatterers but parts of the image seem also overfiltered (the forest and some building blocks); as a result, much geometric content is visible in its ratio image. The PPB and FANS results are visually comparable, although some bright scatterers due to buildings are lost by PPB. FANS is also better at edge preservation. Once again, FANS ratio image resembles pure speckle, as seen in the bottom right image, while the structural contents in the PPB ratio image are noticeable.

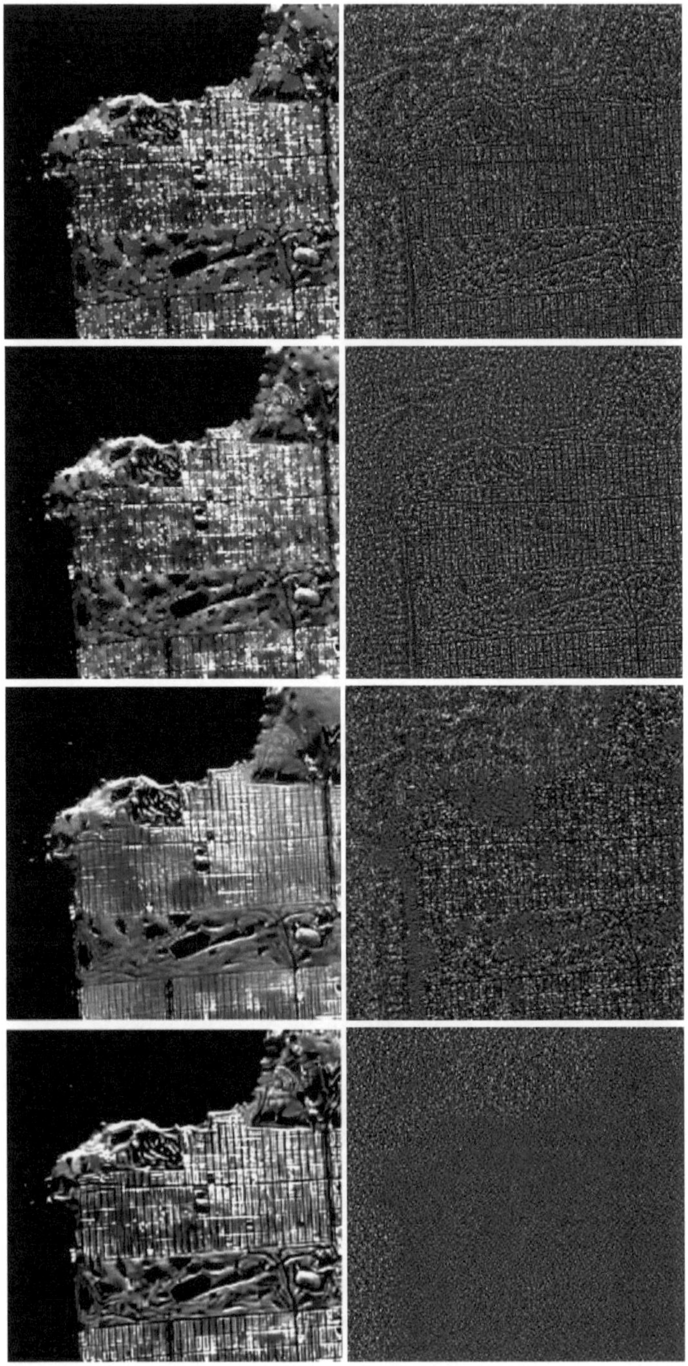

Figure 16. Result for the San Francisco bay image. Top to bottom, (**left**) results of applying SRAD, E-Lee, PPB and FANS. Top to bottom (**right**), their ratio images.

Figure 17 shows a detail of those results. Notice that PPB and FANS results are visually acceptable and quite similar.

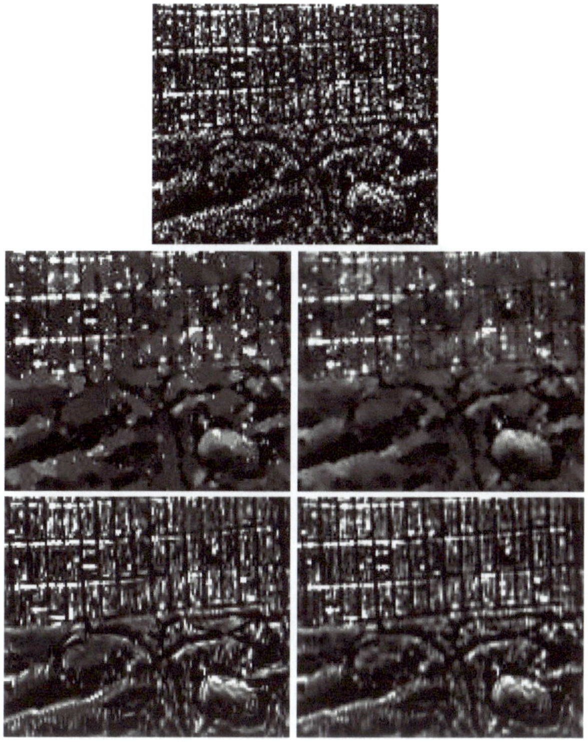

Figure 17. Zoom of the results for San Francisco image: (**top**) Noisy image, (**first row, left**) SRAD filter, (**first row, right**) E-Lee filter, (**second row, left**) PPB filter and, (**second row, right**) FANS filter.

Table 6 presents the mean, standard deviation and ENL estimated in the boxed regions identified in Figure 13b. As with the Flevoland data, no conclusive results stem from those values. However, FANS is consistently the best over ROI-2. A similar conclusion is reached for the estimators measured on the ratio images shown in Table 7.

Table 6. Quantitative assessment of San Francisco bay filtered data in selected ROIs (best values in boldface).

Filter	ROI-1			ROI-2		
	$\widehat{\mu}$	s	ENL	$\widehat{\mu}$	s	ENL
Original	6.8327×10^{-4}	3.8422×10^{-4}	3.1625	0.0018	8.9834×10^{-4}	4.1959
SRAD	7.0597×10^{-4}	3.6942×10^{-5}	365.2071	0.0021	8.1671×10^{-5}	674.1768
E-Lee	$\mathbf{6.8252} \times 10^{-4}$	8.9163×10^{-5}	58.5959	**0.0020**	1.7975×10^{-4}	129.6569
PPB	6.9884×10^{-4}	$\mathbf{3.2459} \times 10^{-5}$	**463.5443**	**0.0020**	1.5831×10^{-4}	163.9516
FANS	6.9156×10^{-4}	4.7095×10^{-5}	215.6278	**0.0020**	$\mathbf{3.5513} \times 10^{-5}$	$\mathbf{3.2947} \times 10^{3}$

Table 7. Quantitative assessment of ratio images for San Francisco bay filtered data in selected ROIs (best values in boldface).

Filter	ROI-1		ROI-2	
	$\widehat{\mu}$	ENL	$\widehat{\mu}$	ENL
SRAD	0.9651	**3.3477**	0.8634	4.7106
E-Lee	**0.9955**	3.5834	0.8948	4.9673
PPB	0.9692	3.4437	0.9004	4.7289
FANS	0.9829	3.3819	**0.9023**	**4.2443**

Table 8 presents the proposed $\mathcal{M}$ metric. Again, FANS obtains the best score as expected from the visual inspection of the ratio images. Although the best result for ENL value and μ preservation is for the SRAD filter, due to the large amount of residual structure within its related ratio image, δh is large enough to rank it to the last position among all despeckling filters discussed in this work.

Table 8. Quantitative evaluation of ratio images for San Francisco bay data (best value in boldface), computed on $n = 10$ automatically detected homogeneous areas.

Filter	h_o	$\overline{h_g}$	δh	$r_{\widehat{\text{ENL}},\widehat{\mu}}$	$\mathcal{M}$
SRAD	0.5643	0.5368	487.26	**0.2216**	5.0942
E-Lee	0.5813	0.5586	390.35	0.3262	4.2297
PPB	0.7449	0.7419	40.65	0.5395	0.9460
FANS	0.7138	0.7141	**5.10**	0.4231	**0.4741**

The above results for actual SAR data support the use of our proposed $\mathcal{M}$ metric.

4.3. Using $\mathcal{M}$ for Filter Design

Next we show the use of $\mathcal{M}$ in fine-tuning the parameters of a despeckling filter on actual data. We use FANS due to its already attested performance, and the Niigata Pi-SAR data as the image to be despeckled.

Figure 18 (left) shows a subimage (300×300 pixels), in intensity format, one look and HH polarization. The resolution of this image is $3\,\text{m} \times 3\,\text{m}$. The selected area includes urban and forest patches.

As indicated in [16] FANS requires more than ten control parameters, although the authors also mentioned that "*All parameters have been set once and for all, obtaining always satisfactory results in the experiments, so the user can forget about them and keep the default values*".

However, we show that some improvement can be achieved by a basic optimization strategy. We selected three control parameters: S (size of rows and columns of neighborhood blocks), P_{FA} (false alarm probability related to wavelet thresholding for the classification process), and W (wavelet transform used in the 2D spatial domain). The default values for these parameters are $S = 16$, $P_{FA} = 10^{-3}$ and, the Daubechies-4 wavelet for the choice of W. These control parameters are extensively discussed in [16], and they seem to have a strong impact on the filter performance.

The filter was optimized by exhaustive search: $S \in [4, 20]$ with steps $h_S = 1$, $P_{FA} \in [0.001, 0.01]$ with steps $h_P = 0.001$, and wavelet transforms from the ones suggested in the author's Matlab implementation: Meyer, DCT (discrete cosine transform), Haar, Daubechies-2, Daubechies-3, Daubechies-4, biorthogonal-1.3, and biorthogonal-1.5.

The optimal values found were $S = 4$, $P_{FA} = 0.0041$ and the Haar wavelet transform. With these, eighteen 15×15 homogeneous areas were detected.

The despeckled result by FANS with default parameters is shown in Figure 18 (middle) and the result by using the optimized parameters is shown in the same figure (right). A seen, some artifacts have been notably reduced and homogeneous areas, which seem more uniform with the optimized filter. The ratio images are depicted in Figure 19. A visual inspection suggests that there remains less geometrical structure within the ratio image by filtering with the optimized parameters.

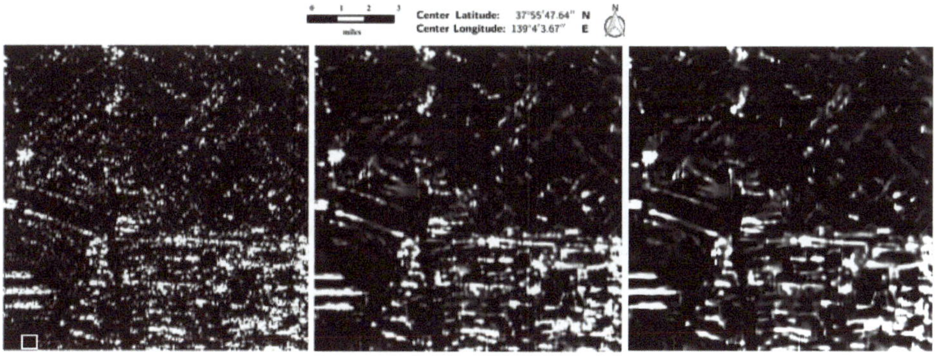

Figure 18. Intensity Pi-SAR, HH one look Niigata image (**left**); Results of applying FANS filters with default parameters (**middle**) and with optimized parameters (**right**).

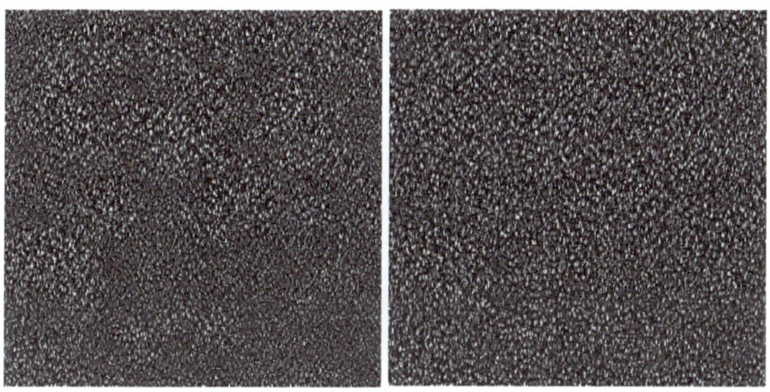

Figure 19. Ratio images for Niigata data; FANS with default parameters (**left**) and with optimized parameters (**right**).

Table 9 presents the mean, standard deviation and ENL estimated in the boxed region identified in Figure 18 (left). Best results for the three estimators are for the optimized FANS filter.

Table 9. Quantitative assessment of San Francisco bay filtered data in selected ROIs (best values in boldface).

Filter	$\widehat{\mu}$	s	ENL
Original	0.0283	0.0261	1.1757
FANS (default parameters)	0.0302	0.0106	8.1171
FANS (optimized parameters)	**0.0295**	**0.0083**	**12.6325**

Similar conclusion is reached for the estimators measured on the ratio images shown in Table 10.

Table 10. Quantitative evaluation of ratio images for Niigata data (best values in boldface), computed on $n = 18$ automatically detected homogeneous areas.

Filter	$\widehat{\mu}$	ENL
FANS (default parameters)	0.8627	2.2270
FANS (optimized parameters)	**0.9006**	**1.8745**

The proposed $\mathcal{M}$ metric is presented in Table 11.

Table 11. Quantitative evaluation of ratio images for Niigata data (best value in boldface), computed on $n = 18$ automatically detected homogeneous areas.

Filter	$r_{\widehat{\text{ENL}},\widehat{\mu}}$	δh	$\mathcal{M}$
FANS (default parameters)	0.4833	20.89	10.6867
FANS (optimized parameters)	**0.3794**	**6.50**	**3.4397**

From these results, it is clear that $\mathcal{M}$ can be applied to design a despeckling filter working on actual data without the need of ground references.

5. Conclusions

We proposed a new image-quality index, $\mathcal{M}$, to objectively evaluate despeckling filters. The proposal operates only in the ratio image and requires no reference. The evaluation relies on measuring deviations from the ideal statistical properties of the ratio image and their residual structural contents. The last component is computed by comparing a textural measure in the ratio image with random permutations of the data.

We have shown the expressiveness and adequacy of $\mathcal{M}$ using both simulated data and SAR images, and we verified that it is consistent with widely used image-quality indices as well as with the visual inspection of both filtered and ratio images. It has been also shown that the proposed unassisted image quality index can also be embedded into the design of despeckling filters. Additionally, the computational cost related to the proposed estimator is comparable to state-of-the art indexes.

The proposal is valid as long as the multiplicative model holds and provided that at least one (even small) region can be detected as textureless. The user is required to input an estimate of the number of looks. The index employs a random component, but it is reproducible once fixed the platform, the random number generator, and the seed.

Acknowledgments: The third author (ACF) wishes to express his gratitude to Xin Li (Chinese Academy of Sciences, Lanzhou) for the two-month stay during which this work was concluded.

Author Contributions: All authors contributed equally to this manuscript.

Conflicts of Interest: The authors declare no conflict of interest. The founding sponsors had no role in the design of the study; in the collection, analyses, or interpretation of data; in the writing of the manuscript, and in the decision to publish the results.

Appendix A. Computational Platform

The code and data for reproducing the results here reported are available here http://www.de.ufpe.br/raydonal/ReproducibleResearch/UNASSISTED/UNASSISTED-QUANTITATIVE.html. The Matlab [23] language was used to simulate and analyze the data. Haralick's textural features were also computed by the available libraries in Matlab. The computational cost for the 500×500 synthetic data shown in this work (see Figure 10) with the parameters setting as mentioned in Section 4, is around 20 s in an Intel Core i7 Q740 1.73 GHz machine.

Appendix B. Extension to the Gaussian Additive Noise Model

The idea of using the residual image as a proxy for filter quality can be also used for the Gaussian additive model. If the observed image is $Z = X + Y$, with X and Y independent fields, and Y a collection of iid zero-mean Gaussian random variables, then the ideal filter will produce $\widehat{X} = X$, and the residual image $I = Z - \widehat{X} = Y$ should bear no structure and be formed by Gaussian deviates with zero mean and the same variance. This idea was used by Hale [24] to attest to the superiority of a new filter for seismic images. The analysis is visual, so there is room for research using, for instance, the Anderson-Darling test for normality. Peng et al. [25] also analyze residuals as a measure of quality of subspace clustering.

References

1. Lee, J.S. Digital Image Enhancement and Noise Filtering by Use of Local Statistics. *IEEE Trans. Pattern Anal. Mach. Intell.* **1980**, *PAMI-2*, 165–168.
2. Lee, J.S. Speckle Suppression and Analysis for Synthetic Aperture Radar Images. *Opt. Eng.* **1986**, *25*, 636–643.
3. Argenti, F.; Lapini, A.; Bianchi, T.; Alparone, L. A Tutorial on Speckle Reduction in Synthetic Aperture Radar Images. *IEEE Geosci. Remote Sens. Mag.* **2013**, *1*, 6–35.
4. Oliver, C.; Quegan, S. *Understanding Synthetic Aperture Radar Images*; Artech House: Boston, MA, USA, 1998.
5. Achim, A.; Kuruoglu, E.; Zerubia, J. SAR image filtering based on the heavy-tailed Rayleigh model. *IEEE Trans. Image Process.* **2006**, *15*, 2686–2693.
6. Parrilli, S.; Poderico, M.; Angelino, C.V.; Verdoliva, L. A nonlocal SAR image denoising algorithm based on LLMMSE wavelet shrinkage. *IEEE Trans. Geosci. Remote Sens.* **2012**, *50*, 606–616.
7. Martino, G.D.; Poderico, M.; Poggi, G.; Riccio, D.; Verdoliva, L. Benchmarking framework for SAR despeckling. *IEEE Trans. Geosci. Remote Sens.* **2014**, *52*, 1596–1615.
8. Gomez, L.; Buemi, M.E.; Jacobo-Berlles, J.; Mejail, M. A new image quality index for objectively evaluating despeckling filtering in SAR images. *IEEE J. Sel. Top. Appl. Earth Obs. Remote Sens.* **2016**, *9*, 1297–1307.
9. Yu, Y.; Acton, S.T. Speckle reducing anisotropic diffusion. *IEEE Trans. Image Process.* **2002**, *11*, 1260–1270.
10. Lee, J.S.; Jurkevich, I.; Dewaele, P.; Wambacq, P.; Oosterlinck, A. Speckle filtering of synthetic aperture radar images: A review. *Remote Sens. Rev.* **1994**, *8*, 313–340.
11. Feng, H.; Hou, B.; Gong, M. SAR Image Despeckling Based on Local Homogeneous-Region Segmentation by Using Pixel-Relativity Measurement. *IEEE Trans. Geosci. Remote Sens.* **2011**, *49*, 2724–2737.
12. Haralick, R.M.; Shanmugam, K.; Dinstein, I. Textural Features for Image Classification. *IEEE Trans. Syst. Man Cybern.* **1973**, *SMC-3*, 610–621.
13. Gomez, L.; Alvarez, L.; Pinheiro, R.L.; Frery, A.C. A Benchmark for Despeckling Filters. In Proceedings of the 11th European Conference on Synthetic Aperture Radar, Hamburg, Germany, 6–9 June 2016; pp. 451–454.
14. Lee, J.S.; Wen, J.H.; Ainsworth, T.L.; Chen, K.; Chen, A.J. Improved sigma filter for speckle filtering of SAR imagery. *IEEE Trans. Geosci. Remote Sens.* **2009**, *47*, 202–213.
15. Deledalle, C.A.; Denis, F.; Tupin, F. Iterative weighted maximum likelihood denoising with probabilistic patch-based weights. *IEEE Trans. Image Process.* **2009**, *18*, 2661–2672.
16. Cozzolino, D.; Parrilli, S.; Scarpa, G.; Poggi, G.; Verdoliva, L. Fast adaptive nonlocal SAR despeckling. *IEEE Geosci. Remote Sens. Lett.* **2014**, *11*, 524–528.
17. Deledalle, C. Probabilistic Patch-Based Filter (PPB). 2015. Avaliable online: http://www.math.u-bordeaux1.fr/~cdeledal/ppb.php (accessed on 19 April 2017).
18. Image Processing Research Group. 2015. Avaliable online: http://www.grip.unina.it/web-download.html?dir=JSROOT/FANS (accessed on 19 April 2017).
19. Zhong, H.; Li, Y.; Jiao, L. SAR image despeckling using Bayesian nonlocal means filter with sigma preselection. *IEEE Geosci. Remote Sens. Lett.* **2011**, *8*, 809–813.
20. Wang, Z.; Bovik, A.; Sheikh, H.; Simoncelli, E. Image quality assessment: from error visibility to structural similarity. *IEEE Trans. Image Process.* **2004**, *13*, 600–611.

21. Sattar, F.; Floreby, L.; Salomonsson, G.; Lovstrom, B. Image enhancement based on a nonlinear multiscale method. *IEEE Trans. Image Process.* **1997**, *6*, 888–895.
22. Chabrier, S.; Laurent, H.; Rosenberger, C.; Emile, B. Comparative study of contour detection evaluation criteria based on dissimilarity measures. *EURASIP J. Image Video Process.* **2008**, *2008*, 1–13.
23. MATLAB. *Version 8.3.0.532 (R2014a)*; The MathWorks Inc.: Natick, MA, USA, 2014.
24. Hale, D. Structure-oriented bilateral filtering. In Proceedings of the 2011 SEG Annual Meeting, Society of Exploration Geophysicists, San Antonio, TX, USA, 18–23 September 2011; pp. 239–244.
25. Peng, X.; Tang, H.; Zhang, L.; Yi, Z.; Xiao, S. A Unified Framework for Representation-Based Subspace Clustering of Out-of-Sample and Large-Scale Data. *IEEE Trans. Neural Netw. Learn. Syst.* **2016**, *27*, 2499–2512.

remote sensing

MDPI

Article

Nonlocal Tensor Sparse Representation and Low-Rank Regularization for Hyperspectral Image Compressive Sensing Reconstruction

Jize Xue [1], Yongqiang Zhao [2,*], Wenzhi Liao [3] and Jonathan Cheung-Wai Chan [4]

1 School of Automation, Northwestern Polytechnical University, Xi'an 710072, China; xuejize900507@mail.nwpu.edu.cn
2 Research & Development Institute of Northwestern Polytechnical University in Shenzhen, Shenzhen 518057, China
3 Department of Telecommunications and Information Processing, Ghent University-TELIN-IMEC, 9000 Ghent, Belgium; Wenzhi.Liao@UGent.be
4 Department of Electronics and Informatics, Vrije, Universiteit Brussel, 1050 Brussel, Belgium; jchengw@etro.vub.ac.be
* Correspondence: zhaoyq@nwpu.edu.cn; Tel.: +86-1599-175-1747

Received: 7 December 2018; Accepted: 17 January 2019; Published: 19 January 2019

Abstract: Hyperspectral image compressive sensing reconstruction (HSI-CSR) is an important issue in remote sensing, and has recently been investigated increasingly by the sparsity prior based approaches. However, most of the available HSI-CSR methods consider the sparsity prior in spatial and spectral vector domains via vectorizing hyperspectral cubes along a certain dimension. Besides, in most previous works, little attention has been paid to exploiting the underlying nonlocal structure in spatial domain of the HSI. In this paper, we propose a nonlocal tensor sparse and low-rank regularization (NTSRLR) approach, which can encode essential structured sparsity of an HSI and explore its advantages for HSI-CSR task. Specifically, we study how to utilize reasonably the l_1-based sparsity of core tensor and tensor nuclear norm function as tensor sparse and low-rank regularization, respectively, to describe the nonlocal spatial-spectral correlation hidden in an HSI. To study the minimization problem of the proposed algorithm, we design a fast implementation strategy based on the alternative direction multiplier method (ADMM) technique. Experimental results on various HSI datasets verify that the proposed HSI-CSR algorithm can significantly outperform existing state-of-the-art CSR techniques for HSI recovery.

Keywords: hyperspectral image; compressive sensing; structured sparsity; tensor sparse decomposition; tensor low-rank approximation

1. Introduction

Hyperspectral image (HSI) is a three-dimension data cube by simultaneously capturing the information over two spatial and one spectral dimensions. The abundant spatial-spectral information is able to provide more accurate and reliable signature features on distinct materials, which contributes to various applications such as scene classification [1], object detection [2], environmental monitoring [3], etc. However, due to the large data sizes of HSI, the storage and transmission on limited resource platform become a challenge problem. Although various methods, mainly including wavelet transform [4–6], TDLT + KLT [7], DPCM [8] and JPEG2000 [9,10], have been proposed to compress HSI effectively, they treat the HSI as a collection of single band images and neglect the spatial-spectral knowledge redundancy. Thus, how to build rational and powerful HSI compressive reconstruction models is still a worthy research issue.

Recently, the compressive sensing (CS) [11–13] theory offers a brand-new field for HSI acquisition or compression, which only needs to capture a small number of incoherent measurements in the imaging stage. Then, the acquired measurements can be employed to reconstruct the whole HSI. For convenient application of CS on HSI, many well-known techniques [14–41] have been presented to convert an HSI into a sparse signal. Although HSI CS can greatly reduce the resource consumption on imaging, storage and transmission compared with those conventional compression methods, how to reconstruct precisely the HSI from fewer measurements is still a challenging problem.

One of the main concerns to the ill-posed reconstruction problem is to convert HSI into sparse description form via imposing some proper sparsity priors. For example, some effective sparsity terms with l_0, l_1 and l_p $(0 < p < 1)$ norms [13–16] have been presented to characterize the sparsity for signal recovery, but those methods neglect the underlying structure information. Regularization-based approaches usually incorporate the prior knowledge into the observation model and develop a united framework [17–20]. For those methods, one key issue is how to design a proper regularization term to characterize the sparsity of HSI. The works in [21–23] mainly consider the sparsity of abundance matrix by the linear unmixing of an HSI, and then HSI CS models are built using spectral unmixing procedures. By introducing structured sparsity across spatial or spectral dimension, Zhang et al. [24–28] extended the compression method based sparse representation/dictionary learning to HSI compression. More recently, Meza et al. [29,30] explored the group sparsity based spatial/spectral redundancy structure to achieve HSI compressive sensing reconstruction (HSI-CSR). The HSI CS model proposed by Golbabaee et al. [31–34] utilized the piecewise smooth structure to explain the underlying gradient sparsity of an HSI. However, as those techniques depict the HSI sparsity in vector space, the description form of sparsity is treated as one vector without considering its multidimensional structure. It will inevitably induce losses and distortions of useful structure information.

Tensor-based HSI-CSR approaches can improve remarkably the HSI recovery quality, since the existing methods jointly take into account the spatial-spectral information, and reduce the losses and distortions caused by HSI reshaping [35–44]. Karami et al. [35,36] exploited discrete wavelet transform and Tucker decomposition (DWT-TD) to encode the spatial-spectral information of HSI. The core idea behind those techniques is first to use DWT to effectively separate an HSI into different sub-images, and then to apply TD on the DWT coefficients of HSI bands to compact the energy of sub-images. Zhang et al. [37,38] compressed an HSI to the core tensor and the HSI could be reconstructed by the multi-linear projection of the factor matrices. Those methods only consider an HSI as a whole 3D tensor while they are short of more potent constraints on spatial-spectral structure of an HSI. Yang [39] employed nonlinear tensor sparse representation to recover an HSI from small number of measurements, and some training examples are required. Wang [40] used the global spatial-spectral correlation and local smoothness properties underlying in an HSI to enhance the HSI-CSR task, in which the tensor Tucker decomposition and 3-D total variation jointly characterize the sparsity of an HSI. Du [41] proposed a patch-based low-rank tensor decomposition for HSI-CSR algorithm that combined the nonlocal similarity across the spatial domain and the low-rank property over spectral domain in a united framework.

Although methods reported in [37,38,40,41] are considerably effective for HSI-CSR compared with vector based approaches, it is difficult to estimate the accurate rank under tensor decomposition and further acquire unique decomposition. Thus, the methods based on tensor decomposition cannot provide an elaborate characterization on spatial-spectral information in HSI-CSR problem. In [42,43], this reasonable usage of the global correlation across spectrum (GCS) and nonlocal self-similarity over space (NSS) prior knowledge have led to quite powerful HSI denoising algorithms, and the effectiveness of GCS and NSS for HSI-CSR has not been reported in the public literature. Such facts inspire us to solve the challenging HSI-CSR problem by the structured sparsity based on GCS and NSS in this paper, and a unified framework combining nonlocal tensor sparse representation and low-rank regularization is proposed for HSI-CSR, as shown in Figure 1. The main contributions of this paper are listed as follows.

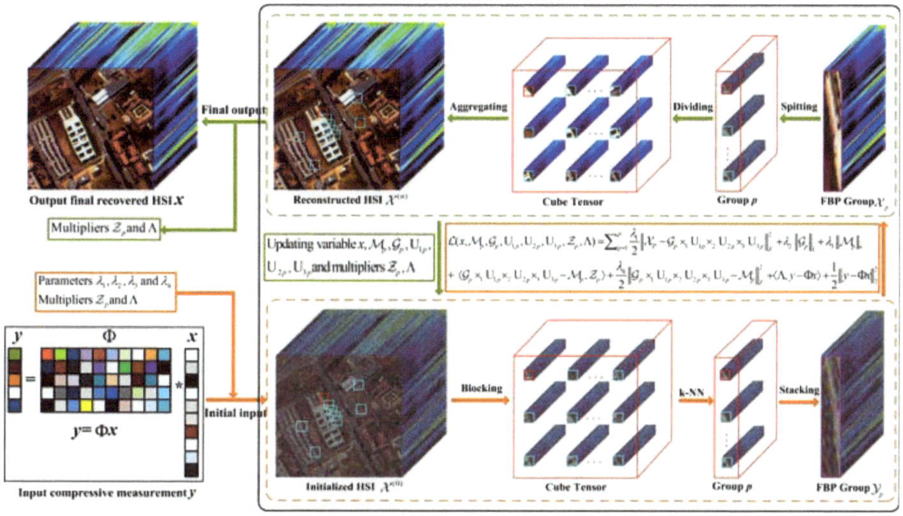

Figure 1. Flowchart of the proposed HSI-CSR algorithm, which consists of two steps: sensing and reconstruction. First, it acquires the compressive measurement y by a random sampling matrix Φ. Second, NTSRLR recovers an HSI from the measurements $y = \Phi x$.

1. To the best of our knowledge, we are the first to exploit GCS and NSS to construct the nonlocal structure sparsity of HSI that is a faithfully structured sparsity representation form for HSI-CSR task.
2. For each cube that is formed by grouping nonlocal similar cubes, the tensor representation based on tensor sparse and low-rank approximation is introduced to encode the intrinsic spatial-spectral correlation.
3. The HSI-CSR task is treated as an optimization problem; we resort to alternative direction multiplier method (ADMM) [44] to solve it.

A preliminary version of this work has appeared in [45], which presents the basic approach. In [45], we established the nonlocal structured sparsity from the perspective of the tensor low-rank property, which adopts the two most commonly used tensor low-rank representation forms: tensor low-rank approximation and tensor low-rank decomposition. In this paper, we depict the nonlocal structured sparsity via the tensor low-rank approximation and sparse representation. Although the tensor low-rank decomposition and sparse representation are derived from the Tucker decomposition model, the former needs to preset the ranks along all dimension while the latter introduces an l_1-based sparse term on core tensor. In practical application, the latter possesses the reliable capability to represent the high-dimension data by mitigating the tensor rank overfitting or underfitting. In addition, this paper adds: (1) the detailed background of HSI-CSR; (2) the theoretical analysis of NTSRLR; and (3) additional HSI-CSR experiments.

The remainder of this paper is organized as follows. Section 2 introduces the tensor notations and operations commonly used in this paper, and background of CS. In Section 3, a novel algorithm for HSI-CSR based on the NTSRLR model is proposed. Section 4 demonstrates the results of extensive experiments and Section 5 draws the conclusion.

2. Notations and Background of HSI-CS

2.1. Notations

Throughout the paper, we denote scalars, vectors, matrices and tensors by non-bold letters, bold lower case letters, bold upper case letters and calligraphic upper case letters, respectively. Besides, we introduce some necessary notations and preliminaries about tensor as follows. A tensor of order N, which corresponds to a N-dimensional data array, is denoted as $\mathcal{X} \in \mathbb{R}^{I_1 \times \cdots \times I_n \times \cdots \times I_N}$. Elements of $\mathcal{X}$ are denoted as $a_{i_1 \cdots i_n \cdots i_N}$, where $1 \leq i_n \leq I_n$. Definitions of tensor terminologies in the paper follow exactly the same description in [46]. Denote $\|\mathcal{X}\|_F = \langle \mathcal{X}, \mathcal{X} \rangle \, (\sum_{i_1 i_2, \ldots, i_N} |a_{i_1 i_2, \ldots, i_N}|^2)^{1/2}$, $\|\mathcal{X}\|_1 = \sum_{i_1 i_2, \ldots, i_N} |a_{i_1 i_2, \ldots, i_N}|$ and $\|\mathcal{X}\|_0$ as the F-norm, l_1 norm and l_0 norm of a tensor $\mathcal{X}$, respectively. $\|\mathcal{X}\|_0 \leq K$ means that K is the number of non-zero entries of $\mathcal{X}$. It is convenient to unfold a tensor into a matrix during the algorithm. The "unfold" operation along the mode-n on a tensor $\mathcal{X}$ is defined as $\text{unfold}_n(\mathcal{X}) := X_{(n)} \in \mathbb{R}^{I_n \times (I_1 \times \cdots \times I_{n-1} I_{n+1} \times \cdots \times I_N)}$, and its opposite operation "fold" is defined as $\text{fold}_n(X_{(n)}) := \mathcal{X}$. The Kronecker product of matrices $A \in \mathbb{R}^{I \times J}$ and $B \in \mathbb{R}^{K \times L}$ is a matrix of size $IK \times JL$, denoted by $A \otimes B$. The multiplication of a tensor $\mathcal{X}$ with a matrix $Y \in \mathbb{R}^{I_k \times J_k}$ on mode-k is denoted by $\mathcal{X} \times_k Y = \mathcal{Z}$, which also can be defined in terms of mode-k unfolding as $Z_k = Y X_k$.

Definition 1. *(Tucker decomposition) [46]: The Tucker decomposition form of a tensor $\mathcal{X}$ is:*

$$\mathcal{X} = \mathcal{G} \times_1 \mathrm{U}_1 \times_2 \cdots \times_N \mathrm{U}_\mathrm{N} \tag{1}$$

where $\mathcal{G} \in \mathbb{R}^{J_1 \times J_2 \times \cdots \times J_N}$ is the core tensor and it reflects the interaction between components along different modes, and $\mathrm{U}_n \in \mathbb{R}^{I_n \times J_n}$ is the orthogonal factor matrix in each mode. Thus, we can achieve the k-unfolding form of Tucker decomposition in Equation (1)

$$\mathrm{X}^{(n)} = \mathrm{U}_\mathrm{n} \mathrm{G}^{(n)} (\mathrm{U}_\mathrm{N} \otimes \cdots \otimes \mathrm{U}_{\mathrm{n}+1} \otimes \mathrm{U}_{\mathrm{n}-1} \otimes \cdots \otimes \mathrm{U}_1) \tag{2}$$

2.2. Background of HSI-CS

For a given HSI $\mathcal{X} \in \mathbb{R}^{W \times H \times S}$ ($W \times H$ spatial resolution and S spectral bands), $x \in \mathbb{R}^{WHS}$ denotes the vector form of $\mathcal{X}$. Let $N = WHS$, then the compressive measurement $y \in \mathbb{R}^M$ can be obtained from the following CS model:

$$y = \Phi x \tag{3}$$

where $\Phi \in \mathbb{R}^{M \times N} (M < N)$ denotes the compressive operator. The CS theory indicates that a sufficiently sparse signal x can be exactly reconstructed from only a few observation y when the compressive operator Φ satisfies the restricted isometry property (RIP) [11]. Under the RIP, the ill-posed recovery problem can be formulated into following form by pursuing the sparsest signal x, i.e.,

$$x = \min_x \|x\|_0, \ \text{ s.t. } y = \Phi x \tag{4}$$

where $\|\cdot\|_0$ denotes l_0 norm as a sparsity constraint. However, the l_0 norm minimization in Equation (4) is combinatorially NP-hard and unstable with the noise. For this reason, a feasible strategy is to replace nonconvex l_0 norm as a convex l_1 counterpart [15,47] as follows:

$$x = \min_x \|x\|_1, \ \text{ s.t. } y = \Phi x \tag{5}$$

The optimization for above l_1-minimization CS problem can resort to iterative shrinkage algorithm [48] and Bregman Split algorithm [49].

Since an HSI can be sparsely represented in a certain domain, many CS models have been proposed for an HSI. Zhang et al. [21–23] unmixed the HSI into a spatially sparse abundance matrix with an endmember matrix. Meza et al. [29–31] extracted the spatial/spectral redundancy structure and then applied the group sparsity constraint. Golbabaee [34] used a wavelet basis to transform the HSI into a sparse matrix, and then adopted the low-rankness and l_1 norm to jointly encode sparsity of the matrix. Zhang et al. [37,38] depicted the sparsity of an HSI in the core tensor domain, instead of reshaped vector domain. Further works [39–41] employ the sparse tensor decomposition to characterize sparsity of an HSI. However, those sparsity constraint terms are incapable of capturing the underlying structure in an HSI or handling the unwanted noise and artifacts in the CSR procedure. In our method, we try to cope with those problems by introducing more refined prior knowledge of an HSI to perfectly promote HSI-CSR performance.

3. The Proposed HSI-CSR via NTSRLR

Structured sparsity is of great importance to the HSI-CSR model that often reveals the rich self-repetitive structures over spatial domain and the highly correlated bands across the spectral domain. Several previous works exploiting nonlocal prior have indicated that the structured sparsity based on nonlocal self-similarity is fairly effective for image restoration [18,19]. However, the research works in HSI-CSR fields have not been documented. In this paper, we present a unified framework for HSI-CSR using the structured sparsity via nonlocal tensor sparse representation and low-rank approximation.

3.1. Non-Local Tensor Formula for Structure Sparsity

The proposed regularization model for structured sparsity consists of two steps: cube grouping for characterizing GCS and NSS and tensor formulation for sparsity enforcement.

3.1.1. Non-Local Structure Sparsity Analysis

Concerning the GCS and NNS underlying an HSI, we provide an analysis for nonlocal tensor sparsity and low-rankness, as illustrated in Figure 2. To begin with, for an initial third-order tensor HSI $\mathcal{X} \in \mathbb{R}^{W \times H \times S}$ (e.g., *PaviaU* dataset), we divide the HSI into a group of 3D full-band cubes (FBC) $\{\mathcal{P}_{i,j}\}_{1 \leq i \leq W-w+1, 1 \leq j \leq H-h+1} \in \mathbb{R}^{w \times h \times S} (w < W, h < H)$ with overlaps. For the exemplar cube $\mathcal{P}_{i,j}$ of size $8 \times 8 \times 60$ located at spatial position (i, j) in Figure 2a marked in red, we first search K-1 (here, we set K = 80) similar cubes by k-NN within a local window (e.g., 70×70), shown as k-NN clustering in Figure 2b. Then, to avoid destroying the high spectral correlation, we unfold a series of 3D cubes into corresponding 2D matrices along the spectral modes (Figure 2c), and obtain a new third-order tensor $\mathcal{Y}_p$ of size $64 \times 80 \times 60$ by stacking a series of similar items (Figure 2d), where $p = 1, \ldots P$, and P denotes the group number. Such constructed third-order tensor simultaneously employ the spatial local sparsity (mode-1), the non-local similarity between cubes (mode-2) and strong spectral correlation (mode-3). The outcome of such arrangement maximizes the benefit from nonlocal tensor representation form. Next, we give a visual interpretation for the nonlocal tensor sparsity and low-rank property.

First, by Tucker decomposition for a nonlocal similar cube group from *PaviaU* dataset, Figure 2e shows the location of singular values in the core tensor, where redder and bluer colors of elements represent large values and smaller values, respectively. To further understand the sparsity of tensor core, Figure 2(e2)–(e4) present three typical slices of core tensor. It is easy to find that the core tensor satisfies sparse property, with 82.59% of its elements being zeroes. Second, the low-rank analysis is performed along its local spatial, nonlocal spatial, and global spectral modes, as shown in Figure 2f. Evidently, the decaying trends of singular values on three curves (pink, blue and green curves correspond to local spatial, nonlocal spatial, and global spectral modes, respectively) indicate there are strong correlations in the three modes. Comparatively, the decaying trend of the curve in mode-2 is most drastic, which is consistent with the nonlocal spatial low-rank theory of an HSI given in [50]. According to the definition of the accumulation energy ratio (Aer) of top k singular values in [50], we calculate Top 10 singular values of three modes and attain the Aers of 0.8029, 0.9031 and 0.8186. The quantitative values (i.e., Aers) also indicate that each cube by grouping nonlocal similar cubes can possess strong low-rank correlation along the mode-2.

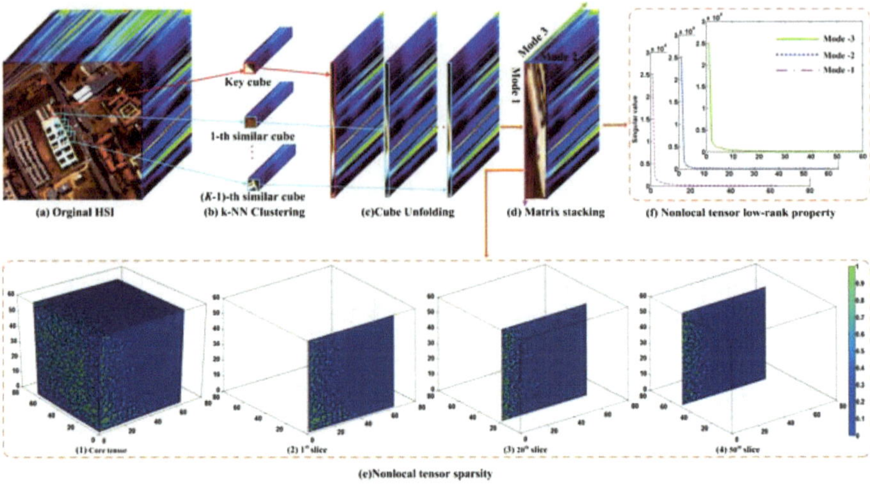

Figure 2. Nonlocal tensor sparsity and low-rank property analysis in HSI.

3.1.2. Non-Local Structure Sparsity Modeling

In Figure 2f, we can observe that the formed FBCs possess the low-rank property, and a tractable strategy is to use the mode-n rank$(r_1, \ldots, r_n)$ to estimate tensor rank by Tucker decomposition [46]. For an Nth-order tensor $\mathcal{X}$, the Tucker rank is defined as rank$(\mathcal{X})$: $=$ $[\text{rank}(X_{(1)}), \text{rank}(X_{(2)}), \ldots, \text{rank}(X_{(N)})]$, where $X_{(i)}$ is the mode-i unfolding of $\mathcal{X}$ [51]. Motivated by the practical applications that the nuclear norm is the convex envelope of the matrix rank within the unit ball of the spectral norm, further tensor nuclear norm, $\|\mathcal{X}\|_* = \sum_{n=1}^{N} \alpha_n \left\|X_{(n)}\right\|_*$ is defined as weighting the unfolding matrix nuclear norm along each mode. Thus, we resort to the following relaxation form for each $\mathcal{X}_p$ to characterize the low-rank property based on GCS and NSS:

$$\mathcal{L}(\mathcal{X}_p) = \sum_i^3 \alpha_i \|\mathcal{X}_{p_{(i)}}\|_* \tag{6}$$

where $\|\mathcal{X}_{p(i)}\|_* = \sum_{k=1}^{\min(m,n)} \sigma_k(\mathcal{X}_{p(i)})$ denotes the nuclear norm of matrix $\mathcal{X}_{p(i)}$ of size $m \times n$.

In practice, $\{\mathcal{Y}_p\}_{p=1}^{P}$ may contain some noise, the data $\mathcal{Y}_p$ can be modeled as: $\mathcal{Y}_p = \mathcal{X}_p + \mathcal{W}_p$, where $\mathcal{X}_p$ and $\mathcal{W}_p$ denote the low-rank component and the noise component, respectively. Hence, we can estimate the low-rank tensor $\mathcal{X}_p$ via the following optimization problem:

$$\mathcal{X}_p = \min_{\mathcal{X}_p} \mathcal{L}(\mathcal{X}_p), \quad \text{s.t. } \left\|\mathcal{Y}_p - \mathcal{X}_p\right\|_F^2 \le \varepsilon \tag{7}$$

where ε is associated with the noise level. The model in Equation (7) is similar to the matrix cases in [18], the difference primarily reflected in that we consider the combination with the correlations along local-nonlocal spatial modes and spectral mode, and measure the low-rankness of a third-order tensor $\mathcal{X}_p$ by a weighted sum of the rank along each unfolding. Besides, considering the strong nonlocal spatial low-rankness along mode-2 than two other modes, we set a larger weight for mode-2 in our experiments.

In addition, as shown in Figure 2e, we give a detailed analysis for another notable representation form for the sparsity prior based on tensor sparse decomposition, which suggests that we can depict the structured sparsity of an HSI from the perspective of core tensor. Some pioneering works are presented in [42,43,52–54]. Here, we draw attention to the structured sparsity formulation of an HSI under tensor sparse representation framework, thus each third-order tensor $\mathcal{X}_p$ can be approximated by following problem:

$$\min_{\mathcal{G}_p,\mathrm{U}_{1p},\mathrm{U}_{2p},\mathrm{U}_{3p}} \mathcal{S}(\mathcal{G}_p), \text{s.t. } \mathcal{X}_p = \mathcal{G}_p \times_1 \mathrm{U}_{1p} \times_2 \mathrm{U}_{2p} \times_3 \mathrm{U}_{3p}, \mathrm{U}_{ip}^T \mathrm{U}_{ip} = I(i = 1,2,3) \tag{8}$$

where $\mathrm{U}_{1p}, \mathrm{U}_{2p}$, and U_{3p} are factor matrices and $\mathcal{S}(\mathcal{G}_p)$ is sparse constraint term, and we assume $\mathcal{S}(\mathcal{G}_p) = \|\mathcal{G}_p\|_0$ as suggested in [42,43,52]. However, the optimization problem based on l_0 constraint deduced by Equation (8) is non-convex, the research in [53,54] further relaxes the l_0-based core sparsity to l_1 case as $\mathcal{S}(\mathcal{G}_p) = \|\mathcal{G}_p\|_1$. The convex optimization problem corresponding to l_1 case can be represented in Lagrangian form as following:

$$\min_{\mathcal{G}_p,\mathrm{U}_{1p},\mathrm{U}_{2p},\mathrm{U}_{3p}} \frac{\lambda_1}{2} \|\mathcal{X}_p - \mathcal{G}_p \times_1 \mathrm{U}_{1p} \times_2 \mathrm{U}_{2p} \times_3 \mathrm{U}_{3p}\|_F^2 + \lambda_2 \|\mathcal{G}_p\|_1, \text{ s.t. } \mathrm{U}_{ip}^T \mathrm{U}_{ip} = I(i = 1,2,3) \tag{9}$$

where λ_1 and λ_2 are the trade-off parameters. Essentially, all factor matrices are orthogonal dictionaries along local–nonlocal spatial modes and spectral mode. It can be seen that the tensor sparse representation model explores the GCS and NSS of HSIs in different dimensions by adaptive multi-dictionaries learning. Compared with the matrix sparse representation technique [19,20], the advantage of tensor modeling is that it not only characterizes the spatial-spectral correlation but also the correlation over nonlocal similar cubes in an HSI.

3.2. Proposed Model

Based on the previous analysis, we now derive the following model for solving the HSI-CSR problem:

$$\begin{gathered}\min_{x,\mathcal{G}_p,\mathrm{U}_{1p},\mathrm{U}_{2p},\mathrm{U}_{3p}} \sum_{p=1}^{P} \frac{\lambda_1}{2} \|\mathcal{X}_p - \mathcal{G}_p \times_1 \mathrm{U}_{1p} \times_2 \mathrm{U}_{2p} \times_3 \mathrm{U}_{3p}\|_F^2 + \lambda_2 \mathcal{S}(\mathcal{G}_p) + \lambda_3 \mathcal{L}(\mathcal{X}_p), \\ \text{s.t. } y = \Phi x,\ \mathrm{U}_{ip}^T \mathrm{U}_{ip} = I(\mathrm{i} = 1,2,3)\end{gathered} \tag{10}$$

where λ_3 is the regularization parameter. It is worth noting that the proposed model can fully exploit the underlying prior over spatial-spectral domain in an HSI, and thus is expected to have a strong ability to enhance HSI-CRS task.

3.3. Optimization Algorithm

For the proposed HSI-CSR model, we apply the ADMM [44], an effective strategy for solving large scale optimization problems, to solve Equation (10). Firstly, we replace $\mathcal{S}(\mathcal{G}_p)$ and $\mathcal{L}(\mathcal{X}_p)$ with the $\|\mathcal{G}_p\|_1$ and $\|\mathcal{X}_p\|_*$, respectively, and introduce P auxiliary tensors $\{\mathcal{M}_p\}_{p=1}^P$ and equivalently reformulate Equation (10) as follows:

$$\begin{gathered}\min_{x,\mathcal{M}_p,\mathcal{G}_p,\mathrm{U}_{1p},\mathrm{U}_{2p},\mathrm{U}_{3p}} \sum_{p=1}^{P} \frac{\lambda_1}{2} \|\mathcal{X}_p - \mathcal{G}_p \times_1 \mathrm{U}_{1p} \times_2 \mathrm{U}_{2p} \times_3 \mathrm{U}_{3p}\|_F^2 + \lambda_2 \|\mathcal{G}_p\|_1 + \lambda_3 \|\mathcal{M}_p\|_*, \\ \text{s.t. } y = \Phi x, \mathcal{M}_p = \mathcal{G}_p \times_1 \mathrm{U}_{1p} \times_2 \mathrm{U}_{2p} \times_3 \mathrm{U}_{3p}, \mathrm{U}_{ip}^T \mathrm{U}_{ip} = I(\mathrm{i} = 1,2,3)\end{gathered} \tag{11}$$

Then, its augmented Lagrangian function is:

$$\begin{aligned}&\mathcal{L}(\mathcal{X}_p, \mathcal{M}_p, \mathcal{G}_p, \mathrm{U}_{1p}, \mathrm{U}_{2p}, \mathrm{U}_{3p}, \mathcal{Z}_p, \Lambda) = \sum_{p=1}^{P} \frac{\lambda_1}{2} \|\mathcal{X}_p - \mathcal{G}_p \times_1 \mathrm{U}_{1p} \times_2 \mathrm{U}_{2p} \times_3 \mathrm{U}_{3p}\|_F^2 + \lambda_2 \|\mathcal{G}_p\|_1 \\ &+ \lambda_3 \|\mathcal{M}_p\|_* + \langle \mathcal{G}_p \times_1 \mathrm{U}_{1p} \times_2 \mathrm{U}_{2p} \times_3 \mathrm{U}_{3p} - \mathcal{M}_p, \mathcal{Z}_p \rangle + \frac{\lambda_4}{2} \|\mathcal{G}_p \times_1 \mathrm{U}_{1p} \times_2 \mathrm{U}_{2p} \times_3 \mathrm{U}_{3p} - \mathcal{M}_p\|_F^2 \\ &+ \langle \Lambda, y - \Phi x \rangle + \frac{1}{2} \|y - \Phi x\|_F^2\end{aligned} \tag{12}$$

where $\{\mathcal{Z}_p\}_{p=1}^{P}$ and Λ are the Lagrange multipliers, λ_4 is the positive scalars. We shall break Equation (12) into five sub-problems and iteratively update each variable via fixing the other ones.

(a) U_{1p}, U_{2p}, U_{3p} problem:

$$\begin{aligned}\min_{U_{1p},U_{2p},U_{3p}} \frac{\lambda_1}{2} \|\mathcal{X}_p - \mathcal{G}_p \times_1 U_{1p} \times_2 U_{2p} \times_3 U_{3p}\|_F^2 + \langle \mathcal{G}_p \times_1 U_{1p} \times_2 U_{2p} \times_3 U_{3p} - \mathcal{M}_p, \mathcal{Z}_p \rangle \\ + \frac{\lambda_4}{2} \|\mathcal{G}_p \times_1 U_{1p} \times_2 U_{2p} \times_3 U_{3p} - \mathcal{M}_p\|_F^2, \text{s.t. } U_{ip}^T U_{ip} = I (i = 1, 2, 3)\end{aligned} \tag{13}$$

which is equivalent to the following sub-problem:

$$\min_{U_{1p},U_{2p},U_{3p}} \sum_{p=1}^{P} \|\mathcal{G} \times_1 U_{1p} \times_2 U_{2p} \times_3 U_{3p} - \mathcal{O}_p\|_F^2, \text{s.t. } U_{ip}^T U_{ip} = I (i = 1, 2, 3) \tag{14}$$

where $\mathcal{O}_p = \frac{\lambda_1 \mathcal{X}_p + \sum_{i=1}^{3} (\lambda_4 \mathcal{M}_i - \mathcal{Z}_i)}{\lambda_1 + 3\lambda_4}$ can be easily solved by the method as suggested in [53,54].

(b) $\mathcal{G}_p$ sub-problem:

$$\begin{aligned}\min_{\mathcal{G}_p} \frac{\lambda_1}{2} \|\mathcal{X}_p - \mathcal{G}_p \times_1 U_{1p} \times_2 U_{2p} \times_3 U_{3p}\|_F^2 + \langle \mathcal{G}_p \times_1 U_{1p} \times_2 U_{2p} \times_3 U_{3p} - \mathcal{M}_p, \mathcal{Z}_p \rangle \\ + \frac{\lambda_4}{2} \|\mathcal{G}_p \times_1 U_{1p} \times_2 U_{2p} \times_3 U_{3p} - \mathcal{M}_p\|_F^2 + \lambda_2 \|\mathcal{G}_p\|_1\end{aligned} \tag{15}$$

It can be rewritten as

$$\min_{\mathcal{G}_p} \frac{1}{2} \|\mathcal{O}_p - \mathcal{G}_p \times_1 U_{1p} \times_2 U_{2p} \times_3 U_{3p}\|_F^2 + \lambda_2 \|\mathcal{G}_p\|_1 \tag{16}$$

It can be solved by the Tensor-based Iterative Shrinkage Thresholding Algorithm (TISTA) in [53,54].

(c) $\mathcal{M}_p$ sub-problem:

$$\min_{\mathcal{M}_p} \lambda_3 \|\mathcal{M}_p\|_* + \langle \mathcal{G}_p \times_1 U_{1p} \times_2 U_{2p} \times_3 U_{3p} - \mathcal{M}_p, \mathcal{Z}_p \rangle + \frac{\lambda_4}{2} \|\mathcal{G}_p \times_1 U_{1p} \times_2 U_{2p} \times_3 U_{3p} - \mathcal{M}_p\|_F^2, \tag{17}$$

It can be briefly reformulated as:

$$\min_{\mathcal{M}_p} \sum_{i=1}^{3} \frac{\lambda_3 \alpha_i}{\lambda_4} \left\|\mathcal{M}_{p(i)}\right\|_* + \frac{1}{2} \|\mathcal{B}_p + \frac{\mathcal{Z}_p}{\lambda_4} - \mathcal{M}_p\|_F^2, \tag{18}$$

where $\mathcal{B}_p = \mathcal{G}_p \times_1 U_{1p} \times_2 U_{2p} \times_3 U_{3p}$, its equivalent form is

$$\min_{\mathcal{M}_p} \sum_{i=1}^{3} \frac{\lambda_3 \alpha_i}{\lambda_4} \left\|\mathcal{M}_{p(i)}\right\|_* + \frac{1}{2} \|\mathcal{B}_{p(i)} + \frac{\mathcal{Z}_{p(i)}}{\lambda_4} - \mathcal{M}_{p(i)}\|_F^2, \tag{19}$$

As suggested in [51], its close-form solution is expressed as:

$$\mathcal{M}_{p(i)} = \text{fold}_i [S_{\alpha_i \lambda_3 / \lambda_4} (\mathcal{B}_{p(i)} + \frac{\mathcal{Z}_{p(i)}}{\lambda_4})], \tag{20}$$

For a given matrix X, the singular value shrinkage operator $S_\tau(X)$ is defined as $S_\tau(X) := U_X D_\tau(\Sigma_X) V_X^T$, and where $X = U_X \sigma_X V_X^T$ is the SVD of X and $D_\tau(A) = \text{sgn}(A_{ij})(|A_{ij}| - \tau)_+$.

(d) x sub-problem:

$$\min_{\mathcal{X}} \sum_{p=1}^{P} \frac{\lambda_1}{2} \left\| \mathcal{X}_p - \mathcal{G}_p \times_1 \mathrm{U}_{1p} \times_2 \mathrm{U}_{2p} \times_3 \mathrm{U}_{3p} \right\|_F^2 + \langle \Lambda, y - \Phi x \rangle + \frac{1}{2} \| y - \Phi x \|_F^2, \tag{21}$$

It is easy to observe that optimizing L with respect to x can be treated as solving the following linear system:

$$\lambda_1 x + \Phi^*(\Phi x) = \Phi^*(y - \Lambda) + \lambda_1 vec(\mathcal{X} - \mathcal{G} \times_1 \mathrm{U}_1 \times_2 \mathrm{U}_2 \times_3 \mathrm{U}_3), \tag{22}$$

where $\mathcal{G} \times_1 \mathrm{U}_1 \times_2 \mathrm{U}_2 \times_3 \mathrm{U}_3 = \sum_{p=1}^{P} \mathcal{G}_p \times_1 \mathrm{U}_{1p} \times_2 \mathrm{U}_{2p} \times_3 \mathrm{U}_{3p}$, $vec(\cdot)$ denotes the vectorization operator for a matrix or tensor, and Φ^* indicates the adjoint of Φ. Obviously, this linear system can be solved by well-known preconditioned conjugate gradient technique.

(e) Update the multipliers

$$\begin{cases} \mathcal{Z}_p = \mathcal{Z}_p + \rho\lambda_4(\mathcal{B}_p - \mathcal{M}_p) \\ \Lambda = \Lambda + \rho(y - \Phi x) \end{cases} \tag{23}$$

where ρ is a parameter associated with the convergence rate at values of, e.g., [1.05–1.1]. The whole optimization procedure for the proposed HSI-CSR model can be summarized as Algorithm 1, and we abbreviate the proposed method as NTSRLR.

Algorithm 1. HSI-CSR based NTSRLR.

Input: The compressive measurements y, measurement operator Φ, and the parameters of the algorithm.
Initialization: Initializing an HSI $x^{(0)}$ via a standard CSR method (e.g., DCT based CSR).
For $l = 1 : L$ **do**
  Extract the set of tensor $\{\mathcal{X}_p\}_{p=1}^{P}$ from $x^{(0)}$ via k-NN search the each exemplar cube;
  For $p = 1 : P$ **do**
    Solve the problem (12) by ADMM;
    Updating $\mathrm{U}_{1p}, \mathrm{U}_{2p}, \mathrm{U}_{3p}$ by via Equation (14);
    Updating $\mathcal{G}_p$ via Equation (16);
    Updating $\mathcal{M}_p$ via Equation (20);
    Updating the multipliers $\mathcal{Z}_p$ via Equation (23);
  End for
  Updating $x^{(l)}$ via Equation (22);
  Updating the multiplier Λ via Equation (23);
End for
Output: CS Reconstructed HSI $x^{(L)}$.

4. Experimential Results and Analysis

In this section, various experiments on real HSI datasets are executed to assess the performance of the proposed NTSRLR method. We chose eight popular methods for comparisons, namely the three classic CS methods including StOMP [55], BCS [56] and multidimensional signal based KCS [57]; total variation based methods with LRTV [34] and TVAL3 [58]; structured sparsity based HSI-CSR methods with RLPHCS [24], SRPREC [25] and CSFHR [28]; and the recent joint tensor decomposition regularization and total variation based method (JTRTV) [40]. These methods represent state-of-the-art HSI-CSR, especially LRTV and JTRTV, which fully consider the HSI sparsity priors. In comparison experiments, we used the default parameter settings of those compared methods described in the reference papers. We adopted random measurement matrix as the sampling operator for all methods.

4.1. Quantitative Metrics

To evaluate the HSI-CSR performances of all methods, five quantitative picture quality indices (PQIs) were employed in experiments. The first index is mean peak signal-to-noise ratio (MPSNR), which is defined as the average PSNR of all bands for HSI, e.g.

$$\mathrm{MPSNR}(\mathcal{X},\hat{\mathcal{X}}) = \frac{1}{S}\sum_{s=1}^{S}\mathrm{PSNR}(\mathcal{X}^s,\hat{\mathcal{X}}^s), \tag{24}$$

where $\mathcal{X}^s$ and $\hat{\mathcal{X}}^s$ denote sth band images of ground truth $\mathcal{X} \in \mathbb{R}^{W\times H\times S}$ reconstructed HSI $\hat{\mathcal{X}} \in \mathbb{R}^{W\times H\times S}$, respectively, and both of them are scaled to the range [0; 255].

The second index, mean structure similarity (MSSIM), was used to evaluate the similarity between the reconstructed HSI and the original HSI based on structural consistency, which is defined as average SSIM [59] of all bands for HSI,

$$\mathrm{MSSIM}(\mathcal{X},\hat{\mathcal{X}}) = \frac{1}{S}\sum_{s=1}^{S}\mathrm{SSIM}(\mathcal{X}^s,\hat{\mathcal{X}}^s), \tag{25}$$

The third index, mean feature similarity (MFSIM), emphasizes the perceptual consistency with the original image, which is defined as average FSIM [60] of all bands for HSI,

$$\mathrm{MFSIM}(\mathcal{X},\hat{\mathcal{X}}) = \frac{1}{S}\sum_{s=1}^{S}\mathrm{FSIM}(\mathcal{X}^s,\hat{\mathcal{X}}^s), \tag{26}$$

High values of these three measures MPSNR, MSSIM and MFSIM represent better reconstructed results.

The fourth index is the spectral angle mapper (SAM) [61], which calculates the average angle between spectrum vectors of the CS reconstructed HSI and the reference one across all spatial positions; its definition is as follows:

$$\mathrm{SAM}(\mathcal{X},\hat{\mathcal{X}}) = \cos^{-1}\left(\frac{x^T\hat{x}}{\sqrt{x^Tx}\sqrt{\hat{x}^T\hat{x}}}\right), \tag{27}$$

where x and $\hat{x}$ denote vector form of the ground truth $\mathcal{X}$ reconstructed HSI $\hat{\mathcal{X}}$, respectively.

The fifth index is the Erreur relative globale adimensionnelle desynthèse (ERGAS) [62], which measures fidelity of the CS reconstructed HSI based on the weighted sum of MSE in each band, defined as follows

$$\mathrm{ERGAS}(\mathcal{X},\hat{\mathcal{X}}) = 100\sqrt{\sum_{s=1}^{S}\frac{\mathrm{MSE}(\mathcal{X}^s,\hat{\mathcal{X}}^s)}{\mu^2_{\hat{\mathcal{X}}^s}}}, \tag{28}$$

where $\mathrm{MSE}(\mathcal{X}^s,\hat{\mathcal{X}}^s)$ is the mean square error between $\mathcal{X}^s$ and $\hat{\mathcal{X}}^s$, and $\mu^2_{\hat{\mathcal{X}}^s}$ is the mean value of $\hat{\mathcal{X}}^s$. Different from the former three PQI measures, smaller values of these two measures represent better reconstruction performances.

4.2. Experiments on Noiseless HSI Datasets

All methods are evaluated on three HSIs, namely *Toy* from the CAVE dataset (http://www1.cs.columbia.edu/CAVE/databases/multispectral/), *PaviaU* and corrected *Indian Pines* from hyperspectral remote sensing scenes (http://www.ehu.eus/ccwintco/index.php?title=Hyperspectra-Remote-Sensing-Scenes). The *Toy* is full spectral resolution reflectance data from 400 nm to 700 nm at 10 nm steps (31 bands total), with spatial resolution 512 × 512. The *PaviaU* dataset contains 103 bands, including 610 × 340 pixels. The *Indian Pines* is of size 145 × 145 with 10 m spatial resolution and consists of 200 bands via removing 20 noisy bands polluted by water absorption, which covers the wavelength in the range from 400 to 2500 nm by 10 nm spectral resolution. We conducted experiments on the three HSI datasets mainly for the following reasons. (1) The three HSI datasets possess higher spatial-spectral resolutions and richer non-local similarity, which facilitates that the structured sparsity across spatial-spectral domains is employed in our HSI-CSR model. (2) These HSIs are benchmark testing datasets in HSI reconstruction, as presented in [21,22,24,25,40,42,43,45,50,53,54]. (3) We selected the dataset with classification label, *Indian Pines*, which helps to compare all methods in term of classification accuracy. For the experiment, we cropped a sub-region of 300 × 300 for all bands of *Toy* and *PaviaU*, as shown in Figure 3. To validate the performance of proposed method, five different sampling rates (SR), namely 0.02, 0.05, 0.10, 0.15 and 0.20, were considered.

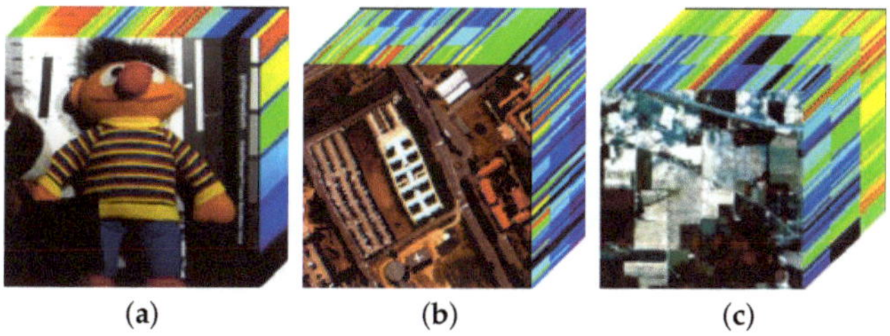

Figure 3. HSIs employed in the compressive sensing experiments: (**a**) *Toy*; (**b**) *PaviaU*; and (**c**) *Indian Pines*.

4.2.1. Visual Quality Evaluation

To visually demonstrate the HSI-CSR performances of the proposed method, we present the pseudocolor images with bands (25,15, 5), bands (55, 30, 5), and bands (23, 13, 3) of reconstructed *Toy*, *PaviaU* and *Indian Pines* obtained by all methods under sampling rates of 0.20, 0.10 and 0.15 in Figures 4–6, respectively. We have the following observations. (1) All the competing methods achieved relatively good reconstructed results. (2) The proposed method outperformed the other methods, as shown by the enlarged subregion (delineated in a red box), where the large-scale sharp edges and small-scale fine texture features are reconstructed well, as shown in Figures 4, 5 and 6j. (3) The method StOMP produced serious noise during reconstruction and the details are blurred in the results of BCS, KCS and CSFHR. Instead of l_1-based sparsity term, the TVAL3 utilizes the TV regularization based on gradient sparsity to preserve the more accurate edges but many details are lost. Although LRTV simultaneously considers the gradient sparsity and low-rankness of the data, the lack of an effective constraint for nonlocal spatial information will generate blurring artifacts. The JTRTV method is a generalization of LRTV for high-dimensional data, although it can deal with the artifacts problem generated by LRTV, it introduces unwanted noises. The RLPHCS and SRPREC consider the structure sparsity based on the reweighted Laplace prior. Nevertheless, their reconstructed results are unsatisfactory and the two methods appear to be virtually powerless for HSI-CSR. We provide following justifications about poor performance of RLPHCS and SRPREC: (1) The two HSI-CSR models use the maximum a posteriori framework to learn the hyperparameters; the accumulation of estimated bias for parameters may lead to a poor HSI-CSR performance. (2) The collected dictionaries in RLPHCS and SRPREC algorithms may not be overcomplete, which do not fully consider the redundant structure over spatial and spectral domain. This demonstrates the effectiveness of NTSRLR technique for HSI-CSR, greatly preserving the local details and structural information of the HSI.

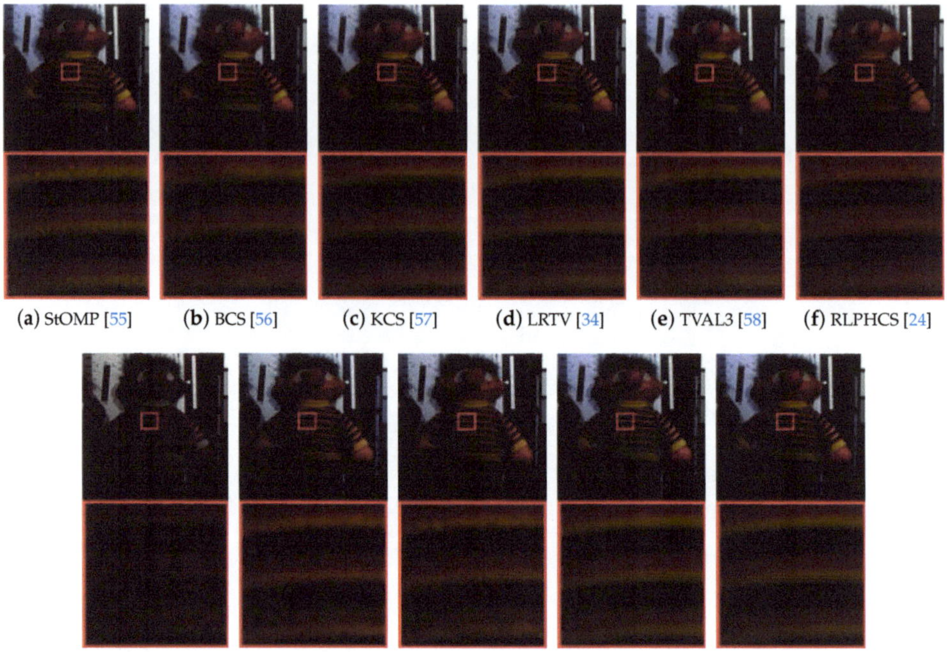

(**a**) StOMP [55] (**b**) BCS [56] (**c**) KCS [57] (**d**) LRTV [34] (**e**) TVAL3 [58] (**f**) RLPHCS [24]

(**g**) SRPREC [25] (**h**) JTRTV [40] (**i**) CSFHR [28] (**j**) NTSRLR (**k**) Original

Figure 4. Compressive sensing reconstructed results on pseudocolor images with bands (25,15, 5) of the *Toy* image from different methods under sampling rate $\rho = 0.20$.

4.2.2. Quantitative Evaluation

In Tables 1 and 2, we provide the performance of all methods using MPSNR, MSSIM, MFSIM, SAM and ERGAS results, over all the spectral bands in *Toy*, *PaviaU* and *Indian Pines*. We highlight the best results for each case in bold in the current and following tables. The proposed method outperforms the other approaches under all sampling rates and in particular the PQIs are better than the recent JTRTV. At sampling rate $\rho = 0.02$, NTSRLR improves the MPSNR at least 10 dB more than JTRTV on the *Toy*, 1.3 dB better on the *PaviaU*, and 2.7 dB better on the *Indian Pines*. For $\rho = 0.20$, the average gain of MPSNR values of NTSRLR are more amplified compared with JTRTV, up to 14 dB on *Toy*, 8 dB on *PaviaU* and 7 dB on *Indian Pines*. MSSIM, MFSIM, SAM and ERGAS values values under three HSI datasets further confirm the robustness of the proposed method at all sampling rates. Although LRTV is second best method, obviously it still is inferior to ours by visual quality evaluation. Since NTSRTR explores the underlying nonlocal structure of an HSI by the tensor sparse representation and low-rank modeling, it gives higher MPNSR, MSSIM, and MFSIM values, and smaller SAM and ERGAS than the other methods, which only consider the local or single sparsity prior.

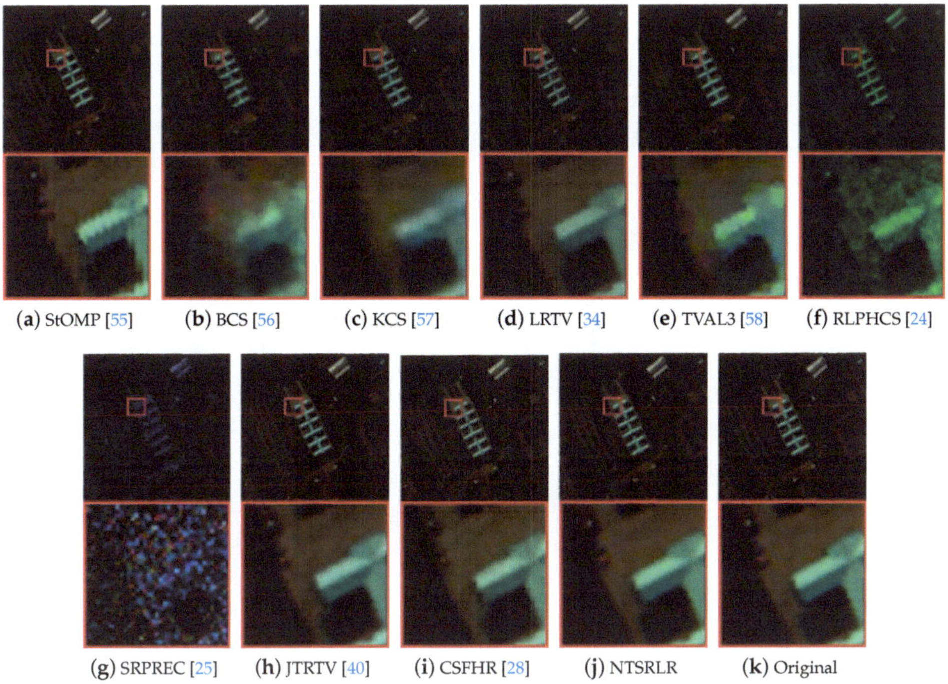

(a) StOMP [55] (b) BCS [56] (c) KCS [57] (d) LRTV [34] (e) TVAL3 [58] (f) RLPHCS [24]

(g) SRPREC [25] (h) JTRTV [40] (i) CSFHR [28] (j) NTSRLR (k) Original

Figure 5. Compressive sensing reconstructed results on pseudocolor images with bands (55, 30, 5) of the *PaviaU* image from different methods under sampling rate $\rho = 0.10$.

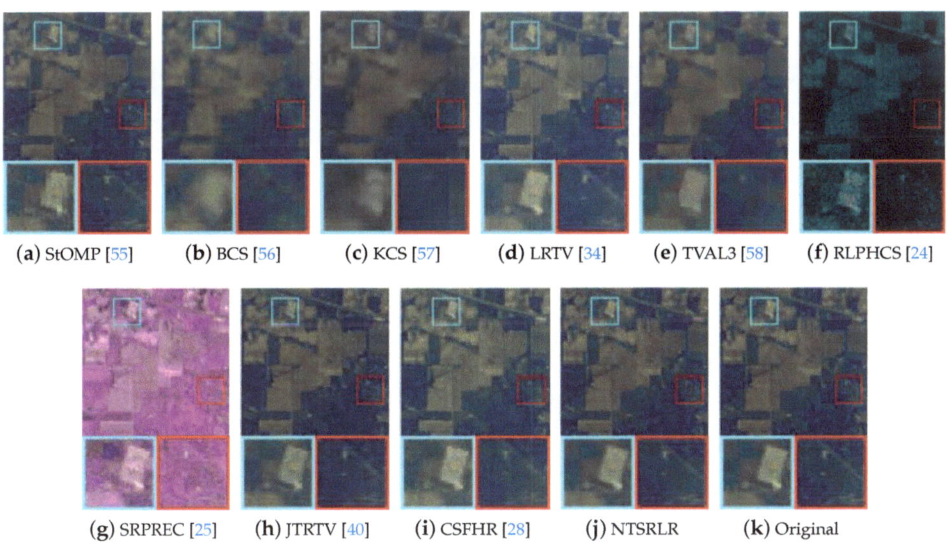

(a) StOMP [55] (b) BCS [56] (c) KCS [57] (d) LRTV [34] (e) TVAL3 [58] (f) RLPHCS [24]

(g) SRPREC [25] (h) JTRTV [40] (i) CSFHR [28] (j) NTSRLR (k) Original

Figure 6. Compressive sensing reconstructed results on pseudocolor images with bands (23, 13, 3) of the *Indian Pines* image from different methods under sampling rate $\rho = 0.15$.

The values of PSNR, SSIM and FSIM across all bands on *Indian Pines* under sampling rate $\rho = 0.10$ are presented in Figure 7. The proposed method achieves the best PSNR, SSIM and FSIM values in most bands of the HSI, which also further validates the robustness of the proposed method over all spectral bands. To further illustrate the superiority of proposed NTSRLR on spectrum reconstruction, we chose four regions in *Toy* and *PaviaU* datasets shown Figure 8a,d; the average reflectance differences were calculated between reconstructed spectra and original spectra across all bands. The curves of those average reflectance differences are plotted in Figure 8b,c for *Toy* and Figure 8e,f for *PaviaU*. It is obvious that the reflectance difference between the reference and the reconstruction by NTSRLR is close to zero—much better than the other comparison methods.

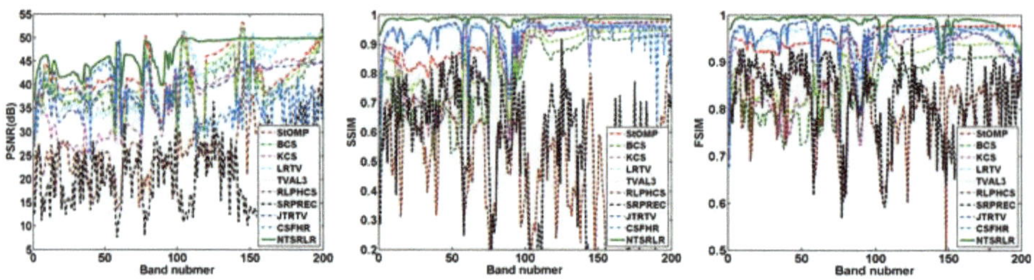

(**a**) PSNR (**b**) SSIM (**c**) FSIM

Figure 7. PSNR, SSIM and FSIM values comparison of different methods for each band on *Indian Pines* dataset under sampling rate $\rho = 0.20$.

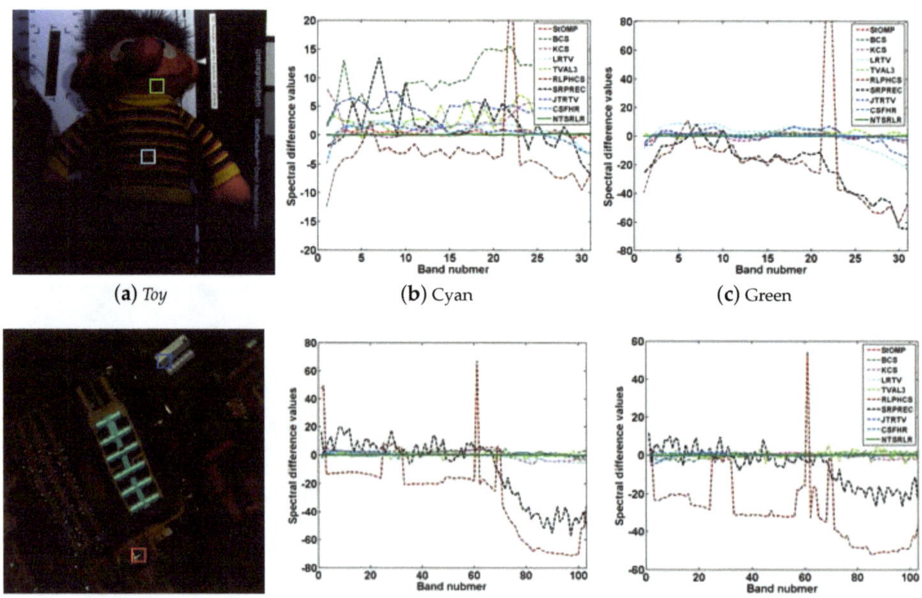

(**a**) *Toy* (**b**) Cyan (**c**) Green

(**d**) *PaviaU* (**e**) Red (**f**) Blue

Figure 8. Comparison of spectra difference on *Toy* and *PaviaU* datasets: (**b**,**c**) the spectra difference curves of different methods corresponding to the region marked by cyan and green rectangles of *Toy* in (**a**) under sampling rate $\rho = 0.05$; and (**e**,**f**) the spectra difference curves of different methods corresponding to the region marked by red and blue rectangles of *PaviaU* in (**d**) under sampling rate $\rho = 0.10$.

Table 1. MPSNRs, MSSIMs, and MFSIMs of different CSR methods on three selected HSIs under different sampling rates.

SRs	PQIs	Methods									
		StOMP [55]	BCS [56]	KCS [57]	LRTV [34]	TVAL3 [58]	RLPHCS [24]	SRPREC [25]	JTRTV [40]	CSFHR [28]	NTSRLR
						Results on *Toy*					
0.02	MPSNR	25.27	18.45	23.39	22.08	22.91	13.19	14.40	17.19	25.87	**27.81**
	MSSIM	0.7040	0.3499	0.6565	0.6651	0.6364	0.2089	0.2786	0.1601	0.6639	**0.7322**
	MFSIM	0.8044	0.6937	0.7820	0.8061	0.7397	0.6651	0.6272	0.5033	0.8389	**0.8484**
0.05	MPSNR	29.35	24.63	26.93	26.51	27.63	13.22	13.89	22.65	29.96	**34.22**
	MSSIM	0.8256	0.6672	0.7811	0.7873	0.7817	0.2372	0.1929	0.3374	0.7462	**0.8930**
	MFSIM	0.9189	0.7837	0.8523	0.8783	0.8273	0.6493	0.5480	0.6233	0.8845	**0.9423**
0.10	MPSNR	29.71	28.24	29.94	32.06	31.81	13.06	15.92	29.93	32.35	**40.12**
	MSSIM	0.8416	0.8072	0.8641	0.9233	0.8871	0.2034	0.1267	0.6860	0.8418	**0.9640**
	MFSIM	0.9261	0.8563	0.8987	0.9517	0.9052	0.6163	0.4505	0.8466	0.9255	**0.9814**
0.15	MPSNR	30.90	29.40	31.88	34.99	33.46	13.69	27.79	31.47	34.99	**44.52**
	MSSIM	0.8982	0.8429	0.9025	0.9427	0.9141	0.1993	0.7492	0.7673	0.8985	**0.9848**
	MFSIM	0.9485	0.8777	0.9232	0.9669	0.9282	0.5642	0.9082	0.8894	0.9527	**0.9928**
0.20	MPSNR	31.75	31.63	33.26	40.54	37.65	13.71	25.74	33.39	38.53	**47.86**
	MSSIM	0.9345	0.8845	0.9236	0.9808	0.9593	0.2495	0.7384	0.8504	0.9541	**0.9925**
	MFSIM	0.9617	0.9094	0.9375	0.9876	0.9664	0.6182	0.8942	0.9307	0.9785	**0.9965**
						Results on *PaviaU*					
0.02	MPSNR	28.11	21.74	23.79	23.08	22.99	15.18	14.84	28.04	25.11	**29.83**
	MSSIM	0.7603	0.4767	0.5486	0.6500	0.5014	0.1562	0.0990	0.6708	0.6923	**0.8000**
	MFSIM	0.8246	0.6825	0.6743	0.7974	0.6429	0.6808	0.5758	0.8593	0.8095	**0.8884**
0.05	MPSNR	30.06	24.26	26.59	27.49	25.29	14.38	15.46	35.73	32.74	**37.96**
	MSSIM	0.8571	0.5572	0.6783	0.8099	0.5914	0.1698	0.1266	0.9235	0.8756	**0.9551**
	MFSIM	0.9371	0.7379	0.7854	0.8863	0.7132	0.7123	0.6379	0.9666	0.9442	**0.9774**
0.10	MPSNR	30.40	26.36	29.14	32.99	27.48	15.73	16.00	37.10	34.36	**42.15**
	MSSIM	0.8223	0.6479	0.7871	0.9158	0.6907	0.1225	0.1157	0.9452	0.9062	**0.9794**
	MFSIM	0.9409	0.7963	0.8606	0.9479	0.7894	0.5930	0.5461	0.9761	0.9583	**0.9905**
0.15	MPSNR	31.59	27.08	30.85	33.81	28.33	26.46	28.29	37.39	36.77	**44.55**
	MSSIM	0.8707	0.6812	0.8422	0.9417	0.7268	0.6771	0.8567	0.9487	0.9417	**0.9872**
	MFSIM	0.9523	0.8137	0.8981	0.9683	0.8165	0.8738	0.9255	0.9778	0.9741	**0.9944**
0.20	MPSNR	32.49	28.54	32.13	40.56	30.46	28.14	35.38	38.03	40.56	**46.55**
	MSSIM	0.9020	0.7445	0.8745	0.9740	0.8057	0.7328	0.9547	0.9548	0.9705	**0.9917**
	MFSIM	0.9594	0.8518	0.9198	0.9862	0.8745	0.8964	0.9800	0.9807	0.9871	**0.9965**
						Results on *Indian Pines*					
0.02	MPSNR	30.45	33.03	31.46	22.81	30.12	19.51	23.58	30.87	30.85	**33.54**
	MSSIM	0.7487	0.7692	0.7385	0.4916	0.7839	0.2234	0.4025	0.8010	0.8089	**0.8202**
	MFSIM	0.8299	0.8128	0.7337	0.8421	0.8026	0.7149	0.8327	0.8102	0.8500	**0.8775**
0.05	MPSNR	35.70	37.23	33.71	26.77	37.28	16.44	21.01	37.07	36.86	**41.15**
	MSSIM	0.8693	0.8153	0.7763	0.8057	0.8221	0.0920	0.2944	0.9240	0.8671	**0.9470**
	MFSIM	0.8639	0.8554	0.7983	0.8936	0.8517	0.4714	0.8125	0.9475	0.9210	**0.9553**
0.10	MPSNR	40.77	38.97	35.38	34.10	39.66	16.06	25.10	39.29	37.38	**44.12**
	MSSIM	0.9395	0.8427	0.8165	0.9153	0.8606	0.0614	0.5336	0.9338	0.8798	**0.9719**
	MFSIM	0.9420	0.8867	0.8491	0.9440	0.8919	0.3846	0.8317	0.9472	0.9439	**0.9750**
0.15	MPSNR	43.71	39.42	36.39	34.65	40.47	19.62	24.05	39.85	39.27	**45.65**
	MSSIM	0.9465	0.8478	0.8417	0.9248	0.8743	0.4756	0.4416	0.9354	0.9197	**0.9810**
	MFSIM	0.9794	0.8942	0.8743	0.9496	0.9056	0.7956	0.7804	0.9476	0.9569	**0.9818**
0.20	MPSNR	44.92	40.72	37.12	41.66	42.36	20.95	26.07	39.67	41.81	**46.96**
	MSSIM	0.9350	0.8740	0.8601	0.9670	0.9052	0.5259	0.4957	0.9367	0.9475	**0.9863**
	MFSIM	0.9772	0.9179	0.8907	0.9748	0.9349	0.8216	0.7966	0.9465	0.9706	**0.9858**

Table 2. SAM and ERGAS comparisons of different CSR methods on three selected HSIs under different sampling rates.

SRs	PQIs	Methods									
		StOMP [55]	BCS [56]	KCS [57]	LRTV [34]	TVAL3 [58]	RLPHCS [24]	SRPREC [25]	JTRTV [40]	CSFHR [28]	NTSRLR
Results on *Toy*											
0.02	SAM	0.3040	0.6548	0.3062	0.5096	0.3888	0.9853	0.9707	0.6599	0.4014	**0.2810**
	ERGAS	165.5	864.5	294.7	362.5	309.7	2411	2740	582.3	178.9	**154.8**
0.05	SAM	0.2500	0.2781	0.2351	0.3967	0.2886	0.9633	0.9210	0.6532	0.3401	**0.2029**
	ERGAS	147.8	257.3	193.9	204.3	181.9	2064	2536	321.4	141.5	**84.65**
0.10	SAM	0.2318	0.1968	0.1894	0.2162	0.2080	0.6234	0.8382	0.4129	0.2750	**0.1031**
	ERGAS	141.94	170.4	136.9	107.0	113.1	1273	1853	140.3	108.1	**35.92**
0.15	SAM	0.2629	0.1654	0.1635	0.1940	0.1828	0.4228	0.4562	0.3582	0.2151	**0.0998**
	ERGAS	123.9	148.6	109.9	78.40	93.82	1262	1620	118.5	79.23	**28.08**
0.20	SAM	0.1123	0.1471	0.1478	0.1112	0.1294	0.3866	0.4250	0.2964	0.1599	**0.0733**
	ERGAS	112.5	116.1	94.29	41.20	58.60	978	1305	95.77	53.85	**20.86**
Results on *PaviaU*											
0.02	SAM	0.1819	0.2223	0.1931	0.1576	0.2460	0.9542	0.9950	0.1722	0.1248	**0.1128**
	ERGAS	137.8	345.6	264.4	329.0	284.3	2537	3585	156.7	153.8	**125.8**
0.05	SAM	0.1542	0.1749	0.1512	0.1347	0.2021	0.8849	0.9646	0.0817	0.1019	**0.0550**
	ERGAS	123.4	245.2	187.6	153.2	213.4	2079	2997	67.56	96.19	**50.98**
0.10	SAM	0.1447	0.1417	0.121	0.0862	0.1701	0.7069	0.8168	0.0725	0.0905	**0.0389**
	ERGAS	118.7	188.0	138.7	90.35	165.2	1858	2425	58.58	80.19	**32.53**
0.15	SAM	0.1116	0.1326	0.1059	0.0708	0.1596	0.2914	0.2368	0.0708	0.0728	**0.0315**
	ERGAS	103.6	173.3	113.96	77.17	149.8	1247	1921	56.68	61.15	**24.90**
0.20	SAM	0.0858	0.1178	0.0957	0.0462	0.1359	0.2407	0.0836	0.0674	0.0521	**0.0260**
	ERGAS	93.40	146.2	98.66	38.73	117.7	1231	1427	52.63	41.26	**19.74**
Results on *Indian Pines*											
0.02	SAM	0.1511	0.1622	0.1383	0.2774	0.1246	0.9166	0.9476	0.1075	0.1087	**0.0821**
	ERGAS	143.2	161.8	138.6	759.7	126.5	1723	2297	129.7	198.7	**116.0**
0.05	SAM	0.1447	0.0830	0.1063	0.0832	0.0911	0.5668	0.8286	0.0553	0.0723	**0.0382**
	ERGAS	89.48	88.69	119.2	233.2	87.85	1558	1988	64.84	152.6	**49.62**
0.10	SAM	0.0434	0.0728	0.0888	0.0587	0.0743	0.4821	0.6523	0.0515	0.0659	**0.0282**
	ERGAS	38.77	74.77	96.91	43.08	68.53	1078	1323	58.37	127.4	**35.96**
0.15	SAM	0.0365	0.0714	0.0799	0.0498	0.0693	0.3914	0.4663	0.0505	0.0549	**0.0229**
	ERGAS	34.98	72.32	86.24	37.45	62.99	917	1258	56.15	78.07	**30.81**
0.20	SAM	0.0295	0.0622	0.0741	0.0344	0.0586	0.2749	0.4590	0.0481	0.0553	**0.0190**
	ERGAS	31.39	61.87	79.43	33.59	51.61	366	982	50.78	59.69	**27.19**

4.2.3. Classification Performance on *Indian Pines* Dataset

The classification accuracy of the HSI with different algorithms was employed to further verify the effectiveness of the proposed method. Under the same circumstance, we chose the support vector machine (SVM) [63] and overall accuracy (OA) as the classifier and evaluation index, respectively. During the classification results with SVM algorithm, we used 16 ground-truth classes in *Indian Pines* and 10% randomly generated training sets from each class to test the classification accuracy. The classification results with different HSI-CSR methods under sampling rate $\rho = 0.20$ are revealed in Figure 9a–j. The OA are given in Table 3. As shown in Figure 9j, the classification results in original HSI appear continuous, and the OA is 86.37%. As shown in Figure 9i, the classification results of NTSRLR still show a continuous phenomenon, and the OA of NTSRLR is closer to the reference value. However, the classification results of other methods are more fragmentary in most regions of the image, with lower OA values.

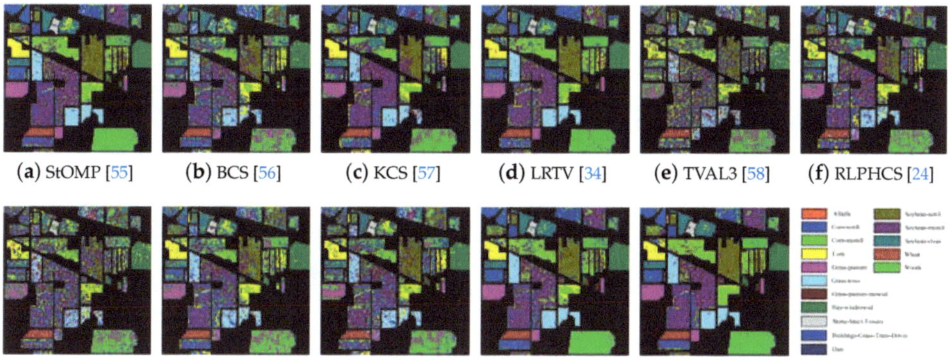

(a) StOMP [55] **(b)** BCS [56] **(c)** KCS [57] **(d)** LRTV [34] **(e)** TVAL3 [58] **(f)** RLPHCS [24]

(g) SRPREC [25] **(h)** JTRTV [40] **(i)** CSFHR [28] **(j)** NTSRLR **(k)** Original **(l)** 16 classes

Figure 9. Classification results for the *Indian Pines* image using SVM before and after CSR under sampling rate $\rho = 0.20$.

Table 3. Classification performance comparison before and after CSR on *Indian Pines* under different sampling rates.

SRs	StOMP [55]	BCS [56]	KCS [57]	LRTV [34]	TVAL3 [58]	RLPHCS [24]	SRPREC [25]	JTRTV [40]	CSFHR [28]	NTSRLR	Original
0.02	71.19%	50.64%	52.37%	60.96%	51.85%	29.61%	10.51%	20.03%	53.21%	**73.69%**	86.37%
0.05	75.70%	57.83%	56.18%	69.64%	57.83%	36.66%	13.32%	54.47%	59.17%	**77.32%**	
0.10	76.32%	59.01%	62.01%	71.24%	60.92%	41.82%	14.62%	55.66%	62.98%	**79.31%**	
0.15	78.41%	63.80%	65.80%	77.03%	62.70%	45.53%	45.53%	56.84%	65.24%	**80.26%**	
0.20	80.28%	68.73%	70.73%	79.19%	65.73%	46.57%	57.83%	58.13%	67.70%	**81.79%**	

4.3. Robustness for Noise Suppression during HSI-CSR

To further evaluate the effectiveness and robustness of proposed HSI-CSR method for noise suppression, we chose the *Urban* dataset (http://www.tec.army.mil/hypercube) contaminated by different degrees of mixture noise, which was with size of 307 × 307 and 4 m spatial resolution, and covers the wavelength in the range from 400 to 2400 nm by 10 nm spectral resolution. Under same competing methods, we removed 24 bands seriously affected by atmospheric attenuations and water absorptions, and finally reserved 186 bands for the dataset.

We present the pseudocolor image with bands (186, 131, 1), in which the input data is polluted by Gaussian noise and stripes, as shown in Figure 10k. The CSR results produced by StOMP, BCS, CSFHR and TVAL3 could neither recover the original HSI nor perform the denoising task well. Instead, the methods RLPHCS and SRPREC amplified the noise. Although the methods KCS, LRTV and JTRTV could suppress the noise to some extent, they lost the edges and textural details when compared to NTSRLR.

Furthermore, we present the quantitative comparisons by showing the horizontal mean profiles of bands 1 and 186 in *Urban* dataset before and after CSR in Figures 11 and 12. The horizontal axis in the figure denotes the row number, and the vertical axis represents the mean gray value of each row. As shown in Figures 11k and 12k, the profiles have huge fluctuation due to the disturbance of noises. After CSR, the fluctuation has been moderately alleviated. Evidently, the profiles with the proposed NTSRLR method are more natural and smoother. The over-smooth profiles corresponding to BCS are mainly due to the image blurring. This further substantiates the efficiency and robustness of the proposed HSI-CSR method for noise suppression.

(a) StOMP [55] **(b)** BCS [56] **(c)** KCS [57] **(d)** LRTV [34] **(e)** TVAL3 [58] **(f)** RLPHCS [24]

(g) SRPREC [25] **(h)** JTRTV [40] **(i)** CSFHR [28] **(j)** NTSRLR **(k)** Original

Figure 10. Compressive sensing reconstructed results on pseudocolor images with bands (186, 131, 1) of the noisy *Urban* image from different methods under sampling rate ρ = 0.10.

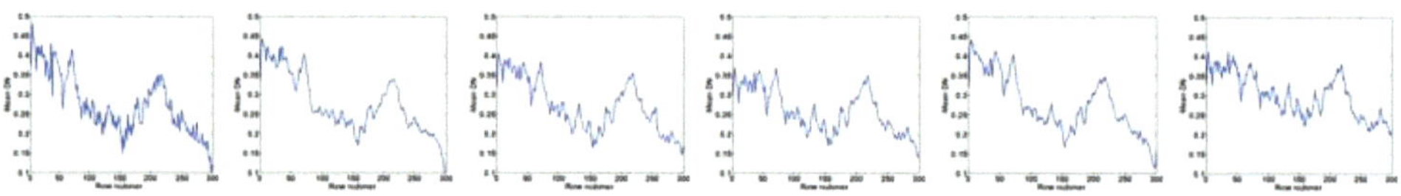

(a) StOMP [55] **(b)** BCS [56] **(c)** KCS [57] **(d)** LRTV [34] **(e)** TVAL3 [58] **(f)** RLPHCS [24]

(g) SRPREC [25] **(h)** JTRTV [40] **(i)** CSFHR [28] **(j)** NTSRLR **(k)** Original

Figure 11. Horizontal mean profiles of compressive sensing reconstructed results on 1st band of real noisy *Urban* HSI data from different methods under sampling rate ρ = 0.10.

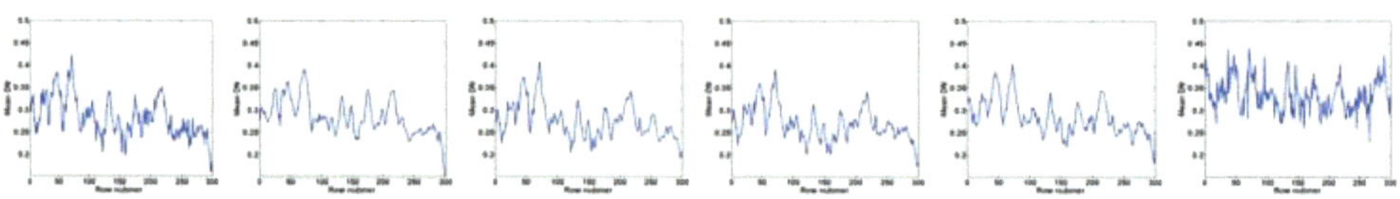

(a) StOMP [55] **(b)** BCS [56] **(c)** KCS [57] **(d)** LRTV [34] **(e)** TVAL3 [58] **(f)** RLPHCS [24]

(g) SRPREC [25] **(h)** JTRTV [40] **(i)** CSFHR [28] **(j)** NTSRLR **(k)** Original

Figure 12. Horizontal mean profiles of compressive sensing reconstructed results on 186th band of real noisy *Urban* HSI data from different methods under sampling rate ρ = 0.10.

Here, we give the theoretical analysis to explain why the proposed HSI-CSR algorithm is able to suppress noise at the same time. The primary cause is that proposed NTSRLR contributes the noise suppression to the joint tensor sparse and low-rank constraint on nonlocal cubes. The work in [64] refers to the fact that the low-rank representation for those nonlocal similar patches to a given patch offer helpful remedy for its better image denoising. For tensor data, one can obtain the same results when unfolding a tensor into a matrix along certain mode, and the nonlocal tensor low-rank term of NTSRLR model can simultaneously provide complementary low-rank structures along all modes to promote the denoising performance of tensor data. Therefore, the noise of HSI can be suppressed to some extent. Besides, the research is [53,54] has demonstrated the effectiveness of tensor sparse models in multi-dimensional signals denoising, which verifies the positive impact of NTSRLR on noise suppression from the perspective of tensor sparse representation.

Note that we removed all noisy bands and preserved only 171 bands for quantitative assessment. Table 4 presents MPSNR, MSSIM, MFSIM ERGAS and SAM of all methods under sampling rates 0.10, 0.15 and 0.20. It can be seen that our method not only recovered the structural and perceptual feature of *Urban* dataset, but also preserved better spectral information.

Table 4. MPSNRs, MSSIMs, MFSIMs ERGAS and SAM of different CSR methods on *Urban* with different sampling rates.

SRs	PQIs	Methods									
		StOMP [55]	BCS [56]	KCS [57]	LRTV [34]	TVAL3 [58]	RLPHCS [24]	SRPREC [25]	JTRTV [40]	CSFHR [28]	NTSRLR
0.10	MPSNR	19.63	16.95	23.63	24.76	17.79	22.04	15.13	27.74	26.76	**30.88**
	MSSIM	0.6523	0.4147	0.8152	0.8705	0.4423	0.8155	0.4245	0.8959	0.8933	**0.9471**
	MFSIM	0.8841	0.6918	0.8916	0.9277	0.6562	0.9088	0.7711	0.9561	0.9279	**0.9746**
	ERGAS	280.2	380.4	184.2	159.6	346.3	261.6	480.9	111.5	109.8	**76.89**
	SAM	0.2884	0.2157	0.1551	0.1197	0.2644	0.2737	0.4775	0.1196	0.1252	**0.0682**
0.15	MPSNR	20.61	17.45	25.78	26.40	18.48	24.16	20.94	27.94	28.27	**33.51**
	MSSIM	0.7088	0.4546	0.8740	0.9134	0.4924	0.8442	0.8306	0.8992	0.9064	**0.9662**
	MFSIM	0.8972	0.7138	0.9242	0.9575	0.6946	0.9284	0.9016	0.9580	0.9582	**0.9845**
	ERGAS	250.4	359.7	145.4	122.9	320.0	202.4	296.3	108.9	91.23	**56.89**
	SAM	0.2461	0.2076	0.1310	0.1024	0.2518	0.2202	0.2885	0.1180	0.1075	**0.0564**
0.20	MPSNR	20.93	18.72	27.37	33.26	20.35	25.99	25.24	28.40	30.11	**35.62**
	MSSIM	0.7274	0.5509	0.9051	0.9664	0.6133	0.8583	0.9034	0.9040	0.9275	**0.9762**
	MFSIM	0.9011	0.7645	0.9418	0.9840	0.7810	0.9459	0.9445	0.9608	0.9705	**0.9896**
	ERGAS	241.4	310.8	122.3	59.45	259.0	165.8	183.7	103.1	67.60	**44.66**
	SAM	0.2323	0.1879	0.1156	0.0592	0.2207	0.1831	0.1859	0.1149	0.0828	**0.0481**

4.4. *Effectiveness Analysis of Single NTSR or NTLR Constraint*

To further demonstrate the effectiveness of nonlocal tensor sparse representation and low-rank regularization in our model, we conducted two more experiments using the *PaviaU* dataset. The first experiment was to perform CSR without the nonlocal tensor low-rank regularization term, and the reconstructed HSI was achieved solely by nonlocal tensor sparse representation (NTSR). The second experiment was a reconstruction with the nonlocal tensor low-rank regularization method, but without NTSR, which is abbreviated as NTLR.

Figure 13 shows the comparison results of MPSNR, MSSIM and SAM of all methods under sampling rates from 0.05 to 0.20 with interval 0.05. Compared with other methods, the proposed NTSRLR obtained larger MPSNRs and MSSIMs, and smaller errors as measured by SAM under different sampling rates. In particular, when the sampling rate is small, the results from NTSRLR are significantly better than the NTSR and NTLR, which are based on a single constraint. This provides additional evidence for the effectiveness of the proposed method from the perspective of having integrated constraints with both non-local sparse representation and low rankness in our model.

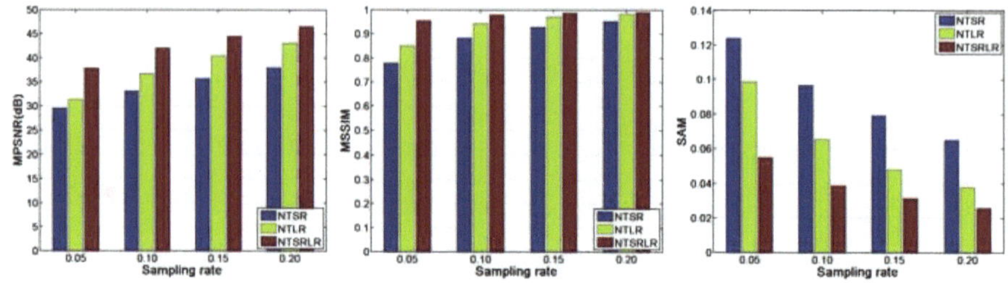

(a) MPSNR (b) MSSIM (c) SAM

Figure 13. MPSNR, MSSIM and SAM bars of different methods under sampling rates 0.05 to 0.20 with interval 0.05 on *PaviaU* dataset.

4.5. Computational Complexity Analysis

For an input HSI $\mathcal{X} \in \mathbb{R}^{W \times H \times S}$, the number of FBCs is $P = O(WH)$, the size of each FBC group is $wh \times s \times S$, where s is number of FBCs in each group. The computation cost seems not very small for quite large P. However, CSR on the P FBCs can be processed in parallel, each with relatively small computational complexity. The computational complexity of the proposed algorithm that mainly lies in the update of $\mathcal{M}_{p(i)}, \mathrm{U}_{ip}(i = 1, 2, 3)$. Updating U_{ip} requires computing an SVD of $I_i \times I_i$ matrix, and updating $\mathcal{M}_{p(i)}$ requires computing an SVD of $I_i \times (\prod_{j \neq i} I_j)$ matrix. Relatively, the other variables $\mathcal{G}_p, x$ and multipliers updating will not consume lots of running time.

4.6. Convergence Analysis

Lastly, we have conducted experiments to show the convergence of our method using the *Toy* and *Indian Pines* dataset as examples under different sampling rates and different initializations. Figure 14 plots the PSNRs versus iteration numbers for the tested HSIs when the sampling rates are at 0.10 for *Toy* and 0.15 for *Indian Pines*, when using initialization $x = \Phi^* y$ and DCT. As can be seen, the different initialization ways can provide quite close solutions, which indicates the performance of proposed algorithm is not sensitive to initialization. However, the two initialization ways possess different rates of convergence, and, by contrast, the initialization via DCT requires only a small number of iterations to get to the final PSNR. Therefore, we adopted the initialization strategy based on DCT to speed up our algorithm. Besides, the value of PSNR will become a constant when the algorithm converges. Thus, in the experiment, we set the maximum number of iterations for termination condition.

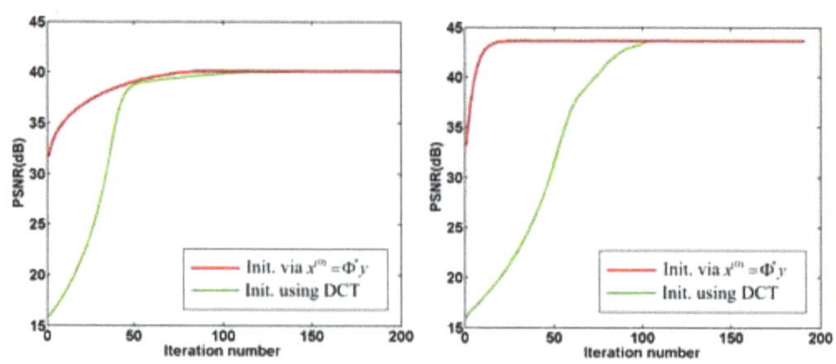

(a) $\rho = 0.10$ (b) $\rho = 0.15$

Figure 14. Verification of the convergence of the proposed method. Progression of the PSNRs for the *Toy* and *Indian Pines* datasets under different sampling rates.

4.7. Parameters Analysis

There are four parameters $\{\lambda_i\}_{i=1}^4$ in the proposed model. Considering the different roles of nonlocal tensor sparseness and low-rankness terms, we conducted two more experiments on *PaviaU* dataset in Section 4.4. The results of MPSNR, MSSIM and SAM demonstrate the nonlocal tensor low-rank regularization term plays a more important role in proposed model than nonlocal tensor sparse representation term. It implies that the nonlocal tensor low-rankness term should be assigned a greater weight to balance the two parts. Therefore, we set $\lambda_2 = 1$ and $\lambda_3 = 10$ in all our experiments. Correspondingly, we can regard the other two parts with λ_1 and λ_4 tradeoff as loyalty terms of the nonlocal tensor sparseness and low-rankness; it is reasonable to obtain a greater value for λ_4, and we set $\lambda_1 = 0.02$ and $\lambda_4 = 250$, as suggested in [42].

Besides, the spatial size of cube and the number of non-local similar cubes are two key parameters. Some research [17,18,30,41] reports that the spatial size of cube and the number of non-local similar cubes are dependent on sampling rates. The higher the sampling rate is, the more detailed information of texture and structure the HSI loses. For this reason, the bigger spatial size and more non-local similar cubes are beneficial to provide extra knowledge to further promote the HSI reconstruction performance. Thus, according to the parameter setting principle in [17,18,30,41], we set spatial size to 6×6, 7×7, 8×8, 9×9 and 10×10 for $\rho = 0.20$, 0.15, 0.10, 0.05 and 0.02, respectively; and the corresponding number of non-local similar cubes are set to 50, 55, 60, 65 and 70.

5. Conclusions

In this paper, we propose a novel method for hyperspectral image compressed sensing reconstruction by non-local tensor sparse representation and low-rank regularization. The proposed method considers intrinsic structured sparsity, where the nonlocal similarity between spatial cubes and the global correlation across all bands are considered fully. Each cube group contains similar structures; its tensor-based sparsity and low-rank properties can be regarded as very valuable priors. Experimental results reveal that the proposed methods outperform the state-of-the-art methods in term of visual inspection, quantitative and classification accuracy assessment. The proposed method is also superior in noise suppression. We also conclude that it is advantageous to have integrated constraints using both non-local tensor sparse representation and low-rankness rather than using only one of them in our model.

Author Contributions: All authors contributed to the design of the methodology and the validation of experiments. J.X. wrote the paper. Y.Z., W.L. and J.C.-W.C. reviewed and revised the paper.

Funding: This work was supported by the National Natural Science Foundation of China (61371152 and 61771391), the Shenzhen Municipal Science and Technology Innovation Committee (JCYJ20170815162956949), and the Fund for Scientific Research in Flanders (FWO) project G037115N Data fusion for image analysis in remote sensing. Wenzhi Liao, a postdoctoral fellow of the Research Foundation Flanders (FWO-Vlaanderen), acknowledges its support.

Conflicts of Interest: The authors declare no conflict of interest.

References

1. Yang, J.; Zhao, Y.; Chan, J.C.-W. Learning and transferring deep joint spectral—Spatial features for hyperspectral classification. *IEEE Trans. Geosci. Remote Sens.* **2017**, *55*, 4729–4742. [CrossRef]
2. Yuan, Y.; Lin, J.; Wang, Q. Hyperspectral image classification via multitask joint sparse representation and stepwise MRF optimization. *IEEE Trans. Cybern.* **2016**, *46*, 2966–2977. [CrossRef]
3. Liu, Y.; Shi, Z.; Zhang, G.; Chen, Y.; Li, S.; Hong, Y.; Shi, T.; Wang, J.; Liu, Y. Application of Spectrally Derived Soil Type as Ancillary Data to Improve the Estimation of Soil Organic Carbon by Using the Chinese Soil Vis-NIR Spectral Library. *Remote Sens.* **2018**, *10*, 1747. [CrossRef]
4. Khelifi, F.; Bouridane, A.; Kurugollu, F. Joined spectral trees for scalable spiht-based multispectral image compression. *IEEE Trans. Multimed.* **2008**, *10*, 316–329. [CrossRef]

5. Christophe, E.; Mailhes, C.; Duhamel, P. Hyperspectral image compression: Adapting spiht and ezw to anisotropic 3-d wavelet coding. *IEEE Trans. Image Process.* **2008**, *17*, 2334–2346. [CrossRef] [PubMed]
6. Töreyın B. U.; Yilmaz O.; Mert Y. M.; Türk F. Lossless hyperspectral image compression using wavelet transform based spectral decorrelation. In Proceedings of the IEEE 7th International Conference on Recent Advances in Space Technologies (RAST), Istanbul, Turkey, 16–19 June 2015; pp. 251–254.
7. Wang, L.; Wu, J.; Jiao, L.; Shi, G. Lossy-to-lossless hyperspectral image compression based on multiplierless reversible integer TDLT/KLT. *IEEE Geosci. Remote Sens. Lett.* **2009**, *6*, 587–591. [CrossRef]
8. Mielikainen, J.; Toivanen, P. Clustered DPCM for the lossless compression of hyperspectral images. *IEEE Trans. Geosci. Remote Sens.* **2003**, *41*, 2943–2946. [CrossRef]
9. Du, Q.; Fowler, J.E. Hyperspectral image compression using JPEG2000 and principal component analysis. *IEEE Geosci. Remote Sens. Lett.* **2007**, *4*, 201–205. [CrossRef]
10. Du, Q.; Ly, N.; Fowler, J.E. An operational approach to PCA+JPEG2000 compression of hyperspectral imagery. *IEEE J. Sel. Top. Appl. Earth Observ. Remote Sens.* **2014**, *7*, 2237–2245. [CrossRef]
11. Donoho, D.L. Compressed sensing. *IEEE Trans. Inf. Theory* **2006**, *52*, 1289–1306. [CrossRef]
12. Boufounos, D.; Liu, D.; Boufounos, P.T. A lecture on compressive sensing. *IEEE Signal Process. Mag.* **2007**, *24*, 1–9.
13. Huang, J.; Zhang, T.; Metaxas, D. Learning with structured sparsity. *J. Mach. Learn. Res.* **2011**, *12*, 3371–3412.
14. Tan, M.; Tsang, I.W.; Wang, L. Matching pursuit LASSO part I: Sparse recovery over big dictionary. *IEEE Trans. Signal Process.* **2015**, *63*, 727–741. [CrossRef]
15. Candes, E.J.; Wakin, M.B.; Boyd, S.P. Enhancing sparsity by reweighted l_1 minimization. *J. Fourier Anal. Appl.* **2008**, *14*, 877–905. [CrossRef]
16. Chartrand, R.; Yin, W. Iterative Reweighted Algorithms for Compressive Sensing. In Proceedings of the IEEE International Conference on Acoust. Speech Signal Process, Las Vegas, NV, USA, 31 March–4 April 2008; pp. 3869–3872.
17. Dong, W.; Wu, X.; Shi, G. Sparsity fine tuning in wavelet domain with application to compressive image reconstruction. *IEEE Trans. Image Process.* **2014**, *23*, 5249–5262. [CrossRef]
18. Dong, W.; Shi, G.; Li, X.; Ma, Y.; Huang, F. Compressive sensing via nonlocal low-rank regularization. *IEEE Trans. Image Process.* **2014**, *23*, 3618–3632. [CrossRef] [PubMed]
19. Dong, W.; Li, X.; Zhang, L.; Shi, G. Sparsity-based image denoising via dictionary learning and structural clustering. In Proceedings of the Computer Vision and Pattern Recognition (CVPR), Colorado Springs, CO, USA, 20–25 June 2011; pp. 457–464.
20. Mairal, J.; Bach, F.; Ponce, J.; Sapiro, G.; Zisserman, A. Non-local sparse models for image restoration. In Proceedings of the IEEE 12th International Conference on Computer Vision, Kyoto, Japan, 29 September–2 October 2009; pp. 2272–2279.
21. Zhang, L. Wei, W.; Zhang, Y.; Yan, H.; Li, F.; Tian, C. Locally similar sparsity-based hyperspectral compressive sensing using unmixing. *IEEE Trans. Comput. Imaging* **2016**, *2*, 86–100. [CrossRef]
22. Wang, L.; Feng, Y.; Gao, Y.; Wang, Z.; He, M. Compressed sensing reconstruction of hyperspectral images based on spectral unmixing. *IEEE J. Sel. Topics Appl. Earth Observ. Remote Sens.* **2018**, *11*, 1266–1284. [CrossRef]
23. Li, C.; Sun, T.; Kelly, K.F.; Zhang, Y. A compressive sensing and unmixing scheme for hyperspectral data processing. *IEEE Trans. Image Process.*, **2012**, *21*, 1200–1210.
24. Zhang, L.; Wei, W.; Tian, C.; Li, F.; Zhang, Y. Exploring structured sparsity by a reweighted laplace prior for hyperspectral compressive sensing. *IEEE Trans. Image Process.* **2016**, *25*, 4974–4988. [CrossRef]
25. Zhang, L.; Wei, W.; Zhang, Y.; Shen, C.; Hengel, A.V.D.; Shi, Q. Dictionary learning for promoting structured sparsity in hyperspectral compressive sensing. *IEEE Trans. Geosci. Remote Sens.* **2016**, *54*, 7223–7235. [CrossRef]
26. Fu, W.; Li, S.; Fang, L.; Benediktsson, J. A. Adaptive spectral—Spatial compression of hyperspectral image with sparse representation. *IEEE Trans. Geosc. Remote Sens.* **2017**, *55*, 671–682. [CrossRef]
27. Lin, X.; Liu, Y.; Wu, J.; Dai, Q. Spatial-spectral encoded compressive hyperspectral imaging. *ACM Trans. Graphics (TOG)* **2014**, *33*, 233. [CrossRef]
28. Zhang, L.; Wei, W.; Zhang, Y.; Shen, C.; Hengel, A.V.D.; Shi, Q. Cluster sparsity field: An internal hyperspectral imagery prior for reconstruction. *Int. J. Comput. Vis.* **2015**, *11*, 1–25. [CrossRef]

29. Meza, P.; Ortiz, I.; Vera, E.; Martinez, J. Compressive hyperspectral imaging recovery by spatial-spectral non-local means regularization. *Opt. Express* **2018**, *26*, 7043–7055. [CrossRef] [PubMed]
30. Wei, J.; Huang, Y.; Lu, K.; Wang, L. Nonlocal low-rank-based compressed sensing for remote sensing image reconstruction. *IEEE Geosci. Remote Sens. Lett.* **2016**, *13*, 1557–1561. [CrossRef]
31. Khan, Z.; Shafait, F.; Mian, A. Joint group sparse pca for compressed hyperspectral imaging. *IEEE Trans. Image Process.* **2015**, *24*, 4934–4942. [CrossRef] [PubMed]
32. Eason, D.T.; Andrews, M. Total variation regularization via continuation to recover compressed hyperspectral images. *IEEE Trans. Image Process.* **2015**, *24*, 284–293. [CrossRef] [PubMed]
33. Jia, Y.; Luo, Z. Weighted total variation iterative reconstruction for hyperspectral pushbroom compressive imaging. *J. Image Process. Theory Appl.*, **2016**, *1*, 6–10.
34. Golbabaee, M.; Vandergheynst, P. Joint trace/TV norm minimization: A new efficient approach for spectral compressive imaging. In Proceedings of the 19th IEEE International Conference on Image Processing (ICIP), Orlando, FL, USA, 30 September–3 October 2012; pp. 933–936.
35. Karami, A.; Yazdi, M.; Mercier, G. Compression of hyperspectral images using discerete wavelet transform and Tucker decomposition. *IEEE J. Sel. Top. Appl. Earth Observ. Remote Sens.* **2012**, *5*, 444–450. [CrossRef]
36. Wang, L.; Bai, J.; Wu, J.; Jeon, G. Hyperspectral image compression based on lapped transform and Tucker decomposition. *Signal Process. Image Commun.* **2015**, *36*, 63–69. [CrossRef]
37. Zhang, L.; Zhang,L.; Tao, D.; Huang, X.; Du, B. Compression of hyperspectral remote sensing images by tensor approach. *Neurocomputing* **2015**, *147*, 358–363. [CrossRef]
38. Fang, L.; He, N.; Lin, H. CP tensor-based compression of hyperspectral images. *J. Opt. Image Sci. Vis.* **2017**, *34*, 252. [CrossRef] [PubMed]
39. Yang, S.; Wang, M.; Li, P.; Jin, L.; Wu, B.; Jiao, L. Compressive hyperspectral imaging via sparse tensor and nonlinear compressed sensing. *IEEE Trans. Geosci. Remote Sens.* **2015**, *53*, 5943–5957. [CrossRef]
40. Wang, Y.; Lin, L.; Zhao, Q.; Yue, T.; Meng, D.; Leung, Y. Compressive sensing of hyperspectral images via joint tensor tucker decomposition and weighted total variation regularization. *IEEE Geosci. Remote Sens. Lett.* **2017**, *14*, 2457–2461. [CrossRef]
41. Du, B.; Zhang, M.; Zhang, L.; Hu, R.; Tao, D. PLTD: Patch-based low-rank tensor decomposition for hyperspectral images. *IEEE Trans. Multimed.* **2016**, *19*, 67–79. [CrossRef]
42. Xie, Q.; Zhao, Q.; Meng, D.; Xu, Z.; Gu, S.; Zuo, W.; Zhang, L. Multispectral images denoising by intrinsic tensor sparsity regularization. In Proceedings of the IEEE Conference on Computer Vision and Pattern Recognition, Las Vegas, NV, USA, 27–30 June 2016; pp. 1692–1700.
43. Peng, Y.; Meng, D.; Xu, Z.; Gao, C.; Yang, Y.; Zhang, B. Decomposable nonlocal tensor dictionary learning for multispectral image denoising. In Proceedings of the IEEE Conference on Computer Vision and Pattern Recognition, Columbus, OH, USA, 23–28 June 2014; pp. 2949–2956.
44. Boyd, S.; Parikh, N.; Chu, E.; Peleato, B.; Eckstein, J. Distributed optimization and statistical learning via the alternating direction method of multipliers. *Found. Trends Mach. Learn.* **2011**, *3*, 1–122. [CrossRef]
45. Xue, J.; Zhao, Y.; Hao, J. Tensor non-local low-rank regularization for recovering compressed hyperspectral images. In Proceedings of the IEEE International Conference on Image Processing (ICIP), Beijing, China, 17–20 September 2017; pp. 3046–3050.
46. Kolda, T.G.; Bader, B.W. Tensor decompositions and applications. *SIAM Rev.* **2009**, *51*, 455–500. [CrossRef]
47. Schwab, H. For most large underdetermined systems of linear equations the minimal l_1 solution is also the sparsest solution. *Commun. Pur Appl. Math.* **2006**, *59*, 797–829.
48. Daubechies, I.; Defrise, M.; Mol, C.D. An iterative thresholding algorithm for linear inverse problems with a sparsity constraint. *Commun. Pure Appl. Math.* **2004**, *57*, 1413–1457. [CrossRef]
49. Zhang, X.; Burger, M.; Bresson, X.; Osher, S. Bregmanized nonlocal regularization for deconvolution and sparse reconstruction. *SIAM J. Imag. Sci.* **2010**, *3*, 253–276. [CrossRef]
50. Xue, J.; Zhao, Y.; Liao, W.; Kong, S.G. Joint spatial and spectral low-rank regularization for hyperspectral image denoising. *IEEE Trans. Geosci. Remote Sens.* **2018**, *56*, 1940–1958. [CrossRef]
51. Liu, J.; Musialski, P.; Wonka, P.; Ye, J. Tensor completion for estimating missing values in visual data. *IEEE Trans. Pattern Anal. Mach. Intell.*, **2013**, *35*, 208–220. [CrossRef] [PubMed]
52. Quan, Y.; Huang, Y.; Ji, H. Dynamic texture recognition via orthogonal tensor dictionary learning. In Proceedings of the IEEE International Conference on Computer Vision (ICCV), Santiago, Chile, 7–13 December 2015; pp. 73–81.

53. Qi, N.; Shi, Y.; Sun, X.; Yin, B. Tensor: Multi-dimensional tensor sparse representation. In Proceedings of the IEEE Conference on Computer Vision and Pattern Recognition, Las Vegas, NV, USA, 27–30 June 2016; pp. 5916–5925.
54. Qi, N.; Shi, Y.; Sun, X.; Wang, J.; Yin, B.; Gao, J. Multi-dimensional sparse models. *IEEE Trans. Pattern Anal. Mach. Intell.* **2017**, *40*, 163–178. [CrossRef]
55. Donoho, D.L.; Tsaig, Y.; Drori, I.; Starck, J.L. Sparse solution of underdetermined systems of linear equations by stagewise orthogonal matching pursuit. *IEEE Trans. Inf. Theory* **2012**, *58*, 1094–1121. [CrossRef]
56. Ji, S; Xue, Y.; Carin, L. Bayesian compressive sensing. *IEEE Trans. Signal Process.* **2008**, *56*, 2346–2356. [CrossRef]
57. Duarte, M.F.; Baraniuk, R.G. Kronecker compressive sensing. *IEEE Trans. Image Process.* **2012**, *21*, 494–504. [CrossRef]
58. Li, C.; Yin, W.; Jiang, H.; Zhang, Y. An efficient augmented lagrangian method with applications to total variation minimization. *Comput. Optim. Appl.* **2013**, *56*, 507–530. [CrossRef]
59. Wang, Z.; Bovik, A.C.; Sheikh, H.R.; Simoncelli, E.P. Image quality assessment: From error visibility to structural similarity. *IEEE Trans. Image Process.* **2004**, *13*, 600–612. [CrossRef]
60. Zhang, L.; Zhang, L; Mou, X.; Zhang, D. Fsim: A feature similarity index for image quality assessment. *IEEE Trans. Image Process.* **2011**, *20*, 2378–2386. [CrossRef]
61. Wald, L. *Data Fusion: Definitions and Architectures: Fusion of Images of Different Spatial Resolutions*. Presses des MINES: Paris, France, 2002.
62. Yuhas, R.H.; Boardman, J.W.; Goetz, A.F. Determination of semi-arid landscape endmembers and seasonal trends using convex geometry spectral unmixing techniques. In *Summaries of the 4th Annual JPL Airborne Geoscience Workshop*; NASA: Washington, DC, USA, 1993; Volume 4, pp. 205–208.
63. Melgani, F.; Bruzzone, L. Classification of hyperspectral remote sensing images with support vector machines, *IEEE Trans. Geosci. Remote Sens.* **2004**, *42*, 1778–1790. [CrossRef]
64. Gu, S.; Xie, Q.; Meng, D.; Zuo, W.; Feng, X.; Zhang, L. Weighted nuclear norm minimization and its applications to low level vision. *Int. J. Comput. Vis.* **2017**, *121*, 183–208. [CrossRef]

remote sensing

MDPI

Article

Multiobjective Optimized Endmember Extraction for Hyperspectral Image

Rong Liu [1], Bo Du [2,*] and Liangpei Zhang [1]

[1] The State Key Laboratory of Information Engineering in Surveying, Mapping, and Remote Sensing, Wuhan University, Wuhan 430079, China; lr@whu.edu.cn (R.L.); zlp62@whu.edu.cn (L.Z.)

[2] School of Computer, Wuhan University, Wuhan 430079, China

* Correspondence: gunspace@163.com; Tel.: +86-138-7146-1059

Academic Editors: Qi Wang, Nicolas H. Younan, Carlos López-Martínez, Lenio Soares Galvao and Prasad S. Thenkabail

Received: 19 March 2017; Accepted: 1 June 2017; Published: 3 June 2017

Abstract: Endmember extraction (EE) is one of the most important issues in hyperspectral mixture analysis. It is also a challenging task due to the intrinsic complexity of remote sensing images and the lack of priori knowledge. In recent years, a number of EE methods have been developed, where several different optimization objectives have been proposed from different perspectives. In all of these methods, only one objective function has to be optimized, which represents a specific characteristic of endmembers. However, one single-objective function may not be able to express all the characteristics of endmembers from various aspects, which would not be powerful enough to provide satisfactory unmixing results because of the complexity of remote sensing images. In this paper, a multiobjective discrete particle swarm optimization algorithm (MODPSO) is utilized to tackle the problem of EE, where two objective functions, namely, volume maximization (VM) and root-mean-square error (RMSE) minimization are simultaneously optimized. Experimental results on two real hyperspectral images show the superiority of the proposed MODPSO with respect to the single objective D-PSO method, and MODPSO still needs further improvement on the optimization of the VM with respect to other approaches.

Keywords: hyperspectral remote sensing; endmember extraction; multi-objective; particle swarm optimization

---

1. Introduction

Each pixel of hyperspectral image (HSI) has tens or hundreds of values corresponding to its spectral bands, which can effectively represent the unique ground objects [1,2]. Hyperspectral images have been successfully applied to a wide range of fields [3]. However, mixed pixels, constituting more than one distinct material, may widely exist in the HSI due to the limited spatial resolution, which makes one single pixel not pure and brings troubles to accurate precision analysis of HSIs [4–6]. Spectral unmixing (SU) is an effective technique to resolve the mixed pixels problem, which decomposes the mixed pixels into a collection of pure materials, named endmembers, as well as the corresponding abundances [7]. SU has two tasks: EE and abundance estimation. It is usually assumed that there are some pixels that contain only one kind of ground object in the image, and EE is to find out such pure pixel for basic ground objects [8]. Abundance estimation is the process to estimate different proportion of each endmember in a mixed pixel. This paper mainly focuses on the task of EE.

The studies of mixed pixels are mostly based on the linear mixture model (LMM) in which each observed pixel in the image can be represented as the linear combination of a set of spectrally pure constituent endmembers, weighted by the corresponding abundance coefficients that establish the proportion of each endmember in the pixel [9]. Under the LMM, assuming that the image scene is dominated by P kinds of distinct materials with L bands, mathematically, a pixel vector $\mathbf{y} = [y_1, y_2, \cdots, y_L]^T$ can be written as:

$$\mathbf{y} = \sum_{i=1}^{P} s_i \mathbf{a}_i + \mathbf{n} = \mathbf{A}\mathbf{s} + \mathbf{e} \quad (1)$$

where $\mathbf{A} = [\mathbf{a}_1, \mathbf{a}_2, \cdots, \mathbf{a}_P]$ is an $L \times P$ endmember matrix, with each column being an endmember signature vector. The number of endmember P is a pre-defined parameter, which can be estimated by existing methods, and the commonly used ones are the virtual dimensionality (VD) estimation method [10] and the hyperspectral subspace identification (HySime) method [11]. $\mathbf{s} = [s_1, s_2, \cdots, s_P]^T$ is a P-dimensional column vector composed of abundance coefficients of the corresponding endmembers for the pixel, and $\mathbf{e}$ represents the $L \times 1$ additive observation noise and error vector. Generally, there are various kinds of noise in HSIs, and this work assumed that the error is represented by the additive white Gaussian noise [12]. The LMM for all N observed pixels can be expressed by the matrix notation:

$$\mathbf{Y} = \mathbf{A}\mathbf{S} + \mathbf{E} \quad (2)$$

where $\mathbf{Y} = [\mathbf{y}_1, \mathbf{y}_2, \cdots, \mathbf{y}_N]$, $\mathbf{S} = [\mathbf{s}_1, \mathbf{s}_2, \cdots, \mathbf{s}_N]$, and $\mathbf{E} = [\mathbf{e}_1, \mathbf{e}_2, \cdots, \mathbf{e}_N]$. Due to physical constraints, the abundance vector is subject to the nonnegative constraint (ANC, $s_i \geq 0, i = 1, 2, \cdots, P$) and the abundance sum-to-one constraint (ASC, $\mathbf{1}^T\mathbf{s} = 1$).

With the LMM, the geometrical interpretation of the HSI is that if $\mathbf{e} = 0$ and there are pure pixels of all kinds of materials in the image (pure pixel assumption), all the pixels are contained in a simplex whose vertices are corresponding to the endmembers [13]. Based on the convex geometry theory, the EE problem can be converted into finding the simplex vertices. Typical methods include the pixel purity index (PPI) [14], N-FINDR [15], the simplex growing algorithm (SGA) [16], vertex component analysis (VCA) [17], as well as some new algorithms proposed in recent years [18–22]. However, the classic algorithms such as N-FINDR and VCA have been shown easily affected by noise and outliers [23]. One progress in recent years lies on the intelligent optimization methods to enhance the EE results in real HSIs [8,23–27]. Most of these algorithms [8,23–25] consider the EE problem as a combination optimization problem, and seek the optimal endmember combination that minimizes the root-mean-square error (RMSE) between the original image and its remixed image. It is showed that intelligent optimization methods such as D-PSO can get a smaller RMSE compared to N-FINDR and VCA [24]. Different from the above methods, the MOAQPSO method in [26] takes the VM as the objective function, and the experimental results showed the conflicts between the RMSE minimization and the VM objective functions. Specifically, the two objective functions did not achieve their best values for the same endmember combination. If one method got the optimal endmember combination in terms of the volume value, then there would be another method superior to it in terms of the RMSE value. From the previous studies [23,26,28], although the RMSE minimization-based methods can get superior results than the VM-based methods in terms of the RMSE value (or the VM-based methods can get superior results than the RMSE minimization-based methods in terms of the volume value), neither of them can prove completely superior to the other when comparing each one of those endmember spectra with the reference. The VM-based methods have an obvious advantage over the RMSE minimization-based methods when extracting rare endmembers, while in VM-based methods the noises and outliers located within the bounds of the data simplex may be identified incorrectly as endmembers; the RMSE minimization-based methods are more robust to noises and outliers, while they usually ignore the rare endmembers. Effective EE results can be achieved if there is a good match between the characteristic expressed by the objective function and the characteristic of the real image. However, no a priori knowledge is provided in practice, and different complex hyperspectral remote sensing images usually have different characteristics. It is concluded that the generalization ability of one single objective function is poor, and it may not be enough to provide satisfactory EE results for various images. Hence, it is natural to simultaneously optimize several objective functions so as to capture the different data characteristics. In this article, the two widely used objective functions, the VM [13] and the RMSE minimization [24] objective functions, are integrated to be simultaneously optimized. In this way, the problem of endmember extraction is transformed into multiobjective optimization (MOO) problem.

Some MOO methods have been suggested to solve various multiobjective problems [29–32], in which the PSO-based MOO methods have attracted a lot of attention, and this kind of method is chosen as the optimization method of this paper due to the simplicity and good search ability of PSO. Although the existing MOO methods have provided us some ideas on how to solve MOO problems, to our knowledge, no previous works are reported for EE problem, and the difficulty lies on that the distribution characteristics of search spaces and solution spaces of different problems are usually different, so existing methods that are effective for other problems may not work while solving the EE problem. In this paper, a multiobjective discrete particle swarm optimization algorithm (MODPSO) is proposed to perform the task of EE for hyperspectral images. The work includes three aspects: (1) The update strategy of particles' velocity and position in D-PSO method [24] is selected as the basic searching strategy for the proposed MOO method. Since the search space and solution space of the EE problem are both discrete, the particle's position and the velocity must also be discrete to ensure the validity of the solution, so the update strategy of the particles should be modified to make it suitable to the EE problem. (2) For the proposed MOO method, the two objective functions often conflict with each other during the process of optimization, which means that finding a solution that optimizes both objective functions at the same time is almost impossible during the process of optimization [33]. This brings a trouble for the acquisition of the particle's personal best position (*pbest*) and the population's global best position (*gbest*) in the multiobjective searching space. The nondominated sorting algorithm [34] is used to determine *pbest* and *gbest* according to the multiobjective function values. (3) Different from the single objective optimization, there is more than one *gbest* for the population in the MOO, and all of the non-dominated solutions are *gbest*s. This brings the problem to determine which *gbest* should be chosen when updating the velocity of each particle. To solve the problem, the Sigma method is utilized to find best local guides for each particle of the population [35]. With all of the above works, the multiobjective discrete particle swarm optimization algorithm (MODPSO) is finally formed to perform the task of endmember extraction for hyperspectral images. Like common EE methods, MODPSO is based on the pure pixel assumption, and needs the number of endmembers as a priori parameter.

As far as we know, this is the first attempt to use MOO for the purpose of EE. The remainder of this paper is organized as follows. Section 2 gives a detailed description of the proposed MODPSO algorithm for EE. Section 3 reports the experimental results of the MODPSO method and several representative single objective optimization EE algorithms. Conclusions are drawn in Section 4.

2. MODPSO

The proposed MODPSO method implements the task of EE through a MOO technique, and it aims at finding the Pareto-optimal solutions for simultaneously optimizing multiple objective functions. Hence, the establishment of the objective functions and the optimization strategy for the multiple objective functions are two key elements of the MODPSO method. In the following, we will introduce them in detail.

2.1. Objective Functions for MODPSO

Two kinds of objective functions are elaborately chosen for the proposed algorithm. One is the maximum volume objective function, which is based on the convex geometry theory, and the other is to minimize the RMSE obtained after reconstructing the hyperspectral scene by only assuming the presence of the additive white Gaussian noise [12], like almost all the unmixing methods do [36].

We have transformed the VM into minimizing the volume inverse, so the two objective functions are both minimization problems. The two objectives are listed below:

$$f_1 = \frac{1}{volume(\mathbf{A})} = \frac{(P-1)!}{\left|\det\begin{bmatrix} 1 & 1 & \cdots & 1 \\ \mathbf{a}_1 & \mathbf{a}_2 & \cdots & \mathbf{a}_P \end{bmatrix}\right|} \tag{3}$$

$$f_2 = RMSE(\mathbf{Y}, \hat{\mathbf{Y}}) = \frac{1}{N}\sum_{i=1}^{N}\sqrt{\frac{1}{L}\|\mathbf{y}_i - \hat{\mathbf{y}}_i\|_2^2} \tag{4}$$

where L is the spectral dimensionality of the HSI, N is the total number of pixels, P is the number of endmembers, $\mathbf{A}$ is the endmember matrix, and $\mathbf{Y}$ and $\hat{\mathbf{Y}}$ are the original image and the remixed image, respectively. The abundances used to calculate $\hat{\mathbf{Y}}$ are estimated by Equation (5) rather than the fully constrained least squares method (FCLS) for the sake of efficiency:

$$\hat{\mathbf{S}}_{ij} = \max\left(0, \left(\left(\mathbf{A}^T\mathbf{A}\right)^{-1}\mathbf{A}^T\mathbf{Y}\right)_{ij}\right), 1 \le i \le P, 1 \le j \le N \tag{5}$$

In most cases, these two objective functions will not obtain their optimal solution for the same combination of endmembers, for considering that there usually exists noise or outlier in real hyperspectral images. The strategy for optimizing the multiple objective functions in MODPSO will help to find a number of endmember combinations, and none of the obtained solutions can be further improved on the objective value without degrading another.

2.2. The Updating Strategy of the Particle's Velocity and Position in MODPSO

MODPSO use particles to search in the feasible solution space. Each particle has two properties: the position and the velocity. A particle moves along a trajectory depicted by its position and velocity in the search space, to find an optimal solution. For the EE problem, the search space is discrete, the particle's position and the velocity must also be discrete to ensure the validity of the solution. The Binary coding method used in the D-PSO method is employed here to make particles be able to search in the discrete feasible solution space. The position of the ith particle at iteration time t can be written as:

$$X_i^t = \left\{(x_1, \cdots, x_j, \cdots, x_N) \,|\, x_j \in \{0,1\}, \sum_{j=1}^{N} x_j = P\right\} \tag{6}$$

where $x_j = 1$ if $\mathbf{y}_i \in \mathbf{A}$ and $x_j = 0$ if not. Explicitly, for the position of the ith particle X_i^t, all the elements of it are composed of 0 and 1, and each element $x_j (j = 1, \ldots, N)$ in it represents the attribute of the corresponding pixel $\mathbf{y}_j (j = 1, \ldots, N)$, if the value of x_j is 1, the pixel $\mathbf{y}_j$ is selected as an endmember; otherwise, the pixel $\mathbf{y}_j$ is not selected as an endmember. Hence, P elements in each particle's position are 1, and the remaining elements are 0.

V_i^t is used to specify the ith particle's velocity at time t. $pbest_i^t$ and $gbest^t$ are used to specify the ith particle's personal best position and all population's global best position in history before time t. The updating functions of position and velocity are:

$$\begin{aligned} X_i^{t+1} &= X_i^t + V_i^t \\ V_i^{t+1} &= \begin{cases} T\left((pbest_i^t - X_i^t) + (gbest^t - X_i^t)\right), \; rand() \ge p \\ R(X_i^t), \; rand() < p \end{cases} \end{aligned} \tag{7}$$

where T and R are both random selection functions. The velocity obtained by T is based on self-experience and social experience, while R generates velocity without considering past experiences. Both $T(X)$ and $R(X)$ are vectors with the same dimension of X, and the calculation for them can be divided into three steps: (1) Predefine a random selection probability p, and randomly generate a number between 0 and 1. (2) If the generated number is greater than or equal to p, select T to obtain the velocity. First, we find the positive elements and the negative elements of X, respectively. Then, randomly select one element from all the positive elements of X, and set the element of $T(X)$ with the same position of this randomly select one to 1. Next, randomly select another element from all the negative elements of X, and set the element of $T(X)$ with the same position of this randomly select one to -1. The final velocity is obtained by setting the rest of the elements of $T(X)$ to 0. (3) If the generated number is less than p, select R to obtain the velocity. First, we find the zero elements and the positive elements of X respectively. Then, randomly select one element from all the zero elements of X, and set the element of $R(X)$ with the same position of this randomly select one to 1. Next, randomly select another element from all the positive elements of X, and set the element of $R(X)$ with the same position of this randomly select one to -1. The final velocity is obtained by setting the rest of the elements of $R(X)$ to 0. The acquisition of $pbest_i^t$ and $gbest^t$ will be introduced in the following part.

2.3. *Strategy for Updating pbest and gbest for Optimizing the Multiple Objective Functions*

Considering the minimization optimization problem, a MOO problem is of the form:

$$\min f(z) = [f_1(z), f_2(z), \cdots, f_m(z)] \tag{8}$$

where the decision vectors z belong to the feasible space formed by some constraint functions. $m(\geq 2)$ conflicting objective functions are to be minimized simultaneously. A decision vector z_1 is said to dominate z_2 if:

$$\forall i \in [1, 2, \cdots, m] f_i(z_1) \leq f_i(z_2), \exists f_i(z_1) \neq f_i(z_2) \tag{9}$$

A vector z_1 is called Pareto-optimal if another z_2 that dominates it does not exist. Figure 1 shows the Pareto-optimal solutions when $m = 2$. There is no single optimal solution in MOO, but a set of optimal solutions. The set containing all the optimal solutions is known as the Pareto front, and the task of MOO is to achieve the Pareto front. It is obvious that the solutions in the Pareto front are non-dominated solutions.

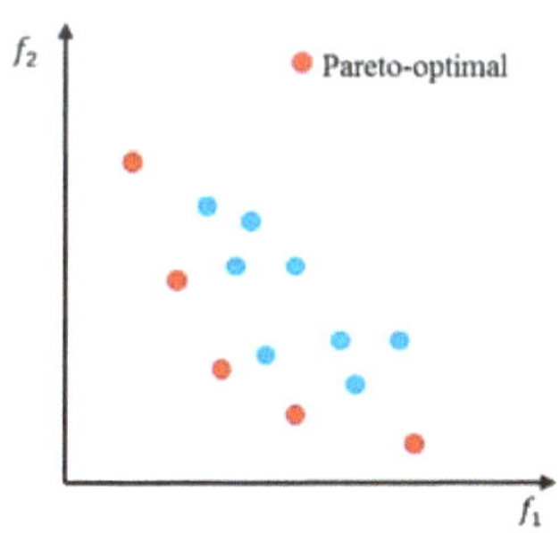

Figure 1. Feasible solutions for minimization optimization. Blue points stand for common solutions, and red points stand for Pareto-optimal solutions.

One main step in MODPSO is to determine the personal and global best positions. They are easy to be determined in single objective optimization by selecting the position best fits the objective function. However, in MOO problems, it is hard to determine which position is better if the solutions represented by two positions are not dominated by each other. To handle this problem, the nondominated sorting algorithm [21] is used to update *pbest* and *gbest*. Among the population, different particles are compared by the concept of Pareto domination. If the solution of one particle is not dominated by that of all the other particles, then it is a Pareto-optimal solution.

It should be noted that there is not only one *gbest* in MOO, all non-dominated solutions in the optimization process are taken as *gbest*. For the update of *gbest*, all the pairwise comparisons of the solutions are conducted by Pareto domination after each iteration, and all the non-dominated solutions are kept as *gbest*. A set named global best archive (GBA) is used to store all these non-dominated solutions (*gbest*).

For the update of *pbest*, the newly generated particle's position X_i^{t+1} is compared with the *pbest* in the history by Pareto domination, if X_i^{t+1} dominates $pbest_i^t$, we set $pbest_i^{t+1} = X_i^{t+1}$; if $pbest_i^t$ dominates X_i^{t+1} $pbest_i^{t+1} = pbest_i^t$; if none of X_i^{t+1} and $pbest_i^t$ dominates the other one, then randomly choose one from them as $pbest_i^{t+1}$, as shown in Figure 2.

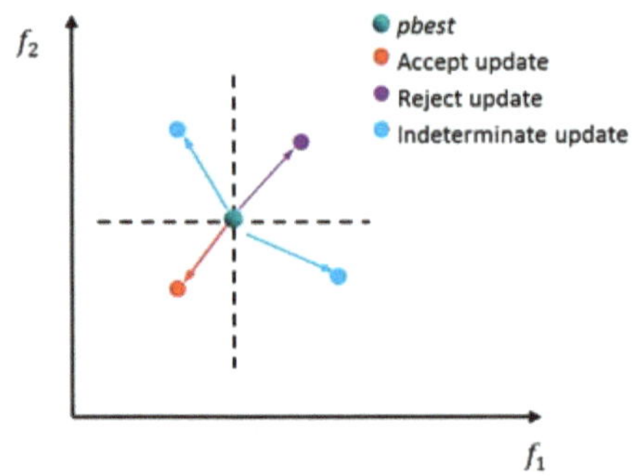

Figure 2. The update of the particle's personal best position. The blue point stands for the current *pbest* of one particle, and other points are possible locations of the particle at the next time. The plane can be divided into four parts centered on the *pbest*. If the particle appears in the area where the purple point located, then the *pbest* of the particle should remain unchanged; if the particle appears in the area where the red point located, then the *pbest* of the particle will be updated by the red point; and if the particle appears in the area where the cyan points located, then randomly select one point as the *pbest*.

2.4. Choose the Best Local Guide for Each Particle

In single objective optimization, there is only one *gbest* for the population, so all of the particles use the same *gbest* to generate the new velocity. However, we have stated that there is not only one *gbest* in MOO, all the solutions in GBA are taken as *gbest*. This brings an additional problem of which *gbest* solution in GBA should be used to generate the velocity for each particle. To solve this problem, The Sigma method [35] is utilized to select one best local guide $gbest_i^t$ from GBA for the ith particle. In the Sigma method, a value σ_i is assigned to each point $(f_{1,i}, f_{2,i})$, and the σ value is defined as:

$$\sigma = \frac{f_1^2 - f_2^2}{f_1^2 + f_2^2} \tag{10}$$

According to Equation (10), all the points on the line $f_2 = af_1$ have the same σ values. By considering the objective space, finding the best local guide $gbest_i^t$ among GBA for the particle i at iteration time t is as follows: in the first step, the σ values of each position in GBA is assigned. In the second step, σ_i for particle i is calculated. Then the distances between σ_i and all the σ values of GBA are calculated. Finally, the kth position in GBA which has the minimum σ value distance with particle i is selected as the best local guide $gbest_i^t$. Figure 3 shows this method for a two objective example.

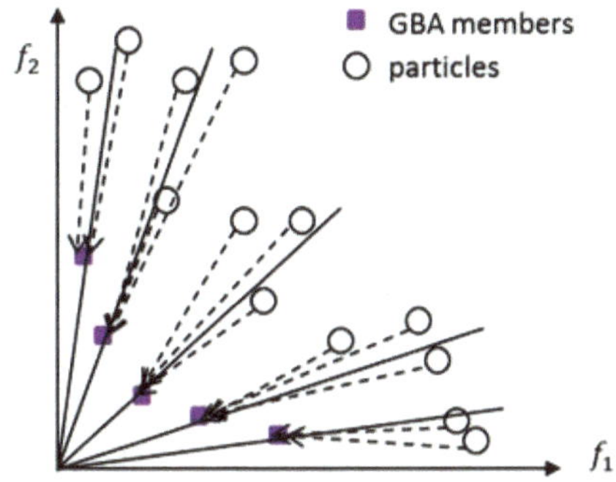

Figure 3. Selection of the best local guide among the global best archive (GBA) for each particle. The squares stand for the GBA members, and circles stand for all the particles. The sigma values of all the GBA members and particles are calculated and compared. For one particle, the GBA member that has the closest sigma value with it is chosen as the best local guide for it.

2.5. The Framework of MODPSO for EE

The overall process of the proposed MODPSO for EE is shown in Figure 4.

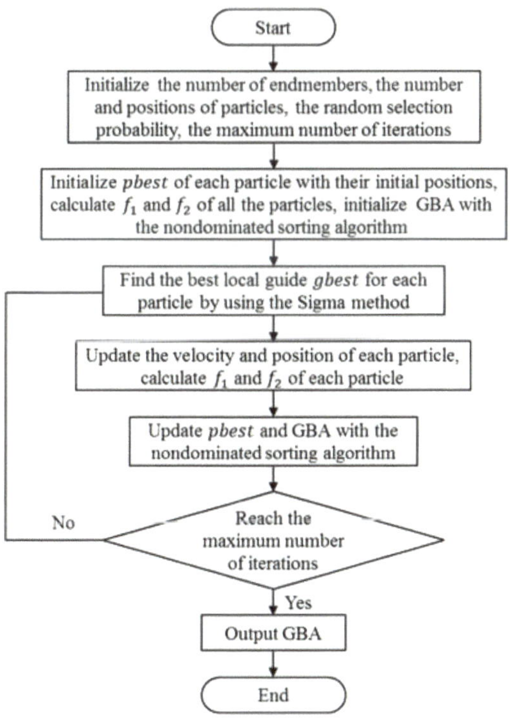

Figure 4. The flowchart of the multiobjective discrete particle swarm optimization (MODPSO) method.

3. Experiments

Two real HSIs are used to test the performance of the proposed method. N-FINDR [15], VCA [17] and D-PSO [24] are comparison algorithms. There are two reasons for selecting these three algorithms as comparing algorithms. One reason is that N-FINDR and VCA are two of the most popular EE methods, and D-PSO is a representative method of the intelligent optimization based methods. Furthermore, the objective functions used in MODPSO have been used in these three methods, so the validity of the proposed method can be checked by comparing the objective values of these methods. For both D-PSO and MODPSO, the maximum iteration number was set to 300, the number of particles was set to 20, the random selection probability was 0.2, and the particles were randomly initialized.

3.1. HYDICE Washington DC Dataset

The first real image dataset was collected by the Hyperspectral Digital Imagery Collection Experiment (HYDICE) sensor over Washington DC, and a subset of 150 × 150 was extracted from the original image for this experiment. In the Washington DC dataset, there are 210 bands, which cover the range of 0.4–2.5 *um*. Low-SNR and water-vapor absorption bands were removed before unmixing, leaving 187 bands for the experiment. Figure 5 shows the false-color image composed of R-band 64, G-band 52, and B-band 36. There are six distinct materials in the image [37], so the endmember number is set to six.

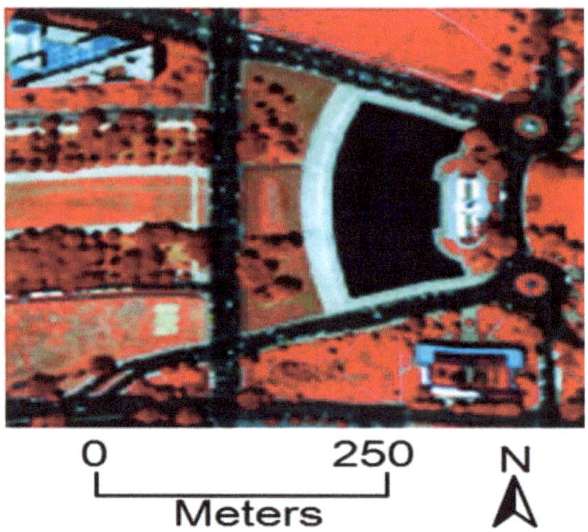

Figure 5. Sub-scene extracted from the Washington DC dataset.

Since there are no standard references of endmembers and abundances for the real image, we cannot directly conduct quantitative evaluation for the extracted endmembers. Considering the following: (1) N-FINDR and VCA try to find the simplex vertices, it is suitable to use the volume of the extracted endmembers to evaluate the searching ability of them; (2) D-PSO searches the endmember combination that minimize the RMSE; and (3) MODPSO try to maximize volume and minimize RMSE simultaneously, two metrics are used to evaluate the performance. (1) The volume inverse: obtained by f_1. (2) RMSE: Obtained by f_2. The smaller f_1 and f_2 are, the better performance the method has.

Figure 6 shows the objective function value as a function of the number of iteration times of MODPSO for the Washington DC dataset. It can be seen that the proposed method can converge to a stationary point when reached the maximum iteration.

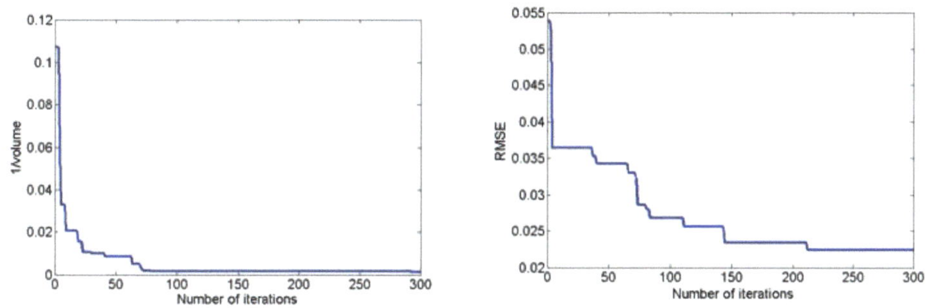

(a) (b)

Figure 6. The objective function value as a function of the number of iteration times for the Washington DC dataset: (**a**) the volume inverse; and (**b**) root-mean-square error (RMSE).

Figure 7a shows the obtained GBA by MODPSO, ten non-dominated solutions are finally remained. The number of solutions in GBA is less than the number of particles, which indicates that some different particles converge to the same solution. We can see in these results that no solution has the minimum f_1 and f_2 values simultaneously. The non-dominated solution with the minimum f_1 has the largest f_2 and vice versa. The ten results are uneven distributed in the objective function value space, they can be easily divided into three parts: three solutions have relatively bigger f_2 and smaller f_1, four solutions have relatively bigger f_1 and smaller f_2, the remaining three solutions have the best tradeoff between the two objective functions. We have also calculated f_1 and f_2 values of the other three methods according to their extracted endmembers. The results are put together with that of MODPSO in Figure 7b, and the numerical results are shown in Table 1 as well as the computation time of them. It can be seen that solutions of MODPSO dominate the result of D-PSO, so the search ability of MODPSO is better than that of D-PSO. VCA and N-FIDNR achieved smaller f_1 and larger f_2 than MODPSO, which indicates that the results with bigger volume are obtained by VCA and N-FINDR, while the RMSE generated by them is larger. Since the results by VCA and N-FINDR are non-dominated solutions compared with the results by MODPSO, it tells us that the Pareto front found by MODPSO is not completed. In terms of the computation time, the two conventional EE methods are more efficient than the two intelligent optimization based methods, especially for the VCA method.

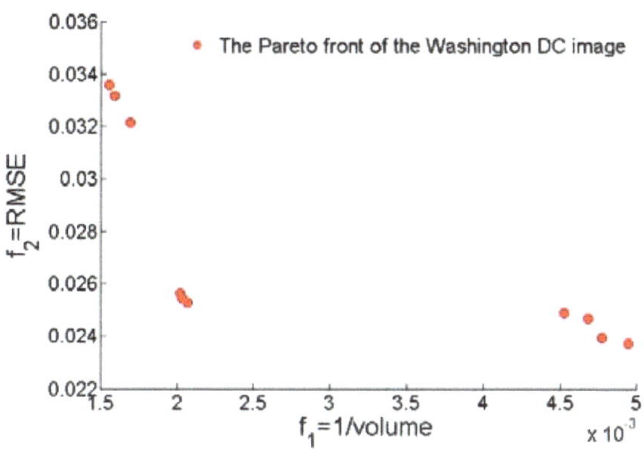

(a)

Figure 7. *Cont.*

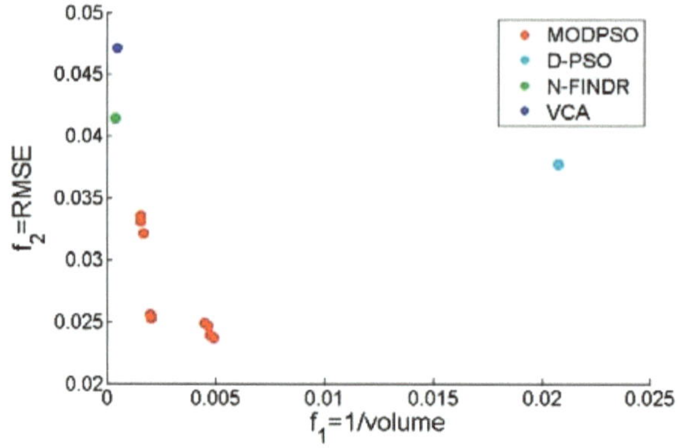

(b)

Figure 7. The results of the Washington DC image: (**a**) the Pareto front obtained by MODPSO; and (**b**) comparison of the results by four methods.

Table 1. The results of objective function values and computation time for the Washington DC Dataset.

	f_1 = 1/volume ($\times10^{-3}$)	f_2 = RMSE	Time (sec)
MODPSO	1.553	0.0335	1556.736
	1.585	0.0331	
	1.692	0.0321	
	2.017	0.0256	
	2.027	0.0254	
	2.064	0.0253	
	4.525	0.0248	
	4.683	0.0246	
	4.774	0.0239	
	4.950	0.0237	
D-PSO	20.762	0.0378	1500.106
N-FINDR	0.414	0.0414	122.118
VCA	0.499	0.0471	0.936

For the HYDICE Washington DC dataset, the ground features are easy to distinguish by visual interpretation; we manually select the endmembers of the six kinds of materials from the image by referring to [37]. These spectra are taken as a rough reference to be shown together with the extracted spectra. The extracted endmembers by the four algorithms and manually selected reference spectra are shown in Figure 8, where the shown endmembers by MODPSO are the union set of the endmembers in GBA. Among the ten sets of results in GBA, there are twelve different endmember spectra. We can see from Figure 8 that N-FINDR and VCA missed the street's spectra and extracted two different paths' spectra. The spectral shapes of the extracted endmembers by the four methods are similar to the shapes of the reference spectra, while there are some differences in the scale, the endmember spectra by D-PSO and MODPSO are more close to the manually selected reference spectra than that of N-FINDR and VCA.

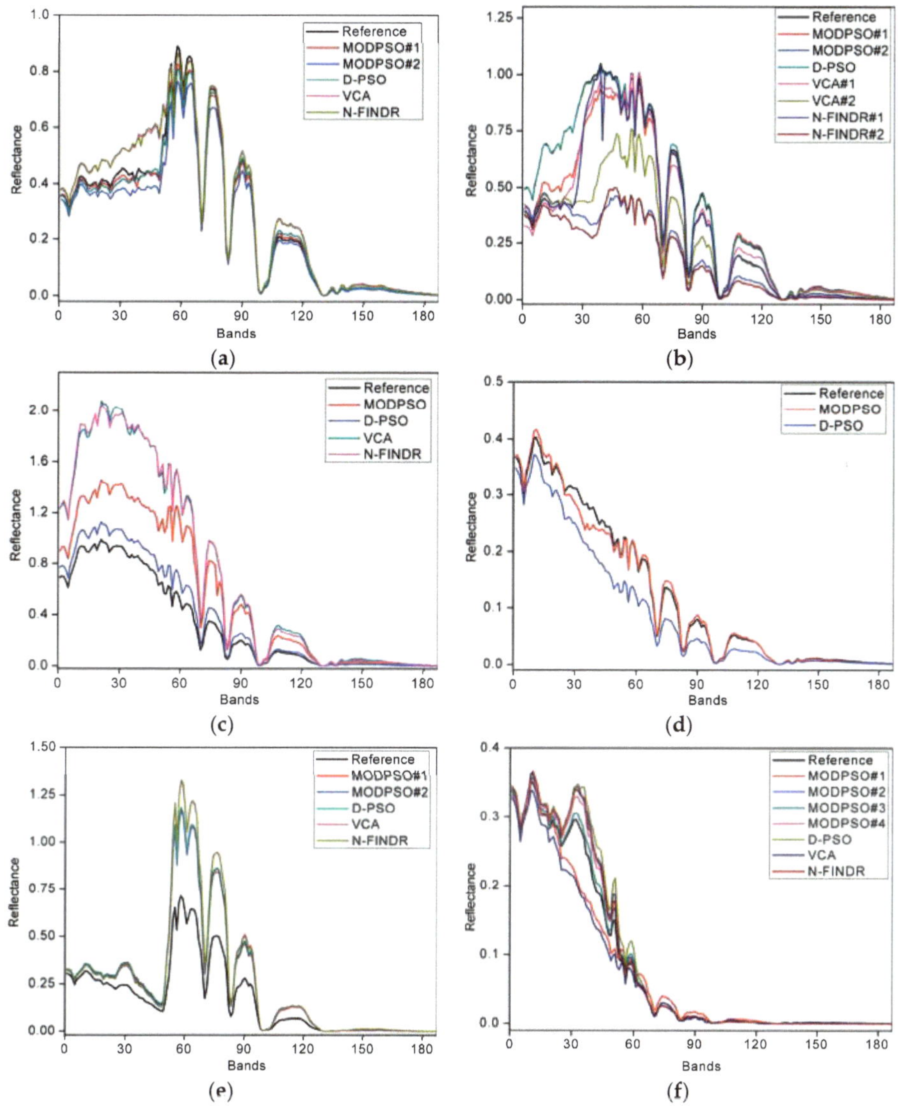

Figure 8. Endmember spectra manually selected from the image and automatically extracted by the four methods for the Washington DC dataset: (**a**) grass; (**b**) path; (**c**) roof; (**d**) street; (**e**) tree; and (**f**) water.

3.2. HYDICE Urban Dataset

The second real dataset was the Urban HYDICE HSI, as shown in Figure 9 by R-band 64, G-band 52, and B-band 36. This image is of size 307 × 307 and has 210 spectral bands in the range of 0.4–2.5 um. A total of 162 bands remained after removing bands 1–4, band 76, band 87, band 111, bands 101–111, bands 136–153 and bands 198–210. The number of endmembers is set to six [38].

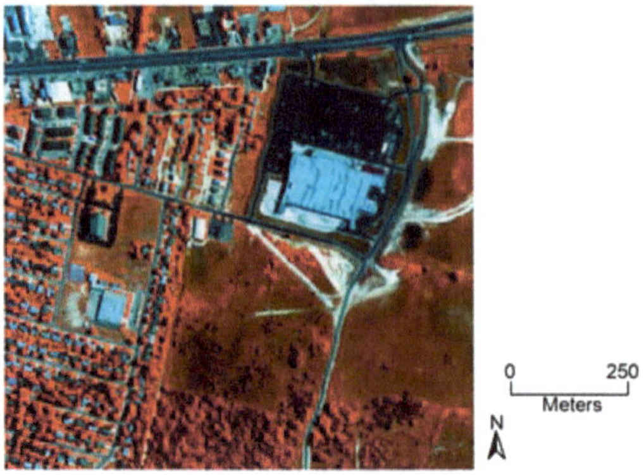

Figure 9. The Urban hyperspectral dataset.

Figure 10 shows the objective function value as a function of the number of iteration times of MODPSO for the Urban dataset. It can be seen that the proposed method can converge to a stationary point when it reached the maximum iteration, and the volume value reached the stationary point earlier than the RMSE value.

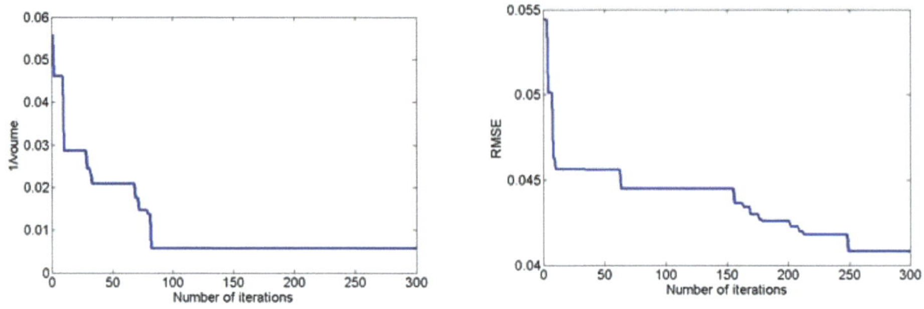

Figure 10. The objective function value as a function of the number of iteration times for the Urban dataset: (**a**) the volume inverse; and (**b**) RMSE.

The Pareto front by MODPSO is displayed in Figure 11a. Eleven non-dominated solutions finally remained, which indicates that some different particles converge to the same solution. We can see that a more uniform distribution of the non-dominated solutions is obtained by the Urban image than the Washington DC image. The comparison results of four methods in Figure 11b and Table 2 are similar to that of the Washington DC image: most of solutions of MODPSO dominate the result of D-PSO; MODPSO and D-PSO have results with smaller RMSE than VCA and N-FINDR; and VCA and N-FIDNR obtained bigger volume than MODPSO and D-PSO, which demonstrate the validity of the MODPSO method. Meanwhile, the MOO result can be further improved. Seen from the computation time, VCA is the most efficient method, while MODPSO and D-PSO are both time consuming.

The extracted endmembers and manually selected reference spectra of the Urban image are shown in Figure 12. Half of the endmembers extracted by N-FINDR and VCA and one endmember extracted by MODPSO cannot be matched with the manually selected reference spectra. Only the N-FINDR algorithm extracted the sixth endmember, and the spectrum of the endmember is not so close to that of the reference endmember. In general, the endmembers extracted by D-PSO and MODPSO are better matched than that of N-FINDR and VCA.

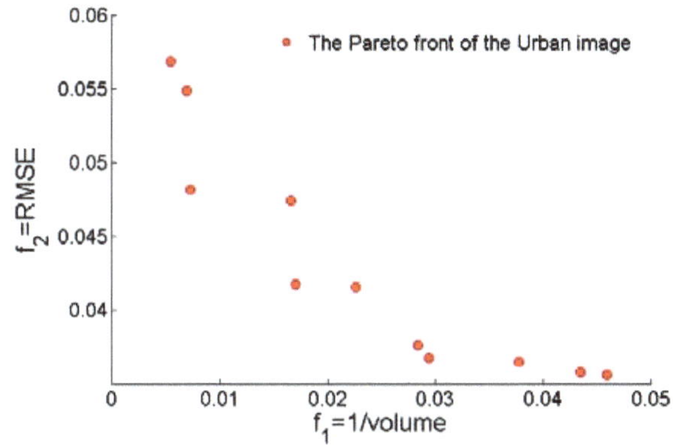

(a)

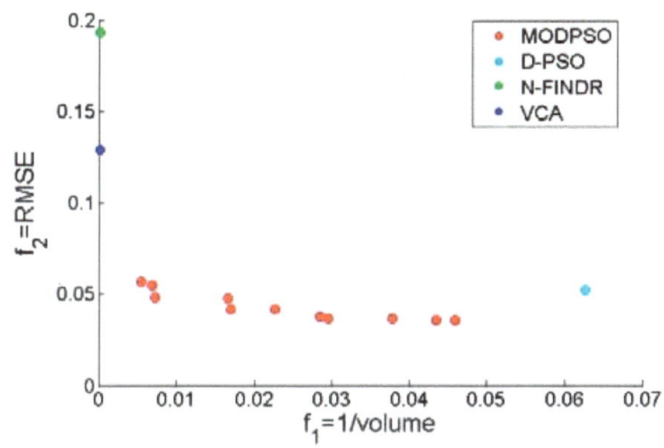

(b)

Figure 11. The results of the Urban image: (**a**) the Pareto front obtained by MODPSO; and (**b**) comparison of the results by four methods.

Table 2. The results of objective function values and computation time for the Urban Dataset.

	f_1 = 1/volume	f_2 = RMSE	Time (sec)
MODPSO	0.0054	0.0568	5544.243
	0.0069	0.0548	
	0.0073	0.0482	
	0.0165	0.0474	
	0.0170	0.0417	
	0.0225	0.0415	
	0.0284	0.0376	
	0.0294	0.0367	
	0.0378	0.0365	
	0.0435	0.0358	
	0.0459	0.0356	
D-PSO	0.0625	0.0522	5324.252
N-FINDR	0.0001	0.1934	502.744
VCA	0.0002	0.1291	3.588

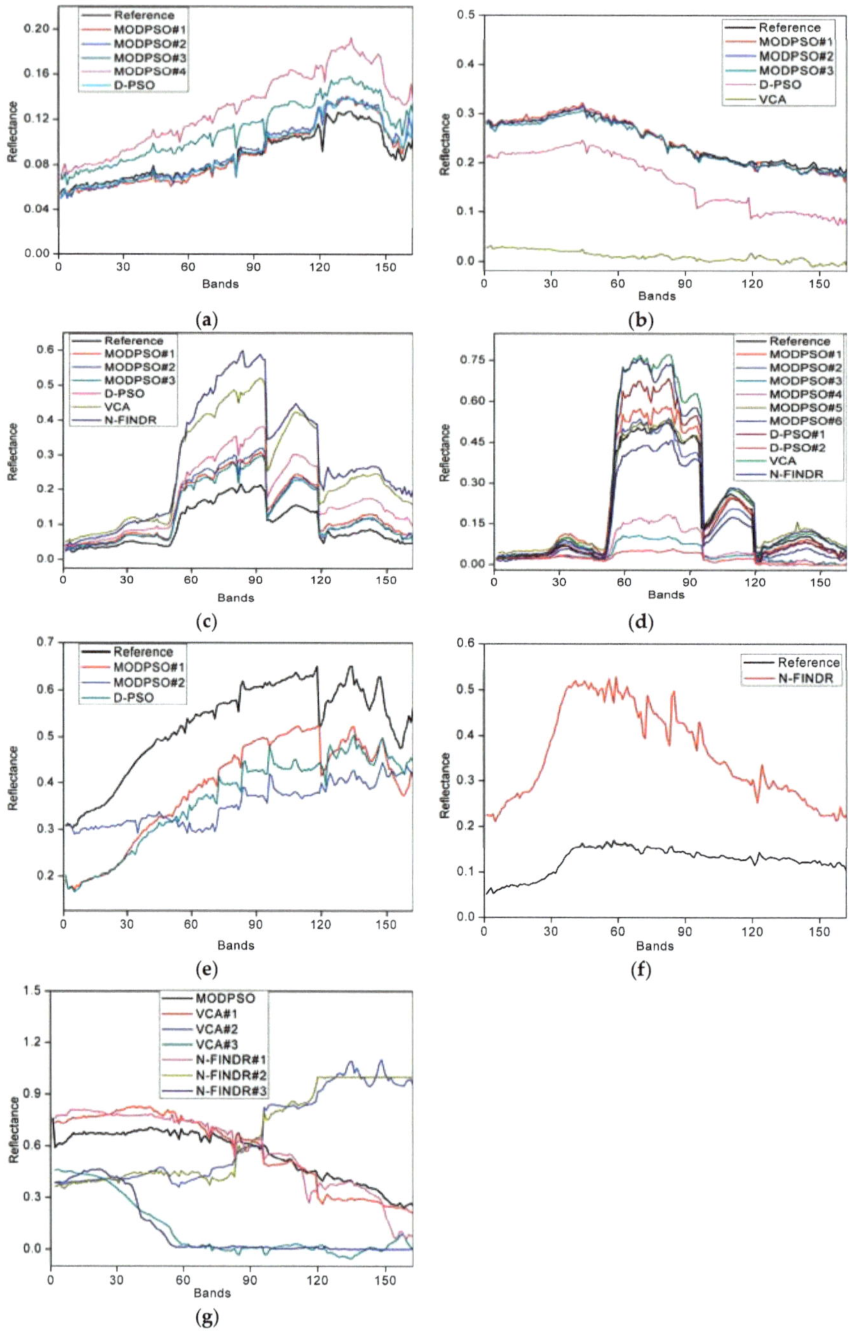

Figure 12. Endmember spectra manually selected from the image and automatically extracted by the four methods for the Urban dataset: (**a**) Road#1; (**b**) Roof#1; (**c**) Grass; (**d**) Tree; (**e**) Road#2; (**f**) Roof#2; and (**g**) spectra unmatched with the reference endmembers.

4. Discussion

4.1. Review of Experimental Results

Experimental results of the Washington dataset showed that N-FINDR and VCA failed to extract the fourth endmember, which resulted in a larger RMSE than the other two methods, while the volumes obtained by N-FINDR and VCA were much larger than the other two methods. Considering the fact that one of the objective functions of MODPSO is VM, we can infer that the searching ability of MODPSO needs further improvement. In term of RMSE, MODPSO was superior to the other methods. Experimental results of the Urban dataset showed that N-FINDR and VCA extracted several outliers, which indicated that they were easy to be affected by outliers. MODPSO and D-PSO were more robust to these interferences. We can infer that the RMSE objective function can play a key role when there are interferences in the image. In both experiments, MODPSO can find better solution than D-PSO. This may because the MOO mechanism of MODPSO (several *gbest* in MODPSO compared to one in D-PSO) increased the diversity of particles and alleviated the premature convergence problem of D-PSO, thus leading to a better optimization result. Time costs of the methods showed that the two intelligent optimization based methods were time consuming, which mainly resulted from the calculation of the RMSE objective function.

4.2. Generalization of MODPSO

In this work, MODPSO assumed that the error is represented by the additive white Gaussian noise. In fact, there may have mixed noise in the HSI such as impulse noise, multiplicative noise or vertical line strips [36,39]. It should be noted that the MODPSO method can also be applied when considering other types of noise, as long as an objective function is built according to a certain type of noise or mixed noise, the RMSE function can be replaced by the newly built one.

5. Conclusions and Future Work

This paper proposed a multiobjective optimization method MODPSO for endmember extraction. In MODPSO, the volume maximization and RMSE minimization objective functions are simultaneously optimized, and the multiobjective optimization framework is especially designed to solve the multiobjective endmember extraction problem. Instead of obtaining one unique solution for one implementation like other endmember extraction methods, the result by MODPSO is a set of non-dominated solutions, and they can be regarded as solutions with different tradeoffs between two objective functions. The experimental results show that the search ability of MODPSO method is superior to that of the D-PSO method, and it can obtain result with smaller RMSE than N-FINDR and VCA. However, the results of N-FINDR and VCA are not dominated by that of MODPSO for the reason that the volume obtained by them is bigger than that of MODPSO, which indicates that the Pareto front obtained by MODPSO is not complete, a part of non-dominated solutions are not founded by them, which revealed the limitation of the MODPSO's search ability.

Considering the future work, in our opinion, two contents are worthy of study. One is that there exist other characteristics of the hyperspectral image that are not considered in this work, so the objective functions can be replaced by others to study the effect of different combinations of objective functions on the endmember extraction result. The other is that the multiobjective optimization method with better search ability can be studied to achieve a Pareto front with higher quality.

Acknowledgments: This work was supported in part by the National Natural Science Foundation of China under Grants 61471274, 41431175, the Natural Science Foundation of Hubei Province under Grants 2014CFB193 and the Fundamental Research Funds for the Central Universities under Grants 2042016kf0152.

Author Contributions: Rong Liu and Bo Du conceived and designed the experiments; Rong Liu performed the experiments; Rong Liu and Bo Du analyzed the data; Liangpei Zhang contributed reagents/materials/analysis tools; Rong Liu wrote the paper.

Conflicts of Interest: The authors declare no conflict of interest. The founding sponsors had no role in the design of the study; in the collection, analyses, or interpretation of data; in the writing of the manuscript, and in the decision to publish the results.

References

1. Wang, Q.; Lin, J.; Yuan, Y. Salient Band Selection for Hyperspectral Image Classification via Manifold Ranking. *IEEE Trans. Neural Netw. Learn. Syst.* **2016**, *27*, 1279–1289. [CrossRef] [PubMed]
2. Yuan, Y.; Lin, J.; Wang, Q. Hyperspectral Image Classification via Multi-Task Joint Sparse Representation and Stepwise MRF Optimization. *IEEE Trans. Cybern.* **2016**, *46*, 2966–2977. [CrossRef] [PubMed]
3. Yuan, Y.; Lin, J.; Wang, Q. Dual Clustering Based Hyperspectral Band Selection by Contextual Analysis. *IEEE Trans. Geosci. Remote Sens.* **2016**, *54*, 1431–1445. [CrossRef]
4. Keshava, N.; Mustard, J.F. Spectral unmixing. *IEEE Signal Process. Mag.* **2002**, *19*, 44–57. [CrossRef]
5. Zhang, L.; Du, B.; Zhong, Y. Hybrid Detectors Based on Selective Endmembers. *IEEE Trans. Geosci. Remote Sens.* **2010**, *48*, 2633–2646. [CrossRef]
6. Du, B.; Zhang, Y.; Zhang, L.; Zhang, L. A hypothesis independent subpixel target detector for hyperspectral Images. *Signal. Process.* **2015**, *110*, 244–249. [CrossRef]
7. Liu, R.; Du, B.; Zhang, L. Hyperspectral Unmixing via Double Abundance Characteristics Constraints Based NMF. *Remote Sens.* **2016**, *8*. [CrossRef]
8. Zhang, B.; Sun, X.; Gao, L.; Yang, L. Endmember Extraction of Hyperspectral Remote Sensing Images Based on the Ant Colony Optimization (ACO) Algorithm. *IEEE Trans. Geosci. Remote Sens.* **2011**, *49*, 2635–2646. [CrossRef]
9. Marrero, R.; Lopez, S.; Callico, G.M.; Veganzones, M.A.; Plaza, A.; Chanussot, J.; Sarmiento, R. A Novel Negative Abundance Oriented Hyperspectral Unmixing Algorithm. *IEEE Trans. Geosci. Remote Sens.* **2015**, *53*, 3772–3790. [CrossRef]
10. Chang, C.I.; Du, Q. Estimation of number of spectrally distinct signal sources in hyperspectral imagery. *IEEE Trans. Geosci. Remote Sens.* **2004**, *42*, 608–619. [CrossRef]
11. Bioucas-Dias, J.M.; Nascimento, J.M. Hyperspectral subspace identification. *IEEE Trans. Geosci. Remote Sens.* **2008**, *46*, 2435–2445. [CrossRef]
12. Eches, O.; Dobigeon, N.; Tourneret, J.Y. Enhancing hyperspectral image unmixing with spatial correlations. *IEEE Trans. Geosci. Remote Sens.* **2011**, *49*, 4239–4247. [CrossRef]
13. Geng, X.; Ji, L.; Zhao, Y.; Wang, F. A New Endmember Generation Algorithm Based on a Geometric Optimization Model for Hyperspectral Images. *IEEE Geosci. Remote Sens. Lett.* **2013**, *10*, 811–815. [CrossRef]
14. Boardman, J.W. Automating Spectral Unmixing of AVIRIS Data Using Convex Geometry Concepts. In Proceedings of the 4th Annual JPL Airborne Geoscience Workshop, Washington, DC, USA, 25–29 October 1993; Volume 1, pp. 11–14.
15. Winter, M.E. N-FINDR: an algorithm for fast autonomous spectral end-member determination in hyperspectral data. In Proceedings of the SPIE Conference Imaging Spectrometry V, Denver, CO, USA, 18 July 1999; pp. 266–275.
16. Chang, C.I.; Wu, C.C.; Liu, W.M.; Ouyang, Y.C. A new growing method for simplex-based endmember extraction algorithm. *IEEE Trans. Geosci. Remote Sens.* **2006**, *44*, 2804–2819. [CrossRef]
17. Nascimento, J.M.; Bioucas Dias, J.M. Vertex component analysis: A fast algorithm to unmix hyperspectral data. *IEEE Trans. Geosci. Remote Sens.* **2005**, *43*, 898–910. [CrossRef]
18. Tits, L.; Heylen, R.; Somers, B.; Scheunders, P.; Coppin, P. A Geometric Unmixing Concept for the Selection of Optimal Binary Endmember Combinations. *IEEE Geosci. Remote Sens. Lett.* **2015**, *12*, 82–86. [CrossRef]
19. Xu, M.; Zhang, L.; Du, B.; Zhang, L. An image-based endmember bundle extraction algorithm using reconstruction error for hyperspectral imagery. *Neurocomputing* **2016**, *173*, 397–405. [CrossRef]
20. Du, Q.; Raksuntorn, N.; Younan, N.H.; King, R.L. Variants of N-FINDR algorithm for endmember extraction. In Proceedings of the SPIE Image and Signal Processing for Remote Sensing XIV, Cardiff, UK, 15–18 September 2008.
21. Liu, J.; Zhang, J. A New Maximum Simplex Volume Method Based on Householder Transformation for Endmember Extraction. *IEEE Trans. Geosci. Remote Sens.* **2012**, *50*, 104–118. [CrossRef]

22. Mei, S.; Du, Q.; He, M.; Wang, Y. Spatial preprocessing for spectral endmember extraction by local linear embedding. In Proceedings of the 2015 IEEE International Geoscience and Remote Sensing Symposium (IGARSS), Milan, Italy, 26–31 July 2015; pp. 5027–5030.
23. Gao, L.; Gao, J.W.; Li, J.; Plaza, A.; Zhuang, L.; Sun, X.; Zhang, B. Multiple Algorithm Integration Based on Ant Colony Optimization for Endmember Extraction From Hyperspectral Imagery. *IEEE J. Sel. Top. Appl. Earth Obs. Remote Sens.* **2015**, *8*, 2569–2582. [CrossRef]
24. Zhang, B.; Sun, X.; Gao, L.; Yang, L. Endmember Extraction of Hyperspectral Remote Sensing Images Based on the Discrete Particle Swarm Optimization Algorithm. *IEEE Trans. Geosci. Remote Sens.* **2011**, *49*, 4173–4176. [CrossRef]
25. Yang, L.; Sun, X.; Peng, L.; Yao, X.; Chi, T. An Agent-Based Artificial Bee Colony (ABC) Algorithm for Hyperspectral Image Endmember Extraction in Parallel. *IEEE J. Sel. Top. Appl. Earth Obs. Remote Sens.* **2015**, *8*, 4657–4664. [CrossRef]
26. Xu, M.; Zhang, L.; Du, B.; Zhang, L.; Fan, Y.; Song, D. A Mutation Operator Accelerated Quantum-Behaved Particle Swarm Optimization Algorithm for Hyperspectral Endmember Extraction. *Remote Sens.* **2017**, *9*, 197. [CrossRef]
27. Luo, W.F.; Gao, L.; Plaza, A.; Marinoni, A.; Yang, B.; Zhong, L.; Gamba, P.; Zhang, B. A New Algorithm for Bilinear Spectral Unmixing of Hyperspectral Images Using Particle Swarm Optimization. *IEEE J. Sel. Top. Appl. Earth Obs. Remote Sens.* **2016**, *9*, 5776–5790. [CrossRef]
28. Liu, R.; Zhang, L.; Du, B. A Novel Endmember Extraction Method for Hyperspectral Imagery Based on Quantum-Behaved Particle Swarm Optimization. *IEEE J. Sel. Top. Appl. Earth Obs. Remote Sens.* **2017**, *10*, 1610–1631. [CrossRef]
29. Hu, W.; Tan, Y. Prototype Generation Using Multiobjective Particle Swarm Optimization for Nearest Neighbor Classification. *IEEE Trans. Cybern.* **2016**, *46*, 2719–2731. [CrossRef] [PubMed]
30. Zheng, Y.; Ling, H.; Xue, J.; Chen, S. Population Classification in Fire Evacuation: A Multiobjective Particle Swarm Optimization Approach. *IEEE Trans. Evol. Comput.* **2014**, *18*, 70–81. [CrossRef]
31. Zhou, D.; Li, X.; Pan, Q.; Zhang, K.; Zeng, L. Multiobjective weapon-target assignment problem by two-stage evolutionary multiobjective particle swarm optimization. In Proceedings of the 2016 IEEE International Conference on Information and Automation (ICIA), Ningbo, China, August 2016; pp. 921–926.
32. Tang, L.; Wang, X. A Hybrid Multiobjective Evolutionary Algorithm for Multiobjective Optimization Problems. *IEEE Trans. Evol. Comput.* **2013**, *17*, 20–45. [CrossRef]
33. Trawiński, K.; Chica, M.; Pancho, D.P.; Damas, S.; Cordòn, O. moGrams: A Network-Based Methodology for Visualizing the Set of Nondominated Solutions in Multiobjective Optimization. *IEEE Trans. Cybern.* **2017**, 1–12. [CrossRef] [PubMed]
34. Deb, K.; Pratap, A.; Agarwal, S.; Meyarivan, T. A fast and elitist multiobjective genetic algorithm: NSGA-II. *IEEE Trans. Evol. Comput.* **2002**, *6*, 182–197. [CrossRef]
35. Mostaghim, S.; Teich, J. Strategies for finding good local guides in multi-objective particle swarm optimization (MOPSO). In Proceedings of the 2003 IEEE Swarm Intelligence Symposium, Indianapolis, IN, USA, 26 April 2003; pp. 26–33.
36. Li, C.; Chen, X.; Jiang, Y. On Diverse Noises in Hyperspectral Unmixing. *IEEE Trans. Geosci. Remote Sens.* **2015**, *53*, 5388 5402.
37. Wang, N.; Du, B.; Zhang, L. An Endmember Dissimilarity Constrained Non-Negative Matrix Factorization Method for Hyperspectral Unmixing. *IEEE J. Sel. Top. Appl. Earth Obs. Remote Sens.* **2013**, *6*, 554–569. [CrossRef]
38. Qu, Q.; Nasrabadi, N.M.; Tran, T.D. Subspace Vertex Pursuit: A Fast and Robust Near-Separable Nonnegative Matrix Factorization Method for Hyperspectral Unmixing. *IEEE J. Sel. Top. Signal Process.* **2015**, *9*, 1142–1155. [CrossRef]
39. Aggarwal, H.K.; Majumdar, A. Hyperspectral Unmixing in the Presence of Mixed Noise Using Joint-Sparsity and Total Variation. *IEEE J. Sel. Top. Appl. Earth Obs. Remote Sens.* **2016**, *9*, 4257–4266. [CrossRef]

remote sensing

MDPI

Article

Quantifying Sub-Pixel Surface Water Coverage in Urban Environments Using Low-Albedo Fraction from Landsat Imagery

Weiwei Sun [1,2,*], Bo Du [3] and Shaolong Xiong [2]

1 Department of Geography and Spatial Information Techniques, 818 Fenghua Road, Ningbo University, Ningbo 315211, China

2 State Key Lab of Information Engineering in Surveying, Mapping and Remote Sensing, Wuhan University, Wuhan 430079, China; xiong_shaolong@163.com

3 School of Computer, Wuhan University, Wuhan 430079, China; gunspace@163.com

* Correspondence: sunweiwei@nbu.edu.cn; Tel.: +86-182-5879-6120

Academic Editors: Qi Wang, Nicolas H. Younan, Carlos López-Martínez, Magaly Koch and Prasad S. Thenkabail
Received: 14 February 2017; Accepted: 16 April 2017; Published: 1 May 2017

Abstract: The problem of mixed pixels negatively affects the delineation of accurate surface water in Landsat Imagery. Linear spectral unmixing has been demonstrated to be a powerful technique for extracting surface materials at a sub-pixel scale. Therefore, in this paper, we propose an innovative low albedo fraction (LAF) method based on the idea of unconstrained linear spectral unmixing. The LAF stands on the "High Albedo-Low Albedo-Vegetation" model of spectral unmixing analysis in urban environments, and investigates the urban surface water extraction problem with the low albedo fraction map. Three experiments are carefully designed using Landsat TM/ETM+ images on the three metropolises of Wuhan, Shanghai, and Guangzhou in China, and per-pixel and sub-pixel accuracies are estimated. The results are compared against extraction accuracies from three popular water extraction methods including the normalized difference water index (NDWI), modified normalized difference water index (MNDWI), and automated water extraction index (AWEI). Experimental results show that LAF achieves a better accuracy when extracting urban surface water than both MNDWI and AWEI do, especially in boundary mixed pixels. Moreover, the LAF has the smallest threshold variations among the three methods, and the fraction threshold of 1 is a proper choice for LAF to obtain good extraction results. Therefore, the LAF is a promising approach for extracting urban surface water coverage.

Keywords: urban surface water extraction; threshold stability; sub-pixel; linear spectral unmixing; Landsat imagery

1. Introduction

Worldwide mass migration to urban areas results in the land use/cover changes, changes in climate and intensifying anthropogenic modifications to urban environments [1]. This directly brings about more unexpected variations in urban surface water, especially in external morphological features of the coverage. The urban surface water changes further impact relevant aquatic biodiversity, healthy human life and even urban ecological balance [2]. Urban surface water deficiencies would aggravate the urban heat island effect and disrupt the living environments of urban vegetation; conversely, surface water inundation would result in flooding and even high fatality because of associated waterborne diseases [3]. Therefore, figuring out the coverage of urban surface water is a crucial issue for urban environments.

Remote sensing is a powerful data source for acquiring prior and comprehensive knowledge of urban surface water [4,5]. It allows synoptic, permanent, and dynamic urban surface water monitoring and is clearly superior to conventional in-situ measurements [6,7]. Among current remote sensing sensors, Landsat sensors have the greatest reputation in urban monitoring because of its advantages in terms of free availability, and moderate spectral, temporal, and spatial resolutions. Therefore, in our study, we implement Landsat imagery to investigate the urban surface water coverage problem.

Many studies have previously reported urban surface water extraction achievements using Landsat images. Regular water extraction methods can be categorized into three main groups [8,9]: (1) thematic classification methods [10–12]; (2) single-band thresholding methods [13,14]; and (3) water index methods [15–17].

Thematic classification methods formulate urban surface water extraction into a regular binary unsupervised or supervised classification problem on urban land cover types, and select surface water as the exclusive thematic class for mapping [10]. The methods easily bring about low accuracy in areas where the background land cover includes low albedo surfaces, such as asphalt roads and building shadows in urban areas [11]. Moreover, they utilize a Boolean set to classify each pixel as either water or non-water, and fail to achieve the desired accuracy, especially at the water-land (i.e., non-water) interface [12]. Single-band thresholding methods select a single diagnostic spectral band from Landsat images (e.g., band 5 from TM/ETM+) and delineate the urban surface water coverage with a manually-defined threshold [18]. Accordingly, the subjectivity of the threshold selection can lead to an overestimated or underestimated result and, moreover, the extracted surface water is affected by shadow noise [16].

Different from the above two methods, water index methods combine two or more spectral bands using algebraic operations to enlarge the divergence between water and non-water areas. McFeeters proposed the normalized difference water index (NDWI) to delineate urban surface water. The NDWI is implemented with a ratio model using the green band (i.e., band 2) and the near-infrared band (i.e., band 4) from Landsat TM/ETM+ data [15]. An empirical value of 0 is set as the threshold for extracting surface water from the raw Landsat images, and pixels with positive NDWI values are regarded as belonging to surface water. Unfortunately, the obtained NDWI surface water suffers from noise in built-up areas, and the threshold of 0 always results in an over-estimation of the surface water [16]. Subsequently, Xu presented another surface water index called modified normalized difference water index (MNDWI) [16]. MNDWI improves NDWI by replacing the near-infrared band (i.e., band 4) with the middle-infrared band (i.e., band 5) from Landsat TM/ETM+ images. MNDWI reduces the built-up area noise in NDWI, and it performs better than NDWI in extracting urban surface water where built-up areas dominate in the image scene. Nevertheless, the threshold of MNDWI is difficult to estimate because of their scene-driven features, and the problem adversely impacts its realistic performance of MNDWI [8]. To address the instability of MNDWI, the automated water extraction index (AWEI) was presented by combining multi-band Landsat images (i.e., bands 2, 4, 5, and 7 of Landsat TM/ETM+ images) [9]. The AWEI argues that the threshold of 0 is a good initialization for urban surface water extraction in the method.

The above three types of methods greatly benefit the studies of urban surface water extraction. However, one big problem of mixed pixels still exists in the urban surface water extraction procedure when using moderate spatial resolution Landsat images. In particular, the problem becomes more pronounced when extracting accurate boundaries of surface water. A simple cause for this problem is that the scale of urban land cover is often smaller than the field of view in the Landsat TM/ETM+ sensor (30 m) [19,20]. Subsequently, a few sub-pixel classifiers were presented to handle the mixed pixel problem. Sethre proposed a sub-pixel classifier named analysis spectral analytical process (AASAP), which aimed to expand the regular classifier into the sub-pixel field to detect the size and shape of ponds [21]. The classifier focuses on sub-pixel wetlands or ponds and requires careful verifications when implemented in the case of urban water extraction. Sun optimized the training samples with mixed training samples and then combined them with the support vector machine (SVM) classifier to improve the urban surface water extraction results [22]. However, the scheme suffers from slow computational speed and complicated manual operations, which seriously restricts its real-word applications in other urban areas.

Spectral unmixing is an alternative technique that can be used to solve the mixed pixel problem encountered in urban environments. It can be classified into linear spectral unmixing (LSU) and nonlinear spectral unmixing (NLSU), according to different mathematical assumptions in mixing patterns of urban land covers in the study area [23]. Numerous applications exploit the powerful performance of LSU in converting spectral information into physical abundances of materials on the earth's surface [23]. Previously, researchers have made some trials related to the surface water extraction problem using spectral unmixing. Zhou integrated a multiscale extraction scheme with spectral mixture analysis techniques to improve water extraction in urban environments from moderate spatial resolution satellite images [24]. The feature of this work is to adopt the multiscale scheme that conducts surface water extraction in multiscale local regions in order to refine the result. Xie combined the water index NDWI with LSU and proposed an automatic subpixel water mapping (ASWM) method to map urban surface water at the sub-pixel scale [25]. Pure water extracted from NDWI and water fractions of mixed water-land pixels estimated from LSU constitute the final urban surface water map. As distinct from previous research, we propose a low albedo fraction (LAF) method based on LSU to extract urban surface water from Landsat imagery. In comparison to all of the above methods, our LAF methods have three major advantages, in the following:

(1) The LAF method stands on the H-L-V [23] (i.e., high albedo-low albedo-vegetation) spectral mixture analysis of urban surface reflectances, and investigates the urban surface water extraction problem with the low albedo fraction map. Accordingly, our idea is different from above water extraction methods, especially sub-pixel classifiers and spectral unmixing methods by Zhou [24] and Xie [25].
(2) The LAF method implements a steady initial threshold at 1 and that significantly reduces the work of parameter tuning in LAF. By contrast, current spectral unmixing-based methods by Zhou and Xie could not provide a stable threshold for fraction segmentation. The water index methods also suffer from the unstable initial threshold problem. Therefore, the LAF is easier to implement in real-word applications than other methods, such as spectral unmixing methods and water index methods.
(3) The LAF method obtains high extraction accuracies of urban surface water, and it significantly improves the accuracy of sub-pixel surface water extraction when compared against MNDWI and AWEI.

2. Test Sites and Datasets

The test sites utilized in the study are located in three representative metropolises of China: Wuhan, Shanghai, and Guangzhou. Different surface features of the urban environments (e.g., different spatial patterns of land covers and different urban backgrounds) of the three sites render them good candidates for testing the proposed LAF method. The Wuhan metropolis lies in one of the fastest-growing regions in central China, and it is becoming a significant strategic center for the rejuvenation of the Chinese nation. Wuhan is centered at the confluence of the Yangzi River and Han River, as shown in Figure 1a. Shanghai is a famous international metropolis, and it is known for advanced economics, shipping, and finance. The Huangpu River in Figure 1b is very important for the health and wellbeing of people in Shanghai. Guangzhou is an important port in China. The Pearl River in Figure 1c runs around Guangzhou city, and is a vital source of drinking water. Figure 1 illustrates the different surface characteristics of all three metropolises, where it can be seen that they have similar land cover types, including built-up surfaces, tall buildings, rivers, and vegetation.

Landsat images of the three metropolises were acquired from the website of the United States Geological Survey (USGS) (available at http://www.glovis.usgs.gov) [26], and the subsets cover the main urban background types and surface water for extraction. The downloaded Landsat imagery belongs to a Level-1 precision- and terrain-corrected product (L1T). The utilized Landsat images are free of clouds in order to avoid any negative effects from cloud. A reference image was utilized to determine the ground truth of water pixels in Landsat images, and it greatly helped in evaluating the accuracies of extracted surface water, at either the pixel level or sub-pixel level. The original sources of the reference data were high spatial-resolution pan-sharpened Quickbird images from the Digital Globe Company, and the JPEG format image at 4m spatial resolution was exported from Google Earth Pro (available at www.google.com). We selected high spatial-resolution images (HSRI) with acquisition times as close as possible to the Landsat images, and tried our best to ensure that the land-cover classes of the Landsat images and the Google Earth images were the same for the same site. Table 1 lists detailed information about the reference data and Landsat images. Geo-referencing HSRI data with Landsat images was implemented to unify spatial references of the corresponding pixels in both datasets. The manual co-registration was carefully undertaken with a Root Mean Square Error (RMSE) of no more than 0.3 pixels, and 19 control points were manually selected from each image. The "true" boundaries of urban surface water at the test sites were manually digitized on screen from the reference data, and were then rasterized at 4 m spatial resolution.

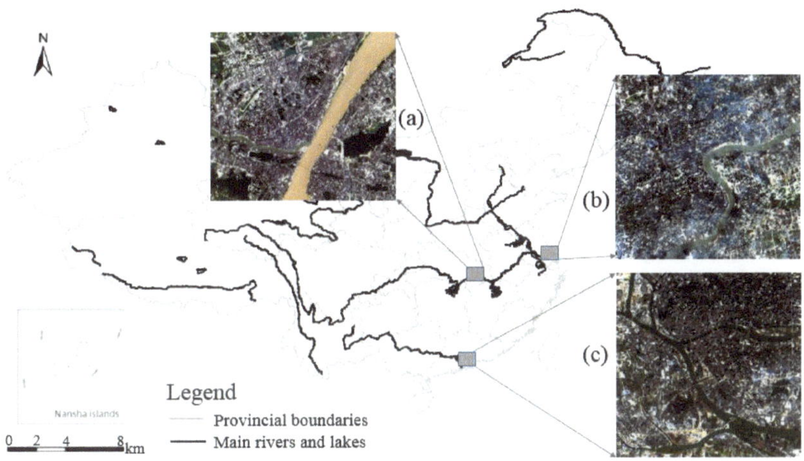

Figure 1. The images of Landsat data on three metropolises: (**a**) Wuhan; (**b**) Shanghai; and (**c**) Guangzhou.

Table 1. Description of Landsat images and their corresponding reference data.

Test Site		Acquisition Date	Sensors	Path	Row	Source
Wuhan	Landsat data	13 September 2000	TM	123	39	USGS
	Reference data	21 September 2000				Google Earth ©Digital globe
Shanghai	Landsat data	27 November 2002	ETM+	118	38	USGS
	Reference data	28 December 2002				Google Earth ©Digital globe
Guangzhou	Landsat data	2 January 2009	TM	122	44	USGS
	Reference data	16 November 2008				Google Earth ©Digital globe

3. Methodology

3.1. The Procedure of LAF Method

The LAF method explores the urban surface water extraction problem from the perspective of linear spectral unmixing and a three-endmember H-L-V (high albedo-low albedo-vegetation) model [23]. It extracts urban surface water coverage through threshold segmentation on the fraction map of the low albedo endmember. The overall procedure of extracting urban surface water using LAF is shown in Figure 2 and includes the following steps:

(1) The Landsat images are preprocessed with radiometric calibration.
(2) The three-endmember H-L-V linear mixture model is implemented to analyze surface reflectances of urban land cover types.
(3) Endmembers covering high albedo, low albedo, and vegetation are carefully selected from Landsat images using our multiple selection scheme.
(4) The unconstrained least square techniques are implemented to unmix Landsat images and to estimate the fractions of all three endmembers at each pixel. Fraction maps of all three endmembers are then obtained.
(5) The binary classification is implemented to segment the fraction map of low albedo endmember, using a given threshold t. The pixels with low albedo fractions no less than t constitute the final surface water map of LAF.

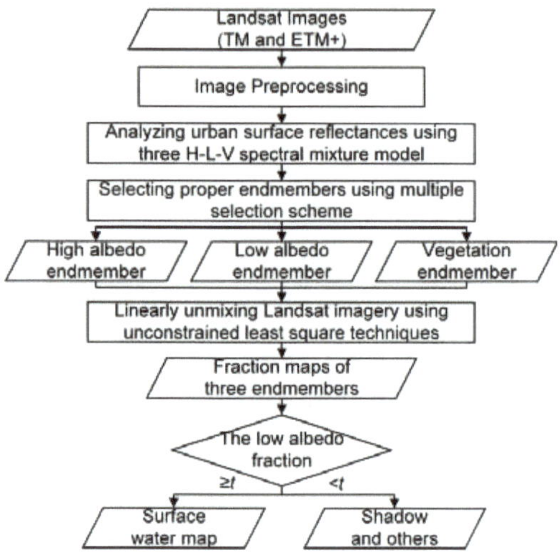

Figure 2. The overall procedure of the LAF method.

3.1.1. Preprocessing of Landsat Images

Radiometric calibration is used to transform the initial digital numbers (DNs) in Landsat images into normalized exo-atmospheric reflectance. The procedure is implemented in ENVI 5.0 [27] with the input of calibrated parameters obtained from the header file of Landsat images. Atmospheric correction is not undertaken because previous studies have shown that the process has an unclear influence on fraction maps when image-based endmembers are used in the LSU method [28,29].

3.1.2. Analyzing Urban Surface Reflectances Using Three-Endmember H-L-V Model

Generally, the three-endmember vegetation-impervious surface-soil (V-I-S) model is utilized for urban landscape analysis from remote sensing data [30]. The model classifies urban land-cover classes into fraction combinations of vegetation, impervious surfaces, and soil; and its typical application is to extract urban vegetation [31]. The V-I-S model is, however, limited in urban surface water extraction because the idea of a single endmember could not represent the complicated land cover types in urban impervious surfaces. As a result, Wu and Murray (2003) separated impervious surfaces into high albedo and low albedo surfaces, and modified the V-I-S model into a four-endmember model [32]. The difference between the four-endmember model and the three-endmember H-L-V model is whether the model includes the soil endmember or not.

In contrast to previous works, we implement the three-endmember H-L-V model. Previous studies have demonstrated that the reflectance properties of land cover in urban environments can be accurately described as linear combinations of three endmembers of high albedo, low albedo and vegetation [33]. Moreover, the three-endmember H-L-V model avoids the misclassification of soil as high albedo that exists in the four-endmember model. Furthermore, our preliminary experimental results showed that the combination of the linear mixture model and H-L-V model is more suitable for urban surface water extraction. The three-endmember H-L-V linear mixture model is represented as follows [23]:

$$R_i = \sum_{j=1}^{3} R_{i,j} f_j + e_i \quad (1)$$

where R_i is the spectral reflectance in band i, $R_{i,j}$ is the reflectance of endmember j in band i, f_j is the fraction of endmember j, and e_i is the bounded approximation error in the model.

3.1.3. Selecting Proper Endmembers Using a Multiple Selection Scheme

The result of endmember selection closely correlates with the success of the linear mixture model in urban surface water extraction. Moreover, a proper three-endmember H-L-V combination helps to robustly estimate a good threshold for extracting urban surface water from the fraction map of the low albedo endmember. In the study, we utilize a combination of different selection schemes to determine the three appropriate H-L-V endmembers from Landsat images. Multiple selection schemes combine the scatter plots of principal component analysis (PCA) transformation, image-based manual selection, and endmember optimization using cross-validation. The image-based selection scheme is adopted because of its advantages in terms of ease of operation and the same spectral response magnitude of selected endmembers with image spectra. The multiple selection schemes are implemented in the following procedures.

The first procedure is to implement PCA transformation to produce covariance-based principal component (PC) rotation and normalize the eigenvalues. The PCA transformation is implemented in ENVI 5.0 software with the input of Landsat images. For the H-L-V model, the two-dimensional normalized eigenvalue distributions of Landsat images could quantify the partitions of reflectance variance among all the PCs and formulate a triangular form with scatter plots of first two PCs [34]. The topology of triangular mixing space in Figure 3 is consistent with the mixing space of Landsat images. The pixels at the vertexes of the triangular topology correspond to high albedo, low albedo and vegetation endmembers [33]. The three vertex endmembers could accurately represent the most important physical properties of the surface reflectance of urban land cover types.

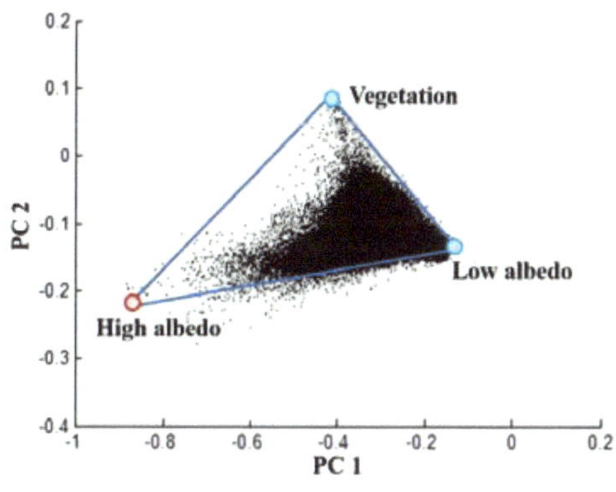

Figure 3. The triangular topology from scatter plots of the first two PCs. The vertexes correspond to three endmembers: high albedo (e.g., concrete), low albedo (e.g., water), and vegetation (e.g., grass).

Meanwhile, because of the limits of spatial and spectral resolution in Landsat sensors, the Landsat images could not discriminate the wide variety of reflectances present in the urban environments. Accordingly, the three vertex endmembers in the triangular form might represent a variety of different ground objects and that might adversely impact the accuracy of estimates for pixels with the three endmember fractions. In particular, the high albedo endmember is the most compositionally variable and the least constrained by the triangular topology. Figure 3 illustrates that a wide variety of spectra exists near the high albedo vertex of the triangular topology of scatter plots. The fraction of the high albedo vertex endmember does not necessarily provide an accurate estimate of the overall albedo because of the non-linearity and dispersion of most mixing spaces near the high albedo vertex; that is, the high albedo vertex endmember in the triangular form could not accurately represent the wide variety of high albedo reflectances observed in the urban Landsat images. In contrast, the vegetation and low albedo endmembers are generally well constrained in the triangular topology. Therefore, the second procedure is to manually select endmembers from Landsat images, compare the endmembers with the vertex endmembers of the triangular form, and optimize the selection result via cross-validation.

The operation rules for three H-L-V endmembers via cross validation are listed in Table 2 and the technique details are as follows:

(1) The low albedo endmember: The low-albedo endmembers correspond to deep dark shadow and water [29]. In this study, water is the most important object. Therefore, we chose the low albedo endmember from the deep dark water pixels, and the endmember has minimal brightness values in the image scene via cross-validation. The low albedo endmember is easy to determine from the image.

(2) The vegetation endmember: The vegetation usually corresponds to grass or dense agriculture. The pixel with maximal normalized difference vegetation index (NDVI) values (dense grass and pasture) in the image scene is chosen as the vegetation endmember, using cross-validation. The vegetation endmember is also easily determined in the LAF method.

(3) The high albedo endmember: The high albedo endmember shows much greater sensitivity to the selection method because it varies most greatly in amplitude within the triangular topology [29]. Therefore, we combine Landsat images with HSRI data to optimize the selection of the high albedo endmember via cross-validation. The initial high albedo endmembers are manually selected from building roofs, airport runways, and highway intersections in Landsat images, with reference to corresponding land covers in the HSRI data. Next, these initial endmembers are compared with the high-albedo vertex endmember in the scatter plots of PC1 and PC2. The endmember located closest to the high albedo vertex of the triangular topology is finally selected as the high albedo endmember [35].

Table 2. The operations of multiple selection schemes in three endmembers.

Endmembers	Difficulty Level	Key Words in Operation	Candidate Sources
Low albedo	Easy	minimum brightness	deep dark water
Vegetation	Easy	maximal normalized difference vegetation index	grass and pasture
High albedo	Difficulty	nearest to the high albedo vertex in the triangular topology	building roofs, airport runway and highway intersections

3.1.4. Spectral Unmixing and Binary Classification of the Low Albedo Fraction Map

Spectral unmixing is utilized to solve the three-endmember H-L-V linear mixture model in Equation (1). Spectral unmixing was initially proposed for calculating land-cover fractions for a pixel [36]. The least square techniques are implemented to estimate the fraction of each endmember at each pixel by minimizing the model errors. The techniques can be grouped into unconstrained and constrained types. The differences between the two types are nonnegativity and sum-to-one constraints in the fractions of each pixel [37].

In the study, we implement the unconstrained least square techniques, for two reasons. The first is that the result of unconstrained least square techniques is only affected by the adopted model, and the second is that our objective is to explore the relations between urban surface water and the fraction map of low albedo endmember, and this purpose differs from current common applications of constrained least square techniques. After spectral unmixing operation with unconstrained least square techniques, the fractions of all three endmembers are estimated and the fraction maps are then obtained.

From the above analysis, the low albedo spectrum dominates in the pixels of urban surface water, and we accordingly extract them from the fraction map of the low albedo endmember. The binary map of urban surface water is achieved by segmenting the low albedo fraction map with a given threshold t, shown as follows:

$$LAF = f_{low-albedo} \geq t \tag{2}$$

where $f_{low-albedo}$ is the fraction or abundance of the low albedo endmember in each pixel.

In LAF, we implement a cross-validation scheme to select an appropriate threshold. The scheme is initialized with a manually-defined threshold, and we then interactively estimate the sub-pixel accuracies (mentioned in Section 3.2) of urban surface water by tuning the threshold parameter from the initial value. Finally, we select an appropriate threshold with the optimal sub-pixel extraction accuracy that best balances over-estimation errors and under-estimation errors. It should be stressed that a good initialization is important for the above scheme. From our trial experiments, we found that, in the low albedo fraction map, pixels with fraction values clearly greater than 1 always belonged to water; pixels with fraction values around 1 were boundary mixed pixels dominated by water; and pixels with fraction values of less than 1 belonged to non-water. We also found that the pixels that were mixed by building shadows and other ground objects had fractions of the low albedo endmember smaller than 1. The shadows belong to non-water and their fractions do not affect the extraction result of LAF in urban surface water. Therefore, we manually select the initial threshold of LAF as 1, and implement the cross-validation scheme to achieve a proper binary classification map of urban surface water. The binary map after thresholding segmentation includes water and non-water, and the image is directly adopted as our final extraction result of urban surface without any filter operations, such as removing isolated or partial water pixels.

3.2. Accuracy Assessment Schemes on the Per-Pixel and Sub-Pixel Levels

Considering the fact that the MNDWI and AWEI obtain a better accuracy of urban surface water extraction than other current water extraction methods [8,9], the two methods are utilized to make comparisons with the proposed LAF. The thresholds in the three methods were estimated via cross-validation, and the best extraction results of urban surface water from all three methods were adopted for the comparison.

The per-pixel accuracy and sub-pixel accuracy were estimated from the binary map to evaluate the performance in extracting urban surface water. The per-pixel accuracy is to evaluate the overall performance of the LAF binary classification map, with pure pixels and boundary mixed pixels of surface water included. The ratio of spatial resolutions between the reference HSRI data and Landsat images is 4:30, meaning that one pixel in Landsat images corresponds to about 50 HSRI pixels. Similar to the idea expressed in [9], we regarded the pixels in Landsat imagery that consist predominantly of water (>50% proportions, i.e., over 25 HSRI pixels) as true water pixels, and vice versa. Using the random sampling scheme, the labels (water and non-water) of testing water pixels for overall per-pixel accuracy evaluation was manually digitized from Landsat imagery, and then compared with their true labels from reference data. The kappa coefficients (KC) were calculated and used to quantify the overall extraction accuracy of all three methods.

Different from overall per-pixel accuracy, the sub-pixel accuracy is to testify the specific performance of all three methods in extracting water from mixed pixels, especially from boundary pixels. The sub-pixel accuracy evaluation implemented the following four main steps.

(1) The actual water fractions of testing boundary pixels were manually estimated via the visual overlay analysis of reference data and Landsat images. By overlaying the binary maps of extracted surface water from all three methods (AWEI, MNDWI and LAF) with the HSRI data, the water fraction of each boundary pixel from each method can be calculated. This was equal to the percentages of water pixels in the total number of HSRI pixels that were fully contained within the area of one pixel of Landsat imagery. For example, within the scene of one pixel from Landsat imagery, if the water occupies 20 of the total 50 HSRI pixels, the water fraction of the targeted boundary pixel is 40%. The process is repeated and the actual water fractions of all testing boundary pixels resulting from the three methods were achieved.
(2) The testing boundary pixels were designated into six categories according to their true water fractions. The true water fractions of all testing boundary pixels in the HSRI data can be classified into six categories, 0–10%, 10–30%, 30–50%, 50–70%, 70–90% and 90–100%. For example, Figure 4 shows six categories of true water proportions in the testing boundary pixels of Shahu lake, and the number of testing boundary water pixels is 106.

Figure 4. The category of true water proportions in testing boundary pixels in Shahu lake, Wuhan.

(3) The estimation errors (EEs) of all three methods on each testing boundary pixel were estimated. The EEs for each testing boundary pixels at the sub-pixel level are the summation of over-estimation error and under-estimation error, defined according to the following two conditions: (a) if a testing boundary pixel in the binary classification map of each method was classified as water, its complement of the true water fraction is regarded as the sub-pixel over-estimation error; (b) in contrast, if the pixel was classified as non-water, its true water fraction is quantified as the under-estimation error at the sub-pixel level.
(4) The average estimation errors (AEEs) in all six categories of testing boundary pixels were calculated and the set of AEEs with six elements for all three methods were obtained to quantify the sub-pixel water extraction accuracy of boundary mixed pixels at different water proportions.

4. Experimental Results and Analysis

4.1. Water Extraction Maps and Per-Pixel Accuracy Assessment in Overall Result

The water extraction results using the three methods of MNDWI, AWEI, and LAF at the three test sites are illustrated in Figure 5. A visual inspection of the figure indicates that LAF results in a better (or at least comparable) accuracy of urban surface water mapping than the AWEI and MNDWI. For the test sites of Wuhan and Shanghai, in particular, the LAF method performs better in suppressing non-water surfaces. Unfortunately, at the test site in Guangzhou, a visual inspection of Figure 5 tells us that the proposed method produces noisy results, as do the other two methods.

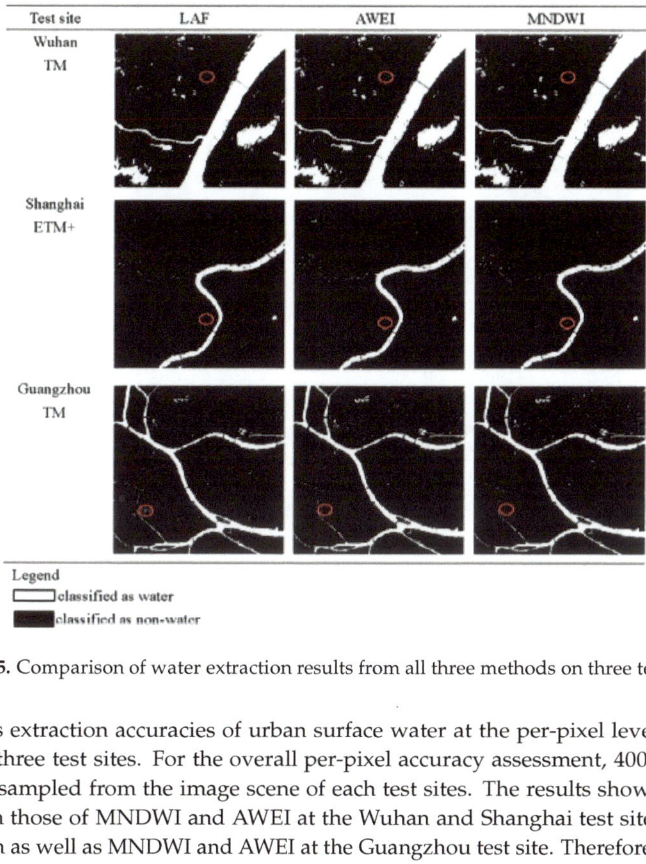

Figure 5. Comparison of water extraction results from all three methods on three test sites.

Table 3 lists extraction accuracies of urban surface water at the per-pixel level from the three methods at the three test sites. For the overall per-pixel accuracy assessment, 400 testing samples were randomly sampled from the image scene of each test sites. The results show that the KCs of LAF outperform those of MNDWI and AWEI at the Wuhan and Shanghai test sites, whereas LAF does not perform as well as MNDWI and AWEI at the Guangzhou test site. Therefore, from the above observations, we can conclude that LAF achieves a better, or at least comparable, per-pixel extraction accuracy for urban surface water than MNDWI and AWEI.

Table 3. List of extraction accuracies at the per-pixel level for the three methods at three test sites.

Water Extraction Methods	Kappa Coefficient (KC)		
	Wuhan	Shanghai	Guangzhou
LAF	0.97	0.93	0.91
AWEI	0.95	0.92	0.93
MNDWI	0.96	0.92	0.92

4.2. Sub-Pixel Accuracy Assessment of LAF in Boundary Mixed Pixels

We also compare extraction accuracies at the sub-pixel level for the three methods. The experiment aims to investigate the performance of LAF in extracting the water from boundary mixed pixels that consist of mixtures of water and non-water components. Table 4 lists extraction errors of the three methods for the boundary mixed pixels at all three test sites. For the sub-pixel accuracy assessment, the testing samples on Wuhan, Shanghai and Guangzhou were randomly chosen along the boundary of Shahu Lake, Huangpu River and Pearl River. The testing samples were mixed by water and concrete pavement, vegetation and soil. The detailed information of three test sites for sub-pixel accuracy assessment is listed in Table 5. The numbers of testing pixels on Wuhan, Shanghai and Guangzhou are 210, 198 and 201, respectively. The accuracies within each water fraction range are the average of AEEs from three test sites.

Table 4. List of extraction errors at the sub-pixel level for the three methods with boundary mixed pixels of all three test sites.

Water Extraction Methods	Extraction Errors of % Water in the Boundary Mixed Pixels					
	0–10%	10–30%	30–50%	50–70%	70–90%	90–100%
LAF	0.04	0.22	0.43	0.41	0.23	0.03
AWEI	0.04	0.30	0.49	0.47	0.34	0.03
MNDWI	0.04	0.33	0.48	0.49	0.37	0.03

Table 5. The detailed information of three test sites for sub-pixel accuracy assessment.

City	Name of Water Bodies	Center Point Coordinate (UTM)	Area (km)	Characteristics of Water Bodies	Topography	Climate
Wuhan	Shahu lake	30°34′04.30″N, 114°19′41.76″E	3.04	Clear lake	flat	Subtropical wet
Shanghai	Huangpu river	31°14′33.18″N, 121°29′21.00″E	6.79	Turbid river	flat	Subtropical wet
Guangzhou	Zhujiang river	23°06′19.23″N, 113°14′17.30″E	13.69	Turbid river	flat	Subtropical wet

The results are in agreement for the three methods in that boundary mixed pixels consisting of 0–10% and 90–100% water are correctly classified as non-water and water, respectively. However, the performance of the three methods varies greatly in extracting water having proportions of 10–90% in the boundary mixed pixels. For the 10–90% boundary pixels, AWEI and MNDWI obtain similar extraction accuracies, with AWEI being slightly superior to MNDWI. The accuracy of LAF clearly surpasses that of AWEI and MNDWI, and it reduces extraction errors by at least 5% in the 10–90% proportion of the boundary pixels. Therefore, we conclude that LAF performs significantly better at the sub-pixel level than AWEI and MNDWI.

4.3. Threshold Analysis

Section 3.1 describes that an initial threshold estimation is essential for the parameter tuning of LAF. A good initialization reduces the computational complexity of threshold estimation in LAF, thereby promoting the feasibility of LAF for real-word applications. This experiment therefore explores the stability of the threshold in LAF.

Table 6 lists the parameter settings of the three water extraction methods that produces the best extraction results in experiments 4.1 and 4.2. The standard deviation (*Std*) is adopted to quantify the variation in threshold parameters of the three methods. The appropriate threshold for MNDWI at the three test sites ranges from 0.35 to 0.515, giving the largest *Std* in parameter estimation. The appropriate threshold for AWEI varies from 0.086 to 0.2, and the *Std* is smaller than that of MNDWI but is higher than that of LAF. The comparison shows that the appropriate threshold of LAF shows the smallest variation across the three test sites, with the narrowest range from 1 to 1.08. The appropriate threshold of LAF is close to its initial value of 1, with only slight tuning work required. Therefore, we conclude from the above that the appropriate threshold in LAF has the smallest variation among all the three methods at the three test sites, and the threshold value at 1 is a good and stable initial value for LAF in extracting urban surface water.

Table 6. Stability analysis for the thresholds of all three water extraction methods.

Water Extraction Methods	Test Site			Threshold Variability
	Wuhan TM	Shanghai ETM+	Guangzhou TM	*Std*
LAF	1.000	1.080	1.000	0.046
AWEI	0.086	0.200	0.156	0.057
MNDWI	0.350	0.515	0.470	0.085

5. Discussion

In the above experiments, we implemented LAF to extract urban surface water from Landsat imagery on three metropolises, Wuhan, Shanghai and Guangzhou. The extraction results were evaluated on the aspects of per-pixel accuracy and sub-pixel accuracy and were compared with two state-of-the-art methods, AWEI and MNDWI. All the experimental results demonstrate the superiority of LAF to other two methods.

First, from per-pixel accuracy estimation experiment on three test sites, our LAF shows better performance in differentiating urban surface water from other ground objects (e.g., building roofs, roads, and vegetation), especially in the image scenes of Wuhan and Shanghai. The better per-pixel accuracy results, in our estimation, from two main causes. The first is that the H-L-V linear mixture model could explain reflectance features of land covers in Landsat imagery, while also avoiding nonnegative effects from soil. The second is that multiple selection schemes maximize the divergence of three endmembers of high albedo, low albedo and vegetation, and it guarantees three vertexes of triangular topology in mixing space of all land covers of urban environments.

Second, with regard to sub-pixel accuracy estimation results on three test sites, our LAF behaves better at recognizing water fractions from boundary mixed pixels. The LSU feature of our method guarantees that it is better able to identify water fractions from boundary mixed pixels, using a fraction threshold of low albedo. On the contrary, the AWEI and MNDWI could not avoid the large uncertainty in boundary water pixels originating from the hard-binary classification of water and non-water at the pixel level.

Finally, the threshold analysis explains that the LAF has a relatively more stable threshold than other two methods. For many water extraction methods, the threshold value for binary classification is difficult to estimate because of its data-driven nature [8]. Our LAF has the smallest variations in the threshold on three test sites among all three methods, making the implementation of the method simpler. It is essential to note that the different endmember selection scheme described in [38] would also greatly affect the stability or value of the fraction threshold.

However, our work has several limitations that require further study. The first is that we could not explain theoretical reasons for good behaviors of empirical threshold value as 1. The fraction relations between water and other urban land covers should be carefully analyzed in further experiments to explain the physical meanings of the recommended initial threshold. The second is that we did not carefully investigate the water extraction problem in the presence of cloud and SLC-gaps. Many algorithms including the multi-temporal linear regression algorithm [39] and the GNSPI algorithm [40] have been proposed to detect the thick clouds and fill gap pixels in SLC-OFF Landsat imagery. The combination of the above algorithms with our LAF would be a promising direction to extend the LAF into urban water extraction of any archived Landsat images. The third is that the H-L-V linear mixture model restricts the applications of LAF into other image scenes. It is not difficult to extend the LAF for the purposes of extracting urban wetlands and identifying water fractions from mixed vegetation-water pixels. Unfortunately, the method would not directly apply to other situations, such as open water or coastal wetlands, because the spectral features of their land covers do not satisfy the H-L-V linear mixture model, especially the unavailability of high albedo reflectance such as building roofs and airports. In such cases, other linear mixture models or nonlinear mixture models might be a good addition to the proposed method. The fourth one is that the endmember selection scheme involves too much manual operations and it might restrict the application of LAF to too large an image scene. The automatic or intelligent scheme should be further investigated to satisfy the demands from its complicated image scenes in massive Landsat datasets. The last one is that most recently proposed methods including the enhanced water index (EWI) [39] and dynamic surface water extent (DSWE) [40] have not been considered in comparisons with the LAF. Further performance contrast with modifications of MNDWI and newly-proposed methods on more Landsat images is essential to promote the LAF in real-word applications.

6. Conclusions

The main purpose of this study was to devise a method that improves the sub-pixel water extraction accuracy and has a stable threshold value. Using Urban Landsat images, we presented the LAF method, and then compared its per-pixel and sub-pixel extraction accuracies and threshold stability with those of two state-of-the-art methods, AWEI and MNDWI, at three test sites including Wuhan, Shanghai, and Guangzhou. The results show that LAF achieves a better sub-pixel water extraction accuracy and reduces errors by at least 5% when compared to AWEI and MNDWI, and obtains better, or at least comparable, extraction results at the per-pixel level than the other two methods. Moreover, the method has the smallest variation in appropriate threshold, and the threshold at 1 is a good and stable initialization for parameter tuning in LAF.

Acknowledgments: This work was funded by the National Natural Science Foundation (41671342, 41401389, U1609203) and the Chinese Postdoctoral Science Foundation (2016T90732, 2015M570668). We greatly appreciate the handling editor and the anonymous reviewers for their hard work in reviewing the paper.

Author Contributions: All coauthors made significant contributions to the paper. Weiwei Sun presented the key idea of the LAF method and carried on the contrast experiments. Bo Du designed the comparison experiments between the proposed LAF and the other two methods. Shaolong Xiong helped to design the procedures of experiments in the paper.

Conflicts of Interest: The authors declare no conflict of interest.

References

1. Liu, Y.; Bai, X.; Shi, P. Realizing China's urban dream. *Nature* **2014**, *509*, 158–160.
2. United States Geological Survey (USGS). *Facing Tomorrow's Challenges—U.S. Geological Survey Science in the Decade 2007–2017*; U.S. Geological Survey: Reston, VA, USA, 2007.
3. Giardino, C.; Bresciani, M.; Villa, P.; Martinelli, A. Application of remote sensing in water resource management: The case study of lake trasimeno, Italy. *Water Resour. Manag.* **2010**, *24*, 3885–3899. [CrossRef]
4. Morss, R.E.; Wilhelmi, O.V.; Downton, M.W.; Gruntfest, E. Flood risk, uncertainty, and scientific information for decision making: Lessons from an interdisciplinary project. *Bull. Am. Meteorol. Soc.* **2005**, *86*, 1593–1601. [CrossRef]
5. Wang, Q.; Lin, J.; Yuan, Y. Salient band selection for hyperspectral image classification via manifold ranking. *IEEE Trans. Neural Netw. Learn. Syst.* **2016**, *27*, 1279–1289. [CrossRef] [PubMed]
6. Zhang, P.; Lu, J.Z.; Feng, L.; Chen, X.L.; Zhang, L.; Xiao, X.W.; Liu, H.G. Hydrodynamic and inundation modeling of China's largest freshwater lake aided by remote sensing data. *Remote Sens.* **2015**, *7*, 4858–4879. [CrossRef]
7. Wang, Q.; Chen, M.; Li, X. Quantifying and Detecting Collective Motion by Manifold Learning. In Proceeding of the AAAI Conference on Artificial Intelligence (AAAI), San Francisco, CA, USA, 4–9 February 2017; pp. 4292–4298.

8. Ji, L.; Zhang, L.; Wylie, B. Analysis of dynamic thresholds for the normalized difference water index. *Photogramm. Eng. Remote Sens.* **2009**, *75*, 1307–1317. [CrossRef]
9. Feyisa, G.L.; Meilby, H.; Fensholt, R.; Proud, S.R. Automated water extraction index: A new technique for surface water mapping using Landsat imagery. *Remote Sens. Environ.* **2014**, *140*, 23–35. [CrossRef]
10. Lira, J. Segmentation and morphology of open water bodies from multispectral images. *Int. J. Remote Sens.* **2006**, *27*, 4015–4038. [CrossRef]
11. Jiang, H.; Feng, M.; Zhu, Y.; Lu, N.; Huang, J.; Xiao, T. An automated method for extracting rivers and lakes from Landsat imagery. *Remote Sens.* **2014**, *6*, 5067–5089. [CrossRef]
12. Yang, Y.; Liu, Y.; Zhou, M.; Zhang, S.; Zhan, W.; Sun, C.; Duan, Y. Landsat 8 OLI image based terrestrial water extraction from heterogeneous backgrounds using a reflectance homogenization approach. *Remote Sens. Environ.* **2015**, *171*, 14–32. [CrossRef]
13. Jain, S.K.; Singh, R.; Jain, M.; Lohani, A. Delineation of flood-prone areas using remote sensing techniques. *Water Resour. Manag.* **2005**, *19*, 333–347. [CrossRef]
14. Jain, S.K.; Saraf, A.K.; Goswami, A.; Ahmad, T. Flood inundation mapping using noaa avhrr data. *Water Resour. Manag.* **2006**, *20*, 949–959. [CrossRef]
15. McFeeters, S. The use of the normalized difference water index (NDWI) in the delineation of open water features. *Int. J. Remote Sens.* **1996**, *17*, 1425–1432. [CrossRef]
16. Xu, H. Modification of normalised difference water index (NDWI) to enhance open water features in remotely sensed imagery. *Int. J. Remote Sens.* **2006**, *27*, 3025–3033. [CrossRef]
17. Rogers, A.; Kearney, M. Reducing signature variability in unmixing coastal marsh thematic mapper scenes using spectral indices. *Int. J. Remote Sens.* **2004**, *25*, 2317–2335. [CrossRef]
18. Verpoorter, C.; Kutser, T.; Tranvik, L. Automated mapping of water bodies using Landsat multispectral data. *Limnol. Oceanogr. Methods* **2012**, *10*, 1037–1050. [CrossRef]
19. Cracknell, A.P. Review article synergy in remote sensing-What's in a pixel? *Int. J. Remote Sens.* **1998**, *19*, 2025–2047. [CrossRef]
20. Yuan, Y.; Lin, J.; Wang, Q. Dual-clustering-based hyperspectral band selection by contextual analysis. *IEEE Trans. Neural Netw. Learn. Syst.* **2016**, *54*, 1431–1445. [CrossRef]
21. Sethre, P.R.; Rundquist, B.C.; Todhunter, P.E. Remote detection of prairie pothole ponds in the devils lake basin, north dakota. *GISci. Remote Sens.* **2005**, *42*, 277–296. [CrossRef]
22. Sun, X.; Li, L.; Zhang, B.; Chen, D.; Gao, L. Soft urban water cover extraction using mixed training samples and support vector machines. *Int. J. Remote Sens.* **2015**, *36*, 3331–3344. [CrossRef]
23. Keshava, N.; Mustard, J.F. Spectral unmixing. *IEEE Signal Process. Mag.* **2002**, *19*, 44–57. [CrossRef]
24. Zhou, Y.; Luo, J.; Shen, Z.; Hu, X.; Yang, H. Multiscale water body extraction in urban environments from satellite images. *IEEE J. Sel. Top. Appl. Earth Obs. Remote Sens.* **2014**, *7*, 4301–4312. [CrossRef]
25. Xie, H.; Luo, X.; Xu, X.; Pan, H.; Tong, X. Automated Subpixel Surface Water Mapping from Heterogeneous Urban Environments Using Landsat 8 OLI Imagery. *Remote Sens.* **2016**, *8*, 584. [CrossRef]
26. United States Geological Survey (USGS). *Landsat Data Archive*; USGS Global Visualization Viewer (GLOVIS): Reston, VA, USA, 2012.
27. EXELIS. *Exelis Visual Information Solutions*; ENVI v5.0; EXELIS: Boulder, CO, USA, 2013.
28. Lu, D.; Batistella, M.; Moran, E.; Mausel, P. Application of spectral mixture analysis to amazonian land-use and land-cover classification. *Int. J. Remote Sens.* **2004**, *25*, 5345–5358. [CrossRef]
29. Small, C. The Landsat ETM+ spectral mixing space. *Remote Sens. Environ.* **2004**, *93*, 1–17. [CrossRef]
30. Ridd, M.K. Exploring a VIS (vegetation-impervious surface-soil) model for urban ecosystem analysis through remote sensing: Comparative anatomy for cities. *Int. J. Remote Sens.* **1995**, *16*, 2165–2185. [CrossRef]
31. Small, C. Estimation of urban vegetation abundance by spectral mixture analysis. *Int. J. Remote Sens.* **2001**, *22*, 1305–1334. [CrossRef]
32. Wu, C.; Murray, A.T. Estimating impervious surface distribution by spectral mixture analysis. *Remote Sens. Environ.* **2003**, *84*, 493–505. [CrossRef]
33. Small, C. A global analysis of urban reflectance. *Int. J. Remote Sens.* **2005**, *26*, 661–681. [CrossRef]
34. Smith, M.O.; Johnson, P.E.; Adams, J.B. Quantitative determination of mineral types and abundances from reflectance spectra using principal components analysis. *J. Geophys. Res. Solid Earth* **1985**, *90*, C797–C804. [CrossRef]

35. Weng, Q.; Hu, X. Medium spatial resolution satellite imagery for estimating and mapping urban impervious surfaces using lsma and ann. *IEEE Trans. Geosci. Remote Sens.* **2008**, *46*, 2397–2406. [CrossRef]
36. Roberts, D.A.; Gardner, M.; Church, R.; Ustin, S.; Scheer, G.; Green, R.O. Mapping chaparral in the santa monica mountains using multiple endmember spectral mixture models. *Remote Sens. Environ.* **1998**, *65*, 267–279. [CrossRef]
37. Heinz, D.C.; Chang, C.I. Fully constrained least squares linear spectral mixture analysis method for material quantification in hyperspectral imagery. *IEEE Trans. Geosci. Remote Sens.* **2001**, *39*, 529–545. [CrossRef]
38. Nascimento, J.M.P.; Dias, J.M.B. Vertex component analysis: A fast algorithm to unmix hyperspectral data. *IEEE Trans. Geosci. Remote Sens.* **2005**, *43*, 898–910. [CrossRef]
39. Wang, S.; Baig, M.H.A.; Zhang, L.; Jiang, H.; Ji, Y.; Zhao, H.; Tian, J. A simple enhanced water index (EWI) for percent surface water estimation using Landsat data. *IEEE J. Sel. Top. Appl. Earth Obs. Remote Sens.* **2015**, *8*, 90–97. [CrossRef]
40. Jones, J. Efficient wetland surface water detection and monitoring via Landsat: Comparison with in situ data from the everglades depth estimation network. *Remote Sens.* **2015**, *7*, 12503–12538. [CrossRef]

remote sensing

MDPI

Article

Online Hashing for Scalable Remote Sensing Image Retrieval

Peng Li [1,2], Xiaoyu Zhang [3], Xiaobin Zhu [4] and Peng Ren [1,2,*]

[1] College of Information and Control Engineering, China University of Petroleum (East China), Qingdao 266580, China; lipeng@upc.edu.cn

[2] State Key Laboratory of Mathematical Engineering and Advanced Computing, Wuxi 214125, China

[3] Institute of Information Engineering, Chinese Academy of Sciences, Beijing 100093, China; zhangxiaoyu@iie.ac.cn

[4] College of Computer and Information Engineering, Beijing Technology and Business University, Beijing 100048, China; brucezhucas@gmail.com

* Correspondence: pengren@upc.edu.cn; Tel.: +86-532-8698-3478

Received: 9 April 2018; Accepted: 3 May 2018; Published: 4 May 2018

Abstract: Recently, hashing-based large-scale remote sensing (RS) image retrieval has attracted much attention. Many new hashing algorithms have been developed and successfully applied to fast RS image retrieval tasks. However, there exists an important problem rarely addressed in the research literature of RS image hashing. The RS images are practically produced in a streaming manner in many real-world applications, which means the data distribution keeps changing over time. Most existing RS image hashing methods are batch-based models whose hash functions are learned once for all and kept fixed all the time. Therefore, the pre-trained hash functions might not fit the ever-growing new RS images. Moreover, the batch-based models have to load all the training images into memory for model learning, which consumes many computing and memory resources. To address the above deficiencies, we propose a new online hashing method, which learns and adapts its hashing functions with respect to the newly incoming RS images in terms of a novel online partial random learning scheme. Our hash model is updated in a sequential mode such that the representative power of the learned binary codes for RS images are improved accordingly. Moreover, benefiting from the online learning strategy, our proposed hashing approach is quite suitable for scalable real-world remote sensing image retrieval. Extensive experiments on two large-scale RS image databases under online setting demonstrated the efficacy and effectiveness of the proposed method.

Keywords: hashing; remote sensing image retrieval; online learning

---

1. Introduction

With the rapid development of satellite and aerial vehicle technologies, we have entered an era of remote sensing (RS) big data. Automatic knowledge discovery from massive RS data has become increasingly urgent. Among emerging RS big data mining efforts, large-scale RS image retrieval has attracted an increasing amount of research interest due to its broad applications in the RS research community. For example, a fast and accurate retrieval of similar satellite cloud images can provide valuable judging information for short-term weather forecasting. Besides, in the disaster rescue scenario, a fast rescue and optimal resources allocating also depend on the real-time and precise retrieval strategies for the photographs of disaster area.

In earlier RS image retrieval systems, RS image retrieval mainly relied on manual tags in terms of sensor types, waveband information, and geographical locations of remote sensing images. However, the manual generation of tags is quite time consuming and becomes especially prohibitive when the volume of remote sensing images is oversized. As an effective method to manage a large number of images, content-based image retrieval (CBIR) can retrieve the interesting images according to their visual content. In recent years, content-based RS image retrieval has been comprehensively studied [1–4], in which the similarity of RS images is measured by different kinds of visual descriptors. More specifically, local invariant [5], morphological [6], textural [7–9], and data-driven features [10–13] have been evaluated in terms of content-based RS image retrieval tasks. To further improve image retrieval performance levels, Li et al. [14] proposed a multiple feature-based remote sensing image retrieval approach by combining handcrafted features and data-driven features via unsupervised feature learning. Wang et al. [15] proposed a multilayered graph model for hierarchically refining retrieval results from coarse to fine. Although some encouraging progress has been made, there remains a great challenge for the content-based RS image retrieval tasks. For the aforementioned visual descriptors, their dimensions can be in the hundreds or even thousands. Exhaustively comparing the high dimensional feature descriptor of an inquiry remote sensing image with each image in the retrieval set is computationally expensive and impossible to achieve on an oversized database. Besides, the storage of the image descriptors is also a bottleneck for large-scale RS image retrieval problems.

Hashing technique is a potential solution to cope with big data retrieval due to its excellent ability in compact feature representation. The hashing methods map the input images from the high dimensional feature space to a low dimensional code space, i.e., hamming space, where each image is represented by a short binary code. It is extremely fast to perform image retrieval over such binary codes, because the hamming distance between binary codes can be efficiently calculated with XOR operation even in a modern CPU. Moreover, binary code representation significantly reduces the amount of memory required for storing the large-scale content information of images. Existing hashing approaches can be broadly categorized as data-independent and data-dependent schemes. Data-independent methods usually adopt random projections as hash functions without using any training data. One representative data-independent method is Locality Sensitive Hashing (LSH) [16–18], which projects data points to a random hyperplane and then conducts random thresholding. Although this data-independent random scheme is quite computationally efficient, it usually cannot achieve satisfactory retrieved results because it totally disregards the image data structure. Moreover, to achieve a reasonable recall rate, the LSH based methods typically require long codes and multiple tables, which degrade the search efficiency in practice. On the contrary, data-dependent hashing methods attempt to learn good data-aware hash codes by utilizing various machine learning techniques, which are usually demonstrated to be more effective than data-independent LSH. Data-dependent hashing can further be divided into unsupervised hashing [19–23] and supervised hashing methods [24–30]. For example, spectral hashing [19] and Principal Component Analysis (PCA) based hashing methods [20] belong to the unsupervised category, which does not utilize the label information of training images when learning the binary codes. Supervised hashing approaches, such as kernel-based supervised hashing [25], supervised discrete hashing [27] and deep hashing methods [29], incorporate the label information to learn semantic hashing functions.

Due to the great success of hashing in the field of natural image retrieval, many efforts have been devoted to develop efficient hashing methods for large-scale RS images retrieval tasks recently. More specifically, kernel-based nonlinear hashing was first introduced into the remote sensing community by Demir and Bruzzone [31]. Then, Li and Ren [32] proposed a novel unsupervised hashing method named partial randomness hashing (PRH) for efficient hash function construction. In [33], a novel large-scale RS image retrieval approach was proposed based on deep hashing neural networks under the supervision of labeled images. Ye et al. [34] proposed a multiple-feature learning framework for large-scale RS image hashing problem, which takes multiple complementary features as the input and learns the hybrid hash functions.

Although the hashing-based RS image retrieval methods have achieved some improvements for large-scale applications, there exist two important problems that are rarely exploited in the existing RS image hashing approaches. (1) The existing RS image hashing methods are based on a batch learning fashion, which assume all training images are available in advance for training and the hash functions keep unchanged once the learning procedure finished. However, in many real-world RS applications, the RS images become available continuously in streaming fashion. For example, the satellite transmits remote sensing images back to the data center every day. In such environments, the RS image database is enriched by time and the new incoming images may have different distribution with the existing images or even belong to a totally new category that has never been seen before. Thus, for the batch-based hashing methods, the pre-learned hash functions may be inappropriate for the new RS images over time. One solution is to accumulate all the available data and repeatedly do batch learning to re-train new hash functions, which is a quite inefficient learning manner, especially for time-consuming hashing methods. (2) The batch-based hashing methods usually have to load all the training RS images into the memory for hash function learning. Thus, these methods make very high demands on the computing hardware such as CPU and memory, which limits their practical application on many mobile remote sensing devices. In addition, for many real large-scale RS image databases, it is even impossible to load the whole training dataset into memory, let alone training hash model. Therefore, batch-based hashing on large-scale data often results in a great deal of computational time and memory cost, which does not satisfy the requirement of the real-world applications.

To overcome the above problems, we propose a novel online hashing method for fast and scalable RS image retrieval in this paper. Online learning approaches are quite efficient for streaming data modeling [35–37]. Specifically, we first formulate our hash model based on a partial random auto-encoder and then develop a novel online hash function learning scheme to continuously update the hash model such that it fits the sequentially arriving new images over time. Our online hashing method only employs the new RS images to optimize the hash functions at each learning round and do not need to revisit all the available data, which has greatly reduced the demands on computing and memory costs. Even for the oversized RS image database that is difficult to handle using batch hashing methods, one can divide the whole big dataset into many small chunks and then implement binary code learning through our proposed online hashing method. As a result, our proposed method is very suitable and efficient for scalable RS image retrieval tasks. The main contributions of this paper are summarized as follows:

(1) A novel online hashing method is developed for scalable RS image retrieval problem. To the best of our knowledge, our work is the first attempt to exploit online hash function learning in the large-scale remote sensing image retrieval literature.
(2) By learning the hash functions in an online manner, the parameters of our hash model can be updated continuously according to the new obtained RS images by time, which in contrast is one main drawback of the existing batch hashing methods.
(3) The proposed online hashing approach reduces the computing complexity and memory cost in the learning process compared with batch hashing methods. Experimental results show the superiority of our online hashing for scalable RS image retrieval tasks.

The rest of the paper is organized as follows. In Section 2, the proposed online RS image hashing method is described in detail. Extensive experiments are conducted in Section 3 to evaluate the performance of our proposed method as well as other compared approaches. Finally, conclusions are given in Section 4.

2. The Proposed Approach

Our proposed hashing approach contains two main steps: (1) hash model formulation, which defines the form of hash model used in the paper; and (2) online hash function learning, which describes how to update the hash functions dynamically based on the sequentially arriving data. The illustration of the proposed online hashing approach for scalable RS image retrieval is shown in Figure 1.

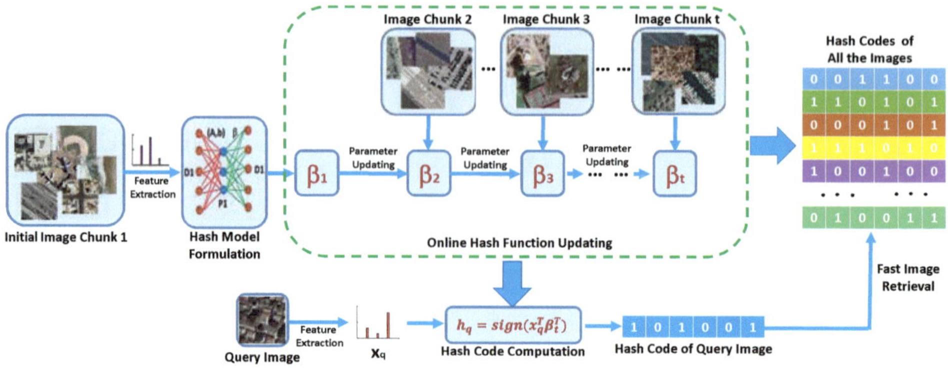

Figure 1. The illustration of the proposed online hashing approach for scalable remote sensing image retrieval.

2.1. Hash Model Formulation

Suppose that the RS image dataset used for training contains n images. Specifically, $\boldsymbol{x}_i \in \mathbb{R}^d$ is a d-dimensional feature vector for the i-th image and the feature vectors for all n images are $\{x_1, x_2, \cdots x_n\}$. Denote $\boldsymbol{X} = [x_1, x_2, \cdots, x_n] \in \mathbb{R}^{d \times n}$ as the whole data matrix. The corresponding binary code matrix for the dataset is $\boldsymbol{H} = [\boldsymbol{h}_1, \boldsymbol{h}_2, \cdots, \boldsymbol{h}_n] \in \{-1, 1\}^{r \times n}$, where r is the code length. The hash code vector for the i-th image is a column of $\boldsymbol{H}$ and is denoted as $\boldsymbol{h}_i$. The goal of hashing is to learn hash functions that encode the original RS images from the d-dimensional feature space into an r-dimensional hamming space.

Our hash model is formulated by a partial random auto-encoder which includes both forward and backward parameters. First, the whole data matrix is randomly projected from the d-dimensional feature space to an r-dimensional relaxed hamming space with sigmoid activation function as follows:

$$\boldsymbol{P} = g(\boldsymbol{X}^T \cdot \boldsymbol{A} + \mathbf{1}_n \boldsymbol{b}) \tag{1}$$

where $\boldsymbol{A} \in \mathbb{R}^{d \times r}$ is a randomly generated projection matrix and $\boldsymbol{b} \in \mathbb{R}^r$ is a randomly generated bias row vector. $g(x) = 1/(1 + e^{-x})$ is the sigmoid activation function and $\mathbf{1}_n$ denotes the $n \times 1$ column vector in which all the elements are equal to 1. $\boldsymbol{P} \in \mathbb{R}^{n \times r}$ is the projected data matrix. This is the forward procedure, whose parameters are randomly generated.

Then, a linear model parameter $\boldsymbol{\beta}$ is employed to fit randomly projected data $\boldsymbol{P}$ back to the original data $\boldsymbol{X}$ and $\boldsymbol{\beta}$ is learned by minimizing the following problem:

$$\hat{\boldsymbol{\beta}} = \arg\min_{\boldsymbol{\beta}} ||\boldsymbol{P} \cdot \boldsymbol{\beta} - \boldsymbol{X}^T||^2 \tag{2}$$

The optimal linear model parameter can be simply computed as follows:

$$\hat{\boldsymbol{\beta}} = \boldsymbol{P}^{\dagger} \boldsymbol{X}^T \tag{3}$$

where the superscript $\dagger$ denotes the Moore–Penrose generalized inverse of a matrix. $\boldsymbol{P}^{\dagger}$ can be given by $\boldsymbol{P}^{\dagger} = (\boldsymbol{P}^T \boldsymbol{P})^{-1} \boldsymbol{P}^T$. This is the backward procedure, whose parameters are optimized based on the training images. Our hash model is inspired by extreme learning machine (ELM) approach [38], however supervised ELM computes forward to a target label matrix while our model computes backward to the original feature data $\boldsymbol{X}^T$. Therefore, our method is in fact an unsupervised data-dependent hashing scheme.

Finally, the hash codes $\boldsymbol{H}$ for all the images in the training database can be simply obtained by $\boldsymbol{H} = \text{sign}\left(\boldsymbol{X}^T \hat{\boldsymbol{\beta}}^T\right)$.

2.2. Online Hash Function Learning

It is easy to observe from Section 2.1 that the defined hash model is a batch-learning based hashing approach, in which all the training images are assumed to be available in advance and the hash model parameters keep fixed once the learning procedure is finished. However, as we have explained in Section 1, such hashing methods are not well adapted to the scalable streaming RS images, which is a common scenario in the real-world applications. For example, as more and more new RS images are available, the pre-learned hash functions may become unsuitable or even fail for hash code generation. Moreover, it is even impossible to load all the images into memory for learning when the training dataset is oversized. In this part, we introduce a novel online hashing scheme which can update the hash functions continuously so that it can fit the sequentially available RS images.

We assume that the new RS images are available in a stream form. Let $\boldsymbol{D}_i$ denote the data chunk received at round i, $i = \{1, 2, ...\}$. One highlight of online learning is that when learning new information at round t, the algorithm should not access the previously seen image data $\boldsymbol{D}_1, ..., \boldsymbol{D}_{t-1}$. Given a chunk of initial training set $\boldsymbol{D}_1$, we can compute its hash code as $\boldsymbol{H}_1 = \text{sign}\left(\boldsymbol{D}_1^T \hat{\boldsymbol{\beta}}_1{}^T\right)$, where $\hat{\boldsymbol{\beta}}_1$ is obtained by the partial random hash model based on Equation (3) as $\hat{\boldsymbol{\beta}}_1 = \boldsymbol{Q}_1^{-1}\boldsymbol{P}_1^T\boldsymbol{D}_1^T$ where $\boldsymbol{Q}_1 = \boldsymbol{P}_1^T\boldsymbol{P}_1$, and $\boldsymbol{P}_1$ is obtained based on Equation (1) as $\boldsymbol{P}_1 = g(\boldsymbol{D}_1^T \cdot \boldsymbol{A} + \mathbf{1}_n\boldsymbol{b})$

Suppose that we are given another chunk of data $\boldsymbol{D}_2$, the proposed method becomes minimizing the following problem if considering both image datasets $\boldsymbol{D}_1$ and $\boldsymbol{D}_2$:

$$\hat{\boldsymbol{\beta}}_2 = \arg\min_{\boldsymbol{\beta}_2} \left\| \begin{bmatrix} \boldsymbol{P}_1 \\ \boldsymbol{P}_2 \end{bmatrix} \boldsymbol{\beta}_2 - \begin{bmatrix} \boldsymbol{D}_1^T \\ \boldsymbol{D}_2^T \end{bmatrix} \right\|^2 \tag{4}$$

where the optimized $\hat{\boldsymbol{\beta}}_2$ can be given by

$$\hat{\boldsymbol{\beta}}_2 = \left(\begin{bmatrix} \boldsymbol{P}_1 \\ \boldsymbol{P}_2 \end{bmatrix}^T \begin{bmatrix} \boldsymbol{P}_1 \\ \boldsymbol{P}_2 \end{bmatrix} \right)^{-1} \begin{bmatrix} \boldsymbol{P}_1 \\ \boldsymbol{P}_2 \end{bmatrix}^T \begin{bmatrix} \boldsymbol{D}_1^T \\ \boldsymbol{D}_2^T \end{bmatrix} \tag{5}$$

If we let $\left(\begin{bmatrix} \boldsymbol{P}_1 \\ \boldsymbol{P}_2 \end{bmatrix}^T \begin{bmatrix} \boldsymbol{P}_1 \\ \boldsymbol{P}_2 \end{bmatrix} \right)$ be denoted by $\boldsymbol{Q}_2$, then

$$\boldsymbol{Q}_2 = \boldsymbol{P}_1^T\boldsymbol{P}_1 + \boldsymbol{P}_2^T\boldsymbol{P}_2 = \boldsymbol{Q}_1 + \boldsymbol{P}_2^T\boldsymbol{P}_2 \tag{6}$$

and $\hat{\boldsymbol{\beta}}_2$ can be rewritten as

$$\begin{aligned} \hat{\boldsymbol{\beta}}_2 &= \boldsymbol{Q}_2^{-1}(\boldsymbol{P}_1^T\boldsymbol{D}_1^T + \boldsymbol{P}_2^T\boldsymbol{D}_2^T) \\ &= \boldsymbol{Q}_2^{-1}(\boldsymbol{Q}_1\boldsymbol{Q}_1^{-1}\boldsymbol{P}_1^T\boldsymbol{D}_1^T + \boldsymbol{P}_2^T\boldsymbol{D}_2^T) \\ &= \boldsymbol{Q}_2^{-1}(\boldsymbol{Q}_1\hat{\boldsymbol{\beta}}_1 + \boldsymbol{P}_2^T\boldsymbol{D}_2^T) \\ &= \boldsymbol{Q}_2^{-1}[(\boldsymbol{Q}_2 - \boldsymbol{P}_2^T\boldsymbol{P}_2)\hat{\boldsymbol{\beta}}_1 + \boldsymbol{P}_2^T\boldsymbol{D}_2^T)] \\ &= \hat{\boldsymbol{\beta}}_1 + \boldsymbol{Q}_2^{-1}\boldsymbol{P}_2^T(\boldsymbol{D}_2^T - \boldsymbol{P}_2\hat{\boldsymbol{\beta}}_1) \end{aligned} \tag{7}$$

From Equations (6) and (7), we can see that $\hat{\boldsymbol{\beta}}_2$ can be expressed as a function of $\hat{\boldsymbol{\beta}}_1$ based on the new data chunk $\boldsymbol{D}_2$.

Without loss of generality, we can easily get a recursive form for the streaming data chunk $\boldsymbol{D}_i$ as new images arrive. When the k-th chunk of image set is received, we have

$$Q_k = Q_{k-1} + \boldsymbol{P}_k^T \boldsymbol{P}_k \tag{8}$$

$$\hat{\boldsymbol{\beta}}_{\boldsymbol{k}} = \hat{\boldsymbol{\beta}}_{\boldsymbol{k-1}} + Q_k^{-1} \boldsymbol{P}_k^T (\boldsymbol{D}_k^T - P_k \hat{\boldsymbol{\beta}}_{\boldsymbol{k-1}}) \tag{9}$$

By recursively applying (8) and (9), the hash model parameter $\hat{\boldsymbol{\beta}}$ is updated with respect to the new available RS images and the learned hash codes for all the images are also improved continuously with the streaming data. More importantly, we only have to handle the current data chunk without needing to access the whole image set at each round. Therefore, our method is less constrained by the computational and space cost limitation compared with the batch hashing approaches.

The learning process of the proposed online partial randomness hashing (OPRH) method is summarized in Algorithm 1.

Algorithm 1 Online Binary Code Learning with OPRH

1: **Input:** Streaming image data chunk $\boldsymbol{D}_1, \boldsymbol{D}_2, ..., \boldsymbol{D}_k$, code length r
2: **Output:** Hash codes $\boldsymbol{H}$ for all the images
3: Randomly generate a projection matrix $\boldsymbol{A} \in \mathbb{R}^{d \times r}$ and a bias row vector $\boldsymbol{b} \in \mathbb{R}^r$
4: Compute $\boldsymbol{P}_1$ by $\boldsymbol{P}_1 = g(\boldsymbol{D}_1^T \cdot \boldsymbol{A} + \boldsymbol{1}_n \boldsymbol{b})$
5: Compute $Q_1 = \boldsymbol{P}_1^T \boldsymbol{P}_1$ and $\hat{\boldsymbol{\beta}}_{\boldsymbol{1}} = Q_1^{-1} \boldsymbol{P}_1^T \boldsymbol{D}_1^T$
6: **for** $i = 2 : k$ **do**
7:     Compute $\boldsymbol{P}_i$ by $\boldsymbol{P}_i = g(\boldsymbol{D}_i^T \cdot \boldsymbol{A} + \boldsymbol{1}_n \boldsymbol{b})$
8:     Update Q_i with Equation (8)
9:     Update $\hat{\boldsymbol{\beta}}_{\boldsymbol{i}}$ with Equation (9)
10: **end for**
11: Compute the hash codes $\boldsymbol{H}$ for the whole database $\mathbf{X} = [\boldsymbol{D}_1, \boldsymbol{D}_2, ..., \boldsymbol{D}_k]$ by $\boldsymbol{H} = \mathrm{sign}\left(\boldsymbol{X}^T \hat{\boldsymbol{\beta}}_k^T\right)$

2.3. Complexity Analysis

We analyze the complexity of our proposed online partial randomness hashing method. Specifically, for a stream of data chunk $\boldsymbol{D}_1, \boldsymbol{D}_2, ..., \boldsymbol{D}_t$, we update the hash model parameters at every round $i = 1, 2, ..., t$. We analyze both the time and space complexity for hash function learning at each round.

Time Complexity: The time complexity of computing $\boldsymbol{P}_k$ at each round is $O(n_k dr)$, where n_k is the number of images in the k-th chunk, d is the dimensionality of the original feature vectors and r is the length for hash codes. The complexity of updating Q_k and $\hat{\boldsymbol{\beta}}_k$ can be $O(n_k r^2)$ and $O(n_k r^3 + n_k dr)$, respectively. Thus, the overall time complexity for each round is $O(n_k r^3 + n_k dr)$.

Space Complexity: In our method, all the operations at each round are conducted on a data chunk $\boldsymbol{D}_k$ without accessing the whole dataset, space overhead of which is $O(n_k d)$. The Q_k and $\hat{\boldsymbol{\beta}}_k$ updating steps occupy a space of $O(n_k r + r^2)$ and $O(dr)$ space is needed to store the final learned hash model parameter $\hat{\boldsymbol{\beta}}_t$. Thus, the overall space complexity is $O(n_k d + r^2 + dr)$.

In the light of the above observations, our OPRH approach is quite suitable for scalable RS image hashing and fast retrieval because the operated data chunk at each round is much smaller than the whole large dataset. Especially when the RS image set is oversized and impossible to be loaded into the memory, we can divide the whole image set into many small chunks and employ our OPRH method for binary code learning, which can be easily finished even on a ordinary computer.

3. Experiments

3.1. Datasets and Settings

In this section, we conduct extensive experiments to evaluate the performance of our proposed OPRH. Two issues are verified in the following experiments: (1) large-scale RS image retrieval performance of our method compared to state-of-the-art batch-based hashing algorithms; and (2) the effectiveness and efficiency of the proposed OPRH method under online setting.

Two public large-scale satellite datasets are used in the experiments, i.e., SAT-4 and SAT-6 airborne datasets [39], which contain 500,000 and 405,000 images, respectively. SAT-4 dataset contains four classes and SAT-6 contains six classes. All the images in these two datasets are normalized to 28×28 pixels in size. Some example images from the two datasets are shown in Figure 2. One thousand images are randomly selected from each dataset as testing queries and the remaining images are used for training and retrieval database. We extract a 512-dimensional GIST descriptor [40] for each image as visual feature representation. Given an input image, a GIST descriptor is computed as follows: (a) convolve the image with 32 Gabor filters at 4 scales and 8 orientations, producing 32 feature maps of the same size of the input image; (b) divide each feature map into 16 regions (by a 4×4 grid), and then average the feature values within each region; and (c) concatenate the 16 averaged values of all 32 feature maps, resulting in a $16 \times 32 = 512$-dimensional GIST descriptor. GIST summarizes the gradient information (scales and orientations) for different parts of an image, which provides a rough description of the scene.

We compare our approach with both batch-based hashing methods and online hashing methods. The batch-based hashing methods include two recent RS image hashing methods, Partial Randomness Hashing (PRH) [32] and Kernel Unsupervised Locality Sensitive Hashing (KULSH) [31], and four hashing approaches, Inductive Hashing on Manifolds (IMH) [23], Isotropic Hashing (IsoHash) [22], Iterative Quantization (ITQ) [20], and Spherical Hashing (SpH) [21], used in computer vision. The compared two online hashing methods are Online Kernel-based Hashing (OKH) [35] and Online Sketch Hashing (OSH) [36], which are used in the natural image processing literature, because our proposed approach is the first online hashing method for large-scale RS image retrieval. For the batch-based hashing methods, all the training images are used to learn the hash functions. For the online hashing methods, we randomly divide the whole training set into 1000 different chunks to simulate the online condition and the hash functions are updated in a streaming way.

To perform fair evaluations, we adopt the hamming ranking search commonly used in the literature. All the images in the database are ranked according to their hamming distance to the query and the desired neighbors are returned from the top of the ranked list. The retrieval performance is measured with average precision of the top K returned examples and the overall precision–recall curves. More specifically, precision and recall are defined as follows:

$$precision = \frac{true\ positive}{true\ positive + false\ positive} \tag{10}$$

$$recall = \frac{true\ positive}{true\ positive + false\ negative} \tag{11}$$

According to Equation (10), we get precision of the top K returned examples for a query image if the correctly predicted samples divided by K. The average precision is obtained by averaging the precision scores over all the test queries.

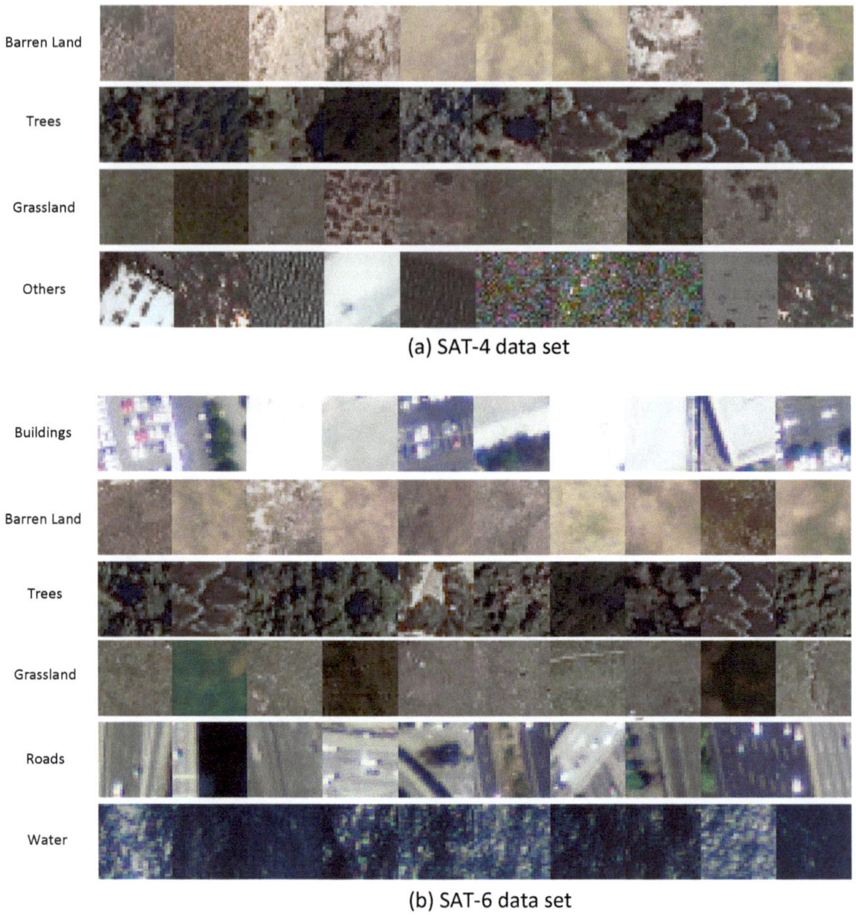

Figure 2. Some sample images from: (**a**) SAT-4 dataset; and (**b**) SAT-6 dataset.

3.2. Results and Analysis

Tables 1 and 2 show the average precision of the Top-10 and Top-100 retrieved image samples by different hashing methods on the two datasets. We can observe that the ITQ and PRH methods achieve relative better results among the batch-based hashing methods under varied hash bits. For the online hashing methods, the proposed OPRH achieves better results compared with the competitors in most cases. By comparing our OPRH method with the batch-based hashing methods, we can find that our OPRH obtains comparable performance to the batch methods on SAT-4 dataset while sometimes achieves even better results than all of the other compared approaches on SAT-6 dataset, which has indicated the effectiveness of the proposed online hashing method. The performance gain our OPRH approach may be attributed to the backward learning procedure, which helps learn more accurate projection parameter to enhance the representational ability of hash codes.

The average precision with respect to different retrieved samples and the precision-recall curves of compared hashing methods on the two datasets are shown in Figure 3. Since too many cures will be overlapped and hard to distinguish, we only keep the online hashing methods and the batch-based RS hashing methods in the figure. In Figure 3a–c,g–i, we can observe that our OPRH method consistently outperforms OSH and OKH methods when the retrieved images increase and the improvements are more notable for long code length. The reason may be that OSH and OKH have large quantization error when generating binary codes while our OPRH can reduce the error in code binarization through the backward decoder learning procedure. Precision–recall curve reflects the overall image retrieval performance of different hashing approaches. In Figure 3d–f,j–l, we also find that our OPRH method achieves the best results among the compared online hashing methods. The proposed OPRH method has comparable overall performance to batch-based PRH method and much better than KULSH approach on the two datasets.

Table 1. The comparison of mean precision of the top *K* returned examples for different methods on the SAT-4 dataset with varied hash bits.

Methods	Top-10			Top-100		
	32-bits	48-bits	64-bits	32-bits	48-bits	64-bits
IMH	0.560	0.538	0.548	0.550	0.524	0.541
IsoHash	0.606	0.640	0.655	0.576	0.594	0.597
ITQ	0.636	0.653	0.662	0.609	0.607	0.610
SpH	0.596	0.623	0.658	0.563	0.588	0.607
KULSH	0.492	0.507	0.553	0.476	0.479	0.526
PRH	0.607	0.621	0.665	0.592	0.595	0.622
OKH	0.439	0.516	0.600	0.418	0.480	0.561
OSH	0.603	0.637	0.647	0.568	0.596	0.596
OPRH	0.608	0.630	0.656	0.598	0.594	0.616

Table 2. The comparison of mean precision of the top *K* returned examples for different methods on the SAT-6 dataset with varied hash bits.

Methods	Top-10			Top-100		
	32-bits	48-bits	64-bits	32-bits	48-bits	64-bits
IMH	0.583	0.626	0.604	0.575	0.614	0.582
IsoHash	0.667	0.680	0.673	0.635	0.645	0.642
ITQ	0.672	0.691	0.681	0.649	0.660	0.653
SpH	0.642	0.664	0.694	0.616	0.631	0.657
KULSH	0.413	0.459	0.452	0.418	0.496	0.520
PRH	0.651	0.682	0.683	0.629	0.658	0.652
OKH	0.541	0.619	0.638	0.521	0.592	0.617
OSH	0.669	0.684	0.680	0.639	0.650	0.647
OPRH	0.645	0.699	0.705	0.631	0.672	0.677

To explicitly compare the online hash function updating process at each round for the online hashing methods, we compute the the average retrieval precision of different methods after each round and show it in Figure 4. It is obvious that the proposed OPRH method outperforms both OKH and OSH approaches on the two datasets. Moreover, OKH has big fluctuations during the online learning process and its performance even deteriorates as the number of received chunks increases on SAT-4 dataset, while our proposed OPRH achieves quite stable improvement when more and more new image chunks are available for training. To show the online updating process of our approach more intuitively, we give an visual example for image retrieval in Figure 5, which shows the first returned 16 samples to the query image by our method after different learning rounds. We can see that the retrieval results become more and more accurate as the learning round increases. The reason is that the learned hash functions improve continuously as new training images are obtained and thus the generated hash codes also become more accurate. This also indicates that our proposed online hashing method can fit the new available streaming data very well, which is the shortcoming of batch-based hashing methods in contrary.

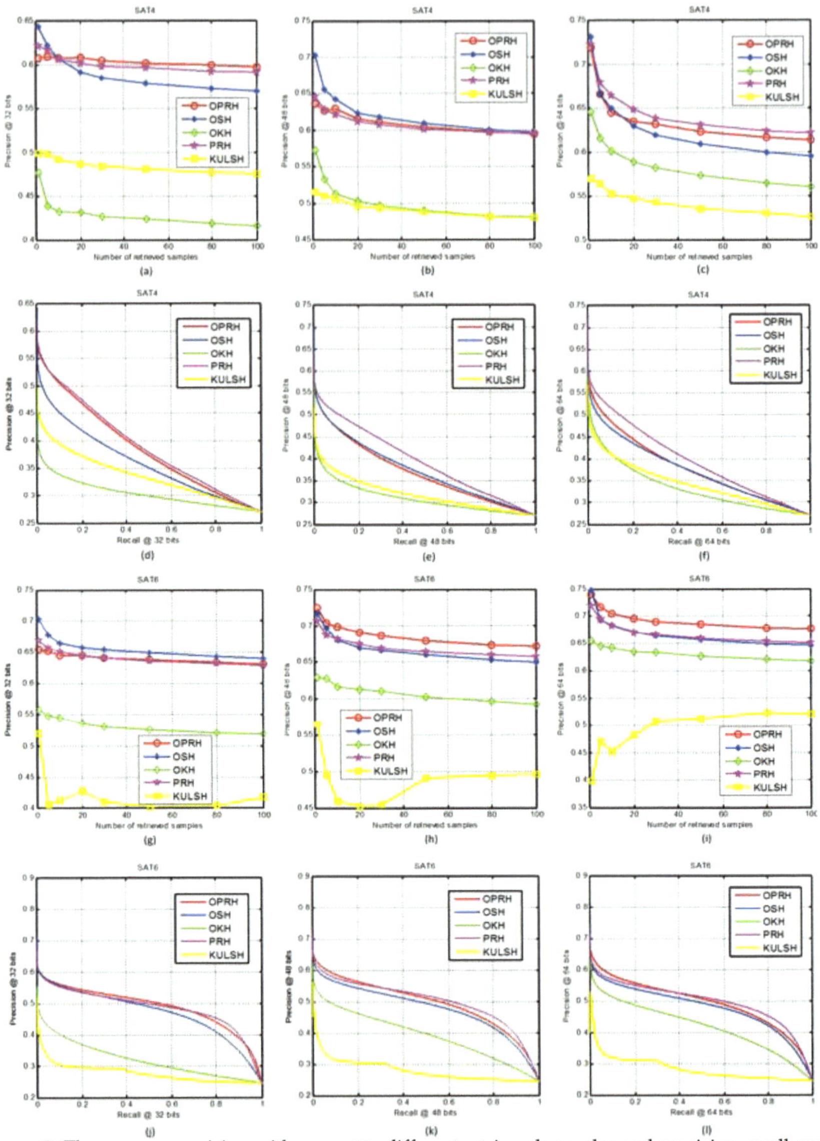

Figure 3. The average precision with respect to different retrieved samples and precision-recall curves for the compared methods on the two datasets: (**a–f**) SAT-4; and (**g–l**) SAT-6.

We also compare the learning efficiency of different hashing methods, which is shown in Table 3. All experiments are implemented with MATLAB code and run on a PC with Intel Core-i5 2.3 GHz CPU, 8 GB RAM. For the batch-based hashing methods, we report their total time on the whole training image set and for the online hashing methods, we show both their average updating time at each round and the accumulated time of total rounds. Among the batch-based methods, PRH and IsoHash are much more efficient than other methods. Among the online hashing methods, our OPRH approach has the fastest updating time at each round and more than 10 times faster than the compared OSH method. The accumulated time of total 1000 rounds of our OPRH is still comparable to the PRH method. For memory cost, the online hashing approaches are much lower than the batch-based hashing methods. This is easy to explain because the online hashing methods only have to handle a small image chunk at each learning round while the batch-based methods have to load all the images into the memory for training. More specifically, the PRH algorithm occupies about 1.2 GB RAM to store the data and parameters in the learning process on SAT-6 dataset with 64-bits in our experiments while only 1.8 MB RAM is needed for our OPRH method. SAT-6 dataset only has 405,000 images. Imagine that, if we are given a RS image dataset consisting of a million or billion images, which is impossible to be loaded into the memory for training, the batch-based hashing methods would not work. However, our proposed online hashing method is still able to do hash function learning by segmenting the whole database into many small chunks. Therefore, the proposed OPRH method is quite suitable for hash code learning and fast image retrieval on oversized RS image sets, which is expected in real-world applications.

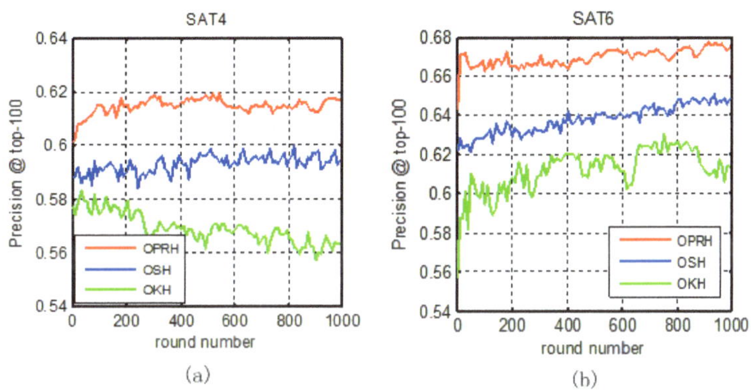

Figure 4. Comparison of average precision at each round of the online hashing methods on: (**a**) SAT-4 dataset; and (**b**) SAT-6 dataset (64-bits).

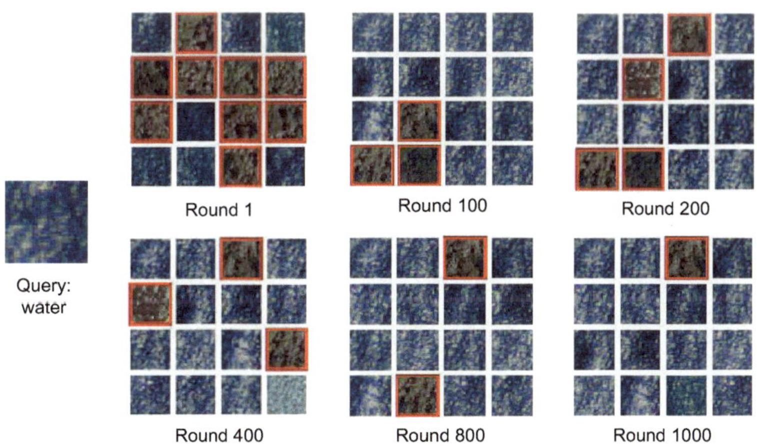

Figure 5. Visualized retrieval example after different rounds by our OPRH method on SAT-6 dataset with 64-bits. Top-16 returned image patches for the query are shown for each round and the false positives are annotated with a red rectangle.

Table 3. The comparison of training time (in seconds) and memory cost (MB) for different kinds of hashing methods.

Methods	SAT-4 Dataset			SAT-6 Dataset		
	Round Time	Total Time	Memory Cost	Round Time	Total Time	Memory Cost
IMH	-	67.6	3696	-	67.7	2990
IsoHash	-	5.5	4915	-	5.8	3942
ITQ	-	47.9	5857	-	61.1	5529
SpH	-	196.3	5109	-	200	4177
KULSH	-	10.3	3901	-	8.2	3143
PRH	-	4.6	1556	-	5.0	1198
OKH	0.32	315.8	10.4	0.27	267	8
OSH	0.11	113.5	4.4	0.11	105.4	3.5
OPRH	0.01	12	2.3	0.009	8.7	1.8

To evaluate the large-scale RS image retrieval performance of our proposed hashing approach and direct linear search strategy, we conduct image retrieval experiments with our OPRH method and ℓ_2 linear scan. For our OPRH method, image retrieval is carried out with learned binary codes in the hamming space. ℓ_2 linear scan directly does image retrieval in the original feature space based on the Euclidean distance of feature vectors. Besides the GIST descriptor used in the previous experiments, CNN feature is also adopted to evaluate the generalizing ability of our OPRH method. We choose AlexNet as the CNN feature extraction model and the output 4096-dimensional feature of the fully connected layer fc7 is extracted for each image. PCA is applied to reduce the dimensionality to 1024 and form the final feature vector for the images. The comparison of average search time per image and mean precision of Top-100 retrieved samples is shown in Table 4. From the results, we can find that, when using CNN feature instead of GIST descriptor, the average retrieval precision can be improved by 30–40% on the two datasets. This is attributed to the powerful representation ability of CNN feature, which is able to learn more high-level semantic information. For different image search strategies, the direct search in the original image feature space can obtain higher accuracy than hashing-based search methods in most cases. However, by sacrificing a little accuracy, the hashing approaches can obtain much faster search speed than the traditional direct search method. For example, compared with direct search in the CNN feature space, OPRH + CNN achieves more than 60 times speed acceleration with only 1% drop in the retrieval accuracy on the SAT-6 dataset. The reason is that our OPRH approach conducts image retrieval based on binary codes and the hamming distance between different codes can be efficiently calculated with XOR operation, which is much more faster that the computation of Euclidean distance in the feature space.

Table 4. The comparison of average search time (in seconds) and accuracy (mean precision of Top-100 retrieved samples) between our proposed hashing method in the hamming space (with 64-bits) and ℓ_2 linear scan in the original feature space based on different feature representations.

	GIST ℓ_2 Scan		CNN ℓ_2 Scan		OPRH+GIST		OPRH+CNN	
	Time	Precision@100	Time	Precision@100	Time	Precision@100	Time	Precision@100
SAT-4	1.93	0.60	4.01	1	0.06	0.61	0.06	0.98
SAT-6	1.67	0.69	3.15	0.98	0.05	0.67	0.05	0.97

Finally, to demonstrate the superiority of the proposed hashing approach for real-world large-scale remote sensing image retrieval, we generate a synthetic dataset consisting of 100 million samples of 1000 dimension. Due to the size of the synthetic dataset, it exceeds the processing ability of traditional batch-based hashing approaches and brute force linear search schemes. However, by dividing the dataset into one million small chunks, our OPRH hashing approach only has to handle 100 samples at each round and can finish hash model training in 33 min on our ordinary PC. With learned binary codes of 64-bits, fast image retrieval from 100 million samples can be carried out at the speed of 5 s per image. These results demonstrate that the proposed OPRH is scalable to massive streaming remote sensing data even on a common computer.

4. Conclusions

In this paper, we have proposed a novel online hashing method, named online partial randomness hashing (OPRH), for retrieving scalable remote sensing image databases. Benefiting from the online learning scheme, the hash model parameters can be updated continuously according to the streaming image data, which is a common scenario in the real-world applications. Therefore, the hash codes learned by our approach have better generalization ability compared with the batch-based hashing approaches. More importantly, the batch-based hashing methods will face difficulties when handling very large database due to the high complexity and space limitation while the proposed method can be easily applied to oversized dataset by dividing it into several small chunks. Thus, our OPRH method is very suitable for large-scale remote sensing image retrieval. Extensive experiments on two public large-scale satellite datasets have demonstrated the effectiveness and efficiency of our approach.

Our proposed online hashing method can be used in many real-time remote sensing applications due to its adapting ability to variations in datasets as they grow and diversify. For example, on-orbit processing of satellite remote sensing images can be conducted through our online hashing method to improve the efficiency of information processing. Real-time retrieval from huge historical satellite cloud pictures with our proposed approach can provide the forecaster more effective information for short-term weather forecasting. At the same time, there are also some challenging issues that need to be addressed for our proposed online hashing approach. The hash functions of our approach are updated gradually according to the changing database, but the updating frequency needs to be well decided in real-world applications. Too high frequency is time-consuming while low updating frequency may lead to unsatisfactory retrieval results. In addition, the hash code indexing must also be frequently updated when hash functions change. This may cause inefficiencies in the real-world systems. Therefore, solutions must be simultaneously developed to alleviate this particular problem in future work.

Author Contributions: P.L. and P.R. conceived and designed the experiments; X.Zhang performed the experiments; X.Zhu analyzed the data; and P.L. wrote the paper. All authors read and approved the final manuscript.

Acknowledgments: This work was supported in part by the National Natural Science Foundation of China (Grant 61602517 and 61501475), the National Key R&D Program of China (Grant SQ2017YFB140187), Qingdao Applied Fundamental Research (Grant 16-5-1-11-jch), the Open Project Program of the State Key Laboratory of Mathematical Engineering and Advanced Computing (Grant 2017A05), the Open Project Program of National Laboratory of Pattern Recognition (Grant 201800018), and the Fundamental Research Funds for Central Universities (Grant 18CX02110A).

Conflicts of Interest: The authors declare no conflict of interest.

References

1. Du, R.; Chen, Y.; Tang, H.; Fang, T. Study on content-based remote sensing image retrieval. In Proceedings of the IEEE International Geoscience and Remote Sensing Symposium (IGARSS), Seoul, South Korea, 25–29 July 2005; pp. 707–710.
2. Wang, Q.; Zhu, G.; Yuan, Y. Statistical quantization for similarity search. *Comput. Vis. Image Underst.* **2014**, *124*, 22–30. [CrossRef]
3. Yang, J.; Liu, J.; Dai, Q. An improved Bag-of-Words framework for remote sensing image retrieval in large-scale image databases. *Int. J. Digit. Earth* **2015**, *8*, 273–292. [CrossRef]
4. Sevilla, J.; Bernabe, S.; Plaza, A. Unmixing-based content retrieval system for remotely sensed hyperspectral imagery on GPUs. *J. Supercomput.* **2014**, *70*, 588–599. [CrossRef]

5. Yang, Y.; Newsam, S. Geographic image retrieval using local invariant features. *IEEE Trans. Geosci. Remote Sens.* **2013**, *51*, 818–832. [CrossRef]
6. Aptoula, E. Remote sensing image retrieval with global morphological texture descriptors. *IEEE Trans. Geosci. Remote Sens.* **2014**, *52*, 3023–3034. [CrossRef]
7. Newsam, S.; Wang, L.; Bhagavathy, S.; Manjunath, B.S. Using texture to analyze and manage large collections of remote sensed image and video data. *Appl. Opt.* **2004**, *43*, 210–217. [CrossRef] [PubMed]
8. Luo, B.; Aujol, J.F.; Gousseau, Y.; Ladjal, S. Indexing of satellite images with different resolutions by wavelet features. *IEEE Trans. Image Process.* **2008**, *17*, 1465–1472. [PubMed]
9. Rosu, R.; Donias, M.; Bombrun, L.; Said, S.; Regniers, O.; Da Costa, J.-P. Structure tensor Riemannian statistical models for CBIR and classification of remote sensing images. *IEEE Trans. Geosci. Remote Sens.* **2017**, *55*, 248–260. [CrossRef]
10. Zhou, W.; Shao, Z.; Diao, C.; Cheng, Q. High-resolution remotesensing imagery retrieval using sparse features by auto-encoder. *Remote Sens. Lett.* **2015**, *6*, 775–783. [CrossRef]
11. Zhou, W.; Newsam, S.; Li, C.; Shao, Z. Learning low dimensional convolutional neural networks for high-resolution remote sensing image retrieval. *Remote Sens.* **2017**, *9*, 489. [CrossRef]
12. Du, Z.; Li, X.; Lu, X. Local structure learning in high resolution remote sensing image retrieval. *Neurocomputing* **2016**, *207*, 813–822. [CrossRef]
13. Wang, Q.; Wan, J.; Yuan, Y. Deep metric learning for crowdedness regression. *IEEE Trans. Trans. Circuits Syst.* **2017**. [CrossRef]
14. Li, Y.; Zhang, Y.; Tao, C.; Zhu, H. Content-based high-resolution remote sensing image retrieval via unsupervised feature learning and collaborative affinity metric fusion. *Remote Sens.* **2016**, *8*, 709. [CrossRef]
15. Wang, Y.; Zhang, L.; Tong, X.; Zhang, L.; Zhang, Z.; Liu, H.; Xing, X.; Takis Mathiopoulos, P. A three-layered graph-based learning approach for remote sensing image retrieval. *IEEE Trans. Geosci. Remote Sens.* **2016**, *54*, 6020–6034. [CrossRef]
16. Andoni, A.; Indyk, P. Near-optimal hashing algorithms for approximate nearest neighbor in high dimensions. In Proceedings of the 47th Annual IEEE Symposium on Foundations of Computer Science (FOCS), Berkeley, CA, USA, 21–24 October 2006; pp. 459–468.
17. Datar, M.; Immorlica, N.; Indyk, P.; Mirrokni, V. Locality-sensitive hashing scheme based on p-stable distributions. In Proceedings of the 20th Annual Symposium on Computational Geometry (SCG), Brooklyn, NY, USA, 8–11 June 2004; pp. 253–262.
18. Kulis, B.; Grauman, K. Kernelized locality-sensitive hashing for scalable image search. In Proceedings of the IEEE 12th International Conference on Computer Vision (ICCV), Kyoto, Japan, 27 September–4 October 2009; pp. 2130–2137.
19. Weiss, Y.; Torralba, A.B.; Fergus, R. Spectral hashing. In Proceedings of the Advances in Neural Information Processing Systems (NIPS), Vancouver, BC, Canada, 8–11 December 2008; pp. 1753–1760.
20. Gong, Y.; Lazebnik, S. Iterative quantization: A procrustean approach to learning binary codes for large-scale image retrieval. *IEEE Trans. Pattern Anal. Mach. Intell.* **2013**, *35*, 2916–2929. [CrossRef] [PubMed]
21. Heo, J.; Lee, Y.; He, J.; Chang, S.; Yoon, S. Spherical hashing: binary code embedding with hyperspheres. *IEEE Trans. Pattern Anal. Mach. Intell.* **2015**, *37*, 2304–2316. [CrossRef] [PubMed]
22. Kong, W.; Li, W. Isotropic hashing. In Proceedings of the Advances in Neural Information Processing Systems (NIPS), Lake Tahoe, NV, USA, 3–6 December 2012; pp. 1655–1663
23. Shen, F.; Shen, C.; Shi, Q.; Hengel, A.; Tang, Z. Inductive hashing on manifolds. In Proceedings of the IEEE Conference on Computer Vision and Pattern Recognition (CVPR), Portland, OR, USA, 23–28 June 2013; pp. 1562–1569.
24. Norouzi, M.; Fleet, D.J. Minimal loss hashing for compact binary codes. In Proceedings of the 28th International Conference on Machine Learning (ICML), Bellevue, WA, USA, 28 June–2 July 2011; pp. 353–360.
25. Liu, W.; Wang, J.; Ji, R.; Jiang, Y.; Chang, S. Supervised hashing with kernels. In Proceedings of the IEEE Conference on Computer Vision and Pattern Recognition (CVPR), Providence, RI, USA, 16-21 June 2012; pp. 2074–2081.
26. Lin, G.; Shen, C.; Shi, Q.; Hengel, A.; Suter, D. Fast supervised hashing with decision trees for high-dimensional data. In Proceedings of the IEEE Conference on Computer Vision and Pattern Recognition (CVPR), Columbus, OH, USA, 23–28 June 2014; pp. 1971–1978.

27. Shen, F.; Shen, C.; Liu, W.; Shen, H. Supervised discrete hashing. In Proceedings of the IEEE Conference on Computer Vision and Pattern Recognition (CVPR), Boston, MA, USA, 7–12 June 2015; pp. 37–45.
28. Xia, R.; Pan, Y.; Lai, H.; Liu, C.; Yan, S. Supervised hashing via image representation learning. In Proceedings of the Twenty-Eighth AAAI Conference on Artificial Intelligence (AAAI), Quebec City, QC, Canada, 27–31 July 2014; pp. 2156–2162.
29. Zhao, F.; Huang, Y.; Wang, L.; Tan, T. Deep semantic ranking based hashing for multi-label image retrieval. In Proceedings of the IEEE Conference on Computer Vision and Pattern Recognition (CVPR), Boston, MA, USA, 7–12 June 2015; pp. 1556–1564.
30. Li, W.; Wang, S.; Kang, W. Feature learning based deep supervised hashing with pairwise labels. In Proceedings of the Twenty-Fifth International Joint Conference on Artificial Intelligence (IJCAI), New York, NY, USA, 7–15 July 2016; pp. 1711–1717.
31. Demir, B.; Bruzzone, L. Hashing-based scalable remote sensing image search and retrieval in large archives. *IEEE Trans. Geosci. Remote Sens.* **2016**, *54*, 892–904. [CrossRef]
32. Li, P.; Ren, P. Partial randomness hashing for large-scale remote sensing image retrieval. *IEEE Geosci. Remote Sens. Lett.* **2017**, *14*, 464–468. [CrossRef]
33. Li, Y.; Zhang, Y.; Huang, X.; Zhu, H.; Ma, J. Large-scale remote sensing image retrieval by deep hashing neural networks. *IEEE Trans. Geosci. Remote Sens.* **2018**, *56*, 950–965. [CrossRef]
34. Ye, D.; Li, Y.; Tao, C.; Xie, X.; Wang, X. Multiple feature hashing learning for large-scale remote sensing image retrieval. *ISPRS Int. J. Geo-Inform.* **2017**, *6*, 364. [CrossRef]
35. Huang, L.; Yang, Q.; Zheng, W. Online hashing. In Proceedings of the International Joint Conference on Artificial Intelligence (IJCAI), Beijing, China, 3–9 August 2013; pp. 1422–1428.
36. Leng, C.; Wu, J.; Cheng, J.; Bai, X.; Lu, H. Online sketching hashing. In Proceedings of the IEEE Conference on Computer Vision and Pattern Recognition (CVPR), Boston, MA, USA, 7–12 June 2015; pp. 2503–2511.
37. Liang, N.; Huang, G.; Saratchandran, P.; Sundararajan, N. A fast and accurate online sequential learning algorithm for feedforward networks. *IEEE Trans. Neural Netw.* **2006**, *17*, 1411–1423. [CrossRef] [PubMed]
38. Huang, G.; Zhu, Q.; Siew, C. Extreme learning machine: Theory and applications. *Neurocomputing* **2006**, *70*, 489–501. [CrossRef]
39. Basu, S.; Ganguly, S.; Mukhopadhyay, S.; Dibiano, R.; Karki, M.; Nemani, R. DeepSat: A learning framework for satellite imagery. In Proceedings of the SIGSPATIAL International Conference on Advances in Geographic Information Systems (SIGSPATIAL/GIS), Bellevue, WA, USA, 3–6 November 2015; pp. 37:1–37:10.
40. Oliva, A.; Torralba, A. Modeling the shape of the sence: A holistic representation of spatial envelope. *Int. J. Comput. Vis.* **2001**, *42*, 145–175. [CrossRef]

remote sensing

Article

Comparative Analysis of Responses of Land Surface Temperature to Long-Term Land Use/Cover Changes between a Coastal and Inland City: A Case of Freetown and Bo Town in Sierra Leone

Musa Tarawally [1], Wenbo Xu [1,*], Weiming Hou [2,*] and Terence Darlington Mushore [3,4]

1 School of Resources and Environment, University of Electronic Science and Technology of China, Chengdu 611731, China; musatarawally28@yahoo.com

2 School of Inoformation Science and Engineering, Hebei University of Science and Technology, Shijiazhuang 050000, China

3 Department of Physics, University of Zimbabwe, P.O. Box MP197, Mount Pleasant, Harare 263, Zimbabwe; tdmushore@gmail.com

4 Discipline of Geography, School of Agricultural, Earth and Environmental Sciences, University of KwaZulu-Natal, P/Bag X01, Scottsville, Pietermaritzburg 3209, South Africa

* Correspondence: xuwenbo@uestc.edu.cn (W.X.); hwm@hebust.edu.cn (W.H.); Tel.: +86-028-6183-0279 (W.X.); +86-311-8166-8796 (W.H.)

Received: 24 November 2017; Accepted: 12 January 2018; Published: 15 January 2018

Abstract: Urban growth and its associated expansion of built-up areas are expected to continue through to the twenty second century and at a faster pace in developing countries. This has the potential to increase thermal discomfort and heat-related distress. There is thus a need to monitor growth patterns, especially in resource constrained countries such as Africa, where few studies have so far been conducted. In view of this, this study compares urban growth and temperature response patterns in Freetown and Bo town in Sierra Leone. Multispectral Landsat images obtained in 1998, 2000, 2007, and 2015 are used to quantify growth and land surface temperature responses. The contribution index (CI) is used to explain how changes per land use and land cover class (LULC) contributed to average city surface temperatures. The population size of Freetown was about eight times greater than in Bo town. Landsat data mapped urban growth patterns with a high accuracy (Overall Accuracy > 80%) for both cities. Significant changes in LULC were noted in Freetown, characterized by a 114 km^2 decrease in agriculture area, 23 km^2 increase in dense vegetation, and 77 km^2 increase in built-up area. Between 1998 and 2015, built-up area increased by 16 km^2, while dense vegetation area decreased by 14 km^2 in Bo town. Average surface temperature increased from 23.7 to 25.5 °C in Freetown and from 24.9 to 28.2 °C in Bo town during the same period. Despite the larger population size and greater built-up extent, as well as expansion rate, Freetown was 2 °C cooler than Bo town in all periods. The low temperatures are attributed to proximity to sea and the very large proportion of vegetation surrounding the city. Even close to the sea and abundant vegetation, the built-up area had an elevated temperature compared to the surroundings. The findings are important for formulating heat mitigation strategies for both inland and coastal cities in developing countries.

Keywords: urban heat island; land surface temperature; climate change; land use; land cover; Landsat; remote sensing

1. Introduction

There has been an increase in the number of urban dwellers, together with an accompanying expansion of built-up area globally [1]. Urban areas are strategic areas economically, as well as from an administrative perspective. They are important for issues such as the improvement of education and health delivery of a nation. Despite their socio-economic importance, urban areas and characteristic complex land use and land cover (LULC) spatial structure also pose a variety of environmental changes [2–5]. According to Acharya et al. [5], the benefits of urban growth in developing countries include opportunities for employment, specialization, and the better production of goods and services. The challenges, however, include air pollution and water pollution in industrialized areas, while flash flooding is prevalent in highly impervious areas. Another notable challenge of urban development is temperature elevation, especially in densely built-up areas [6,7]. Studies have shown that urban areas are comparatively warmer than undisturbed surroundings such as rural areas; a phenomenon called Urban Heat Island (UHI) [8–12]. According to Gusso et al. [8], cities use construction materials such as concrete and asphalt, which do not allow water to penetrate and absorb a large amount of heat, thereby increasing urban temperatures. Elevated temperature results in increased outdoor and indoor human thermal discomfort, as well as increased heat-related health risk [13–16]. Urban heat islands have maximized the number of heat wave days and tropical-like night conditions in several main cities, including Paris, Baltimore, Washington D.C., and Shanghai, during the summer [17–19]. Furthermore, the Intergovernmental Panel on Climate Change (IPCC) [20] stressed that land cover changes have the potential to raise air temperatures of urbanized areas by 4 °C by 2100. The changes and associated adverse impacts seriously threaten the sustainable development of urban areas [21]. Urban land use and land cover heterogeneity, as well as changes, result in the complex and varied spatial structure of heat intensities which also vary from city to city. It is thus important to establish city specific land surface temperature patterns in order to derive relevant mitigation and response strategies.

Remote sensing offers a variety of options for monitoring both LULC and LST spatial structure. Unfortunately, space-borne sensors detect thermal infra-red at either a low (e.g., above 500 m such as METEOSAT) or medium (e.g., 30–500 m such Landsat, ASTER and MODIS), but not high, spatial resolution (e.g., below 30 m such as SPOT). This results in mismatch in the resolution between retrieved LULC and LST maps. High resolution thermal data is often obtained from air-borne missions. Generally, high spatial resolution datasets are expensive to gather, have a low temporal resolution, usually lack a thermal infra-red component, and have very limited historical archives not sufficient for long term analysis [22]. Medium resolution multi-spectral datasets are often reliable for urban LULC and LST analysis. For example, Landsat has large stores of visible, infra-red, and thermal data archives spanning from as early as 1972 to present [6,22]. Recently, studies showed that Landsat data are effective and very accurate in mapping urban LULC distribution, as well as changes thereof [22–25]. For example, using Landsat data, Mushore et al. [9] retrieved LULC spatial and temporal patterns in Harare between 1984 and 2015 at overall accuracies greater than 80%. Studies have also proved the effectiveness of Landsat thermal data in mapping land surface temperature variations, including those in complex urban settings [26–28]. Recently, multi-temporal Landsat data was used to develop a model to predict future urban surface temperatures in Harare [29]. Mushore et al. [29], showed that if historical growth patterns will persist, land surface temperatures will increase by as much as 5 °C by 2045. Therefore, the utility of medium resolution datasets in quantifying the impact of urban growth on LST patterns needs to be continually exploited. This is necessary in cities of low Gross Domestic Product countries such as in Africa, especially where similar studies have not yet been done; for example, in Sierra Leone.

In Africa, the studies have been confined to a few cities mainly in South Africa, Zimbabwe, and Nigeria. For example, Odindi et al. [7] investigated the impact of seasonality of urban greenery on heat island patterns in the Ethkwini municipality in South Africa. However, although they used 30 m multispectral Landsat 7 data for LULC mapping, surface temperatures were retrieved from course resolution (1 km) MODIS thermal data. Other studies in Africa were also confined to a single city; for example, Mushore et al. [9] only focused on Harare in Zimbabwe, while in West Africa, Abegunde and Adedeji [30] focused on Ibadan in Nigeria. Given the projected urban growth which must be faster in developing countries, there is thus a need to understand the implications in other parts of

Africa [19]. While Odindi et al. [31] compared LST patterns in coastal cities of South Africa, there is a general paucity of literature on comparing LST patterns between two cities of an African country. Precisely, there is a lack of literature comparing LST patterns of two cities, especially with one being inland and the other being coastal, such as Freetown and Bo town in Sierra Leone. As such, there is the need for a novel study to understand urban growth patterns, as well as responses of LST, in Sierra Leone, in West Africa. Such analysis is important for understanding both the differential effect of urban growth and of global warming between a coastal and an inland city in West Africa. Adaptation and mitigation strategies derived from such an analysis will take into account the position of a city relative to the ocean. Furthermore, the Contribution Index (CI) has not yet been used to compare growth patterns of two cities, as well as to explain the impacts of growth on surface temperatures in West Africa. To the best of our knowledge, the index has only been successfully tested on the African continent in South Africa [7,23,31] and in Zimbabwe [9]. Odindi et al. [31] used CI to compare LULC and LST patterns between coastal cities of South Africa, but did not compare a coastal city with an inland city. Although Odindi et al. [31] compared LST variations in two cities; they used course resolution MODIS data, leaving a gap on comparison analysis using Landsat data in Africa. Liu and Weng [32] also found the 30 m visible and infrared, as well as the 90 to 120 m resolution thermal infra-red, Landsat data to be optimal in the analysis of the relationship between LULC and LST patterns.

The objectives of this study are thus to (1) use remote sensing to determine urban growth patterns in Sierra Leone; (2) quantify the effect of urban growth on spatial and temporal LST patterns in two major cities of Sierra Leone using the CI; and (3) understand the differences in responses of LST to urban growth and global warming between a coastal city (Freetown) and an inland city (Bo town) in Sierra Leone. The study hypothesizes that urban growth patterns should differ between Freetown and Bo town and thus influence LST spatial and temporal changes to differ between the two cities.

2. Materials and Methods

2.1. Study Area

The study was conducted in the two major cities of Sierra Lone; Freetown and Bo town (Figure 1). Freetown is the major port city on the Atlantic Ocean and is located in the western area of Sierra Leone. Bo town is the second largest city in Sierra Leone (after Freetown) and the biggest city in the Southern Province. Bo town serves as the capital and administrative focus of Bo District in the Southern Province. Freetown has a total area of 357 km^2 and a population of 772,873, constituting 15.53% of the total Sierra Leonean population [33]. From the projected population of local administrative data from 2005 to 2014 (http://statistics.sl), out of 6,348,350 populations in Sierra Leone, 27.14% lives in urban areas, with 16.4% living in Freetown and 4% living in Bo [33]. In Sierra Leone, the national census should be done once every 10 years. To date, five censuses have been conducted in 1963, 1974, 1985, 2004, and 2015. Another census was supposed to be done between 1994 and 1995, but was postponed due to the civil war which commenced in 1991 in the country. In order to ascertain that the population was growing in the study area, we used all the available data from the five censuses, although focus was on the period between 1998 and 2015. Census statistics are obtainable from Statistics Sierra Leone at national, town, and chiefdom levels. Therefore, population statistics for Freetown and Bo town were obtained at the town level.

Climate summaries were obtained from the Sierra Leone Meteorological Department under the Ministry of Transport and Aviation (http://mta.sl/meteorological-department). Freetown and Bo town experience a tropical climate, with a rainy season from May to October and a hot dry season from November to April. Freetown has an average annual precipitation of more than 3500 mm. It receives the highest amount of rainfall in the country due to its proximity to the Peninsula Mountains and Atlantic Ocean. The average annual precipitation of Bo town is around 2616.6 mm. The annual average minimum temperature for Freetown is around 23.8 °C, while the average maximum temperature is 29.9 °C. The annual mean minimum temperature for Bo town is 21.2 °C and the average maximum temperature is 31.3 °C.

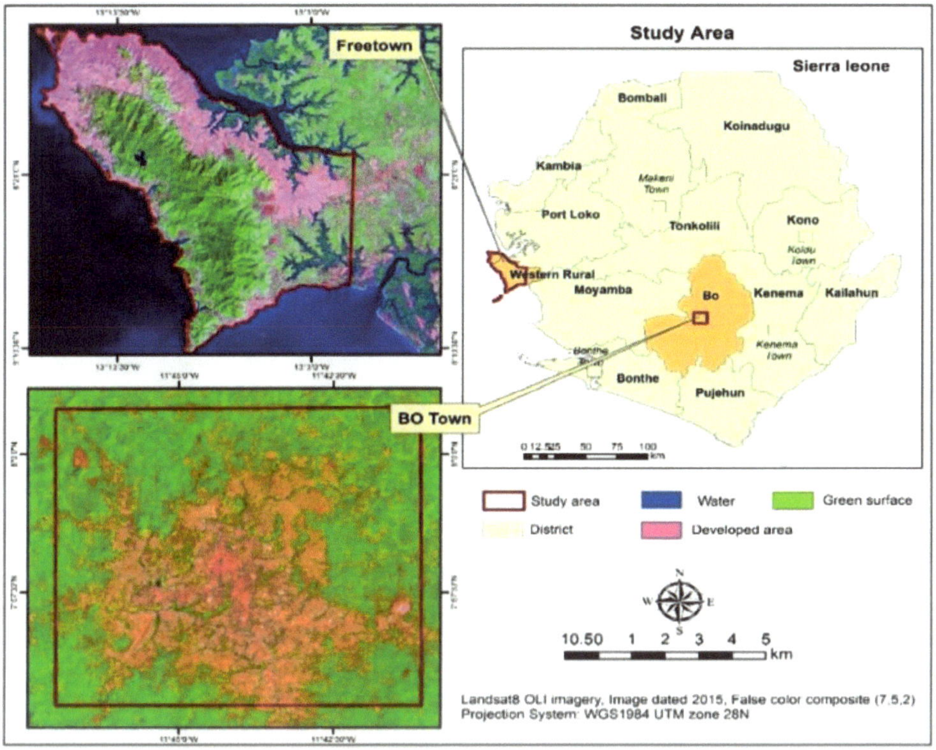

Figure 1. Location of Freetown and Bo town in Sierra Leone, West Africa.

The topography of Freetown is undulated. Elevation ranges between 100 m and 700 m, with slopes exceeding 50 m and Bo town is 104 m above sea level. The prevailing winds are the south west monsoon during the wet season and the northeastern harmattan, which is a dust laden wind from the Sahara Desert, during the dry season. In this research, we have taken the most developed parts of Bo town (10,808.57 ha or 108.08 km^2) and the most developed parts of Freetown (51,896.79 ha or 518.97 km^2), as illustrated in Figure 1. The red bordered area of Freetown and Bo town (study areas) are the rapidly developing areas. The period between November and April was selected for temperature analysis because it is hot and dry, hence posing a threat to human thermal comfort.

2.2. Datasets

This study uses cloud free and geometrically corrected Landsat imagery from the Earth Resources Observation and Science (EROS) center through the United States Geological Survey (USGS) Global Visualization Viewer. The path/row was 202/54 for Freetown and 201/054 for Bo town. The image scenes dated to 27 February 1998 (TM5), 3 February 2000 (ETM), and 23 February 2007 and 28 January 2015 (OLI) for Bo town, and 28 February 1998 (TM5), 3 February 2000 (ETM), 27 February 2007 (ETM), and 4 February 2015 for Freetown. Apart from satellite imagery, several referenced datasets like ground GPS data of different LULC categories, Population census data from the Statistics Sierra Leone, and mean temperature and mean humidity from the Sierra Leone metrological Department were used. High resolution contemporary satellite imagery (GEOEYE-1 and Google Earth historical image of 2015), administrative spatial datasets from the National Tourist Board and Environmental Protection Agency, and ancillary secondary maps were also used as ground truth data for accuracy assessment.

2.3. Image Preprocessing

The remote sensing images are re-projected to the UTM WGS 84 N (UTM zone 29-North) following a third order polynomial fit and nearest neighbor resampling techniques. Digital numbers (DN) of TM5, ETM+, and OLI images are stored as 8 bit and 16 bit, respectively [34–36]. These DNs of each image are converted to the top of atmospheric (TOA) spectral radiance using sensor specific calibration parameters directly obtained from the image MTL (metadata) file following the standard spectral radiance (Equation (1)).

$$L = \left(\frac{A\rho}{1-\rho_e S}\right) + \left(\frac{B\rho_e}{1-\rho_e S}\right) + L_a \quad (1)$$

where, ρ is the pixel surface reflectance, ρ_e is an average surface reflectance for the pixel and a surrounding region, S is the spherical albedo of the atmosphere, L_a is the radiance back scattered by the atmosphere, A and B are coefficients that depend on atmospheric and geometric conditions but not on the surface, and L is the spectral radiance.

The radiance of the reflective bands is then converted to a band interleaved by line (BIL) format to make them efficient for the atmospheric correction process in order to reduce atmospheric effects like water content, dust particles, aerosols, cloud, and varying sun angles, etc., which could significantly influence optical images and thereby degrade their spectral information. Hence, these are subjected to an atmospheric correction process to be applied to minimize those effects and produce corrected surface reflectance. The Fast Line-of-sight Atmospheric Analysis of Hypercube (FLAASH) is applied for the atmospheric correction process [37]. FLAASH is a first principle of atmospheric correction tool which generally corrects wavelengths of visible, near-infrared, and shortwave infrared data. It uses the MODTRAN radiation transfer code [38] for retrieving atmospheric noises like aerosols, dusts, and water vapor content, etc., from dark land pixels in the scene based on a nearly fixed ratio between reflectance from pixels at 660 nm and 2100 nm [39]. The overall FLAASH method takes input from the radiance and provides an atmospherically corrected surface reflectance image output using Equation (2).

$$L_e \approx \left(\frac{(A+B)\rho_e}{1-\rho_e S}\right) + L_a \quad (2)$$

2.4. Urban Growth Assessment Using Remote Sensing and Census Data in Freetown and Bo Town

Land use and land cover (LULC) maps for 1998, 2000, 2007, and 2015 were obtained using supervised image classification of multispectral Landsat data described in Section 2.2 above. Supervised image classification involves the use of ground control points obtained from field surveys or high resolution imagery to assist remote sensing software to assign LULC classes to pixels based on multi-spectral images. In each classification procedure, thermal data were left out since the objective was then to link LULC dynamics with LST derived from these data. The Support Vector Machine (SVM) algorithm was used because it was found to perform better than other common classifiers such as ANN, maximum likelihood, and Mahalanobis distance [22,40,41]. SVM also comparatively requires very little training data. In each year, the ground truth LULC data collected from field work and auxiliary data were split into 70% (for classification) and 30% (for accuracy assessment) following the recommendation of Adelabu et al. [40]. The area is classified into built-up, dense vegetation, sparse vegetation, water/wetlands, and agriculture land. A post classification change detection approach was used to determine the effect of growth on the spatial distribution and areal coverage of LULC types. According to Yu et al. [42], post classification is the most widely used change detection method. Due to simplicity and ease of interpretation, in this study, we detect changes in area per class, as was done by Salvati and Sabbi [43].

In order to link remotely sensed spatial and temporal patterns in LULC with population growth, census data for 1963, 1974, 1985, 2004, and 2015 were used. Although the study focuses on the time interval from 1998 to 2015, the analysis of population dynamics includes time as far back as 1963 in order to take advantage of data availability, as well as to obtain a clearly convincing description of the population trends in the area.

2.5. LST Retrieval from Thermal Infrared Data

The steps as summarised by Weng et al. [26] and described in detail by Weng et al. [44] are followed to retrieve the land surface temperature from Landsat's thermal infrared data. The procedure involved (i) conversion of digital numbers (DN) to spectral radiance; (ii) computation of satellite brightness temperature from spectral radiance; and (iii) retrieval of land surface temperature from brightness temperature (emissivity correction). Full details of the steps are described in the Sections 2.5.1 and 2.5.2 below.

2.5.1. Conversion from Digital Numbers to Brightness Temperature

The DNs of the TIR bands of each year's ETM+ and TM5 images are converted to spectral radiance using the formula adopted by Chander and Markham [45] (Equation (3)) and Landsat 8's thermal infrared images were converted using the USGS standard (Equation (4)).

$$L_\lambda = L_{min} + \frac{L_{max} - L_{min}}{QCAL_{max} - QCAL_{min}} DN \tag{3}$$

$$\mathrm{L}_\lambda = \mathrm{M_L} \times \mathrm{Q_{cal}} + \mathrm{A_L} \tag{4}$$

In the above equations, L_λ is the spectral radiance in $W/(m^2\ sr\mu m)$ received by the sensor from each pixel of the image. M_L and A_L are band specific multiplicative and additive rescaling factors obtained from the image MTL file, *Qcal* is the DN of each image, and $QCAL_{max}$ is the maximum *DN* (65535 for the 16-bit Landsat 8 and 255 for other Landsat missions). L_{max} and L_{min} are the maximum and minimum top of atmospheric (TOA) radiances in $W/(m^2\ sr\mu m)$, respectively.

After the conversion of the DNs to the spectral radiance, the radiant images are converted to the blackbody temperature using (Equation (5)).

$$T_b = \frac{K_2}{\ln\left\{\left(\frac{K_1}{L_\lambda}\right) + 1\right\}} \tag{5}$$

where T_b is the effective at-sensor brightness temperature in Kelvin unit, L_λ is the spectral radiance in $W/(m^2\ sr\mu m)$, and K_1 and K_2 are prelaunch calibration constants in Kelvin unit obtained from the image MTL file.

2.5.2. Surface Emissivity (ε) Retrieval

The land surface emissivity is retrieved using the Normalized Difference Vegetation Index (NDVI) threshold method [45,46]. According to the method, when NDVI < 0.2, the pixels are considered as bare lands and the emissivity is retrieved from the red spectral region. When NDVI > 0.5, the pixels are considered as fully vegetation coverage and the emissivity value is assumed to be 0.99. When NDVI ranges between 0.2 and 0.5, the pixels are considered as a mixture use of soil and vegetation. In this case, emissivity is retrieved using Equation (6), as follows:

$$\varepsilon = \varepsilon_v P_v + \varepsilon_s (1 - P_v) + \Delta\varepsilon \tag{6}$$

where ε_v is the emissivity of vegetation coverage, ε_s is the emissivity of soil surface, and, P_v is the proportion of vegetation calculated from Equation (7),

$$P_{\text{v}} = \left[\frac{NDVI - NDVI_s}{NDVI_v - NDVI_s}\right]^2 \tag{7}$$

where $NDVI_s$ is the NDVI value of pure soil and $NDVI_v$ is the NDVI value of pure vegetation extracted from the NDVI image.

In Equation (6), the term $\Delta\varepsilon$ is the indication of the geometrical distribution of the natural surface, as well as the internal reflection whose value is considered as negligible for the plain and homogenous surfaces. However, in the case of a rough and heterogeneous surface, the value is assumed to be 2% Sobrino et al. [46] and is expressed by the following (Equation (8)):

$$\Delta\varepsilon = (1 - \varepsilon_s)(1 - P_v)F\varepsilon_v \tag{8}$$

where F is the shape factor whose mean value for different geometrical distributions is assumed to be 0.55 [45,46].

By summarizing Equations (6) and (8), the final equation for emissivity estimation is obtained by Equation (9), as follows:

$$\varepsilon = mP_v + n \tag{9}$$

where m and n coefficients are calculated as:

$$m = \varepsilon_v - \varepsilon_s - (1 - \varepsilon_s)F\varepsilon_v \text{ and } n = \varepsilon_s + (1 - \varepsilon_s)F\varepsilon_v \tag{10}$$

Brightness temperatures assume that the earth is a blackbody, which it is not, and this can result in errors in surface temperature. In order to minimize these errors, emissivity correction is necessary and this is done to finally obtain the land surface temperature (*LST*) from T_b using Equation (11) [44].

$$LST = \frac{T_b}{1 + \left\{\lambda T_b\left(\frac{K}{\rho}\right) \times \ln\varepsilon\right\}} \tag{11}$$

In the above equation, λ is the wavelength of emitted radiance (11.5 μm) [47,48], $\rho = hc/\sigma$ (mK), K is the Stefan–Boltzmann's constant (1.38×10^{-23} JK^{-1}), h is the Planck's constant (6.26×10^{-34} Js), c is the velocity of light (2.998×10^8 ms^{-1}), and ε is the surface emissivity.

2.6. Linking Urban Growth to LST

The effect of LULC in the warming or cooling of an area depends on the LULC type and the proportion of the total area occupied by each type. For example, vegetation cover and water/wetlands have a surface cooling effect due to latent heat transfer. However, even though they have a cooling effect, the overall value depends on the proportion of the total area they occupy [49]. The warming or cooling extent of an LULC type taking into account the proportion of the total area it occupies is quantified using the Contribution Index (CI). The CI is used to link spatial structure, as well as long term changes in LULC, to LST intensities. The CI for each LULC type is computed for both cities using Equation (12) for all the periods mentioned in Section 2.2 [7,31,48].

$$CI = D_t \times S \tag{12}$$

D_t is the difference between the average temperature of the entire study area and the average of the LULC class type. Variable S is the proportional area of the LULC type, which is the ratio of the area covered by the class to the total area of the study area. Positive values of CI indicate how much the

LULC type contributes to raising the surface temperatures of an area, while negative values indicate a heat mitigation value.

3. Results

3.1. Remote Sensing Based Urban Growth Assessment in Freetown and Bo Town

Visual inspection of Figure 2A–D indicates the expansion of built-up area in Freetown. This is notable in the northern, eastern, and western parts of the city. Since 1998, the city has been characterized by a tongue of dense vegetation occupying most of the central part of the city. This dense vegetation area is not diminishing, even as built-up area is expanding. The growth of Freetown concentrated along the northwestern and eastern margins is influenced by the ocean (Figure 2A–D). On the other hand, the growth of Bo town since 1998 has been largely characterized by expansion from the central to the southwestern areas of the city (Figure 2E–H). The growth of Bo town also infiltrated into densely vegetated areas between 1998 and 2015.

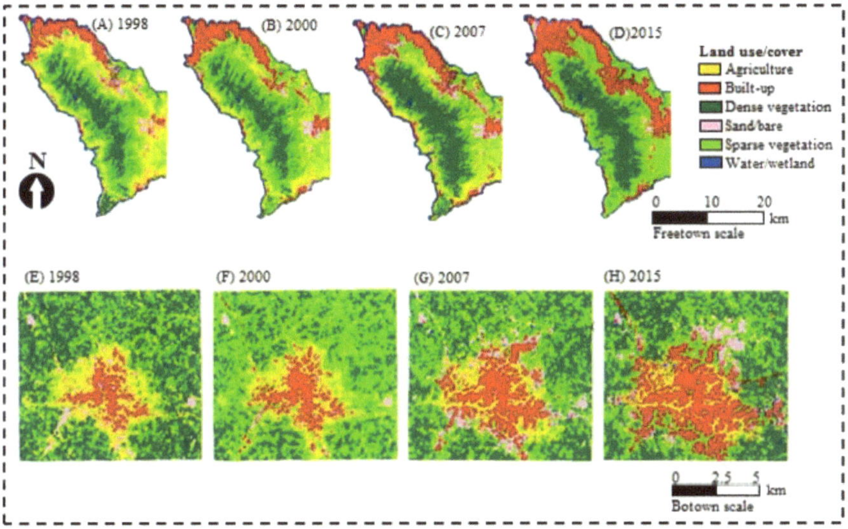

Figure 2. Urban growth induced LULC changes in Freetown (**A–D**) and Bo town (**E–H**) between 1998 and 2015.

Table 1 indicates the overall accuracy (OA) and kappa coefficient (k) obtained in LULC classification for different years in Freetown and Bo town. The overall accuracies were greater than 85% for both cities in all years. Accuracies per individual LULC class (i.e., user accuracy (UA) and producer accuracy (PA)) are shown in Appendix A.

Table 1. Accuracy of multi-temporal LULC classifications in Freetown and Bo town.

Year	Freetown		Bo Town	
	OA	Kappa	OA	Kappa
1998	91.56	0.91	89.87	0.88
2000	95.56	0.95	89.44	0.87
2007	93.33	0.92	87.88	0.85
2015	89.44	0.87	88.33	0.86

Between 1985 and 2015, the agriculture area has decreased by about 84 km^2, while the built-up area increased by almost 80 km^2 in Freetown (Figure 3). Dense vegetation areas increased by 22 km^2, while sparse vegetation areas increased by 40 km^2. The increase in vegetation areas could be part of an explanation of why bare areas reduced in area by 28 km^2. A difference was observed in Freetown because, here, growth occurs along the coast away from the central zone of dense vegetation. Land use and land cover changes in Bo town were not as marked as in Freetown. For example, built-up areas increased by 15 km^2, while areas with sparse vegetation increased by 7 km^2 in Bo town. During the same period, the dense vegetation and agriculture areas decreased by 14 km^2 and 10 km^2, respectively.

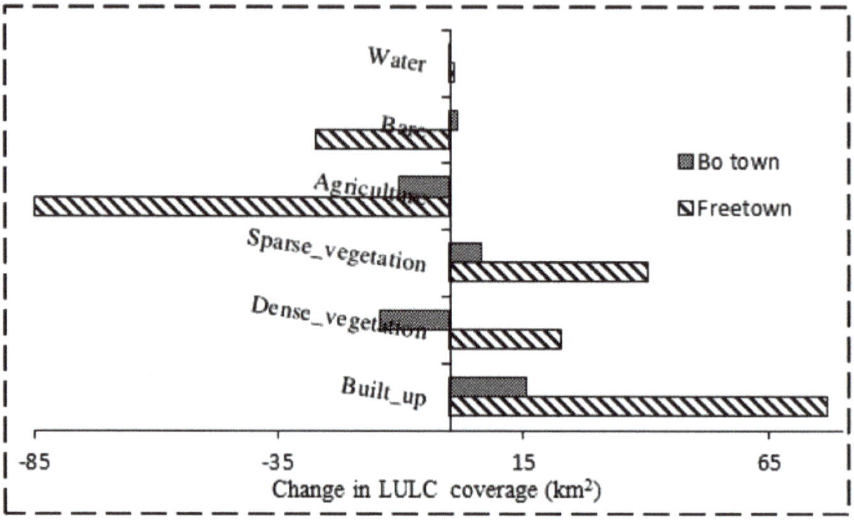

Figure 3. Urban growth induced LULC changes in Freetown and Bo town (1998 to 2015).

3.2. Census Based Urban Growth Patterns in Freetown and Bo town

The population increased by almost ten times in both Freetown and Bo town between 1963 and 2015 (Table 2). Population densities also changed from 246.5 to 2023.8 (people/km^2) and from 246.2 to 1609.0 (people/km^2) in Freetown and Bo town, respectively. The population size of Freetown has always far exceeded that of Bo town, such that in 2015, the sizes were 1,050,301 and 173,905, respectively.

Table 2. Census-based population growth in Freetown and Bo town.

Year	Population Size	
	Freetown	Bo Town
1963	127,917	26,613
1974	276,247	39,741
1985	469,776	59,768
2004	772,873	148,705
2015	1,050,301	173,905

3.3. Responses of LST to Growth Patterns in Freetown and Bo Town

High surface temperatures (above 30 °C) are most notable in the northern and western parts of Freetown in 1998 (Figure 4A). Over the years, high surface temperatures have also been spreading southward along the western margin of the city (Figure 4B–D). Low surface temperatures below 22 °C have remained characteristic of the central and southwestern parts of the city. On the contrary, since 1998, the high surface temperature has spread from the central parts of the city of Bo town, especially towards the southwest (Figure 4E–H). Low temperature areas (below 20 °C) surround this expanding hot spot and are shrinking in size. The shape of high surface temperature areas in both Freetown and Bo town closely mimics that of the built-up area, indicating their strong warming influence. Conversely, low surface temperature patterns also track areas with vegetation cover, being low in dense vegetation areas in both cities. In both cities, average temperatures are rising with time (Table 3).

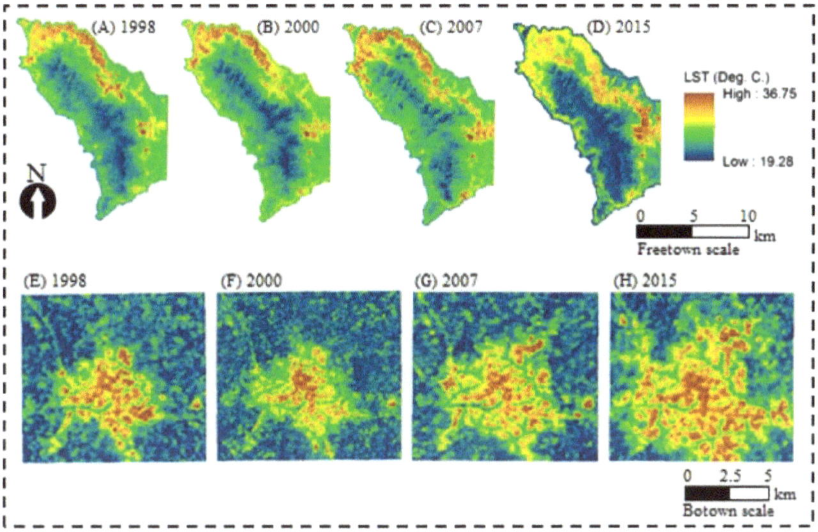

Figure 4. Land surface temperature change in Freetown (**A–D**) and Bo town (**E–H**) between 1998 and 2015.

Table 3. Changes in the heat source/sink role of land use and land cover types in Freetown between 1998 and 2015. Green means vegetation.

	1998			2000			2007			2015		
	DT (°C)	**S (%)**	**CI**	**DT (°C)**	**S (%)**	**CI**	**DT (°C)**	**S (%)**	**CI**	**DT (°C)**	**S (%)**	**CI**
Built-up	2.67	9.92	0.26	3.17	13.16	0.42	2.15	17.84	0.38	2.92	24.73	0.72
Dense green	−2.19	21.43	−0.47	−2.79	21.98	−0.61	−1.51	28.76	−0.43	−2.60	25.82	−0.67
Sparse green	−1.15	29.52	−0.34	−1.28	34.55	−0.44	−0.48	22.57	−0.11	−0.37	37.29	−0.14
Agriculture	0.18	22.93	0.04	−0.48	17.16	−0.08	−0.42	13.85	−0.06	0.97	0.98	0.10
Bare/sand	1.70	10.46	0.18	1.21	7.61	0.09	1.43	10.60	0.15	1.74	5.24	0.09
Water	−1.17	5.73	−0.07	−0.41	5.55	−0.02	−1.15	6.38	−0.07	−2.65	5.95	−0.16

3.4. Link between Long Term Changes in LULC and LST Dynamics

The agriculture area has a positive contribution index (CI) in Freetown, indicating that such places increase heat in the city during the dry season (Table 3). Although the area under agriculture has reduced between 1998 and 2015, the CI has remained positive and increased. The cooling contribution of dense vegetation is increased as indicated by a CI of −0.47 in 1998 followed by −0.85 in 2015. Sparse vegetation also has a significant cooling effect in Freetown, although its value has decreased slightly between 1998 (CI = −0.34) and 2015 (CI = −0.24). The heat mitigation value of vegetation was also noted in Texas, where woodlands were 1.5–3.9 °C cooler than neighboring areas. The cooling effect of dense vegetation was more than that of sparse vegetation indicated for an example by a CI of −0.67 compared to −0.13 in Freetown in 2015 for dense and sparse greenery, respectively. The built-up area in Freetown increased in terms of the warming effect by almost three times, as indicated by the CI of 0.26 in 1998 and 0.72 in 2015.

In Bo town, the surface cooling effect of dense vegetation is increasing significantly (CI = −0.55 in 2000 and −0.85 in 2015). At the same time, bare areas and areas with sparse vegetation are decreasing in terms of their warming and cooling effect, respectively (Table 4). The hot spot area expanding from the centre of the city, mainly to the southeast, can thus be explained by the increasing warming effect of the built-up area between 1998 (CI = 0.26) and 2015 (CI = 0.41). Although water bodies have a cooling effect, their contribution has remained minimal over the years due to the low proportion of the cities they occupy. In both cities, the CI for water has remained less than −0.2 in all the years.

Table 4. Changes in the heat source/sink role of land use and land cover types in Bo town between 1998 and 2015.

	1998			2000			2007			2015		
	DT (°C)	S (%)	CI	DT (°C)	S (%)	CI	DT (°C)	S (%)	CI	DT (°C)	S (%)	CI
Built-up	3.65	5.08	0.19	2.78	7.38	0.21	2.58	12.69	0.33	2.09	19.79	0.41
Dense green	−2.31	47.06	−1.09	−2.21	24.68	−0.55	−1.63	31.09	−0.51	−2.51	33.97	−0.85
Sparse green	−1.44	19.91	−0.29	−1.48	49.14	−0.73	−0.83	31.39	−0.26	−0.93	26.11	−0.24
Agriculture	0.52	18.26	0.10	0.54	14.76	0.08	0.38	10.49	0.04	0.20	8.62	0.02
Bare/sand	1.25	7.96	0.10	1.56	3.72	0.06	0.78	11.15	0.09	1.29	9.69	0.13
Water	−1.70	1.73	−0.03	−1.22	0.31	−0.01	−1.27	3.20	−0.01	−0.13	1.83	−0.01

Urban growth patterns in Freetown are unique, in that they are characterized by the expansion of built-up and dense vegetation areas. Although Freetown is larger in size and growing faster, it was about 2 °C cooler than Bo town in all periods.

4. Discussion

The study obtained a high classification accuracy both in a coastal city (Freetown) and an inland city (Bo town). The overall classification accuracy reached the 85% recommendation by Anderson [49], because even at a 30 m resolution of Landsat optical data, the mixed pixel problem did not significantly affect the quality of the LULC maps produced. Despite the complexity of classification in urban areas due to surface heterogeneity, the mapping accuracies are also higher than the 80% overall accuracy recommended by Omran [50]. The high level of accuracy can be justified by Voogt and Oke [51], who noticed that improvements that have occurred in satellite sensors over the years provide detailed and accurate land surface representation at a low cost. The high classification accuracy could also be attributed to the renowned performance of the Support Vector Machine algorithm [22,40,41]. According to Jia et al. [41], the Support Vector Machine (SVM) algorithm was found to outperform other common classifiers such as ANN, maximum likelihood, and Mahalanobis distance. The algorithm was also used for multi-temporal Landsat-based classification in an urban setting in Harare, where overall accuracies above 80% were also obtained. These findings show the value of freely available medium resolution space-borne remotely sensed datasets for monitoring urban extent and growth, especially in resource-constrained nations.

The population increased by almost ten-fold in both Freetown and Bo town between 1963 and 2015, while the population densities also increased. In all the periods considered, the population size of Freetown has always far exceeded that of Bo town. Most of the economic and administrative activities of Sierra Leone are concentrated in Freetown, hence the larger population size and faster growth than Bo town. Furthermore, the beauty of the sea seems to make residents prefer to concentrate along the coastal margins of Freetown than to spread further inland towards the dense vegetation area. Besides increasing population sizes, built-up areas are also expanding in both cities. Growth patterns observed in both cities agree with earlier observations and predictions that urban population is growing, globally [19,52]. Expansion of the built-up area in Freetown has been mainly concentrated along the coast and is most notable in the northern, eastern, and western parts of the city. This growth along the northern margins of Freetown explains why the dense vegetation area in the central part of the city is not diminishing even as the built-up area is expanding. A different pattern is observed in Bo town, where the built-up area is expanding from central locations outwards. Unlike in Freetown, the growth of Bo town has led to a reduction in the area of the densely vegetated LULC category between 1985 and 2015. As observed in Bo town, in most studies, the proportion of total area occupied by dense vegetation decreases with continuous urban expansion [26,27,53,54]. Kamusoko et al. [54] observed that the expansion of built-up areas in Harare Zimbabwe pushed most dense vegetation locations outwards to the peripheries of the city.

As expected, temperature responded strongly to spatiotemporal dynamics of LULC in both Freetown and Bo town. High temperatures in both cities were observed in built-up areas and their extent increased with time as the cities were expanding. The influence of buildings explains why high surface temperatures (above 30 °C) were recorded in northern and eastern parts of Freetown. Over the years, surface temperatures in this regime have also been spreading southward along the western margin of the city following the expansion of the built-up area. The shape of high surface temperature areas in both Freetown and Bo town closely mimics that of the built-up area, indicating their strong warming influence. This concurs with Sha and Ghauri [28], who observed that surface urban heat island expands with expansion in a built-up area. Buildings reduce heat removal by advection and reduce the sky view factor, thus limiting heat escape to space, while walls and pavements absorb and emit heat [28,53,55,56]. This results in large amounts of stagnant heat and high temperatures, especially in closely packed and high rise buildings. The warming in both cities could also be explained by increased anthropogenic activities supported by an increasing population size in both cities over time, which increases long wave radiation in the lower atmosphere. Nayak and Mandal [3] and Grimmond [57] also attributed urban warming to both LULC changes and other anthropogenic effects such as greenhouse gas emissions. The rising temperature in response to the growth of both cities can be captured by the explanation that, as population grows, urbanization increases and the magnitude of the urban heat island also expands [58]. Similar findings were obtained in Australia between 1951 and 2003, where land cover changes produced statistically significant warming [59].

Vegetation cover has been indicated to be a strong mitigation measure against the elevation of surface temperatures in both cities. For example, in Freetown, low surface temperatures (below 22 °C) remained characteristic of the central and southwestern parts of the city where buildings have not yet replaced vegetation cover. Similarly, low temperature areas (below 20 °C) surround an expanding hot spot in the central parts of the city of Bo town. The heat mitigation value of vegetation was also captured by a strong negative Contribution Index (between −0.5 and −1) in areas with dense and sparse vegetation. This concurs with Odindi et al. [7] who in the EThekwini municipality, South Africa, showed that the temperature reduction effect of vegetation increases with the percentage of total area covered. Although water bodies also have a cooling effect (negative Contribution Index [CI]), their contribution has remained minimal over the years due to the low proportion of the cities they occupy in both cities. Based on CI, the cooling effect of dense vegetation was more than of sparse vegetation, which echoes the suggestion by Zhang et al. [60] that not only vegetation types but also spatial structure affects LST distribution. Vegetation cover promotes surface cooling due to latent heat transfer.

In both Freetown and Bo town, agriculture areas were causing warming of the city, as indicated by a positive Contribution Index (CI) in all periods. This could be because, during the dry seasons, agriculture areas will either be covered by drying crop residue or will be semi-bare/bare, thus absorbing a considerable amount of heat. This is in agreement with the findings of Mushore et al. [61] in Harare, which showed that, during the hot dry season, croplands act as a heat source as they absorb and release large amounts of heat due to negligible evaporation. Although areas under agriculture have reduced between 1998 and 2015, the CI has remained positive and increased, implying an increased warming contribution to the city. This could be because the temperature of these areas has increased over the years with the changes attributed to global warming. Early planting of crops means that by the dry season the residues will be completely dry, resulting in high heat absorption, which could also be another explanation. However, the decrease in area under agriculture may indicate a shift of agriculture to the secondary industry and services in both cities. In other cities such as Harare [9], growth is also characterized by the major replacement of dense vegetation and agriculture areas with building and impervious surfaces, resulting in warming. Therefore, the surface warming mostly of Freetown between 1998 and 2015 can be attributed to global warming, the warming effect of dry agricultural land, and increase in the built-up area which absorbs a significant amount of heat. This agrees with Jiang and Tian [62], who demonstrated that the construction of buildings leads to the transition of an area from a dense vegetation low temperature to sparse vegetation high temperature zone.

Even in coastal cities where the water table is presumed to be high and sea breezes cool the atmosphere, a high density of buildings still causes warming. Although Freetown is larger in population size as well as built-up extent and also growing faster, it was cooler than Bo town in all periods (by about 2 °C). The difference could be a result of surface moisture and cold air advection due to proximity to the sea. Surface wetness reduces the temperature of a surface due to increased evaporation and latent heat transfer [56]. According to Rasul et al. [56], green areas and water bodies act as urban cool islands, hence the low temperature of Freetown despite being larger in size than Bo town. Besides being close to the sea, the proportion of dense vegetation cover is greater in Freetown than Bo town, which reduces the average temperature of the city. According to Sithole and Odindi, green spaces act as heat sinks, tend to be porous, and assimilate heat. Due to the influence of the sea, dense buildings and high surface temperature are found along the coast in Freetown. This has also led to the sustenance and expansion of a tongue of dense green area and low temperature in the central part. Water and vegetation which surround the built-up area of Freetown act as a sink to these gases, which may also explain the lower temperature there than in Bo town. According to Odindi et al. [7], the heat contribution of dense vegetation is similar to that of water, hence Freetown is surrounded by cool areas resulting a in lower mean surface temperature than Bo town.

5. Conclusions

We have compared urban growth and land surface temperature patterns between a coastal city (Freetown) and an inland city (Bo town) in Sierra Leone in this paper. Multi spectral Landsat data are used to quantify land use and land cover, as well as surface temperature, changes between 1998 and 2015. Based on the findings of the study, we conclude that multi-spectral Landsat data and the Support Vector Machine algorithm retrieve LULC spatial patterns and urban growth with a high accuracy. The growth patterns of Freetown are concentrated along city margins at the coast, while Bo town expanded from the center outwards. The abundance of dense vegetation and proximity to ocean makes Freetown cooler, although it is larger in population and is expanding in terms of the built-up area faster than Bo town. However, even in cool areas such as at the coast, built-up areas have warmer surface temperatures than non-built-up areas such as dense vegetation areas. Expansion of the built-up area from the city core pushes out vegetation towards the margin, resulting in a high temperature towards the center, as in Bo town. Overall, the built-up area expansion increases urban temperature, in addition to the effect of global warming, while vegetation has a strong heat mitigation effect. The Freetown-Bo town scenario has indicated that it is possible for a small city to be warmer than larger and faster growing cities within the same country. Temperature patterns depend heavily on position relative to ocean, as well as the size and spatial structure of dense vegetation area. Therefore, even vegetation and water patches around a built-up area (not only those within) have an influence on its temperature. Although the study managed to convincingly link urban growth induced LULC changes with LST dynamics, future efforts must address some limitations which could hamper the reliability of the findings. The study depended on medium spatial resolution Landsat datasets, whose temporal resolution of 16 days is low. This, together with the cloud free image requirement for surface analysis, resulted in a limited amount of data available for the study.

In the presence of sufficient data, averages could have been computed to eliminate the effects of randomness associated with the use of single date images to represent an entire month. Due to the low temporal resolution of Landsat data, it is difficult to obtain in-situ meteorological data at the exact time of satellite overpass for a comparison of temperatures obtained from remote sensing with in-situ observations of air temperature in Sierra Leone. Meteorological operations in Sierra Leone are still manned; taking observations at World Meteorological Organization (WMO) prescribed synoptic hours which do not coincide with the overpass times of Landsat missions. Limited access to in-situ meteorological data inhibited the analysis to test the validity of the findings of this study, although they agreed with global trends. Reflective bands of Landsat are at a higher spatial resolution than the thermal dataset (for example 30 m versus 100 m for Landsat 8). This mismatch has the potential to increase the mixed pixel problem on LST retrievals, thus compromising the link between LULC (30 m resolution) and LST (100 m), even though thermal data is downloaded at a resolution of 30 m after resampling. Other factors which affect thermal properties such as differences in building material and roof types between Freetown and Bo town were not investigated in this study.

Acknowledgments: This work was supported by grants from the National Natural Science Foundation of China (41371398)

Author Contributions: Musa Tawarally, Wenbo Xu, Hou Weiming, and Terence Darlington Mushore designed the study and wrote the paper.

Conflicts of Interest: The authors declare no conflict of interest

Appendix A

Table A1. Accuracy statistic for multi-temporal LULC classification.

Study Area	Year	Bo Town and Freetown Accuracy Assessment				
		LULC Category	Producer Accuracy (%)	User Accuracy (%)	Overall Accuracy (%)	Khat
Freetown	1998	Agricultural land	92.31	89.75	91.56	0.91
		Built-up area	96.57	94.87		
		Dense vegetation	96.77	93.33		
		Exposed land	89.98	91.65		
		Sparse vegetation	85.39	88.71		
		Waterbody	100	100		
	2000	Agricultural land	100	90	95.56	0.95
		Built-up area	100	100		
		Dense vegetation	96.55	93.33		
		Exposed land	100	100		
		Sparse vegetation	87.1	90		
		Waterbody	90.91	100		
	2007	Agricultural land	97.11	86.67	93.33	0.92
		Built-up area	95.33	86.67		
		Dense vegetation	93.33	93.33		
		Exposed land	96.77	100		
		Sparse vegetation	89.57	93.33		
		Waterbody	98.39	100		
	2015	Agricultural land	96.55	93.33	89.44	0.87
		Built-up area	96.55	93.33		
		Dense vegetation	87.88	96.67		
		Exposed land	95.65	73.33		
		Sparse vegetation	85.71	80		
		Waterbody	78.95	100		

Table A1. *Cont.*

Study Area	Year	LULC Category	Producer Accuracy (%)	User Accuracy (%)	Overall Accuracy (%)	Khat
		Bo Town and Freetown Accuracy Assessment				
Botown	1998	Agricultural land	89.78	87.87	89.87	0.88
		Built-up area	93.22	91.33		
		Dense vegetation	95.67	92.89		
		Exposed land	89.89	86.78		
		Sparse vegetation	87.56	83.89		
		Waterbody	100	99.8		
	2000	Agricultural land	93.1	90	89.44	0.87
		Built-up area	96.3	86.67		
		Dense vegetation	100	93.33		
		Exposed land	71.79	93.33		
		Sparse vegetation	92.31	80		
		Waterbody	90.32	93.33		
	2007	Agricultural land	96.15	83.33	87.78	0.85
		Built-up area	96.67	96.67		
		Dense vegetation	100	90		
		Exposed land	68.57	80		
		Sparse vegetation	93.1	90		
		Waterbody	78.79	86.67		
	2015	Agricultural land	96.3	86.67	88.33	0.86
		Built-up area	90.91	100		
		Dense vegetation	100	76.67		
		Exposed land	74.36	96.67		
		Sparse vegetation	86.21	83.33		
		Waterbody	89.66	86.67		

References

1. Owen, T.W.; Carlson, T.N.; Gillies, R.R. An assessment of satellite remotely-sensed land cover parameters in quantitatively describing the climatic effect of urbanization. *Int. J. Remote Sens.* **1998**, *19*, 1663–1681. [CrossRef]
2. Hallegatte, S.; Corfee-Morlot, J. Understanding climate change impacts, vulnerability and adaptation at city scale: An introduction. *Clim. Chang.* **2010**, *104*, 1–12. [CrossRef]
3. Nayak, S.; Mandal, M. Impact of land-use and land-cover changes on temperature trends over Western India. *Curr. Sci.* **2012**, *102*, 1166–1173.
4. Pielke, R.A.; Pitman, A.; Niyogi, D.; Mahmood, R.; McAlpine, C.; Hossain, F.; Goldewijk, K.K.; Nair, U.; Betts, R.; Fall, S.; et al. Land use/land cover changes and climate: Modeling analysis and observational evidence. *Wiley Interdiscip. Rev. Clim. Chang.* **2011**, *2*, 828–850. [CrossRef]
5. Acharya, T.D.; Parajuli, J.; Poudel, D.; Yang, I. Extraction and Modelling of Spatio-Temporal Urban Change in Kathmandu Valley. *Int. J. Eng. Appl. Sci. Res.* **2015**, *4*, 1–11.
6. Shi, T.; Huang, Y.; Wang, H.; Shi, C.-E.; Yang, Y.-J. Influence of urbanization on the thermal environment of meteorological station: Satellite-observed evidence. *Adv. Clim. Chang. Res.* **2015**, *6*, 7–15. [CrossRef]
7. Odindi, J.O.; Bangamwabo, V.; Mutanga, O. Assessing the value of urban green spaces in mitigating multi-seasonal urban heat using MODIS Land Surface Temperature (LST) and Landsat 8 data. *Int. J. Environ. Res.* **2015**, *9*, 9–18.
8. Gusso, A.; Cafruni, C.; Bordin, F.; Veronez, M.R.; Lenz, L. Multitemporal Analysis of Thermal Distribution Characteristics for Urban Heat Island Management. In Proceedings of the 4th World Sustainability Forum, Basel, Switzerland, 1–30 November 2014; pp. 1–17.
9. Mushore, T.D.; Mutanga, O.; Odindi, J.; Dube, T. Linking major shifts in land surface temperatures to long term land use and land cover changes: A case of Harare, Zimbabwe. *Urban Clim.* **2017**, *20*, 120–134. [CrossRef]
10. Zhou, D.; Zhang, L.; Hao, L.; Sun, G.; Liu, Y.; Zhu, C. Spatiotemporal trends of urban heat island effect along the urban development intensity gradient in China. *Sci. Total Environ.* **2016**, *544*, 617–626. [CrossRef] [PubMed]

11. Pal, S.; Devara, P. A wavelet-based spectral analysis of long-term time series of optical properties of aerosols obtained by lidar and radiometer measurements over an urban station in Western India. *J. Atmos. Sol. Terr. Phys.* **2012**, *84–85*, 75–87. [CrossRef]
12. Behrendt, A.; Wagner, G.; Petrova, A.; Shiler, M.; Pal, S.; Schaberl, T.; Wulfmeyer, V. Modular lidar systems for high-resolution 4-dimensional measurements of water vapor, temperature, and aerosols. *SPIE* **2005**, *5653*, 220. [CrossRef]
13. Harlan, S.L.; Declet-Barreto, J.H.; Stefanov, W.L.; Petitti, D.B. Neighborhood effects on heat deaths: Social and environmental predictors of vulnerability in Maricopa county, Arizona. *Environ. Health Perspect.* **2013**, *121*, 197–204. [PubMed]
14. Gronlund, C.J.; Berrocal, V.J.; White-Newsome, J.L.; Conlon, K.C.; O'Neill, M.S. Vulnerability to extreme heat by socio-demographic characteristics and area green space among the elderly in Michigan, 1990–2007. *Environ. Res.* **2015**, *136*, 449–461. [CrossRef] [PubMed]
15. Mushore, T.; Mutanga, O.; Odindi, J.; Dube, T. Outdoor thermal discomfort analysis in Harare, Zimbabwe in Southern Africa. *S. Afr. Geogr. J.* **2017**. [CrossRef]
16. Uejio, C.K.; Wilhelmi, O.V.; Golden, J.S.; Mills, D.M.; Gulino, S.P.; Samenow, J.P. Intra-urban societal vulnerability to extreme heat: The role of heat exposure and the built environment, socioeconomics, and neighborhood stability. *Health Place* **2011**, *17*, 498–507. [CrossRef] [PubMed]
17. Reid, C.E.; Mann, J.K.; Alfasso, R.; English, P.B.; King, G.C.; Lincoln, R.A.; Margolis, H.G.; Rubado, D.J.; Sabato, J.E.; West, N.L.; et al. Evaluation of a Heat Vulnerability Index on Abnormally Hot Days: An Environmental Public Health Tacking Study. *Environ. Health Perspect.* **2012**, *120*, 715–720. [CrossRef] [PubMed]
18. Dousset, B.; Gourmelon, F. Satellite multi-sensor data analysis of urban surface temperatures and landcover. *ISPRS J. Photogramm. Remote Sens.* **2003**, *58*, 43–54. [CrossRef]
19. Simone, D.G.; Janeiro, R.D.; Toronto, T.D.; Jack, D.; York, N.; Toronto, J.P.; Rahman, M. Climate Change and Human Health in Cities. In *Cities and Climate Change—First Assessment Report of the Urban Climate Change Research Network*; Cambridge University Press: Cambridge, UK, 2011; pp. 179–213.
20. Intergovernmental Panel on Climate Change (IPCC). Climate Change 2001: Synthesis Report. In *Acontribution of Working Groups I, II and III to the Third Assessment Report of the Intergovernmental Panel on Climate Change*; Watson, R.T., The Core Writing Team, Eds.; Cambridge University Press: Cambridge, UK, 2001.
21. Wu, H.; Ye, L.-P.; Shi, W.-Z.; Clarke, K.C. Assessing the effects of land use spatial structure on urban heat islands using HJ-1B remote sensing imagery in Wuhan, China. *Int. J. Appl. Earth Obs. Geoinf.* **2014**, *32*, 67–78. [CrossRef]
22. Forkuor, G.; Cofie, O. Dynamics of land-use and land-cover change in Freetown, Sierra Leone and its effects on urban and peri-urban agriculture—A remote sensing approach. *Int. J. Remote Sens.* **2011**, *32*, 1017–1037. [CrossRef]
23. Sithole, K.; Odindi, J.O. Determination of Urban Thermal Characteristics on an Urban/Rural Land Cover Gradient Using Remotely Sensed Data. *S. Afr. J. Geomat.* **2015**, *4*, 384–396. [CrossRef]
24. Mushore, T.D.; Mutanga, O.; Odindi, J.; Dube, T. Assessing the potential of integrated Landsat 8 thermal bands, with the traditional reflective bands and derived vegetation indices in classifying urban landscapes. *Geocarto Int.* **2017**, *32*, 886–899. [CrossRef]
25. Lo, C.P.; Quattrochi, D.A.; Luvall, J.C. Application of high-resolution thermal infrared remote sensing and GIS to assess the urban heat island effect. *Int. J. Remote Sens.* **2010**, *18*, 287–304. [CrossRef]
26. Weng, Q.; Liu, H.; Lu, D. Assessing the effects of land use and land cover patterns on thermal conditions using landscape metrics in city of Indianapolis, United States. *Urban Ecosyst.* **2007**, *10*, 203–219. [CrossRef]
27. Franco, S.; Mandla, V.R.; Rao, K.R.M.; Kumar, M.P.; Anand, P.C. Study of Temperature Profile on Various Land Use and Land Cover for Emerging Heat Island. *J. Urban Environ. Eng.* **2015**, *9*, 32–37. [CrossRef]
28. Sha, B.; Ghauri, B. Mapping Urban Heat Island Effect in Comparison with the Land Use, Land Cover of Lahore District. *Pak. J. Meteorol.* **2015**, *11*, 37–48.
29. Mushore, T.D.; Mutanga, O.; Odindi, J.; Dube, T. Prediction of future urban surface temperatures using medium resolution satellite data in Harare metropolitan city, Zimbabwe. *Build. Environ.* **2017**, *122*, 397–410. [CrossRef]
30. Abegunde, L.; Adedeji, O. Impact of Landuse Change on Surface Temperature in Ibadan, Nigeria. *Int. J. Environ. Ecol. Geol. Geophys. Eng.* **2015**, *9*, 235–241.

31. Odindi, J.; Mutanga, O.; Abdel-Rahman, E.M.; Adam, E.; Bangamwabo, V. Determination of urban land-cover types and their implication on thermal characteristics in three South African coastal metropolitans using remotely sensed data. *S. Afr. Geogr. J.* **2017**, *99*, 52–67. [CrossRef]
32. Liu, H.; Weng, Q.H. Scaling Effect on the Relationship between Landscape Pattern and Land Surface Temperature: A Case Study of Indianapolis, United States. *Photogramm. Eng. Remote Sens.* **2009**, *75*, 291–304. [CrossRef]
33. Statistics Sierra Leone. *Sierra Leone Population and Housing Census 2004*; Final Report; Statistics Sierra Leone: Freetown, Sierra Leone, 2006.
34. Roy, D.P.; Wulder, M.A.; Loveland, T.R.; Woodcock, C.E.; Allen, R.G.; Anderson, M.C.; Helder, D.; Irons, J.R.; Johnson, D.M.; Kennedy, R.; et al. Landsat-8: Science and product vision for terrestrial global change research. *Remote Sens. Environ.* **2014**, *145*, 154–172. [CrossRef]
35. Storey, J.; Scaramuzza, P.; Schmidt, G.; Barsi, J. LANDSAT 7 Scan Line Corrector- off Gap-Filled Product Development. In Proceedings of the Pecora 16 Conference on Global Priorities in Land Remote Sensing, Sioux Falls, SD, USA, 23–27 October 2005.
36. Markham, B.; Goward, S.; Arvidson, T.; Barsi, J.; Scaramuzza, P. Landsat-7 long-term acquisition plan radiometry-evolution over time. *Photogramm. Eng. Remote Sens.* **2006**, *72*, 1129–1135. [CrossRef]
37. Dube, T.; Mutanga, O. Evaluating the utility of the medium-spatial resolution Landsat 8 multispectral sensor in quantifying aboveground biomass in uMgeni catchment, South Africa. *ISPRS J. Photogramm. Remote Sens.* **2015**, *101*, 36–46. [CrossRef]
38. Matthew, M.W.; Adler-Golden, S.M.; Berk, A.; Richtsmeier, S.C.; Levine, R.Y.; Bernstein, L.S.; Acharya, P.K.; Anderson, G.P.; Felde, G.W.; Hoke, M.P. *Status of Atmospheric Correction Using a MODTRAN4-Based Algorithm*; Spectral Sciences Inc.: Burlington, MA, USA, 2000.
39. Kaufman, Y.J.; Wald, A.E.; Remer, L.A.; Gao, B.-C.; Li, R.-R.; Flynn, L. The MODIS 2.1-/SPL mu/m channel-correlation with visible reflectance for use in remote sensing of aerosol. *IEEE Trans. Geosci. Remote Sens.* **1997**, *35*, 1286–1298. [CrossRef]
40. Adelabu, S.; Mutanga, O.; Adam, E.; Cho, M.A. Exploiting machine learning algorithms for tree species classification in a semiarid woodland using RapidEye image. *J. Appl. Remote Sens.* **2013**, *7*, 073480. [CrossRef]
41. Jia, K.; Wei, X.; Gub, X.; Yao, Y.; Xie, X.; Li, B. Land cover classification using Landsat 8 Operational Land Imager data in Beijing, China. *Geocarto Int.* **2014**, *29*, 941–951. [CrossRef]
42. Yu, X.; Zhang, A.; Hou, X.; Li, M. Multi-temporal remote sensing of land cover change and urban sprawl in the coastal city of Yantai, China. *Int. J. Digit. Earth* **2013**, *6*, 37–41. [CrossRef]
43. Salvati, L.; Sabbi, A. Exploring long-Term Land Cover Changes in an Urban Region of Southern Europe. *Int. J. Sustain. Dev. World Ecol.* **2011**, *18*, 273–282. [CrossRef]
44. Weng, Q.; Lu, D.; Schubring, J. Estimation of land surface temperature-vegetation abundance relationship for urban heat island studies. *Remote Sens. Environ.* **2004**, *89*, 467–483. [CrossRef]
45. Sobrino, J.; Caselles, V.; Becker, F. Significance of the remotely sensed thermal infrared measurements obtained over a citrus orchard. *ISPRS J. Photogramm. Remote Sens.* **1990**, *44*, 343–354. [CrossRef]
46. Sobrino, J.A.; Jiménez-Muñoz, J.C.; Paolini, L. Land surface temperature retrieval from LANDSAT TM 5. *Remote Sens. Environ.* **2004**, *90*, 434–440. [CrossRef]
47. Markham, B.L. *Landsat MSS and TM Post-Calibration Dynamic Ranges, Exoatmospheric Reflectances and at-Satellite Temperatures*; Landsat Technical Notes; European Space Agency (ESA): Paris, France, 1986; Volume 1, pp. 3–8.
48. Chen, X.-L.; Zhao, H.-M.; Li, P.-X.; Yin, Z.-Y. Remote sensing image-based analysis of the relationship between urban heat island and land use/cover changes. *Remote Sens. Environ.* **2006**, *104*, 133–146. [CrossRef]
49. Anderson, J.R. *A Land Use and Land Cover Classification System for Use with Remote Sensor Data*; US Government Printing Office: Washington, DC, USA, 1976; Volume 964.
50. Omran, E.-S.E. Detection of Land-Use and Surface Temperature Change at Different Resolutions. *J. Geogr. Inf. Syst.* **2012**, *04*, 189–203. [CrossRef]
51. Voogt, J.A.; Oke, T.R. Thermal remote sensing of urban climates. *Remote Sens. Environ.* **2003**, *86*, 370–384. [CrossRef]
52. Zhang, Q.; Schaaf, C.; Seto, K.C. The Vegetation adjusted NTL Urban Index: A new approach to reduce saturation and increase variation in nighttime luminosity. *Remote Sens. Environ.* **2013**, *129*, 32–41. [CrossRef]

53. Adebowale, B.I.; Kayode, S.E. Geospatial Assessment of Urban Expansion and Land Surface Temperature in Akure, Nigeria. In Proceedings of the ICUC9—9th International Conference on Urban Climate Jointly with 12th Symposium on the Urban Environment, Toulouse, France, 20–24 July 2015; pp. 1–6.
54. Kamusoko, C.; Gamba, J.; Murakami, H. Monitoring Urban Spatial Growth in Harare Metropolitan Province, Zimbabwe. *Adv. Remote Sens.* **2013**, *2*, 322–331. [CrossRef]
55. Blake, R.; Grimm, A.; Ichinose, T.; Horton, R.; Gaffin, S.; Jiong, S.; Bader, D.A.; Cecil, L.D. Urban climate: Processes, trends and projections. In *First Assessment Report of the Urban Climate Change Research Network*; Cambridge University Press: Cambridge, UK, 2011; pp. 43–81.
56. Rasul, A.; Balzter, H.; Smith, C. Spatial variation of the daytime Surface Urban Cool Island during the dry season in Erbil, Iraqi Kurdistan, from Landsat 8. *Urban Clim.* **2015**, *14*, 176–186. [CrossRef]
57. Grimmond, S.U.E. Urbanization and global environmental change: Local effects of urban warming. *Geogr. J.* **2007**, *173*, 83–88. [CrossRef]
58. Zhang, H.; Qi, Z.; Ye, X.; Cai, Y.; Ma, W.; Chen, M. Analysis of land use/land cover change, population shift, and their effects on spatiotemporal patterns of urban heat islands in metropolitan Shanghai, China. *Appl. Geogr.* **2013**, *44*, 121–133. [CrossRef]
59. Thatcher, M.; Hurley, P. Simulating Australian Urban Climate in a Mesoscale Atmospheric Numerical Model. *Bound. Layer Meteorol.* **2012**, *142*, 149–175. [CrossRef]
60. Zhang, Y.; Odeh, I.O.A.; Han, C. Bi-temporal characterization of land surface temperature in relation to impervious surface area, NDVI and NDBI, using a sub-pixel image analysis. *Int. J. Appl. Earth Obs. Geoinf.* **2009**, *11*, 256–264. [CrossRef]
61. Mushore, T.D.; Mutanga, O.; Odindi, J.; Dube, T. Determining extreme heat vulnerability of Harare Metropolitan City using multispectral remote sensing and socio-economic data. *J. Spat. Sci.* **2017**, 1–19. [CrossRef]
62. Jiang, J.; Tian, G. Analysis of the impact of Land use/Land cover change on Land Surface Temperature with Remote Sensing. *Procedia Environ. Sci.* **2010**, *2*, 571–575. [CrossRef]

remote sensing

MDPI

Article

Learning-Based Sub-Pixel Change Detection Using Coarse Resolution Satellite Imagery

Yong Xu [1], **Lin Lin** [2] **and Deyu Meng** [2,*]

1 Institute of Future Cities, The Chinese University of Hong Kong, Hong Kong, China; xuyong@cuhk.edu.hk
2 School of Mathematics and Statistics and Ministry of Education Key Lab of Intelligent Networks and Network Security, Xi'an Jiaotong University, Xi'an 710000, China; linl.xjtu@hotmail.com
* Correspondence: dymeng@xjtu.edu.cn; Tel.: +86-181-0924-1418

Academic Editors: Qi Wang, Nicolas H. Younan, Carlos López-Martínez and Prasad S. Thenkabail
Received: 11 May 2017; Accepted: 5 July 2017; Published: 10 July 2017

Abstract: Moderate Resolution Imaging Spectroradiometer (MODIS) data are effective and efficient for monitoring urban dynamics such as urban cover change and thermal anomalies, but the spatial resolution provided by MODIS data is 500 m (for most of its shorter spectral bands), which results in difficulty in detecting subtle spatial variations within a coarse pixel—especially for a fast-growing city. Given that the historical land use/cover products and satellite data at finer resolution are valuable to reflect the urban dynamics with more spatial details, finer spatial resolution images, as well as land cover products at previous times, are exploited in this study to improve the change detection capability of coarse resolution satellite data. The proposed approach involves two main steps. First, pairs of coarse and finer resolution satellite data at previous times are learned and then applied to generate synthetic satellite data with finer spatial resolution from coarse resolution satellite data. Second, a land cover map was produced at a finer spatial resolution and adjusted with the obtained synthetic satellite data and prior land cover maps. The approach was tested for generating finer resolution synthetic Landsat images using MODIS data from the Guangzhou study area. The finer resolution Landsat-like data were then applied to detect land cover changes with more spatial details. Test results show that the change detection accuracy using the proposed approach with the synthetic Landsat data is much better than the results using the original MODIS data or conventional spatial and temporal fusion-based approaches. The proposed approach is beneficial for detecting subtle urban land cover changes with more spatial details when multitemporal coarse satellite data are available.

Keywords: land cover change; downscaling; sub-pixel change detection; machine learning; MODIS; Landsat

1. Introduction

Timely and accurate information about land cover dynamics is highly important for sustainable urban development and better quality of life in cities. Compared with conventional data collection methods like field surveying and aerial photography, satellite images have proven to be more effective and efficient for land use/cover change monitoring at regional or global scales due to their timely, consistent, repeatable, and cost-effective measurements [1,2]. Until now, a wide variety of change detection approaches have been formulated, ranging from preclassification methods such as image differencing, image ratioing [3], band analysis [4], principal component analysis [5], change vector analysis [6], and composite analysis to postclassification comparisons [7].

The availability of satellite data with improved spatial and temporal resolutions makes it possible to characterize land cover changes (LCCs) at higher spatial and temporal scales [8]. Some multitemporal coarse resolution (CR) sensors, such as the Moderate Resolution Imaging

Spectroradiometer (MODIS), Advanced Very High Resolution Radiometer (AVHRR), the Medium Resolution Imaging Spectrometer (MERIS), and SPOT-Vegetation, have been proven to be suitable for land use/LCC and vegetation dynamics' monitoring [9–11], with which the status and trend of land cover transitions or vegetation dynamics are characterized. Consequently, a variety of multi-temporal change detection approaches have been proposed [12–14].

CR data are effective for phenological change detection due to their high revisit frequencies, but their low spatial resolutions limit their applications for accurate monitoring of urban growth dynamics—especially for rapidly growing areas [7], where dynamic changes commonly occur in sub-pixel scales (like fields, water areas, roads). To enhance the capability of remote sensing for monitoring these dynamics at a sub-pixel scale, researchers have attempted to apply some unmixing approaches to recover high spatial resolution (HR) data directly from CR data [7,8,14–17]. In particular, Le Hégarat-Mascle et al. [8] proposed a statistically-based change detection model in which sub-pixel LCCs are estimated by utilizing previous land cover information as a reminder. Ling et al. [15,16] presented an improved sub-pixel mapping algorithm for change detection using prior land cover percentages, with which temporal contextual information was used to conduct sub-pixel change mapping. However, high-quality land cover percentages are required as input for this approach, which limits its real value. Zurita-Milla et al. [17] presented an unmixing-based approach to downscale multitemporal MERIS data for vegetation dynamics, but it is inappropriate for land cover-type changes. Though soft classification approaches can estimate land cover proportions within a coarse pixel [18], they fail to determine the spatial distribution of each class [19], and needless to say the detection of sub-pixel changes.

Another possible solution for sub-pixel change detection is to explore data-fusion approaches to obtain synthetic data with high spatial and temporal resolutions. These high-resolution synthetic data generated are then used for LCC detection at a HR. In view of conventional data fusion approaches (e.g., pan-sharpening, which integrates both spectral and spatial information rather than spatial and temporal information), they are beneficial for improving spatial resolution, but not suitable for fast change detection. Gao et al. [20] started a pioneering work to develop a spatial and temporal adaptive reflectance data fusion model (STARFM) to obtain high-quality Landsat-like data, but the underlying assumption of having no LCCs over time heavily limits its applications for seasonal change monitoring of vegetation [11,21,22]. In the meantime, Hilker et al. [21,22] proposed an improved spatial and temporal data fusion approach named STAARCH, in which an optimal Landsat was selected with a defined forest disturbance index. It is efficient for detecting forest disturbance, but not very efficient for complex LCCs in cities. Similarly, Zhu et al. [23] proposed an enhanced STARFM to extend the applications of the original approach for complex areas with heterogeneous landscapes. Roy et al. [24] presented a semiphysical fusion approach to characterize surface reflectance variation with the BRDF spectral model parameters and the sun-sensor geometry over time, but it is still not efficient for the fusion task with land cover-type changes.

To address the problem of mixed pixel within the remote sensing community, a superresolution technique long studied by the computer science community has been proposed. Until now, hundreds of superresolution approaches have been proposed, which can be grouped into three categories: interpolation-based [25], construction-based [26], and learning-based [27]. The sparse learning-based superresolution methods outperformed the others and were recognized as an outstanding representative of the learning-based approach. The original one was developed by Yang et al. [28], in which a pair of dictionaries was first learned from prior data, and then applied for downscaling the CR data. Huang et al. [29] extended this approach for spatial and temporal data fusion, and experimental results also show that it outperforms other spatial and temporal data-fusion approaches when compared with actual observations with respect to reflectance fidelity. However, its suitability for actual LCC detection has not been tested. To make use of multisource data fusion for change detection, recent works investigated the use of prior land cover products for generating better change detection results [30,31]. Finally, it is worth mentioning the work in [32], in which a learning-based approach was investigated to allow the achievement of high sub-pixel forest mapping accuracy.

In this study, a novel learning-based approach will be presented to detect LCCs at finer spatial resolution using multitemporal CR data. The proposed approach has two advantages. First, it is well designed to learn the LCC dynamics from previous multisource multitemporal satellite data directly, which indicates that the trained detector has a high capability in detecting high-quality LCC using similar but CR satellite data. Second, the proposed approach makes use of the finer land cover product to provide rich spatial details within a coarse pixel.

The remainder of this paper is organized as follows. In Section 2, the theoretical background and the proposed approach are fully introduced. In Section 3, fused results are validated and applied for LCC detection with actual images in the Guangzhou study area, China. The discussion and conclusions are given in Sections 4 and 5, respectively.

2. Materials and Methods

The proposed approach includes two main steps. First, the CR satellite data at the predicted time (t1) coupled with pairs of coarse and finer resolution satellite data at previous times (e.g., t0) were used to produce a finer resolution synthetic data at the predicted time (t1). Second, the LCC was detected at finer spatial resolution using the obtained finer resolution synthetic data and previous land cover maps.

2.1. Learning-Based Approach for Generating Finer Resolution Synthetic Satellite Data

It is an extremely ill-posed problem to infer the HR data directly from CR data. In this study, we will solve the problem from the perspective of LCC recovery. The recovered changed data were added with the high-resolution satellite data at a previous time to obtain the final downscaled image at the predicted time. Under a mild condition, it can be assumed that actual LCC from bitemporal satellite images can be sparsely represented as a linear combination of different LCC bases. As the following shows, a high-resolution LCC patch can be represented as a linear combination of LCC patterns with respect to a dictionary.

$$\Delta X \approx D_h \alpha \ \ where: \ |\alpha|_0 \leq K \tag{1}$$

where ΔX is an LCC patch with HR, D_h is a high-resolution dictionary, α is the sparse representation coefficient, and K is the number of bases for the dictionary D_h.

It is further assumed that a high-resolution LCC patch can be degraded into a CR LCC patch with respect to a projection matrix. Then, the degraded CR LCC patch can also be inferred to have sparse representations with respect to a low-resolution dictionary, as the following formula shows:

$$\Delta Y \approx A\Delta X = AD_h\alpha = D_l\alpha \ \ where: \ |\alpha|_0 \leq K \tag{2}$$

where ΔY and ΔX are CR and HR LCC patches, respectively, A represents the projection matrix from ΔX to ΔY, D_h and D_l are a pair of dictionaries, and α is the estimated coefficient.

Both HR and CR patches have the same sparse representations, and their co-occurrence can be captured by using a pair of coupled dictionaries. Thus, the downscaling issue for estimating HR data from CR data can be transformed into another issue, where both sparse coefficient and coupled dictionaries need to be estimated. There are two main steps to achieving this target: (1) dictionary learning—a pair of dictionaries was learned by using sample patch data from a pair of CR and HR data, of which each column (base) represents a specific LCC pattern; (2) sparse representation—sparse coefficients are estimated to reconstruct the HR LCC from CR LCC patches. The details are given below.

2.1.1. Dictionary Learning

Because the individual sparse coding problem of LCC in the high-resolution and low-resolution patches can be represented by the sparse linear combinations with respect to D_h or D_l, these two targets (see Formulas (1) and (2)) can be combined to form a unique target as shown below:

$$\min_{\{D_h,\ D_l,\alpha\}} \|\Delta X - D_h\alpha\|_2^2 + \|\Delta Y - D_l\alpha\|_2^2 + \gamma\|\alpha\|_1 \ \ where: \ |\alpha|_0 \leq K \tag{3}$$

where ΔX is a change patch for HR data, D_h is the HR dictionary, α is the sparse representation coefficient, and K is the number of bases for dictionary D_h.

With the same learning strategy as Yang et al. [28], sampled training image patch pairs are first sampled from previously acquired low- and high-resolution data before a pair of dictionaries is jointly trained with these sampled patches using the k-singular value decomposition algorithm [33].

2.1.2. Sparse Representation

Sparse representation was then used to estimate the sparse coefficient and finally recover the HR LCC data. Based on the sparse representation of HR image patch shown in Formula (1), the solution of the sparse coefficients of a specific HR patch (ΔX_s) can be obtained via the following optimization function:

$$\Delta X_s = D_h \times \alpha^* \tag{4}$$

$$\text{where } \alpha^* : \min\|\alpha\|_1 \text{ s.t. } \begin{array}{l} \|D_l\alpha - \Delta Y_s\|_2^2 \leq \varepsilon_1 \\ \|D_h\alpha - W\|_2^2 \leq \varepsilon_2 \end{array} \tag{5}$$

where ΔX_s is a change patch for HR data at location s, D_h and D_l are trained dictionaries for both HR and CR LCC, α is the sparse coefficient that needs to be estimated, and W is the overlap between the current target patch and the previously reconstructed high-resolution patch. As recommended in [27,29], the dictionary size used in this study was set to 256, and the patch size was set to 8 × 8.

The process is operated patch by patch. If the sparse coefficient for each patch is sufficiently sparse, HR LCCs should then be recovered from the patches of CR LCCs with respect to the trained dictionaries. To agree with the previously computed adjacent high-resolution patches, a balance term (seen in the second term of Formula (3)) was used to preserve the fidelity of previous recovered LCC patches. Once the sparse coefficient is estimated, then the finer LCC patch can be recovered with Formula (4). Herein, the orthogonal matching pursuit algorithm was used to estimate the sparse coefficient [34].

The above procedure can be used to estimate the HR LCCs. Finally, the recovered LCCs are added with the HR image at the previous time to obtain the final downscaled image at the predicted time.

2.2. Sub-Pixel Change Detection with Synthetic Satellite Data

In the following, the obtained synthetic satellite data with a finer spatial resolution coupled with the land cover product at a previous time were used to detect LCCs at a finer spatial resolution. Given that the synthetic satellite data are not the real satellite data at the predicted time, the land cover map from the synthetic Landsat data appears to be different from actual land cover patterns. Figure 1a shows the initial land cover map obtained from synthetic data, and it appears to have some incorrect classification results at a sub-pixel level (highlighted with red). Thus, in this step, the obtained land cover map needs to be adjusted to ensure that it is consistent with prior land cover patterns as well as finer land cover products at previous times. The change detection procedure involves the following three steps. For more details, refer to [30].

First, a land cover map at the predicted time was produced from the obtained synthetic data using a supervised classification method, then land cover proportions at a CR were estimated from the obtained finer resolution land cover map.

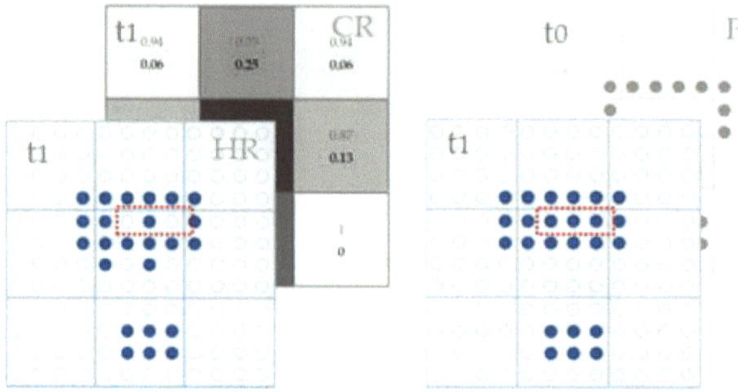

(**a**) (**b**)

Figure 1. Illustration of precise land cover mapping using synthetic Landsat data. (**a**) Synthetic Landsat data and its land cover proportions; (**b**) Land cover map at the predicted time (t1) using the proposed approach.

Second, based on the obtained land cover proportions, sub-pixel labels are initially randomly allocated maintaining the proportions. After random initialization, the labels of sub-pixels are iteratively swapped by counting their spatial correlations with surrounding pixels, and finally, the labels of sub-pixels are consistent with their neighborhood. The surrounding pixels include the nearby pixels at the current predicted time as well as neighboring pixels from land cover maps at previous times. For example, Figure 1a shows the initial land cover map obtained from synthetic data and its land cover proportions, while Figure 1b shows the final obtained land cover map at the predicted time (t1) using both the land cover proportions and a finer resolution land cover product at a previous time (t0).

Third, a refined land cover map at the predicted time was achieved via the above two steps. A change detection result (t1–t0) can be made by comparing the land cover map at the predicted time (t1) with the map from a previous time (t0).

3. Experiments and Result Analysis

The proposed approach was tested using actual data in the study area of Guangzhou, China (23°N, 113°E). This area has experienced a high percentage of land use/LCC during the past several decades, where most of the farmlands and forestlands have been changed into built-up areas due to rapid urbanization. The accurate monitoring of its rapid LCC is beneficial for the scientific management and sustainable development of this area.

Three pairs of medium-resolution Landsat and CR MODIS data for 31 October 2000, 7 November 2002, and 3 October 2004 were acquired for this study area. In this study, the MODIS reflectance products (MOD09GA) provided by NASA were adopted, and these products have been atmospherically corrected to land surface reflectance. For the original Landsat-5 data, they were atmospherically corrected into land surface reflectance using the atmospheric correction tool FLAASH [20]. Moreover, the downloaded MODIS data products were geometrically corrected to the same geographical area as the Landsat data, so both the MODIS and Landsat data cover the same extent. Based on the acquired satellite data, the preprocessed pairs of Landsat and MODIS data for the years 2000 and 2002 were used as training data, while the actual Landsat data for 2004 were used as validation data.

3.1. Synthetic Data Generation and Sub-Pixel Change Detection

Synthetic Landsat-like data for 2004 were predicted via the following main steps. First, some low- and high-resolution LCC patches were randomly sampled from the achieved difference image to train the dictionary, while the difference image reflects LCC from year 2000 to 2002 with the acquired satellite data at years 2000 and 2002. Next, the sparse learning approach introduced in the above section was used to recover HR LCCs from year 2002 to 2004 with respect to the coarse difference image and a pair of dictionaries. Finally, high-quality synthetic Landsat data at the predicted time were recovered by adding the predicted high-resolution difference data to previous Landsat data.

Based on a pair of Landsat and MODIS satellite data for the year 2002 (shown in Figure 2a,b) and MODIS data for the year 2004 (shown in Figure 2d), the finer resolution Landsat-like data for 2004 using the proposed learning-based approach are given in Figure 2e. Using the synthetic Landsat data for 2004, two sets of land cover maps (including the initial and final ones) were generated and are shown in Figure 2f,g. To validate its performance in detecting LCC from year 2002 to 2004, the synthetic Landsat data at year 2004 coupled with the prior land cover product from 2002 were used to generate an LCC map from 2002 to 2004 (Figure 2h). It shows the change detection result using the synthetic satellite data, in which white was used to reflect the correctly predicted LCC classes. In comparison, the MODIS data at year 2004 were also used to generate a change detection result (shown in Figure 2k based on the land cover map at CR (shown in Figure 2j). The actual LCC map from year 2002 to 2004 provided in Figure 2l was used for validation.

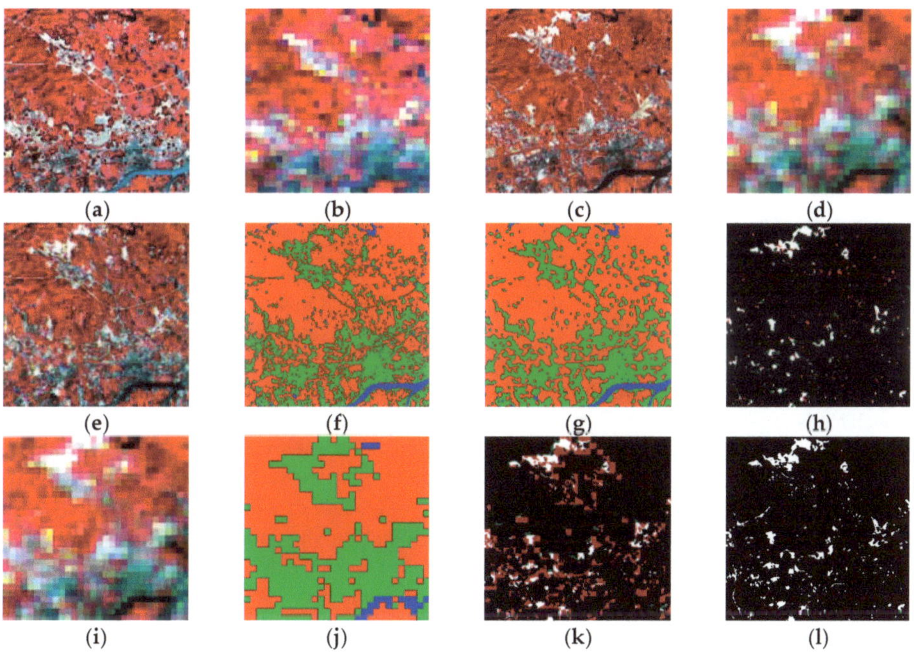

Figure 2. Test with the actual MODIS data for sub-pixel change detection using different downscaling methods: (**a**) Landsat data for the year 2002 and the actual LCC from 2002 to 2004 (highlighted with black); (**b**) MODIS data for 2002; (**c**) Landsat for 2004 as a reference; (**d**) MODIS data for 2004; (**e**) Fused result for 2004 with the proposed approach; (**f**) Initial land cover map for 2004 with the fused result shown in Figure 1e; (**g**) Final land cover map for 2004 with the initial land cover map using the proposed approach; (**h**) Change detection result using the proposed approach; (**i**) MODIS data for 2004; (**j**) Land cover map for 2004 with MODIS data; (**k**) Change detection result with MODIS data from 2002 to 2004; (**l**) Actual LCC from 2002 to 2004 for validation.

3.2. Accuracy Assessment

In the following, the accuracy of different change detection results using different approaches was assessed. Five different scalars—namely, the Kappa statistic, the overall accuracy (OA), the commission error (CE), the omission error (OE), and the correlation coefficient (CC)—were used to assess the change detection accuracy. Other than the omission and commission errors, a higher value of each index reflects a higher change detection accuracy.

Change detection accuracy statistics of the fused results using different approaches are given in Table 1. Results using the simulated MODIS data are also provided for comparison, as shown on the right side of Table 1. It is found that the change detection accuracy with the fused result is much better than using the original MODIS data. Moreover, the proposed approach gives slightly better results than the STARFM method for all scale factors used.

Table 1. Change detection accuracy for the fused result with different methods. STARFM: spatial and temporal adaptive reflectance data-fusion model.

	Actual Data			Simulated Data (S = 16)		
	Soft	**STARFM**	**Proposed**	**Soft**	**STARFM**	**Proposed**
Kappa	0.45	0.46	0.47	0.46	0.49	0.50
OA	83%	84%	85%	83%	85%	86%
CE	19%	18%	17%	18%	17%	17%
OE	32%	38%	37%	31%	36%	30%
CC	0.68	0.69	0.78	0.77	0.86	0.89

4. Discussion

4.1. Strengths

It is apparent that the fusion-based approaches—including STARFM and the proposed one—perform better than the soft classification method when CR data are directly used based on the accuracy statistics provided in Table 1. Let's take the simulated satellite data as an example. Overall accuracies for the results with the proposed and STARFM methods are 86% and 85%, respectively. Varying 83% for the soft classification method is also obtained. Especially, the soft classification approach tends to overestimate the actual LCC, while the fusion-based approach can improve it. When the two downscaling approaches are compared with each other, it is found that the learning-based approach performs slightly better than the conventional STARFM method in terms of all tested indices. Results with the proposed approach, moreover, tend to have better CCs than STARFM. The advantage of the proposed approach is that it can learn the change pattern or spatial texture information from previous satellite data, which is better than the STARFM.

Compared with the conventional unmixing-based fusion approach, an optimized neighborhood size is not required for the proposed approach. Because it is possible to achieve the desired result with a patch covering the whole area of a coarse pixel, a patch size of 8×8 Landsat pixels was used in this study. In addition, the number of land cover types is not required, as a large number of bases (chosen as 256 in this study) is enough to reflect the whole LCC patterns in this study.

4.2. Scale Effect

To assess the impact of spatial scale of the proposed approach in monitoring LCC information, a series of simulated data degraded from the actual Landsat data were used in this study. In our experiment, the scale factors of 4, 8, and 16 were tested. Based on the simulated MODIS and Landsat data, finer resolution fused satellite data and land cover/change maps at different scales were generated. Figure 3h–k shows the results using the proposed approach, while Figure 3a–g show the results using the original MODIS data and the conventional STARFM method for comparison, respectively. Accuracy statistics for all mapping results with different approaches are provided in Table 2.

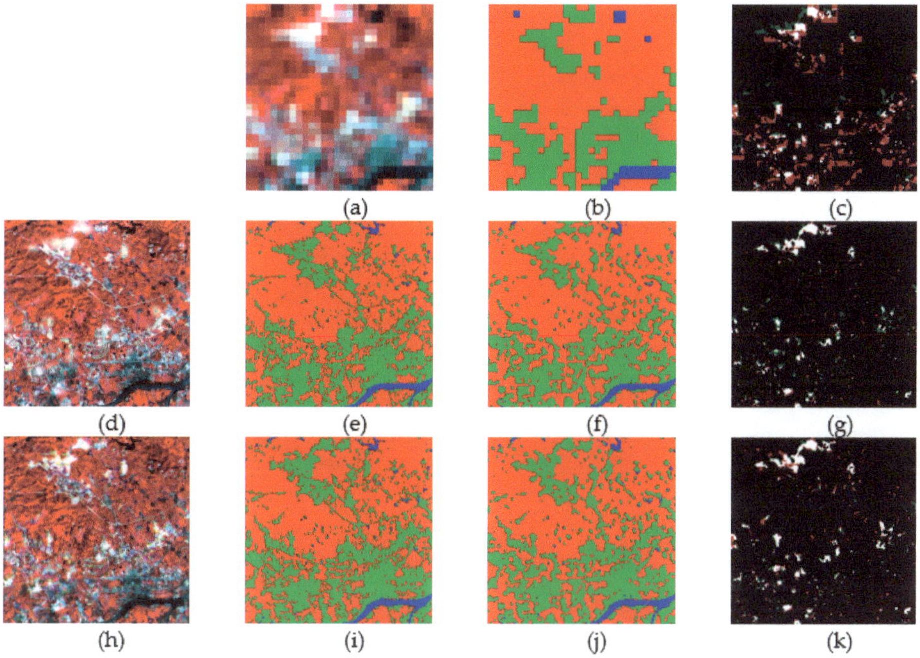

Figure 3. Sub-pixel change detection results with the simulated MODIS data using different methods at a scaling factor of 16. The upper row shows the results using simulated MODIS data (s = 16): (**a**) Simulated MODIS data for the year 2004; (**b**) Land cover map using simulated MODIS data; and (**c**) Change detection result using simulated MODIS data. The middle row shows the results using the conventional fusion-based method. (**d**) Synthetic Landsat data for 2004 using the STARFM method; (**e**) Initial land cover map from the result shown in (**d**); (**f**) Final land cover map from synthetic Landsat data using the STARFM method; and (**g**) Change detection result from 2002 to 2004 using the STARFM method. The lower row shows the results using the proposed approach. (**h**) Synthetic Landsat data for 2004 using the proposed approach; (**i**) Initial land cover map from the result shown in (**h**); (**j**) Final land cover map using the proposed approach; (**k**) Change detection result from 2002 to 2004 using the proposed approach.

Table 2. Change detection accuracy for the fused result with different methods under different scale factors. OA: overall accuracy; CE: commission error; OE: omission error; CC: correlation coefficient.

	Scale Factor = 4			**Scale Factor = 8**			**Scale Factor = 16**		
	Soft	**STARFM**	**Proposed**	**Soft**	**STARFM**	**Proposed**	**Soft**	**STARFM**	**Proposed**
Kappa	0.53	0.60	0.61	0.52	0.53	0.55	0.46	0.49	0.50
OA	86%	89%	90%	85%	87%	88%	83%	85%	86%
CE	15%	13%	15%	16%	15%	15%	18%	17%	17%
OE	23%	27%	16%	23%	30%	26%	31%	36%	30%
CC	0.89	0.92	0.93	0.88	0.89	0.92	0.77	0.86	0.89

According to the accuracy statistics provided in Table 2, three observations can be summarized as below. First, the change detection accuracy decreases significantly as the scale factor increases. Taking the results using the proposed approach as an example, the Kappa index decreased from 0.61

to 0.50 when the scale factor increased from 4 to 16. Second, when the performances of different approaches were compared, it was found that the fused-based approaches achieved better change detection accuracy than results using the original MODIS data. In particular, the use of original MODIS data tended to overpredict the actual LCC, while the fused-based approaches improved it. Third, comparing the performance of STARFM and the proposed method, the proposed approach performed better than STARFM, regardless of which scale factor was used. In particular, a much better CC was achieved by the proposed learning-based approach compared with STARFM, indicating that the learning-based approach is suitable for the downscaling of CR data.

4.3. Limitations

Although the proposed approach has been validated and proven suitable for sub-pixel LCC detection using CR satellite data, there are still some limitations. First, misregistration errors between multisource satellite data of the proposed approach may affect the final change detection results, and thus using the simulated data can achieve better detection accuracy than using actual satellite data. Second, the advantage of the proposed approach is obvious when the predicted LCC percentages are compared with others by referring to the CC index. Nevertheless, the predicted LCCs still have positional errors within a coarse pixel, which may offset its advantage for sub-pixel LCC detection using the actual multisource satellite data. Lastly, it is a computationally expensive approach. Both dictionary training and sparse coefficient estimation processes are computationally expensive.

5. Conclusions

In this paper, a learning-based downscaling method is presented to generate finer resolution LCC results using prior LCC information and one CR data at the predicted time, in which prior LCC patterns are learned and modeled using the popular sparse learning approach. Further experiments demonstrate that it is better than the conventional downscaling approach STARFM when both predicted synthetic data are applied for LCC detection. Experiments conducted at Guangzhou show that the proposed learning-based approach outperforms both the conventional change detection method and the fusion-based change detection method. According to the results with the proposed approach using actual MODIS data, the overall LCC detection accuracy is 85%, which is better than the results using a conventional soft classification method and fusion-based method (83% and 84%, respectively). More importantly, it is found that high-quality LCC percentages—as indicated by the CC index—can be achieved by the proposed approach, as the CC index for the proposed approach is 0.78, which is much better than the results using soft classification and fusion-based methods (0.68 and 0.69, respectively). This finding is meaningful for high-quality LCC detection at a sub-pixel level.

This study also investigated the effect of scale factor on sub-pixel change detection. In particular, results from fusion-based approaches perform much better than when the original coarse satellite data are used directly, regardless of the scale factor used. The soft classification method tends to overestimate the actual LCC area, while the proposed approach tends to miss some actual LCC area. When a large scale factor is adopted, the proposed approach performs slightly better than the STARFM model. When compared with the use of the simulated MODIS data, the use of actual MODIS data achieves a slightly lower LCC detection accuracy. The main reason may be due to the positioning error between MODIS and Landsat data, which needs to be further investigated.

Acknowledgments: This research was supported by the China NSFC project under contract 61373114, 61661166011, and 11690011, the VC's discretionary fund of The Chinese University of Hong Kong, Macau Science and Technology Development Funds with No. 003/2016/AFJ, and 973 Program of China with No. 2013CB329404.

Author Contributions: Y.X. and D.M. conceived the experiments and interpreted the result. Y.X. and L.L. performed the experiments. Y.X. wrote the paper.

Conflicts of Interest: The authors declare no conflict of interest.

References

1. Bruzzone, L.; Prieto, D.F. An adaptive semi-parametric and context-based approach to unsupervised change detection in multitemporal remote-sensing images. *IEEE Trans. Image Process.* **2002**, *11*, 452–466.
2. Xu, Y.; Huang, B. Spatial and temporal classification of synthetic satellite imagery: Land cover mapping and accuracy validation. *Geo-Spat. Inf. Sci.* **2014**, *17*, 1–7. [CrossRef]
3. Singb, A. Digital change detection techniques using remotely sensed data. *Int. J. Remote Sens.* **1989**, *10*, 989–1003. [CrossRef]
4. Wang, Q.; Lin, J.; Yuan, Y. Salient band selection for hyperspectral image classification via manifold ranking. *IEEE Trans. Neural Netw. Learn. Syst.* **2016**, *27*, 1279–1289. [CrossRef] [PubMed]
5. Fung, T.; LeDrew, E. Application of principal components analysis to change detection. *Photogramm. Eng. Remote Sens.* **1988**, *53*, 1649–1658.
6. Chen, J.; Gong, P.; He, C.; Pu, R.; Shi, P. Land-use/land-cover change detection using improved change-vector analysis. *Photogramm. Eng. Remote Sens.* **2003**, *69*, 369–379. [CrossRef]
7. Lu, D.; Mausel, P.; Brondizio, E.; Moran, E. Change detection techniques. *Int. J. Remote Sens.* **2004**, *25*, 2365–2401. [CrossRef]
8. Le Hégarat-Mascle, S.; Ottlé, C.; Guérin, C. Land cover change detection at coarse spatial scales based on iterative estimation and previous state information. *Remote Sens. Environ.* **2005**, *95*, 464–479. [CrossRef]
9. Strugnell, N.C.; Lucht, W.; Schaaf, C. A global albedo data set derived from AVHRR data for use in climate simulations. *Geophys. Res. Lett.* **2001**, *28*, 191–194. [CrossRef]
10. Friedl, M.A.; McIver, D.K.; Hodges, J.C.; Zhang, X.Y.; Muchoney, D.; Strahler, A.H.; Woodcock, C.E.; Gopal, S.; Schneider, A.; Cooper, A.; et al. Global land cover mapping from MODIS: Algorithms and early results. *Remote Sens. Environ.* **2002**, *83*, 287–302. [CrossRef]
11. Walker, J.J.; De Beurs, K.M.; Wynne, R.H.; Gao, F. Evaluation of Landsat and MODIS data fusion products for analysis of dryland forest phenology. *Remote Sens. Environ.* **2012**, *117*, 381–393. [CrossRef]
12. Verbesselt, J.; Hyndman, R.; Zeileis, A.; Culvenor, D. Phenological change detection while accounting for abrupt and gradual trends in satellite image time series. *Remote Sens. Environ.* **2010**, *114*, 2970–2980. [CrossRef]
13. Wu, K.; Du, Q.; Wang, Y.; Yang, Y. Supervised Sub-Pixel Mapping for Change Detection from Remotely Sensed Images with Different Resolutions. *Remote Sens.* **2017**, *9*, 284. [CrossRef]
14. He, D.; Zhong, Y.; Feng, R.; Zhang, L. Spatial-Temporal Sub-Pixel Mapping Based on Swarm Intelligence Theory. *Remote Sens.* **2016**, *8*, 894. [CrossRef]
15. Ling, F.; Li, W.; Du, Y.; Li, X. Land cover change mapping at the subpixel scale with different spatial-resolution remotely sensed imagery. *IEEE Geosci. Remote Sens. Lett.* **2011**, *8*, 182–186. [CrossRef]
16. Ling, F.; Foody, G.M.; Li, X.; Zhang, Y.; Du, Y. Assessing a temporal change strategy for sub-pixel land cover change mapping from multi-scale remote sensing imagery. *Remote Sens.* **2016**, *8*, 642. [CrossRef]
17. Zurita-Milla, R.; Kaiser, G.; Clevers, J.G.P.W.; Schneider, W.; Schaepman, M.E. Downscaling time series of MERIS full resolution data to monitor vegetation seasonal dynamics. *Remote Sens. Environ.* **2009**, *113*, 1874–1885. [CrossRef]
18. Foody, G.M. Approaches for the production and evaluation of fuzzy land cover classifications from remotely-sensed data. *Int. J. Remote Sens.* **1996**, *17*, 1317–1340. [CrossRef]
19. Atkinson, P.M. Sub-pixel target mapping from soft-classified, remotely sensed imagery. *Photogramm. Eng. Remote Sens.* **2005**, *71*, 839–846. [CrossRef]
20. Gao, F.; Masek, J.; Schwaller, M.; Hall, F. On the blending of the Landsat and MODIS surface reflectance: Predicting daily Landsat surface reflectance. *IEEE Trans. Geosci. Remote Sens.* **2006**, *44*, 2207–2218.
21. Hilker, T.; Wulder, M.A.; Coops, N.C.; Linke, J.; McDermid, G.; Masek, J.G.; Gao, F.; White, J.C. A new data fusion model for high spatial-and temporal-resolution mapping of forest disturbance based on Landsat and MODIS. *Remote Sens. Environ.* **2009**, *113*, 1613–1627. [CrossRef]
22. Hilker, T.; Wulder, M.A.; Coops, N.C.; Seitz, N.; White, J.C.; Gao, F.; Masek, J.G.; Stenhouse, G. Generation of dense time series synthetic Landsat data through data blending with MODIS using a spatial and temporal adaptive reflectance fusion model. *Remote Sens. Environ.* **2009**, *113*, 1988–1999. [CrossRef]
23. Zhu, X.; Chen, J.; Gao, F.; Chen, X.; Masek, J.G. An enhanced spatial and temporal adaptive reflectance fusion model for complex heterogeneous regions. *Remote Sens. Environ.* **2010**, *114*, 2610–2623. [CrossRef]

24. Roy, D.P.; Ju, J.; Lewis, P.; Schaaf, C.; Gao, F.; Hansen, M.; Lindquist, E. Multi-temporal MODIS–Landsat data fusion for relative radiometric normalization, gap filling, and prediction of Landsat data. *Remote Sens. Environ.* **2008**, *112*, 3112–3130. [CrossRef]
25. Sun, J.; Xu, Z.; Shum, H.Y. Image super-resolution using gradient profile prior. In Proceedings of the IEEE Conference on Computer Vision and Pattern Recognition (CVPR), Anchorage, AK, USA, 23–28 June 2008; IEEE: Hoboken, NJ, USA, 2008.
26. Baker, S.; Kanade, T. Limits on super-resolution and how to break them. *IEEE Trans. Pattern Anal. Mach. Intell.* **2002**, *24*, 1167–1183. [CrossRef]
27. Gu, S.; Zuo, W.; Xie, Q.; Meng, D.; Feng, X.; Zhang, L. Convolutional sparse coding for image super-resolution. In Proceedings of the IEEE International Conference on Computer Vision, Santiago, Chile, 13–16 December 2015; pp. 1823–1831.
28. Yang, J.; Wright, J.; Huang, T.S.; Ma, Y. Image super-resolution via sparse representation. *IEEE Trans. Image Process.* **2010**, *19*, 2861–2873. [CrossRef] [PubMed]
29. Huang, B.; Song, H. Spatiotemporal reflectance fusion via sparse representation. *IEEE Trans. Geosci. Remote Sens.* **2012**, *50*, 3707–3716. [CrossRef]
30. Xu, Y.; Huang, B. A spatio–temporal pixel-swapping algorithm for subpixel land cover mapping. *IEEE Geosci. Remote Sens. Lett.* **2014**, *11*, 474–478. [CrossRef]
31. Wang, Q.; Shi, W.; Atkinson, P.M.; Li, Z. Land cover change detection at subpixel resolution with a Hopfield neural network. *IEEE J. Sel. Top. Appl. Earth Obs. Remote Sens.* **2015**, *8*, 1339–1352. [CrossRef]
32. Zhang, Y.; Atkinson, P.M.; Li, X.; Ling, F.; Wang, Q.; Du, Y. Learning-Based Spatial–Temporal Superresolution Mapping of Forest Cover with MODIS Images. *IEEE Trans. Geosci. Remote Sens.* **2017**, *55*, 600–614. [CrossRef]
33. Aharon, M.; Elad, M.; Bruckstein, A. K-SVD: An algorithm for designing overcomplete dictionaries for sparse representation. *IEEE Trans. Signal Process.* **2006**, *54*, 4311–4322. [CrossRef]
34. Davis, G.; Mallat, S.; Avellaneda, M. Adaptive greedy approximations. *Construc. Approx.* **1997**, *13*, 57–98. [CrossRef]

remote sensing

MDPI

Article

Urban Change Analysis with Multi-Sensor Multispectral Imagery

Yuqi Tang [1,2,*] and Liangpei Zhang [3,*]

1 School of Geosciences and Info-Physics, Central South University, Changsha 410083, China
2 Key Laboratory of Metallogenic Prediction of Nonferrous Metals and Geological Environment Monitoring (Central South University), Ministry of Education, Changsha 410083, China
3 State Key Laboratory of Information Engineering in Surveying, Mapping, and Remote Sensing and the Collaborative Innovation Center for Geospatial Technology, Wuhan University, Wuhan 430079, China
* Correspondence: yqtang@csu.edu.cn (Y.T.); zlp62@whu.edu.cn (L.Z.); Tel.: +86-139-7311-5356 (Y.T.); +86-139-9555-6225 (L.Z.)

Academic Editors: Qi Wang, Nicolas H. Younan, Carlos López-Martínez, Soe W. Myint and Prasad S. Thenkabail
Received: 5 January 2017; Accepted: 6 March 2017; Published: 9 March 2017

Abstract: An object-based method is proposed in this paper for change detection in urban areas with multi-sensor multispectral (MS) images. The co-registered bi-temporal images are resampled to match each other. By mapping the segmentation of one image to the other, a change map is generated by characterizing the change probability of image objects based on the proposed change feature analysis. The map is then used to separate the changes from unchanged areas by two threshold selection methods and k-means clustering ($k = 2$). In order to consider the multi-scale characteristics of ground objects, multi-scale fusion is implemented. The experimental results obtained with QuickBird and IKONOS images show the superiority of the proposed method in detecting urban changes in multi-sensor MS images.

Keywords: multi-sensor; change feature analysis; object-based; multispectral images

1. Introduction

Change detection involves identifying the changed ground objects between a given pair of multi-temporal (so-called bi-temporal) images observing the same scene at different times [1,2]. The existing change detection methods can be classified into two classes: supervised and unsupervised. Supervised change detection relies on prior information about the ground changes, but unsupervised change detection automatically generates the difference between bi-temporal images to locate [3–6], and even distinguish, changes [5–8].

Most of the unsupervised change detection methods are implemented pixel-wise [9,10], and the classic approach is differencing the bi-temporal images and regarding the pixels with a larger difference as changed [4]. Subsequently, a large number of pixel-based change detection methods have been proposed, including methods based on image transformation [11–17], soft clustering [18–20], and similarity measurement [21]. However, all of these methods presume spatial independence among the image pixels, which is not appropriate for high-resolution images. This is because, in high-resolution images, most of the ground objects cover sets of neighboring pixels, and some information reliance exists among these pixels. Aiming at this drawback of pixel-based change detection in high-resolution images, some researchers have attempted to use the spatial information in a fixed-size image unit, together with the spectrum, to detect ground changes. Examples of such methods include texture extraction [22–24], structural information extraction by Markov random fields (MRFs) [4,25,26], and morphological filtering [27,28].

In order to adapt to the irregular distribution of ground objects, object-based theory has been introduced into change detection for high-resolution images [29]. Object-based theory regards some of the spatially-neighboring and spectrally-similar pixels as a union (a so-called object) to detect whether they are changed. It makes use of the spatial information in the high-resolution image, together with the spectrum, and reduces the salt-and-pepper effect. In recent years, a large number of object-based unsupervised change detection methods [30–33] have been proposed and have improved the accuracy of change detection for high-resolution images. However, most of the existing object-based change detection methods focus on using bi-temporal images acquired by the same sensor. In the case of massive high-resolution images acquired by different sensors, it is necessary to utilize them simultaneously to improve the information extraction. In order to detect changes in multi-sensor remote sensing images, some researchers have addressed change measurement [34,35], and other researchers have focused on the classification of changed features [6,9,36]. Robust change vector analysis (RCVA) was proposed for multi-sensor change detection with very-high-resolution optical satellite data, and this approach improves the robustness of CVA to different viewing geometries or registration noise [37]. Unfortunately, these methods do not consider the incompatibility between different band widths in bi-temporal multispectral (MS) images (Table 1). Moreover, some of the object-based statistical features between bi-temporal images might be affected in the change detection, since changes always arise from ground objects' expansion, reduction, or property variation.

Table 1. Comparison between the bandwidth and spatial resolution of QuickBird and IKONOS images.

	Blue Band (um)	Green Band (um)	Red Band (um)	Near Infrared band (um)	Spatial Resolution (nadir, m)
QuickBird MS image	0.45–0.52	0.52–0.60	0.63–0.69	0.76–0.90	2.44
IKONOS MS image	0.445–0.516	0.506–0.595	0.632–0.698	0.757–0.853	3.28

In this paper, a novel object-based change detection method is proposed for multi-sensor MS imagery. The consistency of bi-temporal image objects is achieved by segmenting one image and mapping this segmentation to the other. Instead of comparing the objects' spectral bands in the bi-temporal images, we summarize the possible distribution between any image object and its relevant changed areas, and we analyze the statistical feature variation of the change-related objects and define a change feature to represent the change probability of the image objects in the bi-temporal MS images. In order to locate the changed areas, binarization of the change map is implemented by thresholding or binary unsupervised classification. In addition, in view of the multi-scale characteristics of the ground objects, multi-scale fusion is carried out.

The rest of this paper is organized as follows. Section 2 describes the proposed method. The experimental results and a discussion are presented in Sections 3 and 4, respectively. Section 5 provides our conclusion and future work directions.

2. Object-Based Change Analysis

The processing flow of the proposed method is shown in Figure 1.

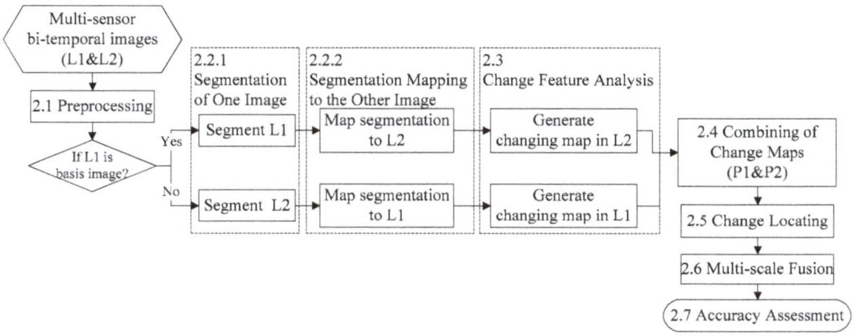

Figure 1. Processing flow of the proposed method.

2.1. Preprocessing

In the preprocessing of the proposed method, image resampling is conducted to unify the size of the multi-sensor bi-temporal images. The bilinear resampling method is adopted to suppress the image heterogeneity, with a reasonable computation cost [38]. When the basis image is the one with a higher spatial resolution, the other image needs to be interpolated by up-sampling. Otherwise, the image is degraded by down-sampling to the lower resolution of the basis image.

2.2. Image Segmentation

Image segmentation is implemented to obtain image objects for the subsequent object-based processes. In this paper, there are three objectives for the image segmentation: (1) the bi-temporal image objects should be in one-to-one correspondence; (2) the spatial distribution between changed objects and their relevant changed areas needs to be preserved for the subsequent change feature analysis (Section 2.3); and (3) the objects obtained from slight under-segmentation are better able to fit the edges of the changed areas in the other image. Therefore, we propose to segment one of the bi-temporal images and map the segmentation to the other. These two segmentation processes are introduced below.

2.2.1. Segmentation of One Image

The segmentation of one image should take into account the spectral and spatial features of the ground objects. In addition, as mentioned above, the image objects should be slightly under-segmented to fit the edges of the changed areas in the other image. In this paper, we use the fractal net evolution approach (FNEA) [39] for the image segmentation. This approach involves calculating the heterogeneity (S_f) between each pair of neighboring objects according to Equation (1), which is a weighted sum of the spectral and spatial criteria:

$$S_f = \omega_{spect.} h_{spect.} + (1 - \omega_{spect.}) h_{spac.} \tag{1}$$

where $0 \leq \omega_{spect.} \leq 1$ is the user-defined weight of the spectral feature. The sum of the weights of the spectral and spatial criteria equals 1. If the spectral feature is emphasized in the segmentation, the value of $\omega_{spect.}$ should be larger. Conversely, the value of $(1 - \omega_{spect.})$, which is the weight of the spatial feature, should be larger when the spatial feature is more important. $h_{spect.}$ and $h_{spac.}$ are, respectively, the spectral and spatial heterogeneity, whose definition can be found in [39].

At the beginning of the segmentation, every pixel is regarded as an individual object. After calculating the heterogeneity (S_f) of each pair of neighboring objects, they are compared to the value of the scale, which can be regarded as the threshold of the heterogeneity:

(1) If S_f < scale, this pair of objects are merged;

(2) Otherwise, the objects are preserved as two individual objects.

This procedure is repeated until no objects can be merged, and the object map is obtained. The scale is critical to the segmentation as it determines the size of the objects.

Using FNEA, only the scale parameter needs to be selected to adjust the size of the image objects. We can make use of Definiens software (Definiens, München, Germany) to simply implement this method. On the premise of efficiency, other segmentation methods [40,41] could also be adopted in the proposed method.

2.2.2. Segmentation Mapping to the Other Image

In this paper, we simply map the segmentation of one image to the other. In this way, the bi-temporal image objects are in one-to-one correspondence. In addition, the spatial distribution between changed objects and their relevant changed areas are also preserved, which is critical for the following change feature analysis.

2.3. Change Feature Analysis

After mapping the segmentation of one image to the other, there will be different spatial distributions between a changed object and its relevant changed area. Figure 2 shows the possible distributions of a changed object and its relevant changed area, in which the bold object represents a changed object, and the object above it is one of its neighboring objects. The shadow area represents the relevant changed area. Through analyzing the six possible distributions in Figure 2, we can deduce the statistical feature variation of the changed objects as follows:

Denoting the bi-temporal images as L1 and L2 and mapping the segmentation of L1 to L2,

(a) if the relevant changed area is contained in the changed object, the standard deviation of the changed object in L2 is larger than L1 (Figure 2a);
(b) if the relevant changed area covers parts of the changed object and its neighborhood, the contrast between the changed object and its neighboring pixels in L2 is less than L1 (Figure 2b);
(c) if the relevant changed area exactly covers the changed object, the ratio of contrast between the changed object and its neighboring pixels in L1 and L2 is not equal to 1 (Figure 2c);
(d) if the relevant changed area covers the whole changed object and parts of its neighborhood, the contrast between the changed object and its neighboring pixels in L2 is less than L1 (Figure 2d);
(e) if the relevant changed area exactly covers the changed object and its neighboring object, the contrast between the changed object and its neighboring pixels in L2 is less than L1 (Figure 2e); and
(f) if the relevant changed area exceeds the changed object and its neighboring object, the contrast between the changed object and its neighboring pixels in L2 is less than L1 (Figure 2f).

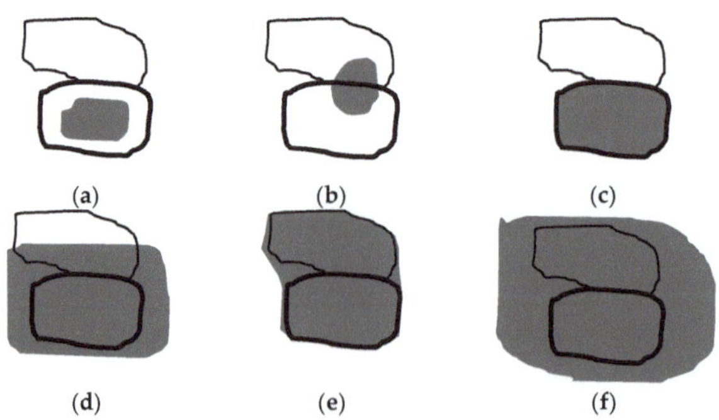

Figure 2. Possible distributions of a changed object and its relevant changed area, whose statistical feature variation is described as above (**a**–**f**).

According to the above statistical feature variations of changed objects, we define a change feature (Equations (2) and (3)) to describe the statistical features of the image objects in bi-temporal MS images. The change feature adequately takes into account the statistical features of the image objects in the bi-temporal images (acquired by the same or different satellites), which is an important innovation of the proposed method.

If $0 < F_{iRatio-Ctr.} < 1$:

$$F_i = \frac{F_{iRatio-Ctr.} \cdot \sum_{\forall(i,j)\in ObjNei_i} F_{ijCtr.}}{F_{iS.D.}} \tag{2}$$

otherwise:

$$F_i = \frac{\sum_{\forall(i,j)\in ObjNei_i} F_{ijCtr.}}{F_{iRatio-Ctr.} \cdot F_{iS.D.}} \tag{3}$$

where F_i is the change feature value for object i, and $F_{iRatio-Ctr.}$ is the ratio of contrast between object i and its neighboring pixels in L1 and L2. $F_{ijCtr.}$ is the contrast between object i and its neighboring pixel (i, j), and $F_{iS.D.}$ is the standard deviation of object i. $ObjNei_i$ is the set of pixels adjacent to object i.

The ratio of contrast between the changed object and its neighboring pixels in L1 and L2 can be defined as:

$$F_{iRatio-Ctr.} = \frac{\sum_{\forall(i,j)\in ObjNei_i} F_{1ijCtr.}}{\sum_{\forall(i,j)\in ObjNei_i} F_{2ijCtr.}} \tag{4}$$

where $F_{1ijCtr.}$ and $F_{2ijCtr.}$ represent the contrast between object i and its neighboring pixel (i, j) in L1 and L2, respectively.

The contrast between the changed object and one of its neighboring pixels can be defined as:

$$F_{ijCtr.} = \frac{|\mu_i - X(i,j)|}{|\mu_i + X(i,j)|} \tag{5}$$

where μ_i is the mean value of the pixels in object i, and $X(i,j)$ is the value of the neighboring pixel (i, j).

The standard deviation of the changed object is defined as:

$$F_{iS.D.} = \sqrt{\frac{1}{n_i} \sum_{\forall(i,j)\in Obj_i} (X(i,j) - \mu_i)^2} \tag{6}$$

where n_i is the number of neighboring pixels in object i, and Ob_{ji} is the set of pixels in object i.

According to the proposed change feature of image objects, there are three statistical factors related to the changes:

(1) the ratio of contrast between any object and its neighboring pixels in L1 and L2;
(2) the sum of contrast between any object and each of its neighboring pixels; and
(3) the standard deviation value of any object.

In other words, if any image object is related to local changes, one of these three factors would vary between the bi-temporal images, and the proposed change feature of this object in L2 would be less in L1. Consequently, the change map in L2 can be generated by representing each object with the change probability:

$$P_{2i} = (F_{1i} - F_{2i})/F_{1i} \tag{7}$$

2.4. Combining the Change Maps

In order to preserve the change information as much as possible, the bi-temporal images take turns to be segmented and mapped to each other. The pair of change maps is combined as:

$$P_{com.i} = \omega_2 \cdot P_{2i} + \omega_1 \cdot P_{1i} \tag{8}$$

where $P_{com.i}$ is the combined change probability of object i. P_{2i} and P_{1i} represent the change probabilities of object i by respectively segmenting L1 and L2 and mapping them to each other. ω_1 and ω_2 are the weights of the change maps. Subsequently, the combined change map can be used for locating the changes. The combination ratio of change maps $R_{com.}$ is an important parameter in this method, which is confirmed in the experiments (Section 3).

$$R_{com.} = \omega_2 / \omega_1 \tag{9}$$

2.5. Change Locating

The changes are located by discriminating them from unchanged areas in the combined change map. Since the combined change map represents the change probability of each gray-level image object, the change locating can be realized by setting a threshold to divide the map into two parts, or applying a binary unsupervised classification method. In this paper, two threshold selection techniques, Otsu's thresholding method [42] and "threshold selection by clustering gray levels of boundary" [43], and k-means clustering [44] (k = 2) are used to extract the changes in the combined change map. These methods could also be replaced by other thresholding or clustering methods [45–47], in which [45] effectively improved the band selection of hyperspectral imagery concerning on dual clustering. However, it is confirmed to have little effect on the proposed method (see Section 3).

(1) Otsu's thresholding method

Otsu's thresholding method is implemented by searching for the optimal threshold to maximize the discrimination criterion and achieve the greatest separability of classes. The criterion is defined as:

$$C = \frac{[\mu_T \cdot \omega(k) - \mu(k)]^2}{\omega(k) \cdot [1 - \omega(k)]} \tag{10}$$

where C is the criterion value of an image unit (pixel or object), and μ_T is the mean of the gray levels in the image. $\omega(k)$ and $\mu(k)$ are the zeroth- and first-order cumulative moments of the histogram up to the k-th gray level, respectively. The optimal threshold is obtained by maximizing the value of C. In this paper, Otsu's thresholding method is used to find the optimal threshold to separate the changes and unchanged areas in the combined change map.

(2) Threshold selection by clustering gray levels of boundary

The threshold selection by clustering gray levels of boundary method involves approximating the mean of the discrete sample pixels lying on the boundary and separating the image into objects and background. The image is divided into square grids, and classified into edge cells intersected by boundary and non-edge cells. Mathematically, the boundary of the image can be represented as:

$$\begin{cases} l(x,y) = 0 \\ \|\Delta f(x,y)\| \geq T_e \end{cases} \tag{11}$$

where $l(x,y)$ and $\|\Delta f(x,y)\|$ are the Laplacian and gradient magnitude functions of pixel (x, y), respectively. If any edge of an edge cell is intersected by the boundary, the edge has the following properties:

(a) its two vertices (p_1 and p_2) are a pair of zero-crossing points, namely, $l(p_1) \cdot l(p_2) < 0$; and
(b) its two vertices (p_1 and p_2) both have high gradient values. For a predefined gradient threshold $\widetilde{T}_e$, $g(p_1) + g(p_2) \geq 2 \cdot \widetilde{T}_e$.

In this way, the intersected pixels of edge cells on the boundary can be obtained. Their positions and gray values are computed by linear interpolation of the two vertices on the edge. These intersected pixels are regarded as the discrete sample pixels on the image boundary. The mean of their gray values is used as the threshold for the image segmentation. In this study, in order to divide the combined change map into changed and unchanged classes, this threshold selection method is used to find a bi-level threshold in the feature map.

(3) *K*-means clustering

K-means clustering is a classical unsupervised classification method. It involves clustering image pixels according to the similarity of their gray levels. The number of clusters depends on the specific application and is defined by the user. In this paper, *k*-means clustering (*k* = 2) is used to classify the combined change map—a gray-level image—into two classes of changed and unchanged areas.

2.6. Multi-Scale Fusion

Considering the multi-scale characteristic of ground objects, multi-scale fusion [30] is applied in the proposed method. The multi-scale fusion is implemented by voting from the single-scale change detection maps. Firstly, we choose an appropriate interval for the segmentation scale, which needs to cover most of the image objects' sizes. We repeat the processes of the proposed method from steps 2.1 to 2.5 (in Figure 1) by increasing the scale with a constant step size, and we obtain a set of single-scale change detection maps. The image objects in these maps only have two values—0 and 1—which, respectively, mean unchanged and changed objects. The sum of the single-scale change detection maps is calculated as:

$$M_i = \sum_{j=1}^{n} S_{ji} \tag{12}$$

where S_{ji} is the value of object i in single-scale change detection map j. M_i is the sum of object i in all of the single-scale change detection maps, and n is the number of single-scale change detection maps. The multi-scale change detection map is defined as:

$$F_i = \begin{cases} 1 & \text{If } M_i > T_f \\ 0 & \text{Otherwise} \end{cases}, \; T_f = 0,\, 1,\, \ldots,\, n-1 \tag{13}$$

where F_i is the value of image object i in the multi-scale change detection map, in which 0 and 1, respectively, mean unchanged and changed objects. T_f is the threshold of the multi-scale fusion.

In this way, if an image object is changed in more than T_f single-scale change detection maps, it is recognized as changed after the multi-scale fusion. Especially, the changed areas after the multi-scale fusion are the sum and the intersection of the changes in all the single-scale change detection maps, when T_f is equal to 0 and 1, respectively.

In the experiments described in Section 3, the optimal result of the multi-scale fusion is the sum of changes in all the single-scale change detection maps, in which T_f is equal to 0.

2.7. Accuracy Assessment

In this paper, false alarms, missed alarms, and overall errors are used to assess the accuracy of the urban change detection. False alarms mean the ratio of unchanged pixels wrongly detected as changed, and missed alarms are the ratio of changed pixels omitted in the change detection. Consequently, overall errors, which is the integrated ratio of the wrongly detected and omitted changed pixels in the image, estimates the effectiveness of the change detection method [30].

Furthermore, in order to validate the effectiveness of the proposed method, it was compared with some of the existing methods. The most important innovations of the proposed method are that it takes into account the incompatibility between different bandwidths and uses an object-based change measure in the multi-sensor MS images. Since there are no other object-based change detection methods for multi-sensor images, we chose to compare the proposed method with the method proposed in [35], which utilizes some features that are invariant to change in the illumination conditions to undertake change detection in multi-sensor images.

3. Experiments

3.1. The First Study Area

The first study area covers the campus of Wuhan University in Hubei province of China. The bi-temporal images were respectively acquired by the QuickBird satellite in April 2005 (L1) and the IKONOS satellite in July 2009 (L2). In order to preserve the spectral information, the MS images were used in the experiments. Although there were four bands in both images, their spectral and spatial characteristics differed as they were acquired by different sensors (Table 1). Either L1 or L2 can be viewed as the basis image in the image resampling preprocessing.

1. L1 as the basis image

With L1 as the basis image, L2 was interpolated to the spatial resolution of L1. Figure 3 shows the bi-temporal images after the interpolation, which are both 400 $\times$ 400 pixels. In order to avoid the effects of vegetation phenology and solar elevation, the vegetation and shadow were extracted and masked out.

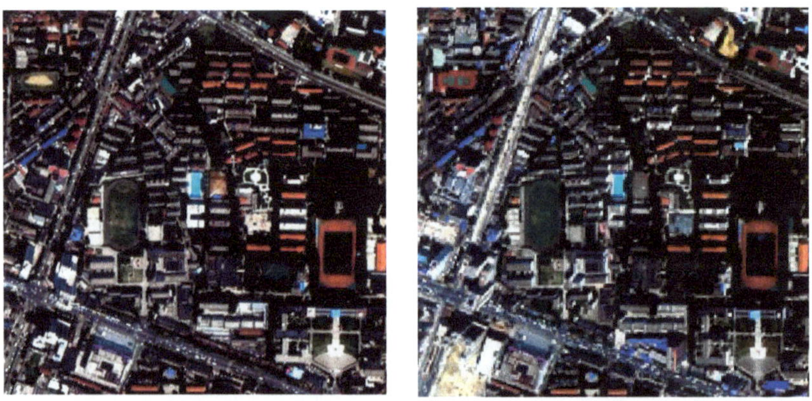

(a) (b)

Figure 3. Interpolated bi-temporal images of the first study area. (**a**) Acquired by QuickBird in April 2005 (L1); and (**b**) acquired by IKONOS in July 2009 (L2).

By mapping the segmentation of L1 to L2, a change map was generated by calculating the value of the change probability (Equation (7)) for each object. The other change map was obtained by exchanging the order of the two images. With different ratios for combining these maps, the characteristics of the combined changed maps varied.

In order to determine the change locations, it is crucial to discriminate the changes from the unchanged areas in the combined change map. The two threshold selection techniques and k-means clustering (k = 2) (introduced in Section 2.5) were used to analyze the combined change map. The results of the three methods are shown in Table 2. In this table, the left, middle, and right parts, respectively, show false, missed alarms, and overall errors among the three methods with different combining ratios of change maps. It can be seen that the overall errors of the three methods are similar. The k-means clustering (k = 2) obtains the smallest number of errors, and the threshold selection by clustering gray levels of boundary method performs a little better than Otsu's thresholding method. Moreover, with the increase of the combination ratio of the change maps, the overall errors of each method decrease. This is because, in Equation (8), P_2 and P_1 represent the change probability of L2 and L1, which was mapped from the segmentation of L1 and L2, respectively. As L2 was interpolated to the spatial resolution of L1, the segmentation of L1 was more accurate than the segmentation of L2. Therefore, a larger weight of P2 leads to a higher accuracy of change feature analysis.

Table 2. Comparison between the change detection results of the three thresholding and clustering methods, with L1 as the basis image in the first study area (scale = 100).

Combination Ratio of Change Maps	False Alarm _Otsu	False Alarm _Edge	False Alarm _K-Means	Missed Alarm _Otsu	Missed Alarm _Edge	Missed Alarm _K-Means	Overall Error s_Otsu	Overall Errors _Edge	Overall Errors _K-Means
1:9	1.47%	1.52%	2.04%	5.57%	4.88%	3.65%	7.04%	6.39%	5.69%
2:8	1.35%	1.41%	1.91%	5.54%	4.82%	3.43%	6.89%	6.23%	5.34%
3:7	1.28%	1.36%	1.89%	5.52%	4.80%	3.38%	6.80%	6.16%	5.27%
4:6	1.24%	1.31%	1.66%	5.38%	4.76%	3.24%	6.62%	6.07%	4.90%
5:5	1.15%	1.25%	1.69%	5.38%	4.70%	2.93%	6.54%	5.94%	4.63%
6:4	1.07%	1.13%	1.70%	4.95%	4.64%	2.88%	6.02%	5.77%	4.58%
7:3	0.98%	1.05%	1.72%	4.73%	4.34%	2.86%	5.70%	5.39%	4.59%
8:2	0.96%	1.08%	1.70%	4.55%	4.20%	2.77%	5.51%	5.27%	4.47%
9:1	0.94%	1.04%	**1.66%**	4.16%	3.86%	**2.42%**	5.10%	4.89%	**4.08%**

The results are visually compared in Figure 4, in which the white and black regions, respectively, represent the changed and unchanged areas. The results of the three methods are similar, but the number of false alarms for k-means clustering (k = 2) is slightly more than for the other two methods, and the missed alarms are fewer in number, especially in the road areas.

According to the spatial resolution and the objects' sizes in the bi-temporal images after preprocessing, the scale interval and step size increase were set as [10, 150] and 10, respectively. The results of the change feature analysis differ with the varying segmentation scales (Figure 5), and the optimal scale is around 100. Considering the multi-resolution characteristics of ground objects, multi-scale fusion is applied in the proposed method, and is realized by voting from the single-scale binary change maps. Figure 6 shows the accuracy of the k-means clustering (k = 2) after the multi-scale fusion. The overall errors are the lowest when T_f in Equation (13) is 0, which means that the optimal multi-scale fusion is the sum of the changes in all of the single-scale change detection maps.

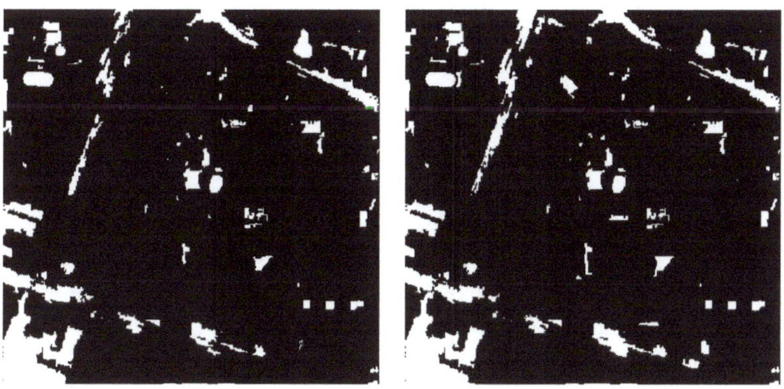

(a) (b)

Figure 4. *Cont.*

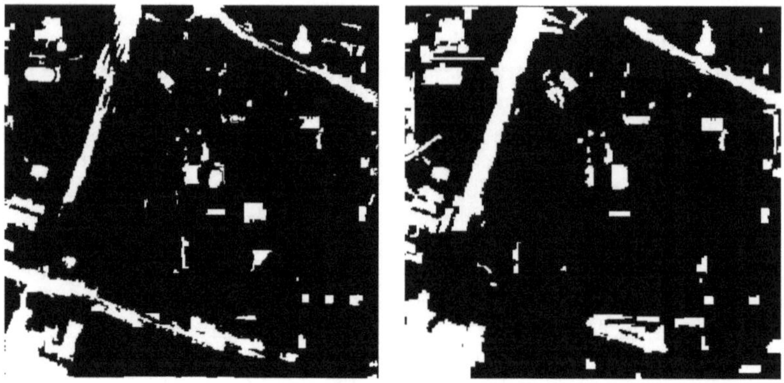

(c) (d)

Figure 4. The change detection maps resulting from: (**a**) Otsu's thresholding method, (**b**) threshold selection by clustering gray levels of boundaries, and (**c**) k-means clustering (k = 2), compared with (**d**) the reference image, with L1 as the basis image in the first study area (scale = 100).

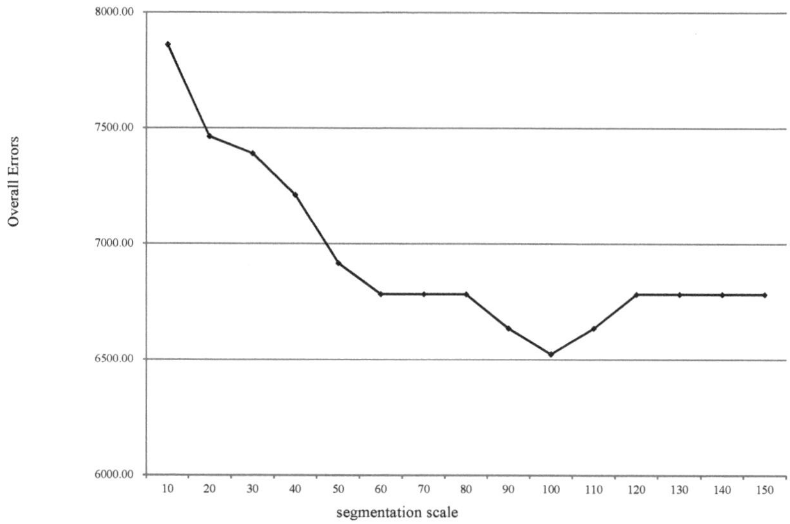

Figure 5. Overall errors of change detection with different segmentation scales, with L1 as the basis image in the first study area.

The accuracies of both the single-scale and multi-scale proposed method are shown in Table 3. As the multi-scale fusion integrates all the single-scale change maps, there are more false alarms but fewer missed alarms than for the optimal single-scale method. Comparing the overall errors, the multi-scale version is more accurate.

Table 3. Comparison between the change detection results of the single-scale and multi-scale proposed method, with L1 as the basis image in the first study area.

	False Alarms_Kmeans	Missed Alarms_Kmeans	Overall Error_Kmeans
The optimal scale = 100	1.66%	2.42%	4.08%
Multi-scale: 10, 20, ... , 150	2.53%	0.81%	3.33%

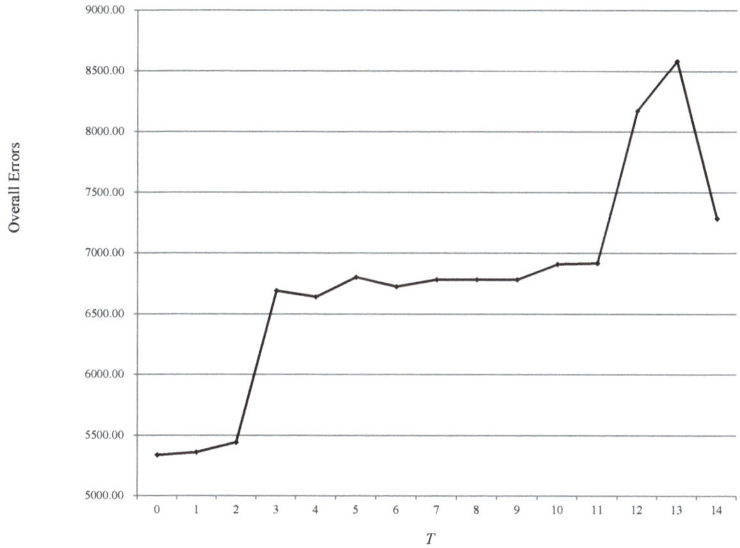

Figure 6. Overall errors of change detection using different multi-scale fusion thresholds, with L1 as the basis image in the first study area.

Moreover, in order to validate the effectiveness of the proposed change detection method for multi-sensor MS imagery, it was compared with the method proposed in [35]. In Figure 7, the white and black regions represent the changed and unchanged areas, respectively. It can be seen that the proposed method effectively decreases the false alarms and suppresses the salt-and-pepper noise in the changed areas. As there are great differences in the visual results, the quantitative assessment and comparison are omitted. The time costs of the two methods were both less than two minutes using MATLAB Software (Mathworks, Natick, MA, USA) on a personal computer with 1.80 GHz CPU and 8.00 GB RAM.

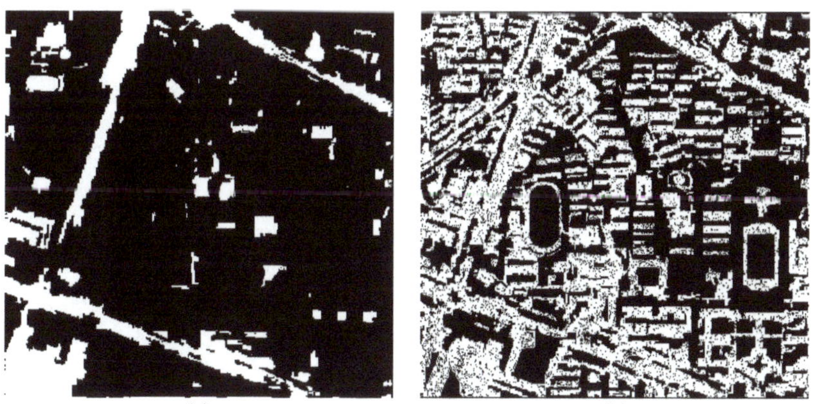

(a) (b)

Figure 7. Change detection maps resulting from (**a**) the proposed multi-scale k-means method and (**b**) the method using varying geometric and radiometric properties [35], with L2 as the basis image in the first study area (scale = 100).

2. L2 as the basis image

In this experiment, L2 was used as the basis image in the preprocessing. Having a higher spatial resolution, L1 was degraded to the same resolution as L2. Figure 8 shows the bi-temporal images after the down-sampling, which are both 240 × 240 pixels. The vegetation and shadow were again masked out.

(a) (b)

Figure 8. Degraded bi-temporal images of the first study area: (**a**) acquired by QuickBird in April 2005 (L1) and (**b**) acquired by IKONOS in July 2009 (L2).

In the analysis of the combined change map, the two threshold selection methods and k-means clustering (k = 2) were again used. The results are shown in Table 4. In this table, the left, middle, and right parts, respectively, show false, missed alarms, and overall errors among the three methods with increasing ratio of P_2. It can be seen that the overall errors of the three methods are again similar. The k-means clustering (k = 2) obtains the least number of errors, and the threshold selection by clustering gray levels of boundary method performs slightly better than Otsu's thresholding method. Figure 9 shows a visual comparison of the results, in which the white and black regions represent the changed and unchanged areas, respectively. The results of the three methods are again similar, and the k-means clustering (k = 2) obtains slightly fewer missed alarms than the two threshold selection methods, which is the same as the result of the experiment with L1 as the basis image.

Table 4. Comparison between the change detection results of the three thresholding and clustering methods, with L2 as the basis image in the first study area (scale = 50).

Combination Ratio of Change Maps	False Alarm _Otsu	False Alarm _Edge	False Alarm _K-means	Missed Alarm _Otsu	Missed Alarm _Edge	Missed Alarm _K-means	Overall Errors _Otsu	Overall Errors _Edge	Overall Errors _K-means
1:9	0.10%	0.10%	0.13%	1.20%	1.09%	0.73%	1.30%	1.19%	0.86%
2:8	0.10%	0.10%	0.13%	1.28%	1.11%	0.74%	1.38%	1.21%	0.87%
3:7	0.10%	0.10%	0.14%	1.28%	1.13%	0.80%	1.38%	1.23%	0.93%
4:6	0.12%	0.11%	0.14%	1.42%	1.28%	0.80%	1.53%	1.39%	0.94%
5:5	0.12%	0.10%	0.14%	1.47%	1.30%	0.81%	1.58%	1.40%	0.94%
6:4	0.12%	0.10%	0.14%	1.50%	1.31%	0.94%	1.62%	1.41%	1.07%
7:3	0.12%	0.11%	0.14%	1.52%	1.33%	1.02%	1.64%	1.44%	1.16%
8:2	0.13%	0.09%	0.17%	1.52%	1.39%	1.06%	1.65%	1.48%	1.23%
9:1	0.13%	0.10%	0.17%	1.52%	1.38%	1.09%	0.00%	1.48%	1.26%

However, it is worth noting that the overall errors increase with the decreasing combination ratio of P1. This is probably because the down-sampling of L1 resulted in the loss of some valuable image information. As a result, the change map of P1, which was generated by the change feature analysis of L1 mapped from the segmentation of L2, was more accurate than the other change map. Therefore, a larger weight of P1 in the combined change map leads to a higher accuracy. From the results of these experiments, we can conclude that the accuracy of the change analysis is improved by increasing the weight of the change map which is generated by mapping the segmentation of the basis image.

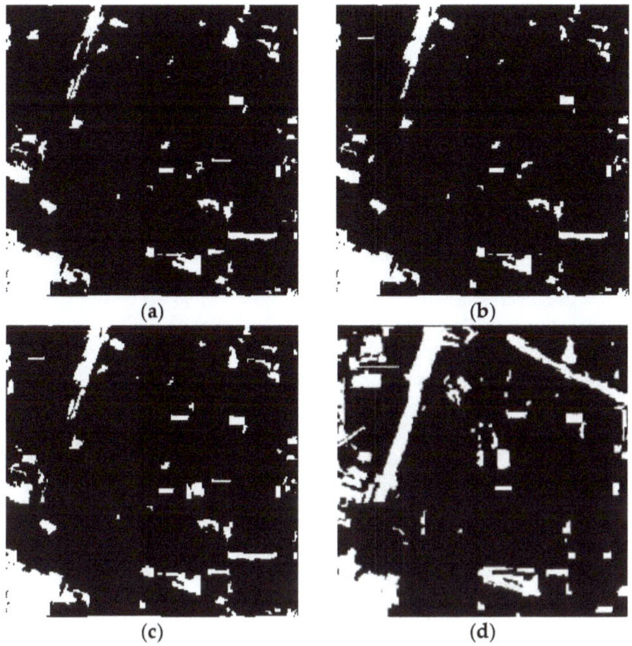

Figure 9. The change detection maps resulting from (**a**) Otsu's thresholding method, (**b**) threshold selection by clustering gray levels of boundary, and (**c**) k-means clustering (k = 2), compared with (**d**) the reference image, with L2 as the basis image in the first study area (scale = 50).

According to the spatial resolution and the objects' sizes in the bi-temporal images after preprocessing, the scale interval and step size increase were set as [10, 100] and 10, respectively. Figure 10 shows the results of the proposed single-scale method using different segmentation scales. The optimal scale is 50. As can be seen in Figure 6, the overall errors are the lowest when T_f in Equation (13) is 0. In addition, Table 5 shows the improvement of the multi-scale fusion with T_f equal to 0, which was realized by k-means clustering (k = 2).

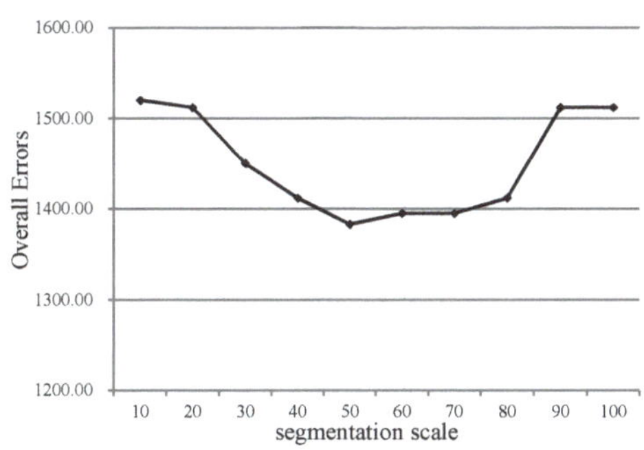

Figure 10. Overall errors of the change detection with different segmentation scales, with L2 as the basis image in the first study area.

Table 5. Comparison between the change detection results of the single-scale and multi-scale proposed method, with L2 as the basis image in the first study area.

	False Alarms_Kmeans	Missed Alarms_Kmeans	Overall Errors_Kmeans
The optimal scale = 50	0.13%	0.73%	0.86%
Multi-scale: 10, 20, . . . , 100	0.15%	0.52%	0.67%

In Figure 11, the proposed method is compared with the method proposed in [35]. The white and black regions represent the changed and unchanged areas, respectively. It can be seen that the proposed method is better able to detect the changes in an urban area with multi-sensor MS images. It suppresses the missed alarms in the changed areas and decreases the false alarms. As there is a significant difference in the visual results, the quantitative assessment and comparison are omitted. The time costs of the two methods were both about one minute using MATLAB Software (Mathworks, Natick, MA, USA) on a personal computer with 1.80 GHz CPU and 8.00 GB RAM.

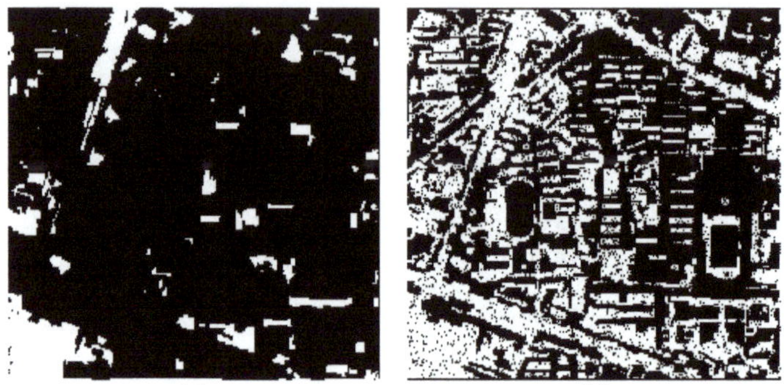

(a) (b)

Figure 11. Change detection maps resulting from (**a**) the proposed multi-scale k-means method and (**b**) the method using varying geometric and radiometric properties [35], with L2 as the basis image in the first study area (scale = 50).

Comparing the two sets of experiments in the first study area, the accuracy is higher in the results with L2 as the basis image. This is probably due to the lower spatial resolution of the basis image.

3.2. The Second Study Area

In order to further verify the proposed method, it was also applied to images from another area in the south of Wuhan, Hubei province, China. The bi-temporal images were respectively acquired by QuickBird in April 2002 (L1) and by IKONOS in July 2009 (L2). L2, with the lower resolution, was regarded as the basis image in the preprocessing, and L1 was degraded by down-sampling. The images after reprocessing, with a size of 240 × 240 pixels, are shown in Figure 12. The vegetation and shadow were, again, masked out to avoid the effects of vegetation phenology and solar elevation.

As the spatial resolutions were the same and the ground objects of the urban area were similar to those of the first study area, the segmentation scale was again set to 50. The results of the two threshold selection methods and k-means clustering (k = 2) are compared in Table 6, with a decreasing P_1 ratio. In this table, the left, middle, and right parts, respectively, show false, missed alarms, and overall errors among the three methods with decreasing ratio of P_1. The accuracies of the three change locating methods are again similar. K-means clustering (k = 2) performs the best, and the threshold selection by clustering gray levels of boundary method performs slightly better than Otsu's thresholding method, which is the same as the first study area. As with the results in the first study area, the accuracy of the proposed method is improved by increasing the weight of P1, which is generated by mapping the segmentation of the basis image of L2. Therefore, it can be concluded that if the weight of the change map, which is mapped from the segmentation of the basis image, is larger than the other, the accuracy of the proposed method increases.

(a)

(b)

Figure 12. Preprocessed bi-temporal images of the second study area: (**a**) acquired by QuickBird in May 2002 (L1) and (**b**) acquired by IKONOS in July 2009 (L2).

Table 6. Comparison between the change detection results of the three thresholding and clustering methods, with L2 as the basis image in the second study area (scale = 50).

Combination Ratio of Change Maps	False Alarm _Otsu	False Alarm _Edge	False Alarm _K-Means	Missed Alarm _Otsu	Missed Alarm _Edge	Missed Alarm _K-Means	Overall Errors _Otsu	Overall Errors _Edge	Overall Errors _K-Means
1:9	0.30%	0.28%	0.36%	1.66%	1.52%	1.00%	1.95%	1.80%	1.37%
2:8	0.31%	0.38%	0.42%	1.67%	1.51%	1.02%	1.98%	1.89%	1.44%
3:7	0.35%	0.39%	0.40%	1.68%	1.50%	1.07%	2.03%	1.90%	1.47%
4:6	0.35%	0.38%	0.40%	1.72%	1.53%	1.08%	2.07%	0.00%	1.48%
5:5	0.36%	0.29%	0.35%	1.76%	1.67%	1.14%	2.11%	1.95%	1.50%
6:4	0.36%	0.36%	0.40%	1.82%	1.70%	1.10%	2.17%	2.06%	1.50%
7:3	0.36%	0.35%	0.40%	1.84%	1.74%	1.15%	2.20%	2.09%	1.54%
8:2	0.38%	0.35%	0.38%	1.89%	1.77%	1.20%	2.27%	2.12%	1.58%
9:1	0.44%	0.34%	0.39%	1.94%	1.81%	1.22%	2.38%	2.16%	1.61%

The binary change maps of the three methods are shown in Figure 13, in which the white and black regions represent the changed and unchanged areas, respectively. Compared with the reference image, the results of the three methods are similar, and the k-means clustering (k = 2) obtains the least number of missed alarms.

As can be seen in Figure 6, the overall errors after the multi-scale fusion are the lowest when T_f in Equation (13) is 0. Table 7 shows the improvement of the multi-scale fusion with T_f equal to 0, which was realized by k-means clustering (k = 2). It can be concluded that the proposed multi-scale method suppresses the missed alarms and keeps the false alarms to an acceptable level.

Table 7. Comparison between the change detection results of the single-scale and multi-scale proposed method, with L2 as the basis image in the second study area.

	False Alarms_Kmeans	Missed Alarms_Kmeans	Overall Errors_Kmeans
The optimal scale = 50	0.36%	1.00%	1.37%
Multi-scale: 10, 20, ... , 100	0.55%	0.22%	0.84%

In Figure 14, the white and black regions represent the changed and unchanged areas, respectively. Compared with the method proposed in [35], the proposed method is shown to be effective in detecting changes in an urban area using multi-sensor MS images. It can effectively decrease the missed alarms in the changed areas while removing the false alarms. As there is a great difference in the visual results, the quantitative assessment and comparison are omitted. The time costs of the two methods were both about one minute using MATLAB Software (Mathworks, Natick, MA, USA) on a personal computer with 1.80 GHz CPU and 8.00 GB RAM.

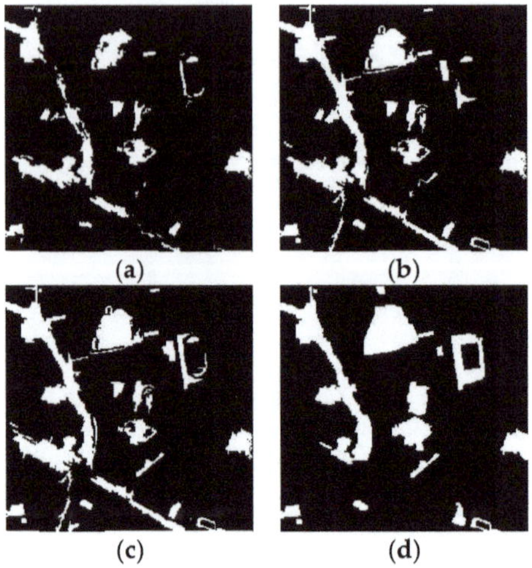

Figure 13. Change detection maps resulting from: (**a**) Otsu's thresholding method, (**b**) threshold selection by clustering gray levels of boundary, and (**c**) *k*-means clustering (k = 2), compared with (**d**) the reference image, with L2 as the basis image in the second study area (scale = 50).

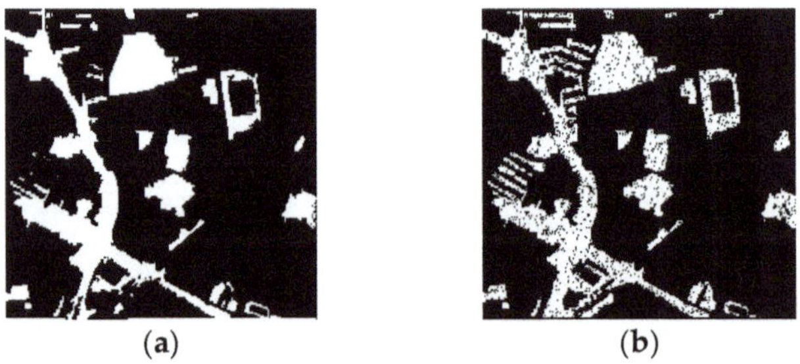

Figure 14. Change detection maps resulting from (**a**) the proposed multi-scale *k*-means method and (**b**) the method using varying geometric and radiometric properties [35], with L2 as the basis image in the second study area (scale = 50).

4. Discussion

In this paper, we have described the experiments conducted with multi-sensor MS images acquired by QuickBird and IKONOS in two different study areas. According to the results of the experiments, the following conclusions can be made:

(1) In the preprocessing of the proposed method, using the image with a lower resolution as the basis image can improve the change detection accuracy. This is probably because some redundant information is removed in the image with lower resolution.

(2) We made use of commercial software (Definiens) to carry out the FNEA and adjust the scale of the image objects to achieve slight under-segmentation. FNEA could be replaced by other segmentation methods, whose results are similar to FNEA.

(3) A change feature is defined to estimate the change possibility of image objects in bi-temporal MS images. The change feature adequately takes into account the statistical features of the image objects in the bi-temporal images (whether acquired by the same or different satellites), which is an important innovation of the proposed method.

(4) In the combining of the change maps, greater precision can be achieved by increasing the ratio of the map which is generated from mapping the segmentation of the basis image to the resampled one. This is probably because the segmentation of the basis image is more precise than the resampled one.

(5) The results of both thresholding and clustering methods for the change locating in gray-level images of the change probability are similar, which confirms that they have little effect on the proposed method.

(6) The multi-scale fusion can effectively improve the accuracy by suppressing the missed alarms and keeping the false alarms to an acceptable level. The overall errors after the multi-scale fusion are the lowest when the changed areas are the sum of the changes in all the single-scale change detection maps.

(7) Compared with the method proposed in [35], the proposed method can effectively detect the changes in multi-sensor MS images by suppressing the missed and false alarms. Instead of utilizing features invariant to different the illumination conditions, the proposed method takes into account the incompatibility between different bandwidths and uses an object-based change measure with the multi-sensor MS images.

5. Conclusions

In this paper, a novel object-based change detection method has been proposed for multi-sensor MS imagery. After the resampling preprocessing, we segment one of the bi-temporal images and map it to the other image, which not only achieves one-to-one correspondence between the bi-temporal images but also preserves the spatial distribution between changed objects and their relevant changed areas. Subsequently, by summarizing the possible distribution between any image object and its relevant changed areas, a change feature is defined to represent the change probability of the image objects in the bi-temporal MS images, whether they are acquired by the same or different satellites. Consequently, thresholding or clustering methods are used to automatically locate the changes in the gray-level image of change probability. Considering the multi-scale feature of ground objects, multi-scale fusion is implemented by voting from the single-scale maps.

According to the experimental results, the urban change analysis method proposed in this paper effectively overcomes the incompatibility between different band widths in bi-temporal (MS) images and utilizes object-based statistical features to describe the changes of ground objects. The overall errors of the proposed method are less than 3.5%. The proposed method makes full use of the spectral and spatial information, and it estimates the change probability of image objects by the use of a novel statistical feature. The object-based change detection method can effectively detect the changes in multi-sensor MS images, and has been confirmed to perform better than the current methods.

Acknowledgments: The authors appreciate the guidance of Prof. Xin Huang from Wuhan University of China. This work was supported by the National 973 Plan of China (grant no. 2012CB719903), the National Natural Science Foundation of China (grant nos. 41401402, 41301453, and 51479215), the Natural Science Foundation of Hunan Province in China (grant no. 2015JJ3150), the Geographical Conditions Monitoring Project of Hunan Province in China (grant no. HNGQJC201503), the Open Research Fund Program of the Key Laboratory of Earth Observation, the State Bureau of Surveying and Mapping (grant no. K201504), the Major Project of High Resolution Earth Observation System of China (Civil Part) (grant no. 03-Y20A11-9001-15/16), and the Demonstration System of High-Resolution Remote Sensing Application in Surveying and Mapping of China (grant no. AH1601-8).

Author Contributions: Yuqi Tang designed the proposed mode, implemented the experiments and drafted the manuscript. Liangpei Zhang provided overall guidance to the project, reviewed and edited the manuscript.

Conflicts of Interest: The authors declare no conflict of interest.

References

1. Ingram, K.; Knapp, E.; Robinson, J.W. Change Detection Technique Development for Improved Urbanized Area Delineation. In *Technical Memorandum, CSC/TM-81/6087*; Computer Sciences Corporation: Silver Springs, MD, USA, 2004.
2. Bruzzone, L.; Bovolo, F. A novel framework for the design of change-detection systems for very-high-resolution remote sensing images. *Proc. IEEE* **2013**, *101*, 609–630. [CrossRef]
3. Tang, Y.; Huang, X.; Zhang, L. Fault-tolerant building change detection from urban high-resolution remote sensing Imagery. *IEEE Geosci. Remote Sens. Lett.* **2013**, *10*, 1060–1064. [CrossRef]
4. Bruzzone, L.; Fernandez-Prieto, D. Automatic analysis of the difference image for unsupervised change detection. *IEEE Geosci. Remote Sens. Lett.* **2000**, *38*, 1170–1182. [CrossRef]
5. Celik, T.; Ma, K.K. Unsupervised change detection for satellite images using dual-tree complex wavelet transform. *IEEE Geosci. Remote Sens. Lett.* **2012**, *48*, 1199–1210. [CrossRef]
6. Bazi, Y.; Melgani, F.; Al-Sharari, H.D. Unsupervised change detection in multispectral remotely sensed imagery with level set methods. *IEEE Geosci. Remote Sens. Lett.* **2010**, *48*, 3178–3187. [CrossRef]
7. Bovolo, F.; Bruzzone, L. A theoretical framework for unsupervised change detection based on change vector analysis in polar domain. *IEEE Geosci. Remote Sens. Lett.* **2007**, *45*, 218–236. [CrossRef]
8. Bovolo, F.; Marchesi, S.; Bruzzone, L. A framework for automatic and unsupervised detection of multiple changes in multitemporal images. *IEEE Geosci. Remote Sens. Lett.* **2012**, *50*, 2196–2212. [CrossRef]
9. Celik, T. Change detection in satellite images using a genetic algorithm approach. *IEEE Geosci. Remote Sens. Lett.* **2010**, *7*, 386–390. [CrossRef]
10. Heng, C.; Celik, T.; Longbotham, N.; Emery, W.J. Gabor feature based unsupervised change detection of multitemporal SAR images based on two-level clustering. *IEEE Geosci. Remote Sens. Lett.* **2015**, *12*, 2458–2462. [CrossRef]
11. Nielsen, A.A. Multiset canonical correlations analysis and multispectral, truly multitemporal remote sensing data. *IEEE Trans. Image Process.* **2002**, *11*, 293–305. [CrossRef] [PubMed]
12. Marchesi, S.; Bovolo, F.; Bruzzone, L. A context-sensitive technique robust to registration noise for change detection in VHR multispectral images. *IEEE Trans. Image Process.* **2010**, *19*, 1877–1889. [CrossRef] [PubMed]
13. Nielsen, A.A.; Conradsem, K.; Simpson, J.J. Multivariate alteration detection(MAD) and MAF postprocessing in multispectral bitemporal image data: New approaches to change detection studies. *Remote Sens. Environ.* **1998**, *64*, 1–19. [CrossRef]
14. Marpu, P.R.; Gamba, P.; Canty, M.J. Improving change detection results of IR-MAD by eliminating strong changes. *IEEE Geosci. Remote Sens. Lett.* **2011**, *8*, 799–803. [CrossRef]
15. Nielsen, A.A. The regularized iteratively reweighted MAD method for change detection in multi- and hyperspectral data. *IEEE Geosci. Remote Sens. Lett.* **2007**, *16*, 463–478. [CrossRef]
16. Wang, Q.; Lin, J.; Yuan, Y. Salient band selection for hyperspectral image classification via manifold ranking. *IEEE Trans. Neural Netw. Learn. Syst.* **2016**, *27*, 1279–1289. [CrossRef] [PubMed]
17. Yuan, Y.; Zhu, G.; Wang, Q. Hyperspectral band selection by multi-task sparsity pursuit. *IEEE Geosci. Remote Sens. Lett.* **2015**, *53*, 631–644. [CrossRef]
18. Luo, W.; Li, H. Soft-change detection in optical satellite images. *IEEE Geosci. Remote Sens. Lett.* **2011**, *8*, 879–883. [CrossRef]
19. Ling, F.; Li, W.; Du, Y.; Li, X. Land cover change mapping at the subpixel scale with different spatial-resolution remotely sensed imagery. *IEEE Geosci. Remote Sens. Lett.* **2011**, *8*, 182–186. [CrossRef]
20. Robin, A.; Moisan, L.; Hegarat-Mascle, S.L. An a-contrario approach for subpixel change detection in satellite imagery. *IEEE Trans. Pattern Anal. Mach. Intell.* **2010**, *32*, 1977–1993. [CrossRef] [PubMed]
21. Gueguen, L.; Soille, P.; Pesaresi, M. Change detection based on information measure. *IEEE Geosci. Remote Sens. Lett.* **2011**, *49*, 4503–4515. [CrossRef]
22. Healey, G.; Slater, D. Computing illumination-invariant descriptors of spatially filtered color image regions. *IEEE Trans. Image Process.* **1997**, *6*, 1002–1013. [CrossRef] [PubMed]
23. Smits, P.C.; Annoni, A. Updating land-cover maps by using texture information from very high-resolution space-borne imagery. *IEEE Geosci. Remote Sens. Lett.* **1999**, *37*, 1244–1254. [CrossRef]
24. Li, L.; Leung, M.K.H. Integrating intensity and texture differences for robust change detection. *IEEE Trans. Image Process.* **2002**, *11*, 105–112. [PubMed]

25. Moser, G.; Angiati, E.; Serpico, S.B. Multiscale unsupervised change detection on optical images by Markov random fields and wavelets. *IEEE Geosci. Remote Sens. Lett.* **2011**, *8*, 725–729. [CrossRef]
26. Yuan, Y.; Lin, J.; Wang, Q. Hyperspectral image classification via multi-task joint sparse representation and stepwise MRF optimization. *IEEE Trans. Cybern.* **2016**, *46*, 2966–2977. [CrossRef] [PubMed]
27. Dalla Mura, M.; Benediktsson, J.A.; Bovolo, F.; Bruzzone, L. An unsupervised technique based on morphological filters for change detection in very high resolution images. *IEEE Geosci. Remote Sens. Lett.* **2008**, *5*, 433–437. [CrossRef]
28. Dalla Mura, M.; Benediktsson, J.A.; Waske, B.; Bruzzone, L. Morphological attribute profiles for the analysis of very high resolution images. *IEEE Geosci. Remote Sens. Lett.* **2010**, *48*, 3747–3762. [CrossRef]
29. Im, J.; Jensen, J.R.; Tullis, J.A. Object-based change detection using correlation image analysis and image segmentation. *Int. J. Remote Sens.* **2008**, *29*, 399–423. [CrossRef]
30. Bovolo, F. A multilevel parcel-based approach to change detection in very high resolution multitemporal images. *IEEE Geosci. Remote Sens. Lett.* **2009**, *6*, 33–37. [CrossRef]
31. Lu, P.; Stumpf, A.; Kerle, N.; Casagli, N. Object-oriented change detection for landslide rapid mapping. *IEEE Geosci. Remote Sens. Lett.* **2011**, *8*, 701–705. [CrossRef]
32. Huo, C.; Zhou, Z.; Lu, H.; Pan, C.; Chen, K. Fast object-level change detection for VHR images. *IEEE Geosci. Remote Sens. Lett.* **2010**, *7*, 118–122. [CrossRef]
33. Tang, Y.; Zhang, L.; Huang, X. Object-oriented change detection based on the Kolmogorov-Smirnov test using high-resolution multispectral imagery. *Int. J. Remote Sens.* **2011**, *32*, 5719–5740. [CrossRef]
34. Mercier, G.; Moser, G.; Serpico, S. Conditional copulas for change detection in heterogeneous remote sensing images. *IEEE Trans. Geosci. Remote Sens.* **2008**, *46*, 1428–1441. [CrossRef]
35. Habib, A.; AI-Ruzouq, R.; Kim, C. Semi-automatic registration and change detection using multi-source imagery with varying geometric and radiometric properties. *Int. Arch. Photogramm. Remote Sens. Spat. Inf. Sci.* **2012**, *1*, 1–6.
36. Bovolo, F.; Bruzzone, L.; Marconcini, M. A novel approach to unsupervised change detection based on a semisupervised SVM and a similarity measure. *IEEE Trans. Geosci. Remote Sens.* **2008**, *46*, 2070–2082. [CrossRef]
37. Thonfeld, F.; Feihauer, H.; Braun, M.; Menz, G. Robust change vector analysis (RVCA) for multi-sensor very high resolution optical satellite data. *Int. J. Appl. Earth Obs. Geoinform.* **2016**, *51*, 131–140. [CrossRef]
38. Richards, J. *Remote Sensing Digital Image Analysis: An Introduction*; Springer: Berlin/Heidelberg, Germany, 1986; pp. 52–55.
39. Baatz, M.; Schape, A. *Multiresolution Segmentation: An Optimization Approach for High Quality Multi-Scale Image Segmentation*; Wichmann-Verlag: Heidelberg, Germany, 2000; pp. 12–23.
40. Pesaresi, M.; Benediktsson, J.A. A new approach for the morphological segmentation of high-resolution satellite imagery. *IEEE Trans. Geosci. Remote Sens.* **2011**, *39*, 309–320. [CrossRef]
41. Huang, X.; Zhang, L. An adaptive mean-shift analysis approach for object extraction and classification from urban hyperspectral imagery. *IEEE Trans. Geosci. Remote Sens.* **2008**, *46*, 4173–4185. [CrossRef]
42. Otsu, N. A threshold selection method from gray-level histograms. *IEEE Trans. Syst. Man Cybern.* **1979**, *9*, 61–66. [CrossRef]
43. Wang, L.; Bai, J. Threshold selection by clustering gray levels of boundary. *Pattern Recognit. Lett.* **2003**, *24*, 1983–1999. [CrossRef]
44. MacQueen, J. Some Methods for Classification and Analysis of Multivariate Observations. *Pro. 5th Berkeley Symp. Mathem. Stat. Probab.* **1967**, *1*, 281–297.
45. Yuan, Y.; Lin, J.; Wang, Q. Dual clustering based hyperspectral band selection by contextual analysis. *IEEE Trans. Geosci. Remote Sens.* **2016**, *54*, 1431–1445. [CrossRef]
46. Zhao, B.; Zhong, Y.; Ma, A.; Zhang, L. A spatial Gaussian mixture model for optical remote sensing image clustering. *IEEE J. Sel. Top. Appl. Earth Obs. Remote Sens.* **2016**, *9*, 5748–5759. [CrossRef]
47. Peng, X.; Tang, H.; Zhang, L.; Yi, Z.; Xiao, S. A unified framework for representation-based subspace clustering of out-of-sample and large-scale data. *IEEE Trans. Neural Netw. Learn. Syst.* **2016**, *27*, 2499–2512. [CrossRef] [PubMed]

MDPI
St. Alban-Anlage 66
4052 Basel
Switzerland
Tel. +41 61 683 77 34
Fax +41 61 302 89 18
www.mdpi.com

Remote Sensing Editorial Office
E-mail: remotesensing@mdpi.com
www.mdpi.com/journal/remotesensing

www.ingramcontent.com/pod-product-compliance
Lightning Source LLC
Chambersburg PA
CBHW041323210926
51847CB00003B/35
* 9 7 8 3 0 3 8 9 7 6 9 8 1 *